Image *and* Video Compression *for* Multimedia Engineering

Fundamentals, Algorithms, and Standards

Second Edition

IMAGE PROCESSING SERIES

Series Editor: Phillip A. Laplante, Pennsylvania State University

Published Titles

Adaptive Image Processing: A Computational Intelligence Perspective
Stuart William Perry, Hau-San Wong, and Ling Guan

Color Image Processing: Methods and Applications
Rastislav Lukac and Konstantinos N. Plataniotis

Image Acquisition and Processing with LabVIEW™
Christopher G. Relf

**Image and Video Compression for Multimedia Engineering
Second Edition**
Yun Q. Shi and Huiyang Sun

Multimedia Image and Video Processing
Ling Guan, S.Y. Kung, and Jan Larsen

Shape Analysis and Classification: Theory and Practice
Luciano da Fontoura Costa and Roberto Marcondes Cesar Jr.

Software Engineering for Image Processing Systems
Phillip A. Laplante

Image *and* Video Compression *for* Multimedia Engineering

Fundamentals, Algorithms, and Standards

Second Edition

Yun Q. Shi
New Jersey Institute of Technology
Newark, New Jersey, USA

Huifang Sun
Mitsubishi Electric Research Laboratories
Cambridge, Massachusetts, USA

CRC Press
Taylor & Francis Group
6000 Broken Sound Parkway NW, Suite 300
Boca Raton, FL 33487-2742

© 2008 by Taylor & Francis Group, LLC
CRC Press is an imprint of Taylor & Francis Group, an Informa business

No claim to original U.S. Government works
Printed in the United States of America on acid-free paper
10 9 8 7 6 5 4 3 2 1

International Standard Book Number-13: 978-0-8493-7364-0 (Hardcover)

This book contains information obtained from authentic and highly regarded sources Reasonable efforts have been made to publish reliable data and information, but the author and publisher cannot assume responsibility for the validity of all materials or the consequences of their use. The Authors and Publishers have attempted to trace the copyright holders of all material reproduced in this publication and apologize to copyright holders if permission to publish in this form has not been obtained. If any copyright material has not been acknowledged please write and let us know so we may rectify in any future reprint

Except as permitted under U.S. Copyright Law, no part of this book may be reprinted, reproduced, transmitted, or utilized in any form by any electronic, mechanical, or other means, now known or hereafter invented, including photocopying, microfilming, and recording, or in any information storage or retrieval system, without written permission from the publishers.

For permission to photocopy or use material electronically from this work, please access www.copyright.com (http://www.copyright.com/) or contact the Copyright Clearance Center, Inc. (CCC) 222 Rosewood Drive, Danvers, MA 01923, 978-750-8400. CCC is a not-for-profit organization that provides licenses and registration for a variety of users. For organizations that have been granted a photocopy license by the CCC, a separate system of payment has been arranged.

Trademark Notice: Product or corporate names may be trademarks or registered trademarks, and are used only for identification and explanation without intent to infringe.

Library of Congress Cataloging-in-Publication Data

Shi, Yun Q.
  Image and video compression for multimedia engineering : fundamentals, algorithms, and standards / Yun Q. Shi and Huifang Sun. -- 2nd ed.
    p. cm. -- (Image processing series)
  Includes bibliographical references and index.
  ISBN 978-0-8493-7364-0 (alk. paper)
  1. Multimedia systems. 2. Image compression. 3. Video compression. I. Sun, Huifang. II. Title.

QA76.575.S555 2008
006.7--dc22   2007048389

Visit the Taylor & Francis Web site at
http://www.taylorandfrancis.com

and the CRC Press Web site at
http://www.crcpress.com

To beloved Kong Wai Shih, Wen Su,

Yi Xi Li, Shu Jun Zheng, and

Xian Hong Li

and

To beloved Xuedong,

Min, Yin, Andrew, Rich, Haixin, and

Allison

Contents

Preface to the Second Edition xxi
Preface to the First Edition xxiii
Content and Organization of the Book xxvii
Authors xxxi

Part I Fundamentals

Chapter 1 Introduction
1.1 Practical Needs for Image and Video Compression 4
1.2 Feasibility of Image and Video Compression 4
    1.2.1 Statistical Redundancy 5
        1.2.1.1 Spatial Redundancy 5
        1.2.1.2 Temporal Redundancy 7
        1.2.1.3 Coding Redundancy 9
    1.2.2 Psychovisual Redundancy 10
        1.2.2.1 Luminance Masking 11
        1.2.2.2 Texture Masking 13
        1.2.2.3 Frequency Masking 15
        1.2.2.4 Temporal Masking 16
        1.2.2.5 Color Masking 16
        1.2.2.6 Color Masking and Its Application in Video Compression 19
        1.2.2.7 Summary: Differential Sensitivity 20
1.3 Visual Quality Measurement 20
    1.3.1 Subjective Quality Measurement 21
    1.3.2 Objective Quality Measurement 22
        1.3.2.1 Signal to Noise Ratio 22
        1.3.2.2 An Objective Quality Measure Based on Human
                  Visual Perception 23
1.4 Information Theory Results 26
    1.4.1 Entropy 27
        1.4.1.1 Information Measure 27
        1.4.1.2 Average Information per Symbol 27
    1.4.2 Shannon's Noiseless Source Coding Theorem 28
    1.4.3 Shannon's Noisy Channel Coding Theorem 29
    1.4.4 Shannon's Source Coding Theorem 29
    1.4.5 Information Transmission Theorem 29
1.5 Summary 30
Exercises 30
References 31

Chapter 2 Quantization
2.1 Quantization and the Source Encoder 33
2.2 Uniform Quantization 35
    2.2.1 Basics 36

            2.2.1.1 Definitions 36
            2.2.1.2 Quantization Distortion 38
            2.2.1.3 Quantizer Design 39
        2.2.2 Optimum Uniform Quantizer 40
            2.2.2.1 Uniform Quantizer with Uniformly Distributed Input 40
            2.2.2.2 Conditions of Optimum Quantization 41
            2.2.2.3 Optimum Uniform Quantizer with Different
                      Input Distributions 44
    2.3 Nonuniform Quantization 45
        2.3.1 Optimum (Nonuniform) Quantization 45
        2.3.2 Companding Quantization 48
    2.4 Adaptive Quantization 50
        2.4.1 Forward Adaptive Quantization 52
        2.4.2 Backward Adaptive Quantization 52
        2.4.3 Adaptive Quantization with a One-Word Memory 53
        2.4.4 Switched Quantization 53
    2.5 Pulse Code Modulation 54
    2.6 Summary 56
    Exercises 57
    References 58

Chapter 3 Differential Coding
    3.1 Introduction to DPCM 59
        3.1.1 Simple Pixel-to-Pixel DPCM 60
        3.1.2 General DPCM Systems 63
    3.2 Optimum Linear Prediction 64
        3.2.1 Formulation 65
        3.2.2 Orthogonality Condition and Minimum Mean Square Error ... 66
        3.2.3 Solution to Yule–Walker Equations 66
    3.3 Some Issues in the Implementation of DPCM 66
        3.3.1 Optimum DPCM System 67
        3.3.2 1-D, 2-D, and 3-D DPCM 67
        3.3.3 Order of Predictor 68
        3.3.4 Adaptive Prediction 68
        3.3.5 Effect of Transmission Errors 68
    3.4 Delta Modulation 70
    3.5 Interframe Differential Coding 73
        3.5.1 Conditional Replenishment 73
        3.5.2 3-D DPCM 74
        3.5.3 Motion Compensated Predictive Coding 75
    3.6 Information-Preserving Differential Coding 76
    3.7 Summary 77
    Exercises 78
    References 78

Chapter 4 Transform Coding
    4.1 Introduction 81
        4.1.1 Hotelling Transform 81
        4.1.2 Statistical Interpretation 83
        4.1.3 Geometrical Interpretation 84

	4.1.4	Basis Vector Interpretation	85
	4.1.5	Procedures of Transform Coding	86
4.2	Linear Transforms		87
	4.2.1	2-D Image Transformation Kernel	87
		4.2.1.1 Separability	87
		4.2.1.2 Symmetry	88
		4.2.1.3 Matrix Form	88
		4.2.1.4 Orthogonality	89
	4.2.2	Basis Image Interpretation	89
	4.2.3	Subimage Size Selection	90
4.3	Transforms of Particular Interest		92
	4.3.1	Discrete Fourier Transform	92
	4.3.2	Discrete Walsh Transform	93
	4.3.3	Discrete Hadamard Transform	93
	4.3.4	Discrete Cosine Transform	95
		4.3.4.1 Background	95
		4.3.4.2 Transformation Kernel	95
		4.3.4.3 Relationship with DFT	96
	4.3.5	Performance Comparison	98
		4.3.5.1 Energy Compaction	98
		4.3.5.2 Mean Square Reconstruction Error	99
		4.3.5.3 Computational Complexity	101
		4.3.5.4 Summary	101
4.4	Bit Allocation		101
	4.4.1	Zonal Coding	101
	4.4.2	Threshold Coding	102
		4.4.2.1 Thresholding and Shifting	103
		4.4.2.2 Normalization and Roundoff	104
		4.4.2.3 Zigzag Scan	106
		4.4.2.4 Huffman Coding	106
		4.4.2.5 Special Code Words	107
		4.4.2.6 Rate Buffer Feedback and Equalization	107
4.5	Some Issues		108
	4.5.1	Effect of Transmission Error	108
	4.5.2	Reconstruction Error Sources	108
	4.5.3	Comparison between DPCM and TC	109
	4.5.4	Hybrid Coding	109
4.6	Summary		110
Exercises			112
References			112

Chapter 5 Variable-Length Coding: Information Theory Results (II)

5.1	Some Fundamental Results		115
	5.1.1	Coding an Information Source	115
	5.1.2	Some Desired Characteristics	116
		5.1.2.1 Block Code	116
		5.1.2.2 Uniquely Decodable Code	116
		5.1.2.3 Instantaneous Codes	118
		5.1.2.4 Compact Code	120
	5.1.3	Discrete Memoryless Sources	120

       5.1.4 Extensions of a Discrete Memoryless Source 120
             5.1.4.1 Definition 120
             5.1.4.2 Entropy 120
             5.1.4.3 Noiseless Source Coding Theorem 121
5.2 Huffman Codes 122
     5.2.1 Required Rules for Optimum Instantaneous Codes 122
     5.2.2 Huffman Coding Algorithm 123
             5.2.2.1 Procedures 124
             5.2.2.2 Comments 125
             5.2.2.3 Applications 125
5.3 Modified Huffman Codes 125
     5.3.1 Motivation 125
     5.3.2 Algorithm 126
     5.3.3 Codebook Memory Requirement 127
     5.3.4 Bounds on Average Code Word Length 128
5.4 Arithmetic Codes 128
     5.4.1 Limitations of Huffman Coding 129
     5.4.2 The Principle of Arithmetic Coding 129
             5.4.2.1 Dividing Interval [0, 1) into Subintervals 130
             5.4.2.2 Encoding 132
             5.4.2.3 Decoding 132
             5.4.2.4 Observations 134
     5.4.3 Implementation Issues 135
             5.4.3.1 Incremental Implementation 135
             5.4.3.2 Finite Precision 136
             5.4.3.3 Other Issues 136
     5.4.4 History 136
     5.4.5 Applications 137
5.5 Summary 137
Exercises 138
References 139

Chapter 6 Run-Length and Dictionary Coding: Information Theory Results (III)

6.1 Markov Source Model 141
     6.1.1 Discrete Markov Source 141
     6.1.2 Extensions of a Discrete Markov Source 143
             6.1.2.1 Definition 143
             6.1.2.2 Entropy 143
     6.1.3 Autoregressive Model 143
6.2 Run-Length Coding 144
     6.2.1 1-D Run-Length Coding 144
     6.2.2 2-D Run-Length Coding 145
             6.2.2.1 Five Changing Pixels 146
             6.2.2.2 Three Coding Modes 147
     6.2.3 Effect of Transmission Error and Uncompressed Mode 148
             6.2.3.1 Error Effect in the 1-D RLC Case 148
             6.2.3.2 Error Effect in the 2-D RLC Case 149
             6.2.3.3 Uncompressed Mode 149

6.3 Digital Facsimile Coding Standards 149
6.4 Dictionary Coding 150
    6.4.1 Formulation of Dictionary Coding 151
    6.4.2 Categorization of Dictionary-Based Coding Techniques 151
        6.4.2.1 Static Dictionary Coding 151
        6.4.2.2 Adaptive Dictionary Coding 151
    6.4.3 Parsing Strategy 152
    6.4.4 Sliding Window (LZ77) Algorithms 152
        6.4.4.1 Introduction 152
        6.4.4.2 Encoding and Decoding 153
        6.4.4.3 Summary of the LZ77 Approach 155
    6.4.5 LZ78 Algorithms 156
        6.4.5.1 Introduction 156
        6.4.5.2 Encoding and Decoding 157
        6.4.5.3 LZW Algorithm 158
        6.4.5.4 Summary 159
        6.4.5.5 Applications 160
6.5 International Standards for Lossless Still Image Compression 160
    6.5.1 Lossless Bilevel Still Image Compression 161
        6.5.1.1 Algorithms 161
        6.5.1.2 Performance Comparison 161
    6.5.2 Lossless Multilevel Still Image Compression 161
        6.5.2.1 Algorithms 161
        6.5.2.2 Performance Comparison 162
6.6 Summary 162
Exercises 163
References 164

Part II Still Image Compression

Chapter 7 Still Image Coding: Standard JPEG
7.1 Introduction 167
7.2 Sequential DCT-Based Encoding Algorithm 169
7.3 Progressive DCT-Based Encoding Algorithm 174
7.4 Lossless Coding Mode 176
7.5 Hierarchical Coding Mode 177
7.6 Summary 178
Exercises 178
References 178

Chapter 8 Wavelet Transform for Image Coding: JPEG2000
8.1 A Review of Wavelet Transform 179
    8.1.1 Definition and Comparison with Short-Time
        Fourier Transform 179
    8.1.2 Discrete Wavelet Transform 183
    8.1.3 Lifting Scheme 185
        8.1.3.1 Three Steps in Forward Wavelet Transform 185
        8.1.3.2 Inverse Transform 186

		8.1.3.3	Lifting Version of CDF (2,2) 186
		8.1.3.4	A Demonstration Example 187
		8.1.3.5	(5,3) Integer Wavelet Transform 188
		8.1.3.6	A Demonstration Example of (5,3) IWT 188
		8.1.3.7	Summary 189
8.2	Digital Wavelet Transform for Image Compression 189
	8.2.1	Basic Concept of Image Wavelet Transform Coding 189
	8.2.2	Embedded Image Wavelet Transform Coding Algorithms 191
		8.2.2.1	Early Wavelet Image Coding Algorithms and Their Drawbacks 191
		8.2.2.2	Modern Wavelet Image Coding 191
		8.2.2.3	Embedded Zerotree Wavelet Coding 192
		8.2.2.4	Set Partitioning in Hierarchical Trees Coding 194
8.3	Wavelet Transform for JPEG2000 195
	8.3.1	Introduction of JPEG2000 195
		8.3.1.1	Requirements of JPEG2000 196
		8.3.1.2	Parts of JPEG2000 197
	8.3.2	Verification Model of JPEG2000 197
	8.3.3	An Example of Performance Comparison between JPEG and JPEG2000 199
8.4	Summary 200
Exercises 200
References 202

Chapter 9 Nonstandard Still Image Coding

9.1 Introduction 203
9.2 Vector Quantization 204
    9.2.1 Basic Principle of Vector Quantization 204
        9.2.1.1 Vector Formation 205
        9.2.1.2 Training Set Generation 205
        9.2.1.3 Codebook Generation 206
        9.2.1.4 Quantization 206
    9.2.2 Several Image Coding Schemes with Vector Quantization 207
        9.2.2.1 Residual VQ 207
        9.2.2.2 Classified VQ 208
        9.2.2.3 Transform Domain VQ 208
        9.2.2.4 Predictive VQ 208
        9.2.2.5 Block Truncation Coding 209
    9.2.3 Lattice VQ for Image Coding 209
9.3 Fractal Image Coding 212
    9.3.1 Mathematical Foundation 212
    9.3.2 IFS-Based Fractal Image Coding 214
    9.3.3 Other Fractal Image Coding Methods 215
9.4 Model-Based Coding 216
    9.4.1 Basic Concept 216
    9.4.2 Image Modeling 216
9.5 Summary 217
Exercises 217
References 217

Part III Motion Estimation and Compensation

Chapter 10 Motion Analysis and Motion Compensation
10.1 Image Sequences 221
10.2 Interframe Correlation 224
10.3 Frame Replenishment 226
10.4 Motion Compensated Coding 227
10.5 Motion Analysis 230
       10.5.1 Biological Vision Perspective 230
       10.5.2 Computer Vision Perspective 230
       10.5.3 Signal Processing Perspective 232
10.6 Motion Compensation for Image Sequence Processing 233
       10.6.1 Motion Compensated Interpolation 233
       10.6.2 Motion Compensated Enhancement 235
       10.6.3 Motion Compensated Restoration 236
       10.6.4 Motion Compensated Down-Conversion 236
10.7 Summary 236
Exercises 237
References 238

Chapter 11 Block Matching
11.1 Nonoverlapped, Equally Spaced, Fixed Size, Small Rectangular
      Block Matching 241
11.2 Matching Criteria 243
11.3 Searching Procedures 244
       11.3.1 Full Search 244
       11.3.2 2-D Logarithm Search 245
       11.3.3 Coarse–Fine Three-Step Search 246
       11.3.4 Conjugate Direction Search 246
       11.3.5 Subsampling in the Correlation Window 248
       11.3.6 Multiresolution Block Matching 248
       11.3.7 Thresholding Multiresolution Block Matching 250
              11.3.7.1 Algorithm 250
              11.3.7.2 Threshold Determination 251
              11.3.7.3 Thresholding 252
              11.3.7.4 Experiments 253
11.4 Matching Accuracy 256
11.5 Limitations with Block Matching Techniques 256
11.6 New Improvements 257
       11.6.1 Hierarchical Block Matching 257
       11.6.2 Multigrid Block Matching 260
              11.6.2.1 Thresholding Multigrid Block Matching 260
              11.6.2.2 Optimal Multigrid Block Matching 262
       11.6.3 Predictive Motion Field Segmentation 263
       11.6.4 Overlapped Block Matching 265
11.7 Summary 267
Exercises 269
References 269

Chapter 12   Pel Recursive Technique
12.1 Problem Formulation 271
12.2 Descent Methods 273
    12.2.1 First-Order Necessary Conditions 273
    12.2.2 Second-Order Sufficient Conditions 273
    12.2.3 Underlying Strategy 274
    12.2.4 Convergence Speed 275
        12.2.4.1 Order of Convergence 275
        12.2.4.2 Linear Convergence 276
    12.2.5 Steepest Descent Method 277
        12.2.5.1 Formulae 277
        12.2.5.2 Convergence Speed 277
        12.2.5.3 Selection of Step Size 277
    12.2.6 Newton–Raphson's Method 277
        12.2.6.1 Formulae 278
        12.2.6.2 Convergence Speed 279
        12.2.6.3 Generalization and Improvements 279
    12.2.7 Other Methods 279
12.3 Netravali–Robbins' Pel Recursive Algorithm 280
    12.3.1 Inclusion of a Neighborhood Area 280
    12.3.2 Interpolation 280
    12.3.3 Simplification 281
    12.3.4 Performance 281
12.4 Other Pel Recursive Algorithms 281
    12.4.1 Bergmann's Algorithm (1982) 281
    12.4.2 Bergmann's Algorithm (1984) 282
    12.4.3 Cafforio and Rocca's Algorithm 282
    12.4.4 Walker and Rao's Algorithm 282
12.5 Performance Comparison 283
12.6 Summary 283
Exercises 284
References 284

Chapter 13   Optical Flow
13.1 Fundamentals 285
    13.1.1 2-D Motion and Optical Flow 286
    13.1.2 Aperture Problem 287
    13.1.3 Ill-Posed Problem 289
    13.1.4 Classification of Optical Flow Techniques 289
13.2 Gradient-Based Approach 290
    13.2.1 Horn and Schunck's Method 290
        13.2.1.1 Brightness Invariance Equation 290
        13.2.1.2 Smoothness Constraint 292
        13.2.1.3 Minimization 292
        13.2.1.4 Iterative Algorithm 293
    13.2.2 Modified Horn and Schunck Method 295
    13.2.3 Lucas and Kanade's Method 296
    13.2.4 Nagel's Method 296
    13.2.5 Uras, Girosi, Verri, and Torre's Method 297

13.3 Correlation-Based Approach 297
   13.3.1 Anandan's Method 298
   13.3.2 Singh's Method 298
      13.3.2.1 Conservation Information 300
      13.3.2.2 Neighborhood Information 300
      13.3.2.3 Minimization and Iterative Algorithm 301
   13.3.3 Pan, Shi, and Shu's Method 302
      13.3.3.1 Proposed Framework 302
      13.3.3.2 Implementation and Experiments 304
      13.3.3.3 Discussion and Conclusion 312
13.4 Multiple Attributes for Conservation Information 315
   13.4.1 Weng, Ahuja, and Huang's Method 316
   13.4.2 Xia and Shi's Method 317
      13.4.2.1 Multiple Image Attributes 317
      13.4.2.2 Conservation Stage 318
      13.4.2.3 Propagation Stage 319
      13.4.2.4 Outline of Algorithm 319
      13.4.2.5 Experimental Results 320
      13.4.2.6 Discussion and Conclusion 321
13.5 Summary 321
Exercises 323
References 324

Chapter 14 Further Discussion and Summary on 2-D Motion Estimation

14.1 General Characterization 327
   14.1.1 Aperture Problem 327
   14.1.2 Ill-Posed Inverse Problem 327
   14.1.3 Conservation Information and Neighborhood Information 328
   14.1.4 Occlusion and Disocclusion 328
   14.1.5 Rigid and Nonrigid Motion 329
14.2 Different Classifications 330
   14.2.1 Deterministic Methods versus Stochastic Methods 330
   14.2.2 Spatial Domain Methods versus Frequency Domain Methods 330
      14.2.2.1 Optical Flow Determination Using Gabor Energy Filters 331
   14.2.3 Region-Based Approaches versus Gradient-Based Approaches 333
   14.2.4 Forward versus Backward Motion Estimation 335
14.3 Performance Comparison between Three Major Approaches 336
   14.3.1 Three Representatives 336
   14.3.2 Algorithm Parameters 336
   14.3.3 Experimental Results and Observations 337
14.4 New Trends 337
   14.4.1 DCT-Based Motion Estimation 338
      14.4.1.1 DCT and DST Pseudophases 338
      14.4.1.2 Sinusoidal Orthogonal Principle 339
      14.4.1.3 Performance Comparison 339
14.5 Summary 340
Exercises 341
References 341

Part IV Video Compression

Chapter 15 Fundamentals of Digital Video Coding

15.1 Digital Video Representation 345
15.2 Information Theory Results: Rate Distortion Function
       of Video Signal 346
15.3 Digital Video Formats 349
       15.3.1 Digital Video Color Systems 349
       15.3.2 Progressive and Interlaced Video Signals 350
       15.3.3 Video Formats Used by Video Industry 351
              15.3.3.1 ITU-R 351
              15.3.3.2 Source Input Format 351
              15.3.3.3 Common Intermediate Format 352
              15.3.3.4 ATSC Digital Television Format 352
15.4 Current Status of Digital Video/Image Coding Standards 352
       15.4.1 JPEG Standard 354
       15.4.2 JPEG2000 354
       15.4.3 MPEG-1 354
       15.4.4 MPEG-2 354
       15.4.5 MPEG-4 354
       15.4.6 H.261 355
       15.4.7 H.263, H.263 Version 2 (H.263+), H.263++, and H.26L 355
       15.4.8 MPEG-4 Part 10 Advanced Video Coding or H.264/AVC 355
       15.4.9 VC-1 355
       15.4.10 RealVideo 355
15.5 Summary 356
Exercises 356
References 357

Chapter 16 Digital Video Coding Standards: MPEG-1/2 Video

16.1 Introduction 359
16.2 Features of MPEG-1/2 Video Coding 360
       16.2.1 MPEG-1 Features 360
              16.2.1.1 Introduction 360
              16.2.1.2 Layered Structure Based on Group of Pictures 360
              16.2.1.3 Encoder Structure 361
              16.2.1.4 Structure of the Compressed Bitstream 365
              16.2.1.5 Decoding Process 365
       16.2.2 MPEG-2 Enhancements 366
              16.2.2.1 Field/Frame Prediction Mode 367
              16.2.2.2 Field/Frame DCT Coding Syntax 369
              16.2.2.3 Downloadable Quantization Matrix and Alternative
                        Scan Order 369
              16.2.2.4 Pan and Scan 370
              16.2.2.5 Concealment Motion Vector 370
              16.2.2.6 Scalability 371
16.3 MPEG-2 Video Encoding 372
       16.3.1 Introduction 372
       16.3.2 Preprocessing 373

	16.3.3	Motion Estimation and Motion Compensation 374
		16.3.3.1 Matching Criterion 374
		16.3.3.2 Searching Algorithm 375
		16.3.3.3 Advanced Motion Estimation 376
16.4	Rate Control 377
	16.4.1	Introduction of Rate Control 377
	16.4.2	Rate Control of Test Model 5 for MPEG-2 378
16.5	Optimum Mode Decision 381
	16.5.1	Problem Formation 381
	16.5.2	Procedure for Obtaining the Optimal Mode 384
		16.5.2.1 Optimal Solution 384
		16.5.2.2 Near-Optimal Greedy Solution 385
	16.5.3	Practical Solution with New Criteria for the Selection of Coding Mode 386
16.6	Statistical Multiplexing Operations on Multiple Program Encoding 387
	16.6.1	Background of Statistical Multiplexing Operation 388
	16.6.2	VBR Encoders in StatMux 389
	16.6.3	Research Topics of StatMux 391
16.7	Summary 393
Exercises 393
References 394

Chapter 17 Application Issues of MPEG-1/2 Video Coding

17.1	Introduction 395
17.2	ATSC DTV Standards 395
	17.2.1 A Brief History 395
	17.2.2 Technical Overview of ATSC Systems 397
	17.2.2.1 Picture Layer 397
	17.2.2.2 Compression Layer 398
	17.2.2.3 Transport Layer 399
	17.2.2.4 Transmission Layer 399
17.3	Transcoding with Bitstream Scaling 400
	17.3.1 Background 400
	17.3.2 Basic Principles of Bitstream Scaling 402
	17.3.3 Architectures of Bitstream Scaling 403
	17.3.3.1 Architecture 1: Cutting AC Coefficients 403
	17.3.3.2 Architecture 2: Increasing Quantization Step 405
	17.3.3.3 Architecture 3: Re-Encoding with Old Motion Vectors and Old Decisions 406
	17.3.3.4 Architecture 4: Re-Encoding with Old Motion Vectors and New Decisions 407
	17.3.3.5 Comparison of Bistream Scaling Methods 407
	17.3.4 MPEG-2 to MPEG-4 Transcoding 409
17.4	Down-Conversion Decoder 410
	17.4.1 Background 410
	17.4.2 Frequency Synthesis Down-Conversion 412
	17.4.3 Low-Resolution Motion Compensation 413
	17.4.4 Three-Layer Scalable Decoder 416
	17.4.5 Summary of Down-Conversion Decoder 418

17.5	Error Concealment		421
	17.5.1	Background	421
	17.5.2	Error Concealment Algorithms	423
		17.5.2.1 Code Word Domain Error Concealment	424
		17.5.2.2 Spatio-Temporal Error Concealment	424
	17.5.3	Algorithm Enhancements	428
		17.5.3.1 Directional Interpolation	428
		17.5.3.2 I-Picture Motion Vectors	429
		17.5.3.3 Spatial Scalable Error Concealment	429
	17.5.4	Summary of Error Concealment	431
17.6	Summary		431
Exercises			432
References			432

Chapter 18 MPEG-4 Video Standard: Content-Based Video Coding

18.1	Introduction		435
18.2	MPEG-4 Requirements and Functionalities		436
	18.2.1	Content-Based Interactivity	436
		18.2.1.1 Content-Based Manipulation and Bitstream Editing	436
		18.2.1.2 Synthetic and Natural Hybrid Coding	436
		18.2.1.3 Improved Temporal Random Access	436
	18.2.2	Content-Based Efficient Compression	436
		18.2.2.1 Improved Coding Efficiency	437
		18.2.2.2 Coding of Multiple Concurrent Data Streams	437
	18.2.3	Universal Access	437
		18.2.3.1 Robustness in Error-Prone Environments	437
		18.2.3.2 Content-Based Scalability	437
	18.2.4	Summary of MPEG-4 Features	437
18.3	Technical Description of MPEG-4 Video		438
	18.3.1	Overview of MPEG-4 Video	438
	18.3.2	Motion Estimation and Compensation	439
		18.3.2.1 Adaptive Selection of 16×16 Block or Four 8×8 Blocks	440
		18.3.2.2 Overlapped Motion Compensation	440
	18.3.3	Texture Coding	441
		18.3.3.1 INTRA DC and AC Prediction	441
		18.3.3.2 Motion Estimation/Compensation of Arbitrary Shaped VOP	442
		18.3.3.3 Texture Coding of Arbitrary Shaped VOP	444
	18.3.4	Shape Coding	445
		18.3.4.1 Binary Shape Coding with CAE Algorithm	446
		18.3.4.2 Gray-Scale Shape Coding	448
	18.3.5	Sprite Coding	448
	18.3.6	Interlaced Video Coding	449
	18.3.7	Wavelet-Based Texture Coding	449
		18.3.7.1 Decomposition of the Texture Information	449
		18.3.7.2 Quantization of Wavelet Coefficients	450
		18.3.7.3 Coding of Wavelet Coefficients of Low–Low Band and Other Bands	450
		18.3.7.4 Adaptive Arithmetic Coder	450

		18.3.8	Generalized Spatial and Temporal Scalability	451

	18.3.9 Error Resilience	451
18.4	MPEG-4 Visual Bitstream Syntax and Semantics	453
18.5	MPEG-4 Visual Profiles and Levels	454
18.6	MPEG-4 Video Verification Model	455
	18.6.1 VOP-Based Encoding and Decoding Process	455
	18.6.2 Video Encoder	455
	18.6.2.1 Video Segmentation	456
	18.6.2.2 Intra/Inter Mode Decision	457
	18.6.2.3 Off-Line Sprite Generation	458
	18.6.2.4 Multiple VO Rate Control	458
	18.6.3 Video Decoder	459
18.7	Summary	460
Exercises		461
References		461

Chapter 19 ITU-T Video Coding Standards H.261 and H.263

19.1	Introduction	463
19.2	H.261 Video Coding Standard	463
	19.2.1 Overview of H.261 Video Coding Standard	463
	19.2.2 Technical Detail of H.261	464
	19.2.3 Syntax Description	466
	19.2.3.1 Picture Layer	466
	19.2.3.2 Group of Blocks Layer	466
	19.2.3.3 Macroblock Layer	466
	19.2.3.4 Block Layer	467
19.3	H.263 Video Coding Standard	467
	19.3.1 Overview of H.263 Video Coding	468
	19.3.2 Technical Features of H.263	468
	19.3.2.1 Half-Pixel Accuracy	468
	19.3.2.2 Unrestricted Motion Vector Mode	469
	19.3.2.3 Advanced Prediction Mode	469
	19.3.2.4 Syntax-Based Arithmetic Coding	472
	19.3.2.5 PB-Frames	472
19.4	H.263 Video Coding Standard Version 2	473
	19.4.1 Overview of H.263 Version 2	473
	19.4.2 New Features of H.263 Version 2	473
	19.4.2.1 Scalability	473
	19.4.2.2 Improved PB-Frames	474
	19.4.2.3 Advanced Intracoding	475
	19.4.2.4 Deblocking Filter	476
	19.4.2.5 Slice-Structured Mode	477
	19.4.2.6 Reference Picture Selection	477
	19.4.2.7 Independent Segmentation Decoding	478
	19.4.2.8 Reference Picture Resampling	478
	19.4.2.9 Reduced-Resolution Update	478
	19.4.2.10 Alternative Inter VLC and Modified Quantization	480
	19.4.2.11 Supplemental Enhancement Information	480
19.5	H.263++ Video Coding and H.26L	481

| 19.6 | Summary | 481 |

Exercises 481
References 482

Chapter 20 A New Video Coding Standard: H.264/AVC
20.1	Introduction	483
20.2	Overview of H.264/AVC Codec Structure	484
20.3	Technical Description of H.264/AVC Coding Tools	487
	20.3.1 Instantaneous Decoding Refresh Picture	487
	20.3.2 Switching I-Slices and Switching P-Slices	488
	20.3.3 Transform and Quantization	490
	20.3.4 Intraframe Coding with Directional Spatial Prediction	491
	20.3.5 Adaptive Block Size Motion Compensation	492
	20.3.6 Motion Compensation with Multiple References	493
	20.3.7 Entropy Coding	493
	20.3.8 Loop Filter	498
	20.3.9 Error-Resilience Tools	500
20.4	Profiles and Levels of H.264/AVC	502
	20.4.1 Profiles of H.264/AVC	502
	20.4.2 Levels of H.264/AVC	503
20.5	Summary	505

Exercises 505
References 505

Chapter 21 MPEG System: Video, Audio, and Data Multiplexing
21.1	Introduction	507
21.2	MPEG-2 System	508
	21.2.1 Major Technical Definitions in MPEG-2 System Document	509
	21.2.2 Transport Streams	510
		21.2.2.1 Structure of Transport Streams	511
		21.2.2.2 Transport Stream Syntax	512
	21.2.3 Transport Streams Splicing	515
	21.2.4 Program Streams	517
	21.2.5 Timing Model and Synchronization	518
21.3	MPEG-4 System	521
	21.3.1 Overview and Architecture	521
	21.3.2 Systems Decoder Model	523
	21.3.3 Scene Description	524
	21.3.4 Object Description Framework	525
21.4	Summary	526

Exercises 526
References 527

Index 529

Preface to the Second Edition

When looking at the preface of the first edition of this book published in 1999, it is observed that most of the presentation, analyses, and discussion made there are still valid. The trend of switching from analog to digital communications continues. Digital image and video, digital multimedia, Internet, and World Wide Web have been continuously and vigorously growing during the past eight years. Therefore, in this second edition of this book, we have retained most of the material of the first edition, but with some necessary updates and new additions. Two major and some minor changes made in this second edition are as follows.

First, the major parts of JPEG2000 have become standards after 1999. Hence, we have updated Chapter 8, which presents fundamental concepts and algorithms on JPEG2000. Second, a new chapter describing the recently developed video coding standard, MPEG-4 Part 10 Advanced Video Coding or H.264, has been added in the second edition as Chapter 20. For this purpose, Chapter 20 in the first edition covering the system part of MPEG, multiplexing/demultiplexing and synchronizing the coded audio, video, and other data has been changed to Chapter 21 in this new edition. Other minor changes have been made wherever necessary, including the addition of new color systems of digital video, profiles, and levels of video coding standards.

Acknowledgments

Both authors acknowledge the great efforts of Dr. Zhicheng Ni in preparing the solution manual of this book. Yun Qing Shi would like to thank Professor Guorong Xuan, Tongji University, Shanghai, China for constructive discussions on wavelet transform. Huifang Sun expresses his appreciation to his colleague, Dr. Anthony Vetro, for fruitful technical discussions and proofreading related to the new chapter (Chapter 20) in this edition. He would like to thank Drs. Ajay Divakaran and Fatih Porikli for their help in many aspects on this edition. He also extends his appreciation to many friends and colleagues of the MPEGers who provided MPEG documents and tutorial materials cited in some revised chapters of this edition. He would like to thank Drs. Richard Waters, Kent Wittenburg, Masatoshi Kameyama, and Joseph Katz for their continuing support and encouragement. He also would like thank Tokumichi Murakami and Kohtaro Asai for their friendly support and encouragement.

Yun Qing Shi
New Jersey Institute of Technology
Newark, New Jersey

Huifang Sun
Mitsubishi Electric Research Laboratories
Cambridge, Massachusetts

Preface to the First Edition

It is well known that in the 1960s the advent of the semiconductor computer and the space program swiftly brought the field of digital image processing into public focus. Since then the field has experienced rapid growth and has entered every aspect of modern technology. Since the early 1980s, digital image sequence processing has been an attractive research area because an image sequence, as a collection of images, may provide more information than a single image frame. The increased computational complexity and memory space required for image sequence processing are becoming more attainable. This is due to more advanced, achievable computational capability, resulting from the continuing progress made in technologies, especially those associated with the VLSI industry and information processing.

In addition to image and image sequence processing in the digitized domain, facsimile transmission has switched from analog to digital since the 1970s. However, the concept of high definition television (HDTV) when proposed in the late 1970s and early 1980s continued to be analog. This has since changed. In the United States, the first digital system proposal for HDTV appeared in 1990. The Advanced Television Standards Committee (ATSC), formed by the television industry, recommended the digital HDTV system developed jointly by the seven Grand Alliance members as the standard, which was approved by the Federal Communication Commission (FCC) in 1997. Today's worldwide prevailing concept of HDTV is digital. The digital television (DTV) provides the signal that can be used in computers. Consequently, the marriage of TV and computers has begun. Direct broadcasting by satellite (DBS), digital video disks (DVD), video-on-demanding (VOD), video game and other digital video related media and services are now, or soon to be, available.

As in the case of image and video transmission and storage, audio transmission and storage through some media have changed from analog to digital. Examples include entertainment audio on compact disks (CD) and telephone transmission over long and medium distance. Digital TV signals discussed above provide another example since they include audio signals. Transmission and storage of audio signals through some other media are about to change to digital. Examples of this include telephone transmission through local area and cable TV.

Although most signals generated from various sensors are analog in nature, the switching from analog to digital is motivated by the superiority of digital signal processing and transmission over their analog counterparts. The principal advantage of being digital is the robustness against various noises. Clearly, this results from the fact that only binary digits exist in digital format and it is much easier to distinguish one state from the other than to handle analog signals.

Another advantage of being digital is ease of signal manipulation. In addition to the development of a variety of digital signal (including image, video and audio) processing techniques and specially designed software and hardware, that may be well-known, the following development is an example of this advantage. The digitized information format, i.e., the bitstream, often in the compressed version, is a revolutionary change in the video industry that enables many manipulations, which are either impossible or very complicated to execute in analog format. For instance, video, audio and other data can be first compressed to separate bitstreams and then combined to a signal bitstream, thus providing a multimedia solution for many practical applications. Information from different sources and to different devices can be multiplexed and demultiplexed in terms of the bitstream. Bitstream conversion in terms of bit rate conversion, resolution conversion and syntax conversion becomes feasible. In digital video, content-based coding, retrieval and manipulation; and editing video in the compressed domain become feasible. All system-timing signals in the digital systems can be included in the bitstream instead of being transmitted separately as in traditional analog systems.

Being digital is well-suited to the recent development of modern telecommunication structures as exemplified by the Internet and World Wide Web (WWW). Therefore, we can see that digital computers, consumer electronics (including television and video games), and telecommunications networks are combined to produce an information revolution. Combining audio, video and other data, multimedia becomes an indispensable element of modern life. While the pace and the future of this revolution cannot be predicted, one thing is certain: this process is going to drastically change many aspects of our world in the next several decades.

One of the enabling technologies in the information revolution is digital data compression, because the digitization of analog signals causes data expansion. In other words, store and/or transmit digitized signals need more bandwidth and/or storage space than the original analog signals do.

The focus of this book is on image and video compression encountered in multimedia engineering. Fundamentals, algorithms and standards are three emphases of the book. It is intended to serve as a graduate level text. Its material is sufficient for a one-semester or one-quarter graduate course on digital image and video coding. For this purpose, at the end of each chapter, there is a section of exercises, containing problems and projects, for practice, and a section of references for further reading.

Based on this book, a short course, entitled "Image and Video Compression for Multimedia," was conducted at Nanyang Technological University, Singapore in March and April, 1999. The response to the short course was overwhelmingly positive.

Acknowledgments

We are pleased to express our gratitude here for the support and help we received in the course of writing this book.

The first author thanks his friend and former colleague Dr. C.Q. Shu for fruitful technical discussion related to some contents of the book. Sincere thanks also are directed to several of his friends and former students, Drs. J.N. Pan, X. Xia, S. Lin and Y. Shi, for their technical contributions and computer simulations related to some subjects of the book. He is grateful to Ms. L. Fitton for her English editing of 11 chapters, and to Dr. Z.F. Chen for her help in preparing many graphics.

The second author expresses his appreciation to his colleagues Anthony Vetro and Ajay Divakaran for fruitful technical discussion related to some contents of the book and for their proofreading of nine chapters. He also extends his appreciation to Dr. Weiping Li and Xiaobing Lee for their help for providing some useful references, and to many friends and colleagues of the MPEGers who provided wonderful MPEG documents and tutorial materials that are cited in some chapters of this book.

Both authors would like to express their deep appreciation to Dr. Z.F. Chen for her great help in formatting all the chapters of the book. They both thank Dr. F. Chichester for his help in preparing the book.

Special thanks go to the editor-in-chief of the Digital Image Processing book series of the CRC Press, Dr. P. Laplante, for his constant encouragement and guidance. The help from the acquisition editor of Electrical Engineering of the CRC Press, Nora Konopka, is appreciated.

The first author acknowledges the support he received associated with the book writing from the Electrical and Computer Engineering Department at New Jersey Institute of Technology, New Jersey, U.S.A. In particular, thanks are directed to the department chairman, Professor R. Haddad, and the associate chairman, Professor K. Sohn. He is also grateful to the Division of Information Engineering and the Electrical and Electronic Engineering School at Nanyang Technological University (NTU), Singapore for the support he received during his sabbatical leave. It was in Singapore that he finished writing the manuscript. In particular, thanks go to the Dean of the School Professor Er Meng Hwa and the Division head Professor A.C. Kot. With pleasure, he expresses his appreciation to many of his colleagues at NTU for their encouragement and help. In particular, his thanks go to Drs. G. Li and J.S. Li, and Dr. G.A. Bi. Thanks are also directed to many colleagues, graduate students and some technical staff from industrial companies in Singapore, who attended the short course, which was based on this book, in March/April 1999 and contributed their enthusiastic support and some fruitful discussion.

The last, but not the least, both authors thank their families for their patient support during the course of the writing. Without their understanding and support, we would not have been able to complete this book.

Yun Qing Shi and Huifang Sun
June 23, 1999

Content and Organization of the Book

This book consists of 21 chapters, grouped into four parts: (1) fundamentals, (2) still image compression, (3) motion estimation and compensation, and (4) video compression. The following paragraphs summarize the aim and content of each chapter, each part, and the relationship between some chapters and the four parts.

Part I includes Chapters 1–6, which provide readers with the fundamentals for understanding the remaining three parts of the book. In Chapter 1, the practical needs for image and video compression are demonstrated. The feasibility of image and video compression is analyzed. Specifically, both statistical and psychovisual redundancies are analyzed and the removal of these redundancies leads to image and video compression. In the course of the analysis, some fundamental characteristics of the human visual system are discussed. Visual quality measurement, another important concept in the compression, is addressed in both subjective and objective quality measures, and the new trend in combining the merits of the two measures is also discussed. Finally, some information theory results are presented as the concluding subject of the chapter.

Chapter 2 discusses quantization, a crucial step in lossy compression. It is known that quantization has a direct impact on both coding bit rate and quality of reconstructed frames. Both uniform and nonuniform quantizations are covered in this chapter. The issues of quantization distortion, optimum quantization, and adaptive quantization are also addressed. The last subject discussed in the chapter is pulse code modulation (PCM), which, as the earliest, best-established, and most frequently applied coding system, normally serves as a standard against which other coding techniques are compared.

Two efficient coding schemes, differential coding and transform coding (TC), are discussed in Chapters 3 and 4, respectively. Both techniques utilize the redundancies discussed in Chapter 1, thus achieving data compression. In Chapter 3, the formulation of general differential pulse code modulation (DPCM) systems is described first, followed by the discussion of optimum linear prediction and several implementation issues. Then, delta modulation (DM), as an important, simple, special case of DPCM is presented. Finally, application of differential coding technique to interframe coding and information preserving differential coding are covered.

Chapter 4 begins with the introduction of the Hotelling transform, the discrete version of the optimum Karhunen and Loeve transform. Through the statistical, geometrical, and basis vector (image) interpretations, this introduction provides a solid understanding of the transform coding technique. Several linear unitary transforms are then presented, followed by performance comparisons between these transforms in terms of energy compactness, mean square reconstruction error, and computational complexity. It is demonstrated that the discrete cosine transform (DCT) performs better than others in general. In the discussion of bit allocation, an efficient adaptive scheme using thresholding coding devised by Chen and Pratt in 1984 is featured, which established a basis for the international still image coding standard, JPEG. The comparison between DPCM and TC is also given. The combination of these two techniques (hybrid transform/waveform coding), and its application in image and video coding are also described.

The last two chapters in Part I cover several coding (e.g., code word assignment) techniques. In Chapter 5, two types of variable-length coding techniques, Huffman coding and arithmetic coding, are discussed. First, an introduction to basic coding theory is presented, which can be viewed as a continuation of the information theory results presented in Chapter 1. Then, the Huffman code, as an optimum and instantaneous code, and a modified version are covered. Huffman coding is a systematic procedure for encoding a source alphabet with each source symbol having an occurrence probability. As a block code (a fixed code word having an integer number of bits is assigned to a source symbol), it is optimum in the sense that it produces minimum coding redundancy. Some limitations of Huffman coding are analyzed. As a stream-based coding technique, arithmetic coding is distinct from and is gaining more popularity than Huffman coding. It maps a string of source symbols into a string of code symbols. Free of the integer-bits-per-source-symbol restriction, arithmetic coding is more efficient. The principle of arithmetic coding and some of its implementation issues are addressed.

While two types of variable-length coding techniques, introduced in Chapter 5, can be classified as fixed-length-to-variable-length coding techniques, run-length coding (RLC) and dictionary coding (as discussed in Chapter 6) can be classified as variable-length-to-fixed-length coding techniques. The discrete Markov source model (another portion of the information theory results), which can be used to characterize 1-D RLC, is introduced at the beginning of Chapter 6. Both 1-D RLC and 2-D RLC are then introduced. The comparison between 1-D and 2-D RLC is made in terms of coding efficiency and transmission error effect. The digital facsimile coding standards based on 1-D and 2-D RLC are introduced. Another focus of Chapter 6 is on dictionary coding. Two groups of adaptive dictionary coding techniques, the LZ77 and LZ78 algorithms, are presented. Their applications are discussed. At the end of the chapter, a discussion of international standards for lossless still image compression is given. For both lossless bilevel and multilevel still image compression, the respective standard algorithms and their performance comparisons are provided.

Part II of the book includes Chapters 7 through 9, which are devoted to still image compression. In Chapter 7, the international still image coding standard JPEG is introduced. Two classes of encoding, i.e., lossy and lossless, and four modes of operation, i.e., sequential DCT-based mode, progressive DCT-based mode, lossless mode, and hierarchical mode, are covered. The discussion in Part I is very useful in understanding what is introduced here for JPEG.

Because of the higher coding efficiency and superior spatial and quality scalability features over the DCT coding technique, discrete wavelet transform (DWT) coding has been adopted by JPEG2000 still image coding standards as the core technology. Chapter 8 begins with an introduction to wavelet transform (WT), which includes a comparison between WT and the short-time Fourier transform (STFT), and presents WT as a unification of several existing techniques, known as filter bank analysis, pyramid coding, and sub-band coding. Then the DWT for still image coding is discussed. In particular, the embedded zerotree wavelet (EZW) technique and set partitioning in hierarchical trees (SPIHT) are discussed. The updated JPEG2000 standard activity is also presented here.

Chapter 9 presents three nonstandard still image coding techniques: vector quantization (VQ), fractal coding, and model-based image coding. All three techniques have several important features such as very high compression ratio for certain kinds of images, and very simple decoding procedures. Owing to some limitations, however, they have not been adopted by the still image coding standards. On the other hand, the facial model and face animation techniques have been adopted by the MPEG-4 video standard.

Part III of this book, consisting of Chapters 10 through 14, addresses motion estimation and motion compensation, which are key issues in modern video compression. Part III is a prerequisite to Part IV, which discusses various video coding standards. The first chapter in Part III, Chapter 10, introduces motion analysis and compensation in general. The chapter begins with the concept of imaging space, which characterizes all images and all image sequences in temporal and spatial domains. Both temporal and spatial image sequences are special proper subsets of the imaging space. A single image becomes merely a specific cross section of the imaging space. Two techniques in video compression utilizing interframe correlation, both developed in the late 1960s and early 1970s, are presented here. Frame replenishment is relatively simpler in modeling and implementation. However, motion compensated coding achieves higher coding efficiency and better quality in reconstructed frames with a 2-D displacement model. Motion analysis is then viewed from a signal processing perspective. Three techniques in motion analysis are briefly discussed. They are block matching, pel recursion, and optical flow, which are presented in detail in Chapters 11 through 13, respectively. Finally, other applications of motion compensation to image sequence processing are discussed.

Chapter 11 addresses the block matching technique, which is presently the most frequently used motion estimation technique. The chapter first presents the original block matching technique proposed by Jain and Jain. Several different matching criteria and search strategies are then discussed. A thresholding multiresolution block matching algorithm is described in some detail so as to provide an insight into the technique. Then, the limitations of block matching techniques are analyzed, from which several new improvements are presented. They include hierarchical block matching, multigrid block matching, predictive motion field segmentation, and overlapped block matching. All of these techniques modify the nonoverlapped, equally spaced, fix-sized, small rectangular block model proposed by Jain and Jain in some way so that the motion estimation is more accurate and has fewer block artifacts and overhead side information.

The pel recursive technique is discussed in Chapter 12. First, determination of 2-D displacement vectors is converted via the use of the displaced frame difference (DFD) concept to a minimization problem. Second, descent methods in optimization theory are discussed. In particular, the steepest descent method and Newton–Raphson method are addressed in terms of algorithm, convergence, and implementation issues such as selection of step-size and initial value. Third, the first pel recursive techniques proposed by Netravali and Robbins are presented. Finally, several improvement algorithms are described.

Optical flow, the third technique in motion estimation for video coding, is covered in Chapter 13. First, some fundamental issues in motion estimation are addressed. They include the difference and relationships between 2-D motion and optical flow, the aperture problem, and the ill-posed nature of motion estimation. The gradient-based and correlation-based approaches to optical flow determination are then discussed in detail. For the former, the Horn and Schunck algorithm is illustrated as a representative technique and some other algorithms are briefly introduced. For the latter, the Singh method is introduced as a representative technique. In particular, the concepts of conservation information and neighborhood information are emphasized. A correlation-feedback algorithm is presented in detail to provide an insight into the correlation technique. Finally, multiple attributes for conservation information are discussed.

Chapter 14, the last chapter in Part III, provides a further discussion and summary of 2-D motion estimation. First, a few features common to all three major techniques discussed in Chapters 11 through 13 are addressed. They are the aperture and ill-posed inverse problems, conservation and neighborhood information, occlusion and disocclusion, and rigid and nonrigid motion. Second, a variety of different classifications of motion estimation techniques are presented. Frequency domain methods are discussed as well. Third, performance comparison between three major techniques in motion estimation is made. Finally, the new trends in motion estimation are presented.

Part IV, containing Chapters 15 through 21, covers various video coding standards. Chapter 15 presents fundamentals of video coding. First, digital video representation is discussed. Second, the rate distortion function of the video signal is covered, the fourth portion of the information theory results presented in this book. Third, various digital video formats are discussed. Finally, the current digital image/video coding standards are summarized. The full names and abbreviations of some organizations, the completion time, and the major features of various image/video coding standards are listed in two tables.

Chapter 16 is devoted to video coding standards MPEG-1/2, which are the most widely used video coding standards at present. The basic technique of MPEG-1/2 is a full-motion compensated DCT and DPCM hybrid coding algorithm. The features of MPEG-1 (including layered data structure) and the MPEG-2 enhancements (including field/frame modes for supporting the interlaced video input and scalability extension) are described. Issues of rate control, optimum mode decision, and multiplexing are discussed.

Chapter 17 presents several application examples of MPEG-1/2 video standards. They are the ATSC DTV standard, approved by the Federal Communications Commission (FCC) in the United States, transcoding, down-conversion decoder, and error concealment. Discussion of these applications can enhance understanding and mastering of MPEG-1/2 standards. Some research work reported may be found helpful for graduate students to broaden their knowledge of digital video processing, an active research field.

Chapter 18 presents the MPEG-4 video standard. The predominant feature of MPEG-4, content-based manipulation, is emphasized. The underlying concept of audio/visual objects (AVOs) is introduced. The important functionalities of MPEG-4, i.e., content-based interactivity (including bitstream editing, synthetic and natural hybrid coding (SNHC)), content-based coding efficiency, and universal access (including content-based scalability), are discussed. Since neither MPEG-1 nor MPEG-2 includes synthetic video and content-based coding, the most important application of MPEG-4 is in a multimedia environment.

Chapter 19 introduces ITU-T video coding standards H.261 and H.263, which are utilized mainly for videophony and videoconferencing. The basic technical detail of H.261, the earliest video coding standard, is presented. The technical improvements with which H.263 achieves high coding efficiency are discussed. Features of H.263+, H.263++, and H.26L are also presented.

Chapter 20 introduces the recently developed video coding standard, MPEG-4 Part 10 Advanced Video Coding [H.264], which is jointly developed by joint video team (JVC) of MPEG and ITU-T VCEG, and it is simply called as H.264/AVC. The H.264/AVC is an efficient and state-of-the-art video compression standard, whose coding efficiency is about two times better than that of MPEG-2 at the expense of increased complexity. The H.264/AVC has been planed for many applications including HD-DVD, DTV for satellite and wireless networks, IPTV, and many others.

Chapter 21 covers the systems part of MPEG, multiplexing/demultiplexing and synchronizing the coded audio, video and other data. Specifically, MPEG-2 systems and MPEG-4 systems are introduced. In MPEG-2 systems, two forms i.e., program stream and transport stream, are described. In MPEG-4 systems, some multimedia application related issues are discussed.

Yun Qing Shi
New Jersey Institute of Technology
Newark, New Jersey

Huifang Sun
Mitsubishi Electric Research Laboratories
Cambridge, Massachusetts

Authors

Yun Qing Shi joined the New Jersey Institute of Technology (NJIT), Newark, New Jersey in 1987, and is currently a professor of Electrical and Computer Engineering. He obtained his BS and MS from Shanghai Jiao Tong University, Shanghai, China, and his MS and PhD from the University of Pittsburgh, Pennsylvania. His research interests include visual signal processing and communications, multimedia data hiding and security, theory of multidimensional systems, and signal processing. Before entering graduate school, he had industrial experience in numerical control manufacturing and electronic broadcasting. Some of his research projects have been funded by several federal and New Jersey state agencies.

Dr. Shi is an author and coauthor of 200 papers, one book, and four book chapters. He holds two U.S. patents, and has 20 U.S. patents pending (all of these pending patents have been licensed to third parties by NJIT). He is the chairman of the Signal Processing Chapter of IEEE North Jersey Section, the founding editor-in-chief of *LNCS Transactions on Data Hiding and Multimedia Security* (Springer), an editorial board member of *Multidimensional Systems and Signal Processing* (Springer), a member of IEEE Circuits and Systems Society's (CASS) three technical committees, the technical chair of IEEE International Conference on Multimedia and Expo 2007 (ICME07), a co-technical chair of International Workshop on Digital Watermarking 2007 (IWDW07), and a fellow of IEEE. He was an associate editor of *IEEE Transactions on Signal Processing*, *IEEE Transactions on Circuits and Systems Part II*, a guest editor of special issues for several journals, a formal reviewer of the *Mathematical Reviews*, a contributing author for *Comprehensive Dictionary of Electrical Engineering* (CRC), an IEEE CASS Distinguished Lecturer, a member of IEEE Signal Processing Society's Technical Committee of Multimedia Signal Processing, a co-general chair of IEEE 2002 International Workshop on Multimedia Signal Processing (MMSP02), a co-technical chair of MMSP05, and a co-technical chair of IWDW06.

Huifang Sun graduated from Harbin Engineering Institute, China, and received his PhD from the University of Ottawa, Canada. He joined the Electrical Engineering Department of Fairleigh Dickinson University in 1986 and was promoted to an associate professor before moving to Sarnoff Corporation in 1990. He joined the Sarnoff laboratory as a member of the technical staff and was later promoted to technology leader of digital video communication. In 1995, he joined Mitsubishi Electric Research Laboratories (MERL) as a senior principal technical staff member, and was promoted as vice president and fellow of MERL and deputy director in 2003. Dr. Sun's research interests include digital video/image compression and digital communication. He has coauthored two books and has published more than 150 journal and conference papers. He holds 48 U.S. patents. He received the technical achievement award in 1994 at the Sarnoff laboratory. He received the 1992 best paper award of *IEEE Transactions on Consumer Electronics*, the 1996 best paper award of ICCE, and the 2003 best paper award of *IEEE Transactions on Circuits and Systems for Video Technology*. Dr. Sun is now an associate editor for *IEEE Transactions on Circuits and Systems for Video Technology* and was the chair of the Visual Processing Technical Committee of IEEE Circuits and System Society. He is an IEEE Fellow.

Part I

Fundamentals

1
Introduction

Image and video data compression refers to a process in which the amount of data used to represent image and video is reduced to meet a bit rate requirement (below or at most equal to the maximum available bit rate), while the quality of the reconstructed image or video satisfies a requirement for a certain application and the complexity of computation involved is affordable for the application. In this book, the terms image and video data compression, image and video compression, and image and video coding are synonymous. Figure 1.1 shows the functionality of image and video data compression in visual transmission and storage. Image and video data compression has been found to be necessary in these important applications, because the huge amount of data involved in these and other applications usually well-exceeds the capability of today's hardware despite rapid advancements in semiconductor, computer, and other industries.

It is noted that both information and data are closely related yet different concepts. Data represents information and the quantity of data can be measured. In the context of digital image and video, data is usually measured in the number of binary units (bits). Information is defined as knowledge, facts, and news according to the Cambridge International Dictionary of English. That is, while data is the representation of knowledge, facts, and news, information is the knowledge, facts, and news. Information, however, may also be quantitatively measured.

Bit rate (also known as coding rate), as an important parameter in image and video compression, is often expressed in a unit of bits per second (bits/s, or bps), which is suitable in visual communication. In fact, an example in Section 1.1 concerning videophony (a case of visual transmission) uses bit rate in terms of bits per second. In the application of image storage, bit rate is usually expressed in a unit of bits per pixel (bpp). The term pixel is an abbreviation for picture element and is sometimes referred to as pel. In information source coding, bit rate is sometimes expressed in a unit of bits per symbol. In Section 1.4.2, when discussing noiseless source coding theorem, we consider bit rate as the average length of code words in the unit of bits per symbol.

The required quality of the reconstructed image and video is application dependent. In medical diagnosis and some scientific measurements, we may need the reconstructed image and video to mirror the original image and video. In other words, only reversible, information-preserving schemes are allowed. This type of compression is referred to as lossless compression. In applications, such as motion picture and television (TV), a certain amount of information loss is allowed. This type of compression is called lossy compression.

From its definition, one can see that image and video data compression involves several fundamental concepts including information, data, visual quality of image and video, and computational complexity. This chapter is concerned with several fundamental concepts in image and video compression. First, the necessity as well as the feasibility of image and video data compression are discussed. The discussion includes the utilization of several types of redundancy inherent in image and video data, and the visual perception of the

FIGURE 1.1
Image and video compression for visual transmission and storage.

human visual system (HVS). As the quality of the reconstructed image and video is one of our main concerns, the subjective and objective measures of visual quality are addressed. Then we present some fundamental information theory results, considering they play a key role in image and video compression.

1.1 Practical Needs for Image and Video Compression

Needless to say, visual information is of vital importance for human beings to perceive, recognize, and understand the surrounding world. With the tremendous progress that has been made in advanced technologies, particularly in very large-scale integrated (VLSI) circuits, increasingly powerful computers and computations, it is becoming more possible than ever for video to be widely utilized in our daily life. Examples include videophony, videoconferencing, high definition TV (HDTV), and digital video disk (also known as digital versatile disk [DVD]), to name a few.

Video as a sequence of video frames, however, involves a huge amount of data. Let us take a look at an illustrative example. Assume the present switch telephone network (PSTN) modem can operate at a maximum bit rate of 56,600 bits/s. Assume each video frame has a resolution of 288×352 (288 lines and 352 pixels/line), which is comparable with that of a normal TV picture and is referred to as common intermediate format (CIF). Each of the three primary colors RGB (red, green, blue) is represented for one pixel with 8 bits, as usual, and the frame rate in transmission is 30 frames/s to provide a continuous motion video. The required bit rate, then, is $288 \times 352 \times 8 \times 3 \times 30 = 72,990,720$ bits/s. Therefore, the ratio between the required bit rate and the largest possible bit rate is about 1289. This implies that we have to compress the video data by at least 1289 times in order to accomplish the transmission described in this example. Note that an audio signal has not been accounted for yet in this illustration.

With increasingly demanding video services, such as three-dimensional (3-D) movies and games, and high video quality, such as HDTV, advanced image, and video data compression is necessary. It becomes an enabling technology to bridge the gap between the required huge amount of video data and the limited hardware capability.

1.2 Feasibility of Image and Video Compression

In this section, we shall see that image and video compression is not only a necessity for rapid growth of digital visual communications, but is also feasible. Its feasibility rests with two types of redundancies, i.e., statistical redundancy and psychovisual redundancy. By eliminating these redundancies, we can achieve image and video compression.

Introduction

1.2.1 Statistical Redundancy

Statistical redundancy can be classified into two types: interpixel redundancy and coding redundancy. By interpixel redundancy we mean that pixels of an image frame, and pixels of a group of successive image or video frames, are not statistically independent. On the contrary, they are correlated to various degrees. (Difference and relationship between image and video sequences are discussed in Chapter 10, when we begin to discuss video compression). This type of interpixel correlation is referred to as interpixel redundancy. Interpixel redundancy can further be divided into two categories: spatial redundancy and temporal redundancy. By coding redundancy, we mean that the statistical redundancy is associated with coding techniques.

1.2.1.1 Spatial Redundancy

Spatial redundancy represents the statistical correlation between pixels within an image frame. Hence it is also called intraframe redundancy.

It is well known that for most properly sampled TV signals the normalized autocorrelation coefficients along a row (or a column) with a one-pixel shift is very close to the maximum value 1. That is, the intensity values of pixels along a row (or a column) have a very high autocorrelation (close to the maximum autocorrelation) with those of pixels along the same row (or the same column) but shifted by a pixel. This does not come as a surprise because most of the intensity values change continuously from pixel to pixel within an image frame except for the edge regions (Figure 1.2). Figure 1.2a is a pretty normal picture: a boy and a girl in a park, and is of a resolution of 883×710. The intensity profiles along the 318th row and the 262th column are depicted in Figure 1.2b and c, respectively. For easy reference, the positions of the 318th row and 262th column in the picture are shown in Figure 1.2d. That is, the vertical axis represents intensity values, while the horizontal axis indicates the pixel position within the row or the column. These two curves indicate that often intensity values change gradually from one pixel to the other along a row and along a column.

The study of the statistical properties of video signals can be traced back to the 1950s. Knowing that we must study and understand redundancy to remove it, Kretzmer designed some experimental devices such as a picture autocorrelator and a probabiloscope to measure several statistical quantities of TV signals and published his outstanding work in [kretzmer 1952]. He found that the autocorrelation in both horizontal and vertical directions exhibits similar behaviors as shown in Figure 1.3. Autocorrelation functions of several pictures with different complexity were measured. It was found that from picture to picture, the shape of autocorrelation curves ranges from remarkably linear to somewhat like exponential. The central symmetry with respect to the vertical axis and the bell-shaped distribution, however, remain generally the same. When the pixel shifting becomes small, it was found that the autocorrelation is high. This local autocorrelation can be as high as 0.97 to 0.99 for one- or two-pixel shifting. For very detailed pictures, it can range from 0.43 to 0.75. It was also found that autocorrelation generally has no preferred direction.

The Fourier transform of autocorrelation, power spectrum, is known as another important function in studying statistical behavior. Figure 1.4 shows a typical power spectrum of a TV signal [fink 1957; connor 1972]. It is reported that the spectrum is quite flat until 30 kHz for a broadcast TV signal. Beyond this line frequency the spectrum starts to drop at a rate of around 6 dB per octave. This reveals the heavy concentration of video signals in low frequencies, considering a nominal bandwidth of 5 MHz.

Spatial redundancy implies that the intensity value of a pixel can be guessed from that of its neighboring pixels. In other words, it is not necessary to represent each pixel in an image frame independently. Instead, one can predict a pixel from its neighbors. Predictive coding, also known as differential coding, is based on this observation and is discussed in Chapter 3. The direct consequence of recognition of spatial redundancy is that by removing a large amount of the redundancy (or utilizing the high correlation) within an image frame, we may save a lot of data in representing the frame, thus achieving data compression.

FIGURE 1.2 (See color insert following page 288.)
(a) A picture of boy and girl. (b) Intensity profile along 318th row.

Introduction

(c)

(d)

FIGURE 1.2 (continued)
(c) Intensity profile along 262th column. (d) Positions of 318th row and 262th column.

1.2.1.2 Temporal Redundancy

Temporal redundancy is concerned with the statistical correlation between pixels from successive frames in a temporal image or video sequence. Therefore, it is also called interframe redundancy.

Consider a temporal image sequence. That is, a camera is fixed in the 3-D world and it takes pictures of the scene one by one as time goes by. As long as the time interval between two consecutive pictures is short enough, i.e., the pictures are taken densely enough, we can

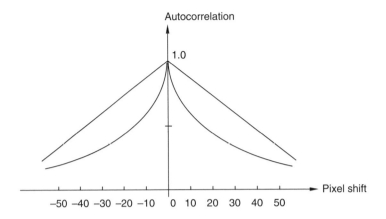

FIGURE 1.3
Autocorrelation in horizontal directions for some testing pictures. (From Kretzmer, E.R., *Bell Syst. Tech. J.*, 31, 751, 1952. With permission.)

imagine that the similarity between two neighboring frames is strong. Figure 1.5a and b shows the 21st and 22nd frames of the "Miss America" sequence, respectively. The frames have a resolution of 176 × 144. Among the total 25,344 pixels, only 3.4% change their gray value more than 1% of the maximum gray value (255 in this case) from the 21st frame to the 22nd frame. This confirms an observation made in [mounts 1969]: For a videophone-like signal with moderate motion in the scene, on average, less than 10% of pixels change their gray values between two consecutive frames by an amount of 1% of the peak signal. The high interframe correlation was reported in [kretzmer 1952]. There, the autocorrelation between two adjacent frames was measured for two typical motion picture films. The measured autocorrelations were 0.80 and 0.86. The concept of frame difference coding of television signals was also reported in [seyler 1962], and the probability density functions of television frame differences was analyzed in [seyler 1965]. In summary, pixels within successive frames usually bear a strong similarity or correlation. As a result, we may predict a frame from its neighboring frames along the temporal dimension. This is referred to as interframe predictive coding and is discussed in Chapter 3. A more precise, hence, more efficient interframe predictive coding

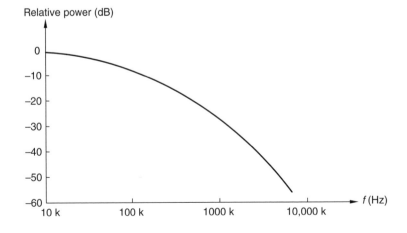

FIGURE 1.4
A typical power spectrum of a TV broadcast signal. (From Fink, D.G., *Television Engineering Handbook*, New York, 1957, Sect. 10.7. With permission.)

FIGURE 1.5
(a) 21st frame and (b) 22nd frame of the Miss America sequence.

scheme, which has been in development since the 1980s, uses motion analysis. That is, it considers that the changes from one frame to the next are mainly due to the motion of some objects in the frame. Taking this motion information into consideration, we refer to the method as motion compensated (MC) predictive coding. Both interframe correlation and MC predictive coding are discussed in detail in Chapter 10.

Removing a large amount of temporal redundancy leads to a great deal of data compression. At present, all the international video coding standards have adopted MC predictive coding, which has been a vital factor to the increased use of digital video in digital media.

1.2.1.3 Coding Redundancy

As we discussed, interpixel redundancy is concerned with the correlation between pixels. That is, some information associated with pixels is redundant. The psychovisual redundancy (Section 1.2.2) is related to the information that is psychovisually redundant, i.e., to which the HVS is not sensitive. It is hence clear that both interpixel and psychovisual redundancies are somehow associated with some information contained in image and video. Eliminating these redundancies, or utilizing these correlations, by using fewer bits to represent the information results in image and video data compression. In this sense, the coding redundancy is different. It has nothing to do with information redundancy but with the representation of information, i.e., coding itself. To see this, let us take a look at the following example.

One illustrative example is provided in Table 1.1. The first column lists five distinct symbols that need to be encoded. The second column contains occurrence probabilities of

TABLE 1.1
An Illustrative Example

Symbol	Occurrence Probability	Code 1	Code 2
a_1	0.1	000	0000
a_2	0.2	001	01
a_3	0.5	010	1
a_4	0.05	011	0001
a_5	0.15	100	001

these five symbols. The third column lists code 1, a set of code words obtained by using uniform-length code word assignment. (This code is known as the natural binary code.) The fourth column lists code 2, in which each code word has a variable length. Therefore, code 2 is called the variable-length code. It is noted that the symbol with a higher occurrence probability is encoded with a shorter length. Let us examine the efficiency of the two different codes. That is, we will examine which one provides a shorter average length of code words. It is obvious that the average length of code words in code 1, $L_{\text{avg},1}$, is 3 bits. The average length of code words in code 2, $L_{\text{avg},2}$, can be calculated as follows:

$$L_{\text{avg},2} = 4 \times 0.1 + 2 \times 0.2 + 1 \times 0.5 + 4 \times 0.05 + 3 \times 0.15 = 1.95 \text{ bits/symbol}. \quad (1.1)$$

Therefore, it is concluded that code 2 with variable-length coding is more efficient than code 1 with natural binary coding.

From this example, we can see that for the same set of symbols, different codes may perform differently. Some may be more efficient than others. For the same amount of information, code 1 contains some redundancy. That is, some data in code 1 is not necessary and can be removed without any effect. Huffman coding and arithmetic coding, two variable-length coding techniques, are discussed in Chapter 5.

From the study of coding redundancy, it is clear that we should search for more efficient coding techniques to compress image and video data.

1.2.2 Psychovisual Redundancy

While interpixel redundancy inherently rests in image and video data, psychovisual redundancy originates from the characteristics of the HVS.

It is known that the HVS perceives the outside world in a rather complicated way. Its response to visual stimuli is a nonlinear function of the strength of some physical attributes of the stimuli such as intensity and color. HVS perception is different from camera sensing. In the VHS, visual information is not perceived equally; some information may be more important than other information. This implies that if we apply less data to represent less important visual information, perception will not be affected. In this sense, we see that some visual information is psychovisually redundant. Eliminating this type of psychovisual redundancy leads to data compression.

To understand this type of redundancy, let us study some properties of the HVS. We may model the human vision system as a cascade of two units [lim 1990], as depicted in Figure 1.6. The first one is a low-level processing unit that converts incident light into a neural signal. The second one is a high-level processing unit that extracts information from the neural signal. Although much research was carried out to investigate low-level processing, high-level processing remains wide open. The low-level processing unit is

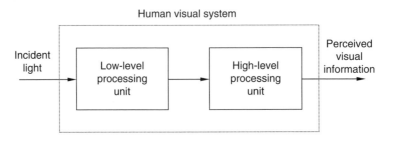

FIGURE 1.6
A two-unit cascade model of the human visual system (HVS).

known as a nonlinear system (approximately logarithmic, as shown below). As a great body of literature exists, we limit our discussion only to video compression-related results. That is, several aspects of the HVS, which are closely related to image and video compression, are discussed in this section. They are luminance masking, texture masking, frequency masking, temporal masking, and color maskings. Their relevance in image and video compression is addressed. Finally, a summary is provided, in which it is pointed out that all of these features can be unified as one: differential sensitivity. This seems to be the most important feature of the human visual perception.

1.2.2.1 Luminance Masking

Luminance masking concerns the brightness perception of the HVS, which is the most fundamental aspect among the five to be discussed here. Luminance masking is also referred to as luminance dependence [connor 1972] and contrast masking [legge 1980; watson 1987]. As pointed in [legge 1980], the term masking usually refers to a destructive interaction or interference among stimuli that are closely coupled in time or space. This may result in a failure in detection, or errors in recognition. Here, we are mainly concerned with the detectability of one stimulus when another stimulus is present simultaneously. The effect of one stimulus on the detectability of another, however, does not have to decrease detectability. Indeed, there are some cases in which a low-contrast masker increases the detectability of a signal. This is sometimes referred to as facilitation, but in this discussion we only use the term masking.

Consider the monochrome image shown in Figure 1.7. There, a uniform disk-shaped object with a gray level (intensity value) I_1 is imposed on a uniform background with a gray level I_2. Now the question is: Under what circumstances can the disk-shaped object be discriminated from the background by the HVS? That is, we want to study the effect of one stimulus (the background in this example, the masker) on the detectability of another stimulus (in this example, the disk). Two extreme cases are obvious. That is, if the difference between the two gray levels is quite large, the HVS has no problem with discrimination, or in other words the HVS notices the object from the background. If, on the other hand, the two gray levels are the same, the HVS cannot identify the existence of the object. What we are concerned with here is the critical threshold in the gray level difference for discrimination to take place.

If we define the threshold ΔI as such a gray level difference $\Delta I = I_1 - I_2$ that the object can be noticed by the HVS with a 50% chance, then we have the following relation, known as contrast sensitivity function, according to Weber's law.

$$\frac{\Delta I}{I} \approx \text{constant}, \tag{1.2}$$

where the constant is approximately 0.02. Weber's law states that for a relatively very wide range of I, the threshold for discrimination, ΔI, is directly proportional to the intensity I. The implication of this result is that when the background is bright, a larger difference in gray levels is needed for the HVS to discriminate the object from the background.

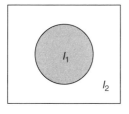

FIGURE 1.7
A uniform object with gray level I_1 imposed on a uniform background with gray level I_2.

On the other hand, the intensity difference required could be smaller if the background is relatively dark. It is noted that Equation 1.1 implies a logarithmic response of the HVS, and Weber's law holds for all other human senses as well.

Further research has indicated that the luminance threshold ΔI increases more slowly than predicted by Weber's law. Some more accurate contrast sensitivity functions have been presented in the literature. In [legge 1980], it was reported that an exponential function replaces the linear relation in Weber's law. The following exponential expression is reported in [watson 1987].

$$\Delta I = I_0 \cdot \max\left\{1, \left(\frac{I}{I_0}\right)^\alpha\right\}, \tag{1.3}$$

where I_0 is the luminance detection threshold when the gray level of the background is equal to zero, i.e., $I = 0$, and α is a constant, approximately equal to 0.7.

Figure 1.8 shows a picture uniformly corrupted by additive white Gaussian noise (AWGN). It can be observed that the noise is more visible in the dark area than in the bright area if comparing, for instance, the dark portion and the bright portion of the cloud above the bridge. This indicates that noise filtering is more necessary in the dark area than in the bright area. The lighter area can accommodate more additive noise before the noise becomes visible. This property has found application in embedding digital watermarks [huang 1998].

The direct impact that luminance masking has on image and video compression is related to quantization, which is covered in detail in Chapter 2. Roughly speaking, quantization is a process that converts a continuously distributed quantity into a set of

(a)

FIGURE 1.8 (See color insert following page 288.)
The bridge in Vancouver: (a) Original [Courtesy of Minhuai Shi].

Introduction

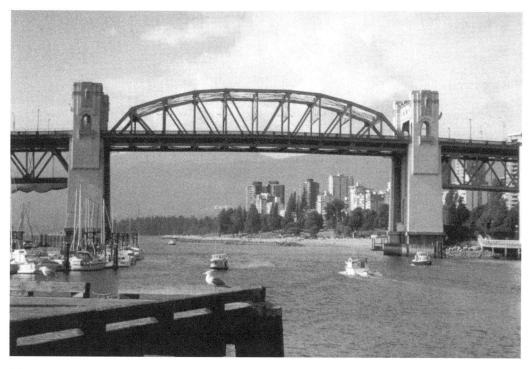

(b)

FIGURE 1.8 (continued)
(b) Uniformly corrupted by additive white Gaussian noise (AWGN).

finitely many distinct quantities. The number of these distinct quantities (known as quantization levels) is one of the keys in quantizer design. It significantly influences the resulting bit rate and the quality of the reconstructed image and video. An effective quantizer should be able to minimize the visibility of quantization error. The contrast sensitivity function provides a guideline in analysis of the visibility of quantization error. Therefore, it can be applied to quantizer design. Luminance masking suggests a nonuniform quantization scheme that takes the contrast sensitivity function into consideration. One such example was presented in [watson 1987].

1.2.2.2 Texture Masking

Texture masking is sometimes also called detail dependence [connor 1972], spatial masking [netravali 1977; lim 1990], or activity masking [mitchell 1996]. It states that the discrimination threshold increases with increasing picture detail. That is, the stronger the texture, the larger the discrimination threshold. In Figure 1.8, it can be observed that the additive random noise is less pronounced in the strong texture area than in the smooth area if comparing, for instance, the dark portion of the cloud (the up-right corner of the picture) with the water area (the bottom-right corner of the picture). This is in confirmation of texture masking.

In Figure 1.9b, the number of quantization levels decreases from 256 (as in Figure 1.9a) to 16. That is, we use only 4 bits, instead of 8 bits, to represent the intensity value for each pixel. The unnatural contours, caused by coarse quantization, can be noticed in the relative uniform regions, compared with Figure 1.9a. This phenomenon was first noted in [goodall 1951] and is called false contouring [gonzalez 1992]. Now we see that the false contouring

14   *Image and Video Compression for Multimedia Engineering*

(a)

(b)

FIGURE 1.9 (See color insert following page 288.)
Christmas in Winorlia: (a) Original, (b) Four-bit quantized.

Introduction

(c)

FIGURE 1.9 (continued)
(c) Improved IGS quantized with 4 bits.

can be explained by using texture masking because texture masking indicates that the human eye is more sensitive to the smooth region than to the textured region, where intensity exhibits a high variation. A direct impact on image and video compression is that the number of quantization levels, which affects bit rate significantly, should be adapted according to the intensity variation of image regions.

1.2.2.3 Frequency Masking

While the above two characteristics are picture dependent in nature, frequency masking is picture independent. It states that the discrimination threshold increases with frequency increase. It is also referred to as frequency dependence.

Frequency masking can be well illustrated by using Figure 1.9. In Figure 1.9c, high-frequency random noise has been added to the original image before quantization. This method is referred to as the improved gray-scale (IGS) quantization [gonzalez 1992, p. 318]. With the same number of quantization levels (16) as in Figure 1.9b, the picture quality of Figure 1.9c is improved dramatically compared with that of Figure 1.9b: the annoying false contours have disappeared despite the increase of the root mean square value of the total noise in Figure 1.9c. This is due to the fact that the low-frequency quantization error is converted to the high-frequency noise, and that the HVS is less sensitive to the high-frequency content. We thus see, as pointed out in [connor 1972], that our human eyes function like a low-pass filter.

Owing to frequency masking, in the transform domain, say, the discrete cosine transform (DCT) domain, we can drop some high-frequency coefficients with small magnitudes to achieve data compression without noticeably affecting the perception of the HVS. This leads to a technique called transform coding, discussed in Chapter 4.

1.2.2.4 Temporal Masking

Temporal masking is another picture-independent feature of the HVS. It states that it takes a while for the HVS to adapt itself to the scene when the scene changes abruptly. During this transition the HVS is not sensitive to details. The masking takes place both before and after the abrupt change. It is called forward temporal masking if it happens after the scene change; otherwise, it is referred to backward temporal masking [mitchell 1996].

This implies that one should take temporal masking into consideration when allocating data in image and video coding.

1.2.2.5 Color Masking

Digital color image processing is gaining increasing popularity due to the wide application of color images in modern life. As mentioned earlier, we are not going to cover all aspects of the perception of the HVS. Instead, we cover only those aspects related to psychovisual redundancy, thus to image and video compression. Therefore, our discussion here on color perception is by no means exhaustive.

In physics, it is known that any visible light corresponds to an electromagnetic spectral distribution. Therefore, a color, as a sensation of visible light, is an energy with an intensity as well as a set of wavelengths associated with the electromagnetic spectrum. Obviously, intensity is an attribute of visible light. The composition of wavelengths is another attribute: chrominance. There are two elements in the chrominance attribute: hue and saturation. The hue of a color is characterized by the dominant wavelength in the composition. Saturation is a measure of the purity of a color. A pure color has a saturation of 100%, whereas a white light has a saturation of 0.

1.2.2.5.1 RGB Model

The RGB primary color system is the most well known among several color systems. This is due to the following feature of the human perception of color. The color sensitive area in the HVS consists of three different sets of cones and each set is sensitive to the light of one of the three primary colors: red, green, and blue. Consequently, any color sensed by the HVS can be considered as a particular linear combination of the three primary colors. Many research results are available, the C.I.E. (Commission Internationale de l'Eclairage) chromaticity diagram being a well-known example. These results can be easily found in many classic optics and digital image processing texts.

The RGB model is used mainly in color image acquisition and display. In color signal processing including image and video compression, however, the luminance–chrominance color system is more efficient and, hence, widely used. This has something to do with the color perception of the HVS. It is known that the HVS is more sensitive to green than to red, and is least sensitive to blue. An equal representation of red, green, and blue leads to inefficient data representation when the HVS is the ultimate viewer. Allocating data only to the information that the HVS can perceive, on the other hand, can make video coding more efficient.

Luminance is concerned with the perceived brightness, while chrominance is related to the perception of hue and saturation of color. That is, roughly speaking, the luminance–chrominance representation agrees more with the color perception of the HVS. This feature makes the luminance–chrominance color models more suitable for color image processing. A good example is presented in [gonzalez 1992], about histogram equalization. It is well known that applying histogram equalization can bring out some details originally in dark regions. Applying histogram equalization to the RGB components separately can certainly achieve the goal. In doing so, however, the chrominance elements hue and saturation have been changed, thus leading to color distortion. With a luminance–chrominance model, histogram equalization can be applied to the luminance component only. Hence, the details in the dark regions are brought out, whereas the chrominance elements remain unchanged, hence no color distortion. With the luminance component Y serving as a black–white signal, a luminance–chrominance color model offers compatibility with black and white TV systems. This is another merit of luminance–chrominance color models.

To be discussed next are several different luminance–chrominance color models: HSI, YUV, YCbCr, and YIQ.

1.2.2.5.2 Gamma-Correction

It is known that a nonlinear relationship (basically a power function) exists between electrical signal magnitude and light intensity for both cameras and CRT-based display monitors [haskell 1996]. That is, the light intensity is a linear function of the signal voltage raised to the power of γ. It is a common practice to correct this nonlinearity before transmission. This is referred to as gamma-correction. The gamma-corrected RGB components are denoted by R', G', and B', respectively. They are used in the discussion on various color models. For the sake of notational brevity, we simply use R, G, and B instead of R', G', and B' in the following discussion, while keeping the gamma-correction in mind.

1.2.2.5.3 HSI Model

In this model, I stands for the intensity component, H for the hue component, and S for saturation component. One merit of this color system is that the intensity component is decoupled from the chromatic components. As analyzed above, this decoupling usually facilitates color image processing tasks. Another merit is that this model is closely related to the way the HVS perceives color pictures. Its main drawback is the complicated conversion between RGB and HSI models. A detailed derivation of the conversion may be found in [gonzalez 1992]. Because of this complexity, the HSI model is not used in any TV systems.

1.2.2.5.4 YUV Model

In this model, Y denotes the luminance component, and U and V are the two chrominance components. The luminance Y can be determined from the RGB model via the following relation:

$$Y = 0.299R + 0.587G + 0.114B. \tag{1.4}$$

It is noted that the three weights associated with the three primary colors, R, G, and B, are not the same. Their different magnitudes reflect different responses of the HVS to different primary colors.

Instead of being directly related to hue and saturation, the other two chrominance components, U and V, are defined as color differences as follows:

$$U = 0.492(B-Y), \tag{1.5}$$

and

$$V = 0.877(R-Y). \tag{1.6}$$

In this way, the YUV model lowers computational complexity. It has been used in PAL (phase alternating line) TV systems. Note that PAL is an analog composite color TV standard and is used in most of the European countries, some Asian countries, and Australia. By composite systems, we mean both the luminance and chrominance components of the TV signals are multiplexed within the same channel. For completeness, an expression of YUV in terms of RGB is given below.

$$\begin{pmatrix} Y \\ U \\ V \end{pmatrix} = \begin{pmatrix} 0.299 & 0.587 & 0.114 \\ -0.147 & -0.289 & 0.436 \\ 0.615 & -0.515 & -0.100 \end{pmatrix} \begin{pmatrix} R \\ G \\ B \end{pmatrix}. \tag{1.7}$$

1.2.2.5.5 YIQ Model

This color space has been utilized in NTSC (National Television Systems Committee) TV systems for years. Note that NTSC is an analog composite color TV standard and is used in the North America and Japan. The Y still acts as the luminance component. The two chrominance components are the linear transformation of the U and V components defined in the YUV model. Specifically,

$$I = -0.545U + 0.839V, \tag{1.8}$$

and

$$Q = 0.839U + 0.545V. \tag{1.9}$$

Substituting the U and V components expressed in Equations 1.4 and 1.5 into Equations 1.8 and 1.9, we can express YIQ directly in terms of RGB. That is,

$$\begin{pmatrix} Y \\ I \\ Q \end{pmatrix} = \begin{pmatrix} 0.299 & 0.587 & 0.114 \\ 0.596 & -0.275 & -0.321 \\ 0.212 & -0.523 & 0.311 \end{pmatrix} \begin{pmatrix} R \\ G \\ B \end{pmatrix} \tag{1.10}$$

1.2.2.5.6 YDbDr Model

The YDbDr model is used in SECAM (Sequential Couleur a Memoire) TV system. Note that SECAM is used in France, Russia, and some eastern European countries. The relationship between YDbDr and RGB appears below.

$$\begin{pmatrix} Y \\ Db \\ Dr \end{pmatrix} = \begin{pmatrix} 0.299 & 0.587 & 0.114 \\ -0.450 & -0.883 & 1.333 \\ -1.333 & 1.116 & -0.217 \end{pmatrix} \begin{pmatrix} R \\ G \\ B \end{pmatrix} \tag{1.11}$$

where

$$Db = 3.059U, \tag{1.12}$$

and

$$Dr = -2.169V. \tag{1.13}$$

Introduction

1.2.2.5.7 YCbCr Model

From the above, we can see that the U and V chrominance components are differences between the gamma-corrected color B and the luminance Y, and the gamma-corrected R and the luminance Y, respectively. The chrominance component pairs I and Q, and Db and Dr are both linear transforms of U and V. Hence they are very closely related to each other. It is noted that U and V may be negative as well. To make chrominance components nonnegative, the Y, U, and V components are scaled and shifted to produce the YCbCr model, which is used in the international coding standards JPEG and MPEG. (These two standards are covered in Chapters 7 and 16, respectively).

$$\begin{pmatrix} Y \\ Cb \\ Cr \end{pmatrix} = \begin{pmatrix} 0.257 & 0.504 & 0.098 \\ -0.148 & -0.291 & 0.439 \\ 0.439 & -0.368 & -0.071 \end{pmatrix} \begin{pmatrix} R \\ G \\ B \end{pmatrix} + \begin{pmatrix} 16 \\ 128 \\ 128 \end{pmatrix} \quad (1.14)$$

1.2.2.6 Color Masking and Its Application in Video Compression

It is well known that the HVS is much more sensitive to the luminance component Y than to the chrominance components U and V. Following Van Ness and Mullen [van ness 1967; mullen 1985], Mitchell, Pennebaker, Fogg, and LeGall included a figure in [mitchell 1996] to quantitatively illustrate the above statement. A modified version is shown in Figure 1.10. There, the abscissa represents spatial frequency in the unit of cycles per degree (cpd), while

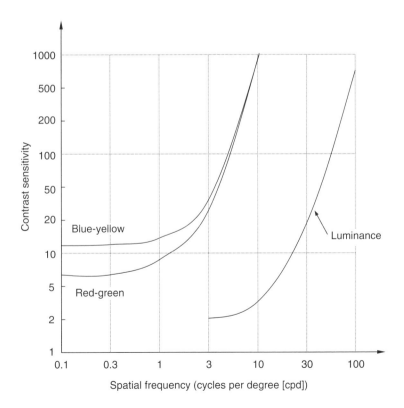

FIGURE 1.10
Contrast sensitivity versus spatial frequency. (From Van Ness, F.I. and Bouman, M.A., *J. Opt. Soc. Am.* 57, 401, 1967; Mullen, K.T., *J. Physiol.*, 359, 381, 1985.)

the ordinate is the contrast sensitivity defined for the sinusoidal testing signal. Two observations are in order. First, for each of the three curves, i.e., curves for the luminance component Y and the chrominance components U and V, the contrast sensitivity increases when spatial frequency increases, in general. This agrees with frequency masking discussed above. Second, for the same contrast sensitivity, we see that the luminance component corresponds to a much higher spatial frequency. This is an indication that the HVS is highly sensitive to luminance than to chrominance. This statement can also be confirmed, perhaps more easily, by examining those spatial frequencies at which all three curves have data available. Then we can see that the contrast sensitivity of luminance is much lower than that of the chrominance components.

The direct impact of color masking on image and video coding is that by utilizing this psychovisual feature, we can allocate more bits to the luminance component than to the chrominance components. This leads to a common practice in color image and video coding: using full resolution for the intensity component, while using a 2×1 subsampling both horizontally and vertically for the two chrominance components. This has been adopted in related international coding standards, discussed in Chapter 16.

1.2.2.7 Summary: Differential Sensitivity

In this section, we have discussed luminance, texture, frequency, temporal, and color maskings. Before we enter Section 1.3, let us summarize what we have discussed so far.

We see that luminance masking, also known as contrast masking, is of fundamental importance among several types of masking. It states that the sensitivity of the eyes to a stimulus depends on the intensity of another stimulus. Thus it is a differential sensitivity. Both texture (detail or activity) and frequency of another stimulus significantly influence this differential sensitivity. The same mechanism exists in color perception, where the HVS is highly sensitive to luminance than to chrominance. Therefore, we conclude that differential sensitivity is the key in studying human visual perception.

These features can be utilized to eliminate psychovisual redundancy, and thus compress image and video data.

It is also noted that variable quantization, which depends on activity and luminance in different regions, seems to be reasonable from a data compression point of view. Its practical applicability, however, is somehow questionable. That is, some experimental work does not support this expectation [mitchell 1996].

It is noted that this differential sensitivity feature of the HVS is common to human perception. For instance, there is also forward and backward temporal masking in human audio perception.

1.3 Visual Quality Measurement

As the definition of image and video compression indicates, image and video quality is an important factor in dealing with image and video compression. For instance, in evaluating two different compression methods, we have to base the evaluation on some definite image and video quality. When both methods achieve the same quality of reconstructed image and video, the one that requires less data is considered to be superior to the other. Alternatively, with the same amount of data, the method providing a higher quality reconstructed image or video is considered the better method. Note that here we have not considered other performance criteria, such as computational complexity.

Introduction

Surprisingly, however, it turns out that the measurement of image and video quality is not straightforward. There are two types of visual quality assessment. One is objective assessment (using electrical measurements), and the other is subjective assessment (using human observers). Each has its own merits and demerits. A combination of these two methods is now widely utilized in practice. In this section, we will first discuss subjective visual quality measurement, followed by objective quality measurement.

1.3.1 Subjective Quality Measurement

It is natural that the visual quality of reconstructed video frames should be judged by human viewers if they are to be the ultimate receivers of the data (see Figure 1.1). Therefore, the subjective visual quality measure plays an important role in visual communications.

In subjective visual quality measurement, a set of video frames are generated with varying coding parameters. Observers are invited to subjectively evaluate the visual quality of these frames. Specifically, observers are asked to rate the pictures by giving some measure of picture quality. Alternatively, observers are requested to provide some measure of impairment to the pictures. A five-scale rating system of the degree of impairment, used by Bell Laboratories, is listed below [sakrison 1979]. It has been adopted as one of the standard scales in CCIR Recommendation 500-3 [CCIR 1986]. Note that CCIR is now ITU-R (International Telecommunications Union-Recommendations).

1. Impairment is not noticeable.
2. Impairment is just noticeable.
3. Impairment is definitely noticeable, but not objectionable.
4. Impairment is objectionable.
5. Impairment is extremely objectionable.

In the subjective evaluation, there are a few things worth mentioning. In most applications, there is a whole array of pictures simultaneously available for evaluation. These pictures are generated with different encoding parameters. By keeping some parameters fixed while making one parameter (or a subset of parameters) free to change, the resulting quality rating can be used to study the effect of the one parameter (or the subset of parameters) on encoding. An example using this method to study the effect of varying numbers of quantization levels on image quality can be found in [gonzalez 1992].

Another possible way to study subjective evaluation is to identify pictures with the same subjective quality measure from the whole array of pictures. From this subset of test pictures, we can produce, in the encoding parameter space, isopreference curves that can be used to study the effect of the parameter(s) under investigation. An example using this method to study the effect of varying both image resolution and numbers of quantization levels on image quality can be found in [huang 1965].

In the rating, a whole array of pictures is usually divided into columns with each column sharing some common conditions. The evaluation starts within each column with a pairwise comparison. This is because pairwise comparison is relatively easy for the eyes. As a result, pictures in one column are arranged in an order according to visual quality, and quality or impairment measures are then assigned to the pictures in the one column. After each column has been rated, unification between columns is necessary. That is, different columns need to have a unified quality measurement. As pointed out in [sakrison 1979], this task is not easy because it means we may need to equate impairment that results from different types of errors.

One thing is understood from the above discussion: Subjective evaluation of visual quality is costly, and it needs large number of pictures and observers. The evaluation takes a long time because human eyes are easily fatigued and bored. Some special measures have to be taken to arrive at an accurate subjective quality measure. Examples in this regard include averaging subjective ratings and taking their deviation into consideration. For further details on subjective visual quality measurement, readers may refer to [sakrison 1979; hidaka 1990; webster 1993].

1.3.2 Objective Quality Measurement

In this section, we first introduce the concept of signal to noise ratio (SNR), which is a popularly utilized objective quality assessment. Then we present a promising new objective visual quality assessment technique based on human visual perception.

1.3.2.1 Signal to Noise Ratio

Consider Figure 1.11, where $f(x, y)$ is the input image to a processing system. The system can be a low-pass filter, a subsampling system, or a compression system.

It can even represent a process in which AWGN corrupts the input image. The $g(x, y)$ is the output of the system. In evaluating the quality of $g(x, y)$, we define an error function $e(x, y)$ as the difference between the input and the output. That is,

$$e(x,y) = f(x,y) - g(x,y). \tag{1.15}$$

The mean square error, E_{ms}, is defined as

$$E_{ms} = \frac{1}{MN} \sum_{x=0}^{M-1} \sum_{y=0}^{N-1} e(x,y)^2, \tag{1.16}$$

where M and N are the dimensions of the image in the horizontal and vertical directions. Note that it is sometimes denoted by MSE. The root mean square error, E_{rms}, is defined as

$$E_{rms} = \sqrt{E_{ms}}. \tag{1.17}$$

It is sometimes denoted by RMSE.

As noted earlier, SNR is widely used in objective quality measurement. Depending whether mean square error or root mean square error is used, the SNR may be called the mean square signal to noise ratio, SNR_{ms}, or the root mean square signal to noise ratio, SNR_{rms}. We have

$$SNR_{ms} = 10 \log_{10} \left(\frac{\sum_{x=0}^{M-1} \sum_{y=0}^{N-1} g(x,y)^2}{MN \cdot E_{ms}} \right), \tag{1.18}$$

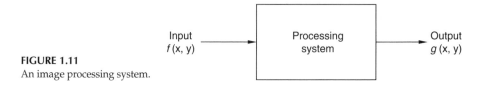

FIGURE 1.11
An image processing system.

and

$$SNR_{rms} = \sqrt{SNR_{ms}}. \qquad (1.19)$$

In image and video data compression, another closely related term, *PSNR* (peak signal to noise ratio), which is essentially a modified version of SNR_{ms}, is widely used. It is defined as follows:

$$PSNR = 10\log_{10}\left(\frac{255^2}{E_{ms}}\right) \qquad (1.20)$$

The interpretation of the SNR is that the larger the SNR (SNR_{ms}, SNR_{rms}, or *PSNR*) the better the quality of the processed image, $g(x, y)$. That is, the closer the processed image $g(x, y)$ is to the original image $f(x, y)$. This seems correct; however, from our earlier discussion about the features of the HVS, we know that the HVS does not respond to visual stimuli in a straightforward way. Its low-level processing unit is known to be nonlinear. Several masking phenomena exist. Each confirms that the visual perception of the HVS is not simple. It is worth noting that our understanding of the high-level processing unit of the HVS is far from complete. Therefore, we may understand that the SNR does not always provide us with reliable assessments of image quality. One good example is presented in Section 1.2.2.3, which uses the IGS quantization technique to achieve high compression (using only four bits for quantization instead of the usual eight bits) without introducing noticeable false contouring. In this case, the subjective quality is high, and the SNR decreases due to low-frequency quantization noise and additive high-frequency random noise. Another example drawn from our discussion about the masking phenomena is that some additive noise in bright areas or in highly textured regions may be masked, whereas some minor artifacts in dark and uniform regions may turn out to be quite annoying. In this case, the SNR cannot truthfully reflect visual quality.

On the one hand, we see that the objective quality measure does not always provide reliable picture quality assessment. On the other hand, however, its implementation is much faster and easier than that of the subjective quality measure. Furthermore, objective assessment is repeatable. Owing to these merits, objective quality assessment is still widely used despite this drawback.

It is noted that combining subjective and objective assessment has been a common practice in international coding-standard activity.

1.3.2.2 *An Objective Quality Measure Based on Human Visual Perception*

Introduced here is a new development in visual quality assessment, which is an objective quality measurement based on human visual perception [webster 1993]. Since it belongs to the category of objective assessment, it possesses merits, such as repeatability, and fast and easy implementation. On the other hand, based on human visual perception, its assessment of visual quality agrees closely to that of subjective assessment. In this sense, the new method attempts to combine the merits of the two different types of assessment.

1.3.2.2.1 *Motivation*

Visual quality assessment is best conducted via the subjective approach because in this case, the HVS is the ultimate viewer. The implementation of subjective assessment is, however, time-consuming, costly, and it lacks repeatability. On the other hand, although not always accurate, objective assessment is fast, easy, and repeatable. The motivation here is to develop an objective quality measurement system such that its quality assessment is very close to that obtained by using subjective assessment. To achieve this goal, this objective system is based on subjective assessment. That is, it uses the rating achieved via subjective assessment as a criterion to search for new objective measurements so as to have the objective rating as close to the subjective one as possible.

1.3.2.2.2 Methodology

The derivation of the objective quality assessment system is shown in Figure 1.12. The input testing video goes through a degradation block, resulting in degraded input video. The degradation block, or impairment generator, includes various video compression codecs (coder–decoder pairs) with bit rates ranging from 56 kbits/s to 45 Mbits/s, and other video operations. The input video and degraded input video form a pair of testing video, which is sent to a subjective assessment block as well as a statistical feature selection block.

A normal subjective visual quality assessment, as introduced in Section 1.3.2.2.2, is performed in the subjective assessment block, which involves a large panel of observers (e.g., 48 observers in [webster 1993]). In the statistical feature selection block, a variety of statistical operations are conducted and various statistical features are selected. Examples cover Sobel filtering, Laplacian operator, first-order differencing, moment calculation, fast Fourier transform, etc. Statistical measurements are then selected based on these statistical operations and features. An objective assessment is formed as follows:

$$\hat{s} = a_0 + \sum_{i=1}^{l} a_i n_i, \qquad (1.21)$$

where $\hat{s}$ denotes the output rating of the object assessment, or simply the objective measure, which is supposed to be a good estimate of the corresponding subjective score. The $n_i, i = 1, \ldots, l$ are selected objective measurements. The $a_0, a_i, i = 1, \ldots, l$ are coefficients in the linear model of the objective assessment.

The results of the objective and subjective assessments are applied to a statistical analysis block. In the statistical analysis block, the objective assessment rating is compared with the subjective assessment rating. The result of the comparison is fed back to the statistical

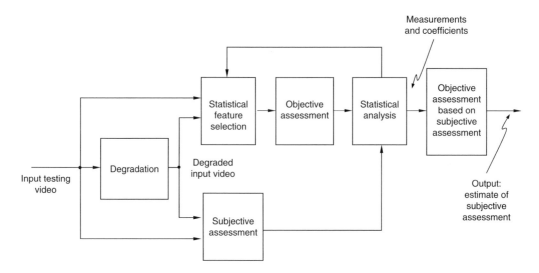

FIGURE 1.12
Block diagram of objective assessment based on subjective assessment.

feature selection block. The statistical measurements obtained in the statistical feature selection block are examined according to their performance in the assessment. A statistical measurement is regarded to be good if it can reduce by a significant amount the difference between the objective and the subjective assessments. The best measurement is determined via an exhaustive search among the various measurements. Note that the coefficients in Equation 1.21 are also examined in the statistical analysis block in a similar manner to that used for the measurements.

The measurements and coefficients determined after iterations result in an optimal objective assessment via Equation 1.21, which is finally passed to the last block as the output of the system. The whole process will become much clearer with the explanation provided below.

1.3.2.2.3 Results

The results reported in [webster 1993] are introduced here.

1.3.2.2.4 Information Features

As mentioned in Section 1.2.2, differential sensitivity is a key in human visual perception. Two selected features, perceived spatial information [SI] (the amount of spatial detail) and perceived temporal information [TI] (the amount of temporal luminance variation), involve pixel differencing. SI is defined as shown:

$$SI(f_n) = STD_s\{Sobel(f_n)\}, \quad (1.22)$$

where
STD_s denotes standard deviation operator in the spatial domain
$Sobel$ denotes the Sobel operation
f_n denotes the nth video frame

Temporal information is defined similarly:

$$TI(f_n) = STD_s\{\Delta f_n\}, \quad (1.23)$$

where $\Delta f_n = f_n - f_{n-1}$, i.e., the successive frame difference.

1.3.2.2.5 Determined Measurements

The parameter l in Equation 1.21 is chosen as three. That is,

$$\hat{s} = a_0 + a_1 n_1 + a_2 n_2 + a_3 n_3. \quad (1.24)$$

The measurements n_1, n_2, and n_3 are formulated based on the above-defined information features: SI and TI.

Measurement n_1

$$n_1 = RMS_t\left(5.81 \left|\frac{SI(\text{of}_n) - SI(\text{df}_n)}{SI(\text{of}_n)}\right|\right), \quad (1.25)$$

where RMS_t represents the root mean square value taken over the time dimension; of_n and df_n denote the original nth frame and the degraded nth frame, respectively. It is observed that n_1 is a measure of the relative change in the SI between the original frame and the degraded frame.

Measurement n_2

$$n_2 = \Im_t\{0.108 \cdot \text{MAX}\{[TI(\text{of}_n) - TI(\text{df}_n)], 0\}\}, \quad (1.26)$$

where

$$\Im_t\{y_t\} = STD_t\{\text{CONV}(y_t, [-1, 2, -1])\}, \quad (1.27)$$

where STD_t denotes the standard deviation operator with respect to time, and CONV indicates the convolution operation between its two arguments. It is understood that TI measures temporal luminance variation (temporal motion) and the convolution kernel, $[-1, 2, -1]$, enhances the variation due to its high-pass filter nature. Therefore, n_2 measures the difference of TI between the original and degraded frames.

Measurement n_3:

$$n_3 = \text{MAX}_t\left\{4.23 \cdot \log_{10}\left[\frac{TI(\text{df}_n)}{TI(\text{of}_n)}\right]\right\}, \quad (1.28)$$

where MAX_t indicates taking of the maximum value over time. Therefore, measurement n_3 responds to the ratio between the TI of the degraded video and that of the original video. Distortion, such as block artifacts (discussed in Chapter 11) and motion jerkiness (discussed in Chapter 10), which occurs in video coding, will cause n_3 to be large.

1.3.2.2.6 Objective Estimator

The least square error procedure is applied to testing video sequences with measurements n_i, $i = 1, 2, 3$, determined above, to minimize the difference between the rating scores obtained from the subjective and the objective assessments, resulting in the estimated coefficients a_0 and a_i, $i = 1, 2, 3$. Consequently, the objective assessment of visual quality $\hat{s}$ becomes

$$\hat{s} = 4.77 - 0.992 n_1 - 0.272 n_2 - 0.356 n_3. \quad (1.29)$$

1.3.2.2.7 Reported Experimental Results

It was reported that the correlation coefficient between the subjective and the objective assessment scores (an estimate of the subjective score) is in the range of 0.92 to 0.94. It is noted that a set of 36 testing scenes containing various amounts of SI and TI were used in the experiment. Hence, it is apparent that quite good performance was achieved.

Although there is surely room for further improvement, this work does open a new and promising way to assess visual quality by combining subjective and objective approaches. It is objective and thus fast and easy; and because it is based on the subjective measure, it is more accurate in terms of the high correlation to human perception. Theoretically, the SI and TI measures defined on differencing are very important. They reflect the most important aspect of human visual perception.

1.4 Information Theory Results

In the beginning of this chapter, it was noted that the term information is considered one of the fundamental concepts in image and video compression. We will now address some

Introduction

information theory results. In this section, measure of information and the entropy of an information source are covered first. We then introduce some coding theorems, which play a fundamental role in studying image and video compression.

1.4.1 Entropy

Entropy is a very important concept in information theory and communications. So is it in image and video compression. We first define the information content of a source symbol. Then we define entropy as average information content per symbol for a discrete memoryless source.

1.4.1.1 Information Measure

As mentioned at the beginning of this chapter, information is defined as knowledge, fact, and news. It can be measured quantitatively. The carriers of information are symbols. Consider a symbol with an occurrence probability p. Its information content (i.e., the amount of information contained in the symbol), I, is defined as follows:

$$I = \log_2 \frac{1}{p} \text{ bits} \quad \text{or} \quad I = -\log_2 p \text{ bits} \tag{1.30}$$

where the bit is a contraction of binary unit. In Equation 1.30, we set the base of the logarithmic function to equal 2. It is noted that these results can be easily converted as follows for the case where the r-ary digits are used for encoding. Hence, from now on, we restrict our discussion to binary encoding.

$$I = -\log_r 2 \cdot \log_2 p \text{ bits.} \tag{1.31}$$

According to Equation 1.30, the information contained within a symbol is a logarithmic function of its occurrence probability. The smaller the probability, the more information the symbol contains. This agrees with common sense. The occurrence probability is somewhat related to the uncertainty of the symbol. A small occurrence probability means large uncertainty. In this way, we see that the information content of a symbol is about the uncertainty of the symbol. It is noted that the information measure defined here is valid for both equally probable symbols and nonequally probable symbols [lathi 1998].

1.4.1.2 Average Information per Symbol

Now consider a discrete memoryless information source. By discreteness, we mean the source is a countable set of symbols. By memoryless, we mean the occurrence of a symbol in the set is independent of that of its preceding symbol. Take a look at a source of this type that contains m possible symbols $\{s_i, i=1,2,\ldots,m\}$. The corresponding occurrence probabilities are denoted by $\{p_i, i=1,2,\ldots,m\}$. According to the discussion in Section 1.4.1.1, the information content of a symbol s_i, I_i, is equal to $I_i = -\log_2 p_i$ bits. Entropy is defined as the average information content per symbol of the source. Obviously, the entropy, H, can be expressed as follows:

$$H = -\sum_{i=1}^{m} p_i \log_2 p_i \text{ bits} \tag{1.32}$$

From this definition, we see that the entropy of an information source is a function of occurrence probabilities. It is straightforward to show that the entropy reaches the maximum when all symbols in the set are equally probable.

1.4.2 Shannon's Noiseless Source Coding Theorem

Consider a discrete, memoryless, stationary information source. In what is called source encoding, a code word is assigned to each symbol in the source. The number of bits in the code word is referred to as the length of the code word. The average length of code words is referred to as bit rate, expressed in the unit of bits per symbol.

Shannon's noiseless source coding theorem states that for a discrete, memoryless, stationary information source, the minimum bit rate required to encode a symbol on average is equal to the entropy of the source. This theorem provides us with a lower bound in source coding. Shannon showed that the lower bound can be achieved when the encoding delay extends to infinity. By encoding delay, we mean the encoder waits and then encodes a certain number of symbols at once. Fortunately, with finite encoding delay, we can already achieve an average code word length fairly close to the entropy. That is, we do not have to actually sacrifice bit rate much to avoid long encoding delay, which involves high computational complexity and a large amount of memory space.

Note that the discreteness assumption is not necessary. We assume a discrete source simply because digital image and video are the focus here. Stationarity assumption is necessary in deriving the noiseless source coding theorem. This assumption may not be satisfied in practice. Hence, Shannon's theorem is a theoretical guideline only. There is no doubt, however, that it is a fundamental theoretical result in information theory.

In summary, the noiseless source coding theorem, Shannon's first theorem published in his celebrated paper [shannon 1948], is concerned with the case where both the channel and the coding system are noise free. The aim under these circumstances is coding compactness. The more compact, the better the coding. This theorem specifies the lower bound, which is the source entropy, and how to reach this lower bound.

One way to evaluate the efficiency of a coding scheme is to determine its efficiency with respect to the lower bound, i.e., entropy. The efficiency η is defined as follows:

$$\eta = \frac{H}{L_{avg}}, \qquad (1.33)$$

where H is entropy, and L_{avg} denotes the average length of the code words in the code. As the entropy is the lower bound, the efficiency never exceeds the unity, i.e., $\eta \leq 1$. The same definition can be generalized to calculate the relative efficiency between two codes. That is,

$$\eta = \frac{L_{avg,1}}{L_{avg,2}}, \qquad (1.34)$$

where $L_{avg,1}$ and $L_{avg,2}$ represent the average code word length for code 1 and code 2, respectively. We usually put the larger of the two in the denominator, and η is called the efficiency of code 2 with respect to code 1. A complementary parameter of coding efficiency is coding redundancy, ζ, which is defined as

$$\zeta = 1 - \eta. \qquad (1.35)$$

1.4.3 Shannon's Noisy Channel Coding Theorem

If a code has an efficiency of $\eta = 1$ (i.e., it reaches the lower bound of source encoding) then coding redundancy is $\zeta = 0$. Now consider a noisy transmission channel. In transmitting the coded symbol through the noisy channel, the received symbols may be erroneous due to the lack of redundancy. On the other hand, it is well known that by adding redundancy (e.g., parity check bits) some errors occurring during the transmission over the noisy channel may be identified. The coded symbols are then resent. In this way, we see that adding redundancy may combat noise.

Shannon's noisy channel coding theorem states that it is possible to transmit symbols over a noisy channel without error if the bit rate is below a channel capacity, C. That is,

$$R < C \tag{1.36}$$

where R denotes the bit rate. The channel capacity is determined by the noise and signal power.

In conclusion, the noisy channel coding theorem, Shannon's second theorem [shannon 1948], is concerned with a noisy, memoryless channel. By memoryless, we mean the channel output corresponding to the current input is independent of the output corresponding to previous input symbols. Under these circumstances, the aim is reliable communication. To be error-free, the bit rate cannot exceed channel capacity. That is, channel capacity sets an upper bound on the bit rate.

1.4.4 Shannon's Source Coding Theorem

As seen in Sections 1.4.2 and 1.4.3, the noiseless source coding theorem defines the lowest possible bit rate for noiseless source coding and noiseless channel transmission; whereas the noisy channel coding theorem defines the highest possible coding bit rate for error-free transmission. Therefore, both theorems work for reliable (no error) transmission. In this section, we continue to deal with discrete memoryless information sources, but we discuss the situation in which lossy coding is encountered. As a result, distortion of the information sources takes place. For instance, quantization, discussed in Chapter 2, causes information loss. Therefore, it is concluded that if an encoding procedure involves quantization, then it is lossy coding. That is, errors occur during the coding process, even though the channel is error-free. We want to find the lower bound of the bit rate for this case.

The source coding theorem [shannon 1948] states that for a given distortion D, there exists a rate distortion function $R(D)$ [berger 1971], which is the minimum bit rate required to transmit the source with distortion less than or equal to D. That is, to have distortion not larger than D, the bit rate, R, must satisfy the following condition:

$$R \geq R(D). \tag{1.37}$$

A more detailed discussion about this theorem and the rate distortion function is given in Chapter 15, which deals with video coding.

1.4.5 Information Transmission Theorem

It is clear that by combining the noisy channel coding theorem and the source coding theorem, we can derive the following relationship:

$$C \geq R(D) \tag{1.38}$$

This is called the information transmission theorem [slepian 1973]. It states that if the channel capacity of a noisy channel, C, is larger than the rate distortion function $R(D)$ then it is possible to transmit an information source with distortion D over a noisy channel.

1.5 Summary

In this chapter, we first discussed the necessity for image and video compression. It is shown that image and video compression becomes an enabling technique in today's exploding number of digital multimedia applications. Then, we show that the feasibility of image and video compression rests in redundancy removal. Three types of redundancy are studied: statistical redundancy, coding redundancy, and psychovisual redundancy. Statistical redundancy comes from interpixel correlation. By interpixel correlation, we mean correlation between pixels either located in one frame (spatial or intraframe redundancy) or pixels located in successive frames (temporal or interframe redundancy). Psychovisual redundancy is based on the features (several types of masking phenomena) of human visual perception. That is, visual information is not perceived equally from human visual point of view. In this sense, some information is psychovisually redundant. Coding redundancy is related to coding technique.

The visual quality of reconstructed image and video is a crucial criterion in the evaluation of the performance of visual transmission or storage systems. Both subjective and objective assessments are discussed. A new and promising objective technique based on subjective assessment is introduced. Because it combines the merits of both types of visual quality assessment, it achieves a quite satisfactory performance. The selected statistical features reveal some possible mechanism of the human visual perception. Further study in this regard would be fruitful.

In the last section, we introduced some fundamental information theory results, relevant to image and video compression. The results introduced include information measurement, entropy, and several theorems. All the theorems assume discrete, memoryless, and stationary information sources. The noiseless source coding theorem points out that the entropy of an information source is the lower bound of coding bit rate that a source encoder can achieve. The source coding theorem deals with lossy coding applied in a noise-free channel. It states that for a given distortion, D, there is a rate distortion function, $R(D)$. Whenever the bit rate in the source coding is greater than $R(D)$, the reconstructed source at the receiving end satisfies the fidelity requirement defined by the D. The noisy channel coding theorem states that, to achieve error-free performance, the source coding bit rate must be smaller than the channel capacity. Channel capacity is a function of noise and signal power. The information transmission theorem combines the noisy channel coding theorem and the source coding theorem. It states that it is possible to have a reconstructed waveform at the receiving end, satisfying the fidelity requirement corresponding to distortion, D, if the channel capacity, C, is larger than the rate distortion function, $R(D)$. Although some of the assumptions on which these theorems were developed, may not be valid in complicated practical situations, these theorems provide important theoretical limits for image and video coding. They can also be used for evaluation of the performance of different coding techniques.

Exercises

1. Using your own words, define spatial and temporal redundancy, and psychovisual redundancy, and state the impact they have on image and video compression.

2. Why is differential sensitivity considered the most important feature in human visual perception?
3. From the description of the newly developed objective assessment technique based on subjective assessment (Section 1.3), what points do you think are related to and support the statement made in problem 2?
4. Using your own words, interpret Weber's law.
5. What is the advantage possessed by color models that decouple the luminance component from chrominance components.
6. Why has the HIS model not been adopted by any TV systems?
7. What is the problem with the objective visual quality measure of *PSNR*?

References

[berger 1971] T. Berger, *Rate Distortion Theory*, Prentice-Hall, Englewood Cliffs, NJ, 1971.

[CCIR 1986] CCIR Recommendation 500-3, Method for the subjective assessment of the quality of television pictures, Recommendations and Reports of the CCIR, 1986, XVIth Plenary Assembly, Volume XI, Part 1.

[connor 1972] D.J. Connor, R.C. Brainard, and J.O. Limb, Interframe coding for picture transmission, *Proceedings of The IEEE*, 60, 7, 779–790, July 1972.

[fink 1957] D.G. Fink, *Television Engineering Handbook*, McGraw-Hill, New York, 1957, Sect. 10.7.

[goodall 1951] W.M. Goodall, Television by pulse code modulation, *Bell System Technical Journal*, 30, 33–49, January 1951.

[gonzalez 1992] R.C. Gonzalez and R.E. Woods, *Digital Image Processing*, Addison Wesley, Reading, MA, 1992.

[haskell 1996] B.G. Haskell, A. Puri and A.N. Netravali, *Digital Video: An Introduction to MPEG-2*, Chapman and Hall, ITP, New York, 1996.

[hidaka 1990] T. Hidaka and K. Ozawa, Subjective assessment of redundancy-reduced moving images for interactive application: Test methodology and report, *Signal Processing: Image Communication*, 2, 201–219, 1990.

[huang 1965] T.S. Huang, PCM picture transmission, *IEEE Spectrum*, 2, 12, 57–63, 1965.

[huang 1998] J. Huang and Y.Q. Shi, Adaptive image watermarking scheme based on visual masking, *IEE Electronics Letters*, 34, 8, 748–750, April 1998.

[kretzmer 1952] E.R. Kretzmer, Statistics of television signal, *Bell System Technical Journal*, 31, 4, 751–763, July 1952.

[lathi 1998] B.P. Lathi, *Modern Digital and Analog Communication Systems*, 3rd edn., Oxford University Press, New York, 1998.

[legge 1980] G.E. Legge and J.M. Foley, Contrast masking in human vision, *Journal of Optical Society of America*, 70, 12, 1458–1471, December 1980.

[lim 1990] J.S. Lim, *Two-Dimensional Signal and Image Processing*, Prentice-Hall, Englewood Cliffs, NJ, 1990.

[mitchell 1996] J.L. Mitchell, MPEG Video: Compression Standard, J.L. Mitchell, W.B. Pennebaker, C.E. Fogg and D.J. LeGall, (Eds.), Chapman and Hall, ITP, New York, 1996.

[mounts 1969] F.W. Mounts, A video encoding system with conditional picture-element replenishment, *Bell System Technical Journal*, 48, 7, 2545–2554, September 1969.

[mullen 1985] K.T. Mullen, The contrast sensitivity of human color vision to red-green and blue-yellow chromatic gratings, *Journal of Physiology*, 359, 381–400, 1985.

[netravali 1977] A.N. Netravali and B. Prasada, Adaptive quantization of picture signals using spatial masking, *Proceedings of the IEEE*, 65, 536–548, April 1977.

[sakrison 1979] D.J. Sakrison, Image coding applications of vision model, in *Image Transmission Techniques*, W.K. Pratt (Ed.), Academic Press, New York, 1979, pp. 21–71.

[seyler 1962] A.J. Seyler, The coding of visual signals to reduce channel-capacity requirements, *Proceedings of the I.E.E.*, 109C, 676–684, 1962.

[**seyler 1965**] A.J. Seyler, Probability distributions of television frame difference, *Proceedings of IREE (Australia)*, 26, 355–366, November 1965.

[**shannon 1948**] C.E. Shannon, A mathematical theory of communication, *Bell System Technical Journal*, 27, 379–423 (Part I), July 1948, 623–656 (Part II), October 1948.

[**slepian 1973**] D. Slepian (Ed.), *Key Papers in the Development of Information Theory*, IEEE Press, New York, 1973.

[**van ness 1967**] F.I. Van Ness and M.A. Bouman, Spatial modulation transfer in the human eye, *Journal of Optical Society of America*, 57, 3, 401–406, March 1967.

[**watson 1987**] A.B. Watson, Efficiency of a model human image code, *Journal of Optical Society of America A*, 4, 12, 2401–2417, December 1987.

[**webster 1993**] A.A. Webster, C.T. Jones, and M.H. Pinson, An objective video quality assessment system based on human perception, *Proceedings of Human Vision, Visual Processing and Digital Display IV*, J.P. Allebach and B.E. Rogowitz (Eds.), *SPIE Proceedings*, 1913, 15–26, September 1993.

2
Quantization

After an introduction to image and video compression was presented in Chapter 1, we address several fundamental aspects of image and video compression in the remaining chapters of Part I. Chapter 2, as the first chapter in the series, deals with quantization. Quantization is a necessary component in lossy coding and has direct impact on the bit rate and the distortion of reconstructed image or video. We will discuss concepts, principles, and various quantization techniques, which include uniform and nonuniform quantization, optimum quantization, and adaptive quantization.

2.1 Quantization and the Source Encoder

The functionality of image and video compression in the applications of visual communications and storage is depicted in Figure 1.1. In the context of visual communications, the whole system may be illustrated as shown in Figure 2.1. In the transmitter, the input analog information source is converted to a digital format in the A/D converter block. The digital format is compressed through the image and video source encoder. In the channel encoder, some redundancy is added to help combat noise and, hence, transmission error. Modulation makes digital data suitable for transmission through the analog channel, such as air space in the application of a TV broadcast. At the receiver, the counterpart blocks reconstruct the input visual information. As far as storage of visual information is concerned, the blocks of channel, channel encoder, channel decoder, modulation, and demodulation may be omitted (Figure 2.2). If input and output are required to be in the digital format in some applications then the A/D and D/A converters are omitted from the system. If they are required, however, other blocks, such as encryption and decryption can be added to the system [sklar 1988]. Hence, what is depicted in Figure 2.1 is a conceptually fundamental block diagram of a visual communication system.

This book is mainly concerned with source encoding and decoding. To this end, we take a step further. That is, we show block diagrams of a source encoder and decoder (Figure 2.3). As shown in Figure 2.3a, there are three components in the source encoding: transformation, quantization, and code word assignment. After the transformation, some form of an input information source is presented to a quantizer. In other words, the transformation block decides which types of quantities from the input image and video are to be encoded. It is not necessary that the original image and video waveform be quantized and coded: we will show that some formats obtained from the input image and video are more suitable for encoding. An example is the difference signal. From the discussion of interpixel correlation in Chapter 1, it is known that a pixel is normally highly correlated with its immediately horizontal or vertical neighboring pixel. Therefore, a better strategy is to encode the difference of gray level values between a pixel and its neighbor. Since these

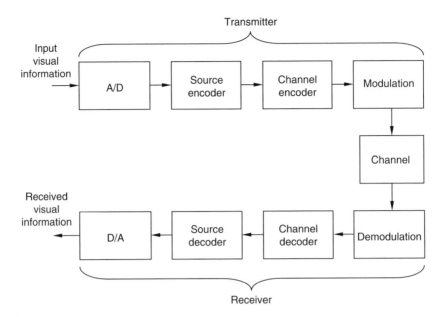

FIGURE 2.1
Block diagram of a visual communication system.

data are highly correlated, the difference usually has a smaller dynamic range. Consequently, the encoding is more efficient. This idea is discussed in detail in Chapter 3.

Another example is what is called transform coding (Chapter 4). There, instead of encoding the original input image and video, we encode a transform of the input image and video. Because the redundancy in the transform domain is reduced greatly, the coding efficiency is much higher compared with directly encoding the original image and video.

Note that the term transformation in Figure 2.3a is sometimes referred to as mapper and signal processing in the literature [gonzalez 1992; li 1995]. Quantization refers to a process that converts input data into a set of finitely many different values. Often, the input data to a quantizer is continuous in magnitude.

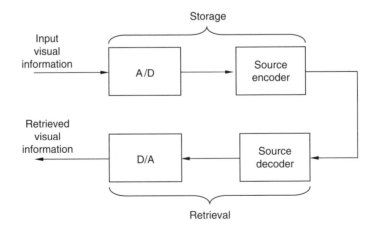

FIGURE 2.2
Block diagram of a visual storage system.

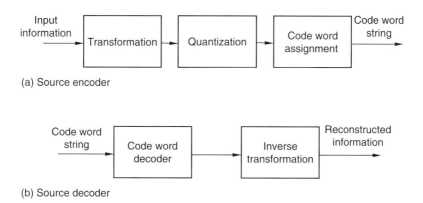

FIGURE 2.3
Block diagram of (a) a source encoder and (b) a source decoder.

Hence, quantization is essentially discretization in magnitude, which is an important step in the lossy compression of digital image and video. (The reason that the term lossy compression is used here is shown shortly.) The input and output of quantization can be either scalars or vectors. The quantization with scalar input and output is called scalar quantization, whereas that with vector input and output is referred to as vector quantization. In this chapter, we discuss scalar quantization. Vector quantization is addressed in Chapter 9.

After quantization, code words are assigned to the finitely many different values, the output of the quantizer. Natural binary code (NBC) and variable-length code (VLC), introduced in Chapter 1, are two examples of this. Other examples are the widely utilized entropy code (including Huffman code and arithmetic code), dictionary code, and run-length code (RLC) (frequently used in facsimile transmission), which are covered in Chapters 5 and 6.

The source decoder, as shown in Figure 2.3b, consists of two blocks: code word decoder and inverse transformation. They are counterparts of the code word assignment and transformation in the source encoder. Note that there is no block that corresponds to quantization in the source decoder. The implication of this observation is the following. First, quantization is an irreversible process. That is, in general, there is no way to find the original value from the quantized value. Second, quantization is, therefore, a source of information loss. In fact, quantization is a critical stage in image and video compression. It has significant impact on the distortion of reconstructed image and video as well as the bit rate of the encoder. Obviously, coarse quantization results in more distortion and lower bit rate than fine quantization.

In this chapter, uniform quantization, which is the simplest yet the most important case, is discussed first. Nonuniform quantization is covered after that, followed by optimum quantization for both uniform and nonuniform cases. Then a discussion of adaptive quantization is provided. Finally, pulse code modulation (PCM) is described as the best established and most frequently implemented digital coding method involving quantization is described.

2.2 Uniform Quantization

Uniform quantization is the simplest, yet very popular quantization technique. Conceptually, it is of great importance. Hence, we start our discussion on quantization with uniform quantization. Several fundamental concepts of quantization are introduced in this section.

2.2.1 Basics

This section concerns several basic aspects of uniform quantization. They are some fundamental terms, quantization distortion, and quantizer design.

2.2.1.1 Definitions

In Figure 2.4, the horizontal axis denotes the input to a quantizer, while the vertical axis represents the output of the quantizer. The relationship between the input and the output best characterizes this quantizer; this type of curve is referred to as the input–output characteristic of the quantizer. From the curve, it can be seen that there are nine intervals along the x-axis. Whenever the input falls in one of the intervals, the output assumes a corresponding value. The input–output characteristic of the quantizer is staircase-like and, hence, clearly nonlinear.

The end points of the intervals are called decision levels, denoted by d_i with i being the index of intervals. The output of the quantization is referred to as the reconstruction level (also known as quantizing level [musmann 1979]), denoted by y_i with i being its index. The length of the interval is called the step size of the quantizer, denoted by Δ. With the above terms defined, we can now mathematically define the function of the quantizer in Figure 2.4 as follows:

$$y_i = Q(x), \quad \text{if } x \in (d_i, d_{i+1}), \tag{2.1}$$

where $i = 1, 2, \ldots, 9$ and $Q(x)$ is the output of the quantizer with respect to the input x.

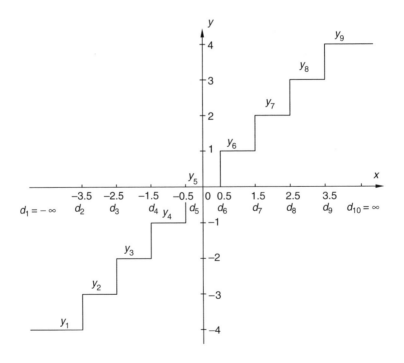

FIGURE 2.4
Input–output characteristic of a uniform midtread quantizer.

Quantization

It is noted that in Figure 2.4, $\Delta = 1$. The decision levels and reconstruction levels are evenly spaced. It is a uniform quantizer because it possesses the following two features:

1. Except possibly the right-most and left-most intervals, all intervals (hence, decision levels) along the x-axis are uniformly spaced. That is, each inner interval has the same length.
2. Except possibly the outer intervals, the reconstruction levels of the quantizer are also uniformly spaced. Furthermore, each inner reconstruction level is the arithmetic average of the two decision levels of the corresponding interval along the x-axis.

The uniform quantizer depicted in Figure 2.4 is called midtread quantizer. Its counterpart is called a midrise quantizer, in which the reconstructed levels do not include the value of zero. A midrise quantizer having step size $\Delta = 1$ is shown in Figure 2.5. While midtread quantizers are usually utilized for an odd number of reconstruction levels, midrise quantizers are used for an even number of reconstruction levels.

Note that the input–output characteristic of both the midtread and midrise uniform quantizers as depicted in Figures 2.4 and 2.5, respectively, is odd symmetric with respect to the vertical axis $x = 0$. In the rest of this chapter, our discussion develops under this symmetry assumption. The results thus derived will not lose generality since we can always subtract the statistical mean of input x from the input data and thus achieve this symmetry. After quantization, we can add the mean value back.

The total number of reconstruction levels of a quantizer is denoted by N. Figures 2.4 and 2.5 reveal that if N is even, then the decision level $d_{(N/2)+1}$ is located in the middle of the input x-axis, i.e., $d_{(N/2)+1} = 0$. If N is odd, on the other hand, then the reconstruction level $y_{(N+1)/2} = 0$. This convention is important in understanding the design tables of quantizers in the literature.

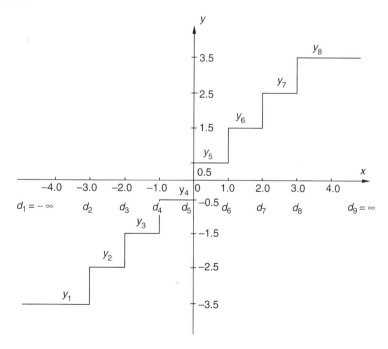

FIGURE 2.5
Input–output characteristic of a uniform midrise quantizer.

2.2.1.2 Quantization Distortion

The source coding theorem presented in Chapter 1 states that for a certain distortion D, there exists a rate distortion function $R(D)$, such that as long as the bit rate used is larger than $R(D)$ it is possible to transmit the source with a distortion smaller than D. Since we cannot afford an infinite bit rate to represent an original source, some distortion in quantization is inevitable. We can also say that since quantization causes information loss irreversibly, we encounter quantization error and, consequently, an issue: how to evaluate the quality or, equivalently, the distortion of quantization. According to our discussion on visual quality assessment in Chapter 1, we know that there are two ways to do so: subjective evaluation and objective evaluation.

In terms of subjective evaluation, in Section 1.3.1 we introduced a five-scale rating adopted in CCIR Recommendation 500-3. We also described the false contouring phenomenon, which is caused by coarse quantization. That is, our human eyes are more sensitive to the relatively uniform regions in an image plane. Therefore, an insufficient number of reconstruction levels results in annoying false contours. In other words, more reconstruction levels are required in relatively uniform regions than in relatively nonuniform regions.

In terms of objective evaluation, in Section 1.3.2 we defined mean square error (MSE) and root mean square error (RMSE), signal to noise ratio (SNR) and peak signal to noise ratio (PSNR). In dealing with quantization, we define quantization error, e_q, as the difference between the input signal and the quantized output:

$$e_q = x - Q(x), \tag{2.2}$$

where x and $Q(x)$ are input and quantized output, respectively. Quantization error is often referred to as quantization noise. It is a common practice to treat input x as a random variable with a probability density function (pdf) $f_x(x)$. Mean square quantization error, MSE_q, can thus be expressed as

$$\text{MSE}_q = \sum_{i=1}^{N} \int_{d_i}^{d_{i+1}} (x - Q(x))^2 f_x(x) dx \tag{2.3}$$

where N is the total number of reconstruction levels. Note that the outer decision levels may be $-\infty$ or ∞ (Figures 2.4 and 2.5). It is clear that, when the pdf, $f_x(x)$, remains unchanged, fewer reconstruction levels (smaller N) result in more distortion. That is, coarse quantization leads to large quantization noise. This confirms the statement that quantization is a critical component in a source encoder, which significantly influences both bit rate and distortion of the encoder. As mentioned earlier, the assumption that the input–output characteristic is odd symmetric with respect to the $x=0$ axis implies that the mean of the random variable, x, is equal to zero, i.e., $E(x) = 0$. Therefore, MSE_q is the variance of x, i.e., $\text{MSE}_q = \sigma_q^2$.

The quantization noise associated with the midtread quantizer depicted in Figure 2.4 is shown in Figure 2.6. It is clear that the quantization noise is signal dependent. It is observed that, associated with the inner intervals, the quantization noise is bounded by $\pm 0.5\Delta$. This type of quantization noise is referred to as granular noise. The noise associated with the right-most and the left-most intervals are unbounded as the input x approaches either $-\infty$ or ∞. This type of quantization noise is called overload noise. Denoting the mean square granular noise and overload noise by $\text{MSE}_{q,g}$ and $\text{MSE}_{q,o}$, respectively, we then have the following relations:

$$\text{MSE}_q = \text{MSE}_{q,g} + \text{MSE}_{q,o} \tag{2.4}$$

Quantization

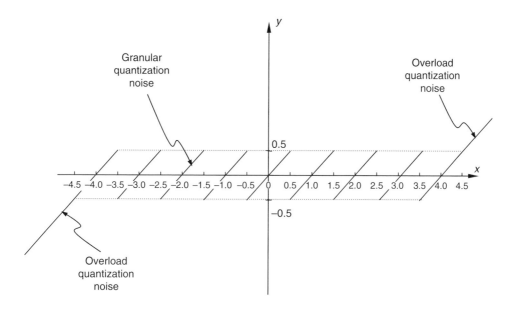

FIGURE 2.6
Quantization noise of the uniform midtread quantizer shown in Figure 2.4.

and

$$\text{MSE}_{q,g} = \sum_{i=2}^{N-1} \int_{d_i}^{d_{i+1}} (x - Q(x))^2 f_x(x) dx \quad (2.5)$$

$$\text{MSE}_{q,o} = 2 \int_{d_1}^{d_2} (x - Q(x))^2 f_x(x) dx \quad (2.6)$$

2.2.1.3 Quantizer Design

The design of a quantizer (either uniform or nonuniform) involves choosing the number of reconstruction levels, N (hence, the number of decision levels, $N+1$), and selecting the values of decision levels and reconstruction levels (deciding where to locate them). In other words, the design of a quantizer is equivalent to specifying its input–output characteristic.

The optimum quantizer design can be stated as follows. For a given pdf of the input random variable, $f_X(x)$, determine the number of reconstruction levels, N, choose a set of decision levels $\{d_i, i=1,\ldots,N+1\}$ and reconstruction levels $\{y_i, i=1,\ldots,N\}$ such that the MSE_q, defined in Equation 2.3, is minimized.

In the uniform quantizer design, the total number of reconstruction levels, N, is usually given. According to the two features of uniform quantizers described in Section 2.2.1.1, we know that the reconstruction levels of a uniform quantizer can be derived from the decision levels. Hence, only one of these two sets is independent. Furthermore, both decision levels and reconstruction levels are uniformly spaced except possibly the outer intervals. These constraints together with the symmetry assumption lead to the following observation: In fact, there is only one parameter that needs to decide in uniform quantizer design, which is the step size Δ. As to the optimum uniform quantizer design, a different pdf leads to a different step size.

2.2.2 Optimum Uniform Quantizer

In this section, we first discuss optimum uniform quantizer design when the input x obeys uniform distribution. Then, we cover optimum uniform quantizer design when the input x has other types of probabilistic distributions.

2.2.2.1 Uniform Quantizer with Uniformly Distributed Input

Let us return to Figure 2.4, where the input–output characteristic of a nine reconstruction level midtread quantizer is shown. Now, consider that the input x is a uniformly distributed random variable. Its input–output characteristic is shown in Figure 2.7. We notice that the new characteristic is restricted within a finite range of x, i.e., $-4.5 \leq x \leq 4.5$. This is due to the definition of uniform distribution. Consequently, the overload quantization noise does not exist in this case, which is shown in Figure 2.8.

The mean square quantization error, MSE_q, is found to be

$$MSE_q = N \int_{d_1}^{d_2} (x - Q(x))^2 \frac{1}{N\Delta} dx$$

$$= \frac{\Delta^2}{12}. \qquad (2.7)$$

This result indicates that if the input to a uniform quantizer has a uniform distribution and the number of reconstruction levels is fixed then the MSE_q is directly proportional to the square of the quantization step size. Or, in other words, the root MSE_q (the standard

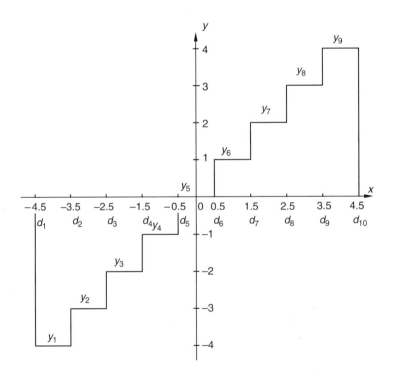

FIGURE 2.7
Input–output characteristic of a uniform midtread quantizer with input x having uniform distribution in $[-4.5, 4.5]$.

Quantization

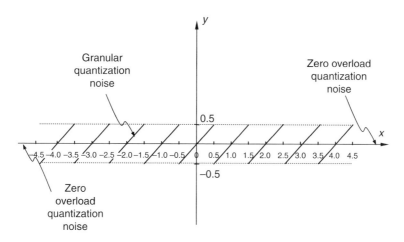

FIGURE 2.8
Quantization noise of the quantizer shown in Figure 2.7.

deviation of the quantization noise) is directly proportional to the quantization step. The larger the step size, the larger (according to square law) the MSE_q. This agrees with our earlier observation: coarse quantization leads to large quantization error.

As mentioned above, the MSE_q is equal to the variance of the quantization noise, i.e., $\text{MSE}_q = \sigma_q^2$. To find the SNR of the uniform quantization in this case, we need to determine the variance of the input x. Note that we assume the input x to be a zero mean uniform random variable. So, according to probability theory, we have

$$\sigma_x^2 = \frac{(N\Delta)^2}{12}. \tag{2.8}$$

Therefore, the mean square signal to noise ratio SNR_{ms}, defined in Chapter 1, is equal to

$$\text{SNR}_{ms} = 10 \log_{10} \frac{\sigma_x^2}{\sigma_q^2} = 10 \log_{10} N^2. \tag{2.9}$$

Note that here we use the subscript ms to indicate the SNR in the mean square sense, as defined in the Chapter 1. If we assume $N = 2^n$, we then have

$$\text{SNR}_{ms} = 20 \log_{10} 2^n = 6.02n \text{ dB}. \tag{2.10}$$

The interpretation of the above result is as follows. If we use the NBC to code the reconstruction levels of a uniform quantizer with a uniformly distributed input source, then every increased bit in the coding brings out a 6.02 dB increase in the SNR_{ms}. An equivalent statement can be derived from Equation 2.7. That is, whenever the step size of the uniform quantizer decreases by a half, the MSE_q decreases four times.

2.2.2.2 Conditions of Optimum Quantization

The conditions under which the MSE_q is minimized were derived [lloyd 1957, 1982; max 1960] for a given pdf of the quantizer input, $f_X(x)$.

TABLE 2.1
Optimal Symmetric Uniform Quantizer for Uniform, Gaussian, Laplacian, and Gamma Distributions (Having Zero Mean and Unit Variance)

N	Uniform d_i	y_i	MSE	Gaussian d_i	y_i	MSE	Laplacian d_i	y_i	MSE	Gamma d_i	y_i	MSE
2	−1.000	−0.500	8.33×10^{-2}	−1.596	−0.798	0.363	−1.414	−0.707	0.500	−1.154	−0.577	0.668
	0.000	0.500		0.000	0.798		0.000	0.707		0.000	0.577	
	1.000			1.596			1.414			1.154		
4	−1.000	−0.750	2.08×10^{-2}	−1.991	−1.494	0.119	−2.174	−1.631	1.963×10^{-1}	−2.120	−1.590	0.320
	−0.500	−0.250		−0.996	−0.498		−1.087	−0.544		−1.060	−0.530	
	0.000	0.250		0.000	0.498		0.000	0.544		0.000	0.500	
	0.500	0.750		0.996	1.494		1.087	1.631		1.060	1.590	
	1.000			1.991			2.174			2.120		
8	−1.000	−0.875	5.21×10^{-3}	−2.344	−2.051	3.74×10^{-2}	−2.924	−2.559	7.17×10^{-2}	−3.184	−2.786	0.132
	−0.750	−0.625		−1.758	−1.465		−2.193	−1.828		−2.388	−1.990	
	−0.500	−0.375		−1.172	−0.879		−1.462	−1.097		−1.592	−1.194	
	−0.250	−0.125		−0.586	−0.293		−0.731	−0.366		−0.796	−0.398	
	0.000	0.125		0.000	0.293		0.000	0.366		0.000	0.398	
	0.250	0.375		0.586	0.879		0.731	1.097		0.796	1.194	
	0.500	0.625		1.172	1.465		1.462	1.828		1.592	1.990	

t_i	x_i	t_i	x_i	t_i	x_i	t_i	x_i
0.750	0.875	1.758	2.051	2.193	2.559	2.388	2.786
1.000		2.344		2.924		3.184	
−1.000	−0.938	−2.680	−2.513	−3.648	−3.420	−4.320	−4.050
−0.875	−0.813	−2.345	−2.178	−3.192	−2.964	−3.780	−3.510
−0.750	−0.688	−2.010	−1.843	−2.736	−2.508	−3.240	−2.970
−0.625	−0.563	−1.675	−1.508	−2.280	−2.052	−2.700	−2.430
−0.500	−0.438	−1.340	−1.173	−1.824	−1.596	−2.160	−1.890
−0.375	−0.313	−1.005	−0.838	−1.368	−1.140	−1.620	−1.350
−0.250	−0.188	−0.670	−0.503	−0.912	−0.684	−1.080	−0.810
−0.125	−0.063	−0.335	−0.168	−0.456	−0.228	−0.540	−0.270
0.000 ($N=16$)	0.063	0.000	0.168	0.000	0.228	0.000	0.270
1.30×10^{-3}		1.15×10^{-2}		2.54×10^{-2}		5.01×10^{-2}	
$\boxed{0.125}$	0.188	$\boxed{0.335}$	0.503	$\boxed{0.456}$	0.684	$\boxed{0.540}$	0.810
0.250	0.313	0.670	0.838	0.912	1.140	1.080	1.350
0.375	0.438	1.005	1.173	1.368	1.596	1.620	1.890
0.500	0.563	1.340	1.508	1.824	2.052	2.160	2.430
0.625	0.688	1.675	1.843	2.280	2.508	2.700	2.970
0.750	0.813	2.010	2.178	2.736	2.964	3.240	3.510
0.875	0.938	2.345	2.513	3.192	3.420	3.780	4.050
1.000		2.680		3.648		4.320	

Sources: From Max, J., *IRE Trans. Inf. Theory*, IT-6, 7, 1960; Paez, M.D. and Glisson, T.H., *IEEE Trans. Commun.*, COM-20, 225, 1972.

Note: The numbers enclosed in rectangles are the step sizes.

The MSE$_q$ was given in Equation 2.3. The necessary conditions for optimum (minimum MSE) quantization are as follows. That is, the derivatives of MSE$_q$ with respect to the d_i and y_i have to be zero.

$$(d_i - y_{i-1})^2 f_x(d_i) - (d_i - y_i)^2 f_x(d_i) = 0, \quad i = 2, \ldots, N \tag{2.11}$$

$$-\int_{d_i}^{d_{i+1}} (x - y_i) f_x(x) dx = 0, \quad i = 1, \ldots, N \tag{2.12}$$

The sufficient conditions can be derived accordingly by involving the second order derivatives [max 1960; fleischer 1964]. The symmetry assumption of the input–output characteristic made earlier holds here as well. These sufficient conditions are listed below:

1. $x_1 = -\infty$ and $x_{N+1} = +\infty$ \quad (2.13)

2. $\int_{d_i}^{d_{i+1}} (x - y_i) f_X(x) dx = 0, \quad i = 1, 2, \ldots, N$ \quad (2.14)

3. $d_i = \dfrac{1}{2}(y_{i-1} + y_i), \quad i = 2, \ldots, N$ \quad (2.15)

Note that the first condition is for an input x whose range is $-\infty < x < \infty$. The interpretation of the above conditions is that each decision level (except for the outer intervals) is the arithmetic average of the two neighboring reconstruction levels, and each reconstruction level is the centroid of the area under the pdf $f_X(x)$ between the two adjacent decision levels.

Note that the above conditions are general in the sense that there is no restriction imposed on the pdf. In the next section, we discuss the optimum uniform quantization when the input of quantizer assumes different distributions.

2.2.2.3 Optimum Uniform Quantizer with Different Input Distributions

Let us return to our discussion on the optimum quantizer design whose input has uniform distribution. Because the input has uniform distribution, the outer intervals are also finite. For uniform distribution, Equation 2.14 implies that each reconstruction level is the arithmetic average of the two corresponding decision levels. Considering the two features of a uniform quantizer, presented in Section 2.2.1.1, we see that a uniform quantizer is optimum (minimizing the MSE$_q$) when the input has uniform distribution.

When the input x is uniformly distributed in $[-1, 1]$, the step size Δ of the optimum uniform quantizer is listed in Table 2.1 for the number of reconstruction levels, N, equal to 2, 4, 8, 16, and 32. From this table, we note that the MSE$_q$ of the uniform quantization with a uniformly distributed input decreases four times as N doubles. As mentioned in Section 2.2.2.1, this is equivalent to an increase of SNR$_{ms}$ by 6.02 dB as N doubles.

The derivation above is a special case, i.e., the uniform quantizer is optimum for a uniformly distributed input. Normally, if the pdf is not uniform, the optimum quantizer is not a uniform quantizer. Due to the simplicity of uniform quantization, however, it may sometimes be desirable to design an optimum uniform quantizer for an input with an other-than-uniform distribution.

Under these circumstances, however, Equations 2.13 through 2.15 are not a set of simultaneous equations one can hope to solve with any ease. Numerical procedures were suggested to solve for design of optimum uniform quantizers. Max derived uniform quantization step size Δ for an input with a Gaussian distribution [max 1960]. Paez and Glisson found step size Δ for Laplacian and Gamma-distributed input signals [paez 1972]. These results are listed in Table 2.1. Note that all three distributions have a zero mean and unit standard deviation. If the mean is not zero, only a shift in input is needed while applying these results. If the standard deviation is not unit, the tabulated step size needs to be multiplied by the standard deviation. The theoretical MSE is also listed in Table 2.1. Note that the subscript q associated with MSE has been dropped for the sake of notational brevity from now on in the chapter as long as it does not cause confusion.

2.3 Nonuniform Quantization

It is not difficult to see that, except for the special case of the uniformly distributed input variable x, the optimum (minimum MSE, also denoted sometimes by MMSE) quantizers should be nonuniform. Consider a case in which the input random variable obeys the Gaussian distribution with a zero mean and unit variance, and the number of reconstruction levels is finite. We naturally consider that having decision levels more densely located around the middle of the x-axis, $x = 0$ (high-probability density region), and choosing decision levels more coarsely distributed in the range far away from the center of the x-axis (low-probability density region) will lead to less MSE. The strategy adopted here is analogous to the superiority of VLC over fixed-length code (FLC) discussed in Chapter 1.

2.3.1 Optimum (Nonuniform) Quantization

Conditions of optimum quantization were discussed in Section 2.2.2.2. With some constraints, these conditions were solved in a closed form [panter 1951]. The equations characterizing these conditions, however, cannot be solved in a closed form in general. Lloyd and Max proposed an iterative procedure to numerically solve the equations. The optimum quantizers thus designed are called Lloyd–Max quantizers.

When input x obeys Gaussian distribution, the solution to optimum quantizer design for finitely many reconstruction levels N was obtained [lloyd 1957, 1982; max 1960]. That is, the decision and reconstruction levels together with theoretical minimum MSE and optimum SNR have been determined. Following this procedure, the design for Laplacian and Gamma distribution were tabulated in [paez 1972]. These results are contained in Table 2.2. As stated before, we see in the table once again that uniform quantization is optimal if the input x is a uniform random variable.

Figure 2.9 [max 1960] gives a performance comparison between optimum uniform quantization and optimum quantization for the case of a Gaussian-distributed input with a zero mean and unit variance. The abscissa represents the number of reconstruction levels, N, and the ordinate the ratio between the error of the optimum quantizer and the error of the optimum uniform quantizer. It can be seen that when N is small, the ratio is close to one. That is, the performances are close. When N increases, the ratio decreases. Specifically, when N is large the nonuniform quantizer is about 20% to 30% more efficient than the uniform optimum quantizer for the Gaussian distribution with a zero mean and unit variance.

TABLE 2.2

Optimal Symmetric Quantizer for Uniform, Gaussian, Laplacian, and Gamma Distributions (The Uniform Distribution Is between [−1, 1]; the Other Three Distributions Have Zero Mean and Unit Variance.)

N	Uniform			Gaussian			Laplacian			Gamma		
	d_i	y_i	MSE	d_i	y_i	MSE	d_i	y_i	MSE	d_i	y_i	MSE
2	−1.000	−0.500	8.33×10^{-2}	−∞	−0.799	0.363	−∞	−0.707	0.500	−∞	−0.577	0.668
	0.000	0.500		0.000	0.799		0.000	0.707		0.000	0.577	
	1.000			∞			∞			∞		
4	−1.000	−0.750	2.08×10^{-2}	−∞	−1.510	0.118	−∞	−1.834	1.765×10^{-1}	−∞	−2.108	0.233
	−0.500	−0.250		−0.982	−0.453		−1.127	−0.420		−1.205	−0.302	
	0.000	0.250		0.000	0.453		0.000	0.420		0.000	0.302	
	0.500	0.750		0.982	1.510		1.127	1.834		1.205	2.108	
	1.000			∞			∞			∞		
8	−1.000	−0.875	5.21×10^{-3}	−∞	−2.152	3.45×10^{-2}	−∞	−3.087	5.48×10^{-2}	−∞	−3.799	7.12×10^{-2}
	−0.750	−0.625		−1.748	−1.344		−2.377	−1.673		−2.872	−1.944	
	−0.500	−0.375		−1.050	−0.756		−1.253	−0.833		−1.401	−0.859	
	−0.250	−0.125		−0.501	−0.245		−0.533	−0.233		−0.504	−0.149	
	0.000	0.125		0.000	0.245		0.000	0.233		0.000	0.149	
	0.250	0.375		0.501	0.756		0.533	0.833		0.504	0.859	
	0.500	0.625		1.050	1.344		1.253	1.673		1.401	1.944	

Quantization

	0.750	0.875	1.748	2.152	2.377	3.087	2.872	3.799
	1.000	∞	∞	∞	∞	∞	∞	∞
−1.000			−∞	−2.733	−∞	−4.316	−∞	−6.085
−0.875	−0.938		−2.401	−2.069	−3.605	−2.895	−5.050	−4.015
−0.750	−0.813		−1.844	−1.618	−2.499	−2.103	−3.407	−2.798
−0.625	−0.688		−1.437	−1.256	−1.821	−1.540	−2.372	−1.945
−0.500	−0.563		−1.099	−0.942	−1.317	−1.095	−1.623	−1.300
−0.375	−0.438		−0.800	−0.657	−0.910	−0.726	−1.045	−0.791
−0.250	−0.313		−0.522	−0.388	−0.566	−0.407	−0.588	−0.386
−0.125	−0.188		−0.258	−0.128	−0.266	−0.126	−0.229	−0.072
	−0.063	1.30×10^{-3}						
16				9.50×10^{-3}		1.54×10^{-2}		1.96×10^{-2}
0.000	0.063		0.000	0.128	0.000	0.126	0.000	0.072
0.125	0.188		0.258	0.388	0.266	0.407	0.229	0.386
0.250	0.313		0.522	0.657	0.566	0.726	0.588	0.791
0.375	0.438		0.800	0.942	0.910	1.095	1.045	1.300
0.500	0.563		1.099	1.256	1.317	1.540	1.623	1.945
0.625	0.688		1.437	1.618	1.821	2.103	2.372	2.798
0.750	0.813		1.844	2.069	2.499	2.895	3.407	4.015
0.875	0.938		2.401	2.733	3.605	4.316	5.050	6.085
1.000			∞		∞		∞	

Sources: From Lloyd, S.P., Least squares quantization in PCM, Institute of Mathematical Statistics Meeting, Atlantic City, NJ, September 1957; Lloyd, S.P., Least squares quantization in PCM, *IEEE Trans. Inf. Theory*, IT-28, 129, 1982; Paez, M.D. and Glisson, T.H., *IEEE Trans. Commun.*, COM-20, 225, 1972; Max, J., *IRE Trans. Inf. Theory*, IT-6, 7, 1960.

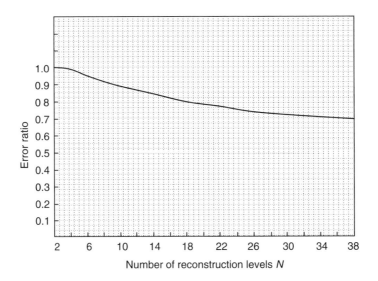

FIGURE 2.9
Ratio of error for optimal quantizer to error for optimum uniform quantizer vs. number of reconstruction levels N. (minimum mean square error [MSE] for Gaussian distributed input with a zero mean and unit variance). (From Max, J., *IRE Trans. Inf. Theory*, IT-6, 7, 1960. With permission.)

2.3.2 Companding Quantization

It is known that a speech signal usually has a large dynamic range. Moreover, its statistical distribution reveals that very low speech volumes predominate most voice communications. Specifically, by a 50% chance, the voltage characterizing detected speech energy is less than 25% of the root mean square (RMS) value of the signal. Large amplitude values are rare: only by a 15% chance does the voltage exceed the RMS value [skalr 1988]. These statistics naturally lead to the need for nonuniform quantization with relatively dense decision levels in the small magnitude range and relatively coarse decision levels in the large magnitude range.

When the bit rate is 8 bits/sample, the following companding technique [smith 1957], which realizes nonuniform quantization, is found to be extremely useful. Though speech coding is not the main focus of this book, we briefly discuss the companding technique here as an alternative way to achieve nonuniform quantization.

The companding technique, also known as logarithmic quantization, consists of the following three stages: compressing, uniform quantization, and expanding [gersho 1977] as shown in Figure 2.10. First, it compresses the input signal with a logarithmic

FIGURE 2.10
Companding technique in achieving quantization.

Quantization

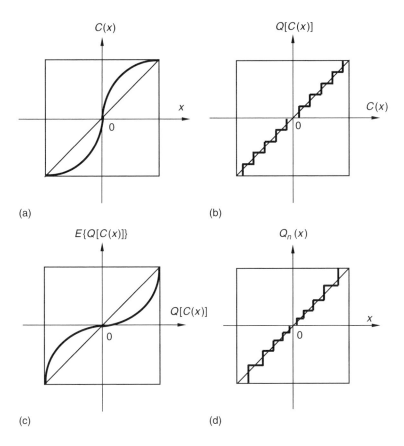

FIGURE 2.11
Characteristics of companding techniques. (a) Compressing characteristic, (b) uniform quantizer characteristic, (c) expanding characteristic, and (d) nonuniform quantizer characteristic.

characteristic, and second, it quantizes the compressed input using a uniform quantizer. Finally, the uniformly quantized results are expanded inversely. An illustration of the characteristics of these three stages and the resultant nonuniform quantization are shown in Figure 2.11.

In practice, a piecewise linear approximation of the logarithmic compression characteristic is used. There are two different ways. In North America, a μ-law compression characteristic is used, which is defined as follows.

$$c(x) = x_{\max} \frac{\ln[1 + \mu(|x|/x_{\max})]}{\ln(1+\mu)} \operatorname{sgn} x, \tag{2.16}$$

where sgn is a sign function defined as

$$\operatorname{sgn} x = \begin{cases} +1 & \text{if } x \geq 0 \\ -1 & \text{if } x < 0 \end{cases} \tag{2.17}$$

The μ-law compression characteristic is shown in Figure 2.12a. The standard value of μ is 255. Note from the figure that the case of $\mu = 0$ corresponds to uniform quantization.

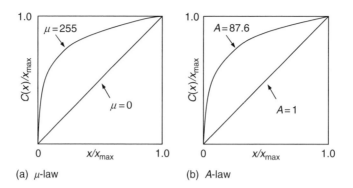

FIGURE 2.12
Compression characteristics.

In Europe, the A-law characteristic is used. The A-law characteristic is depicted in Figure 2.12b, and is defined as follows.

$$c(x) = \begin{cases} x_{max} \dfrac{A(|x|/x_{max})}{1 + \ln A} \text{sgn } x & 0 < \dfrac{|x|}{x_{max}} \leq \dfrac{1}{A} \\ x_{max} \dfrac{1 + \ln[A(|x|/x_{max})]}{1 + \ln A} \text{sgn } x & \dfrac{1}{A} < \dfrac{|x|}{x_{max}} < 1 \end{cases} \quad (2.18)$$

It is noted that the standard value of A is 87.6. The case of $A=1$ corresponds to uniform quantization.

2.4 Adaptive Quantization

In the last section, we studied nonuniform quantization, whose motivation is to minimize MSE_q. We found that nonuniform quantization is necessary if the pdf of the input random variable x is not uniform. Consider an optimum quantizer for a Gaussian-distributed input when the number of reconstruction levels N is eight. Its input–output characteristic can be derived from Table 2.2 and is shown in Figure 2.13. This curve reveals that the decision levels are densely located in the central region of the x-axis and coarsely elsewhere. In other words, the decision levels are densely distributed in the region having a higher probability of occurrence and coarsely distributed in other regions. A logarithmic companding technique also allocates decision levels densely in the small magnitude region, which corresponds to a high occurrence probability, but in a different way. We conclude that nonuniform quantization achieves minimum MSE_q by distributing decision levels according to the statistics of the input random variable.

These two types of nonuniform quantizers are both time-invariant. That is, they are not designed for nonstationary input signals. Moreover, even for a stationary input signal, if its pdf deviates from that with which the optimum quantizer is designed then a mismatch will take place and the performance of the quantizer will deteriorate. There are two main types of mismatch. One is called variance mismatch. That is, the pdf of input signal is matched, while the variance is mismatched. Another type is pdf mismatch. Noted that these two kinds of mismatch also occur in optimum uniform quantization, because there the optimization is also achieved based on the input statistics assumption. For a detailed

Quantization

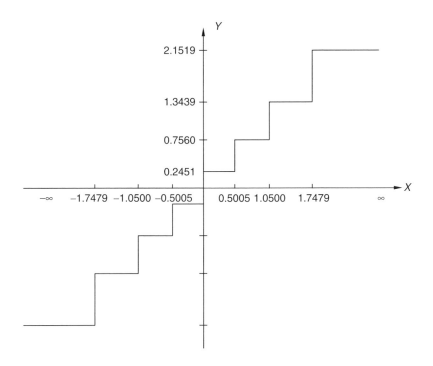

FIGURE 2.13
Input–output characteristic of the optimal quantizer for Gaussian distribution with zero mean, unit variance, and $N=8$.

analysis of the effects of the two types of mismatch on quantization, readers are referred to [Jayant 1984].

Adaptive quantization attempts to make the quantizer design adapt to the varying input statistics to achieve better performance. It is a means to combat the mismatch problem discussed above. By statistics, we mean the statistic mean, variance (or the dynamic range), and type of input pdf. When the mean of the input changes, differential coding (discussed in the next chapter) is a suitable method to handle the variation. For other types of cases, adaptive quantization is found to be effective. The price paid in adaptive quantization is processing delay and an extra storage requirement as seen below.

There are two different types of adaptive quantization: forward and backward adaptations. Before we discuss these, however, let us describe an alternative way to define quantization [Jayant 1984]. Quantization can be viewed as a two-stage process (Figure 2.14). The first stage is the quantization encoder and the second stage is the quantization decoder. In the encoder, the input to quantization is converted to the index of an interval into which the

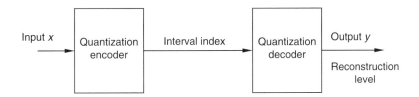

FIGURE 2.14
A two-stage model of quantization.

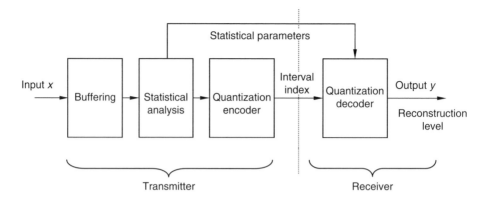

FIGURE 2.15
Forward adaptive quantization.

input x falls. This index is mapped to (the code word that represents) the reconstruction level corresponding to the interval in the decoder. Roughly speaking, this definition considers a quantizer as a communication system in which the quantization encoder is in the transmitter side while the quantization decoder is in the receiver side. In this sense, this definition is broader than that of quantization defined in Figure 2.3a.

2.4.1 Forward Adaptive Quantization

A block diagram of forward adaptive quantization is shown in Figure 2.15. There, the input to the quantizer, x, is first split into blocks, each with a certain length. Blocks are stored in a buffer one at a time. A statistical analysis is then carried out with respect to the block in the buffer. Based on the analysis, the quantization encoder is set up, and the input data within the block are assigned indexes of respective intervals. In addition to these indexes, the encoder setting parameters are sent to the quantization decoder as side information. The term side comes from the fact that the amount of bits used for coding the setting parameter is usually a small fraction of the total amount of bits used.

The selection of block size is a critical issue. If the size is small, the adaptation to the local statistics will be effective, but the side information needs to be sent frequently. That is, more bits are used for sending the side information. If the size is large, the bits used for side information decrease. On the other hand, the adaptation becomes less sensitive to changing statistics, and both processing delay and storage required increase. In practice, a proper compromise between quantity of side information and effectiveness of adaptation produces a good selection of the block size.

Examples of using forward manner to adapt quantization to a changing input variance (to combat variance mismatch) can be found in [jayant 1984; sayood 1996].

2.4.2 Backward Adaptive Quantization

Figure 2.16 shows a block diagram of backward adaptive quantization. A close look at the block diagram reveals that in both the quantization encoder and decoder the buffering and the statistical analysis are carried out with respect to the output of quantization encoder. In this way, there is no need to send side information. The sensitivity of adaptation to the

Quantization

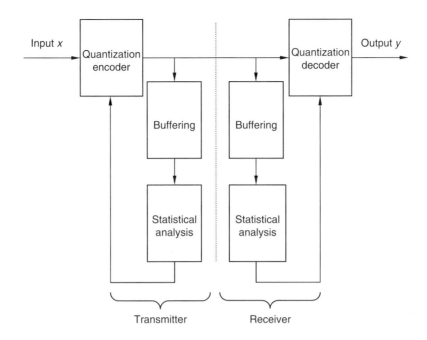

FIGURE 2.16
Backward adaptive quantization.

changing statistics will be degraded, however, since, instead of the original input, only is the output of the quantization encoder used in the statistical analysis. That is, the quantization noise is involved in the statistical analysis.

2.4.3 Adaptive Quantization with a One-Word Memory

Intuitively, it is expected that observing a sufficient large number of input or output (quantized) data is necessary to track the changing statistics and then adapt the quantizer setting in adaptive quantization. Through an analysis, Jayant showed that effective adaptations can be realized with an explicit memory of only one word. That is, either one input sample, x, in forward adaptive quantization or a quantized output, y, in backward adaptive quantization is sufficient [jayant 1973].

In [jayant 1984], examples on step size adaptation (with the number of total reconstruction levels larger than four) were given. The idea is as follows. If at moment t_i the input sample x_i falls into the outer interval, then the step size at the next moment t_{i+1} will be enlarged by a factor of m_i (multiplying the current step size by m_i, $m_i > 1$). On the other hand, if the input x_i falls into an inner interval close to $x = 0$ then, the multiplier is less than 1, i.e., $m_i < 1$. That is, the multiplier m_i is small in the interval near $x = 0$ and monotonically increases for an increased x. Its range varies from a small positive number less than 1 to a number larger than 1. In this way, the quantizer adapts itself to the input to avoid overload as well as underload to achieve better performance.

2.4.4 Switched Quantization

This is another adaptive quantization scheme. A block diagram is shown in Figure 2.17. It consists of a bank of L quantizers. Each quantizer in the bank is fixed, but collectively they

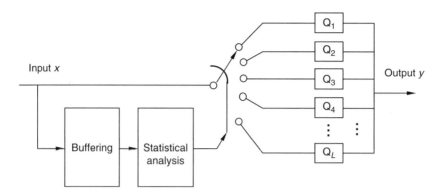

FIGURE 2.17
Switched quantization.

form a bank of quantizers with a variety of input–output characteristics. Based on a statistical analysis of recent input or output samples, a switch connects the current input to one of the quantizers in the bank such that the best possible performance may be achieved. It is reported that in both video and speech applications, this scheme has shown improved performance even when the number of quantizers in the bank, L, is two [Jayant 1984]. Interestingly, it is noted that as $L \to \infty$, the switched quantization converges to the adaptive quantizer discussed above.

2.5 Pulse Code Modulation

Pulse code modulation is closely related to quantization, the focus of this chapter. Furthermore, as pointed in [Jayant 1984], PCM is the earliest, best-established, and most frequently applied coding system despite the fact that it is the most bit-consuming digitizing system (since it encodes each pixel independently) as well as a very demanding system in terms of bit error rate on the digital channel. Therefore, we discuss the PCM technique in this section.

PCM is now the most important form of pulse modulation. The other forms of pulse modulation are pulse amplitude modulation (PAM), pulse width modulation (PWM), and pulse position modulation (PPM), which are covered in most communication texts. Briefly speaking, pulse modulation links an analog signal to a pulse train in the following way. The analog signal is first sampled (a discretization in time domain). The sampled values are used to modulate a pulse train. If the modulation is carried out through the amplitude of the pulse train, it is called PAM. If the modified parameter of the pulse train is the pulse width, we then have PWM. If the pulse width and magnitude are constant—only the position of pulses is modulated by the sample values—we then encounter PPM. An illustration of these pulse modulations is shown in Figure 2.18.

In PCM, an analog signal is first sampled. The sampled value is then quantized. Finally, the quantized value is encoded, resulting in a bit steam. Figure 2.19 provides an example of PCM. We see that through a sampling and a uniform quantization the PCM system converts the input analog signal, which is continuous in both time and magnitude, into a digital signal (discretized in both time and magnitude) in the form of a NBC sequence. In this way, an analog signal modulates a pulse train with a NBC.

Quantization

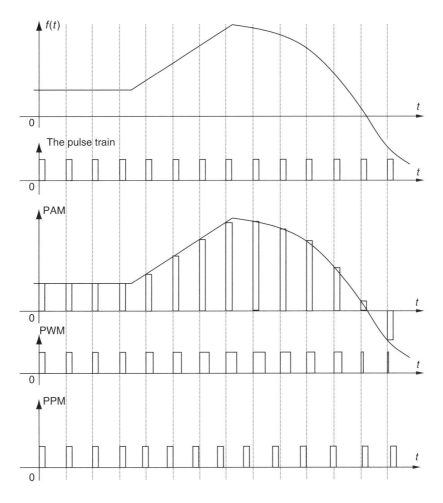

FIGURE 2.18
Pulse modulation.

By far, PCM is more popular than other types of pulse modulation because the code modulation is much more robust against various noises than amplitude modulation, width modulation and position modulation. In fact, almost all coding techniques include a PCM component. In digital image processing, given digital images usually appear in PCM format. It is known that an acceptable PCM representation of monochrome picture requires 6–8 bits/pixel [huang 1965]. It is used so commonly in practice that its performance normally serves as a standard against which other coding techniques are compared.

Let us recall the false contouring phenomenon, discussed in texture masking (Chapter 1). It states that our eyes are more sensitive to relatively uniform regions in an image plane. If the number of reconstruction levels is not large enough (coarse quantization) then some unnatural contours will appear. When frequency masking was discussed, it was noted that by adding some high frequency signal before quantization, the false contouring can be eliminated to a great extent. This technique is called dithering. The high frequency signal used is referred to as a dither signal. Both false contouring and dithering were first reported in [goodall 1951].

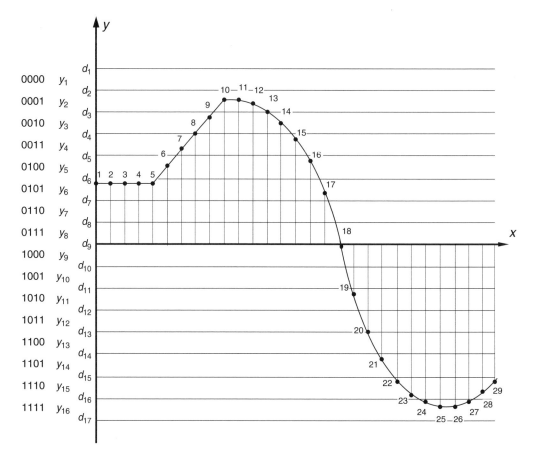

FIGURE 2.19
Pulse code modulation (PCM).

2.6 Summary

Quantization is a process in which a quantity having possibly infinitely many values is converted to another quantity having only finitely many values. It is an important element in source encoding that has significant impact on both bit rate and distortion of reconstructed images and video in visual communication systems. Depending on whether the quantity is a scalar or a vector, quantization is called either scalar or vector quantization. In this chapter, we considered only scalar quantization.

Uniform quantization is the simplest and yet the most important case. In uniform quantization, except for outer intervals, both decision levels and reconstruction levels are uniformly spaced. Moreover, a reconstruction level is the arithmetic average of the two corresponding decision levels. In uniform quantization design, the step size is the only parameter that needs to be specified.

Optimum quantization implies minimization of the MSE_q. When the input has a uniform distribution, uniform quantization is optimum. For the sake of simplicity, a uniform optimum quantizer is sometimes desired even when the input does not obey uniform distribution. The design under these circumstances involves an iterative procedure. The design problem in cases where the input has Gaussian, Laplacian, or Gamma distribution was solved and the parameters are available.

When the constraint of uniform quantization is removed, the conditions for optimum quantization are derived. The resultant optimum quantizer is normally nonuniform. An iterative procedure to solve the design is established and the optimum design parameters for Gaussian, Laplacian, and Gamma distribution are tabulated.

The companding technique is an alternative way to implement nonuniform quantization. Both nonuniform quantization and companding are time-invariant and hence not suitable for nonstationary input. Adaptive quantization deals with nonstationary input and combats the mismatch that occurs in optimum quantization design.

In adaptive quantization, buffering is necessary to store some recent input or sampled output data. A statistical analysis is carried out with respect to the stored recent data. Based on the analysis, the parameters of the quantizer are adapted to changing input statistics to achieve better quantization performance. There are two types of adaptive quantization: forward and backward adaptive quantizations. With the forward type, the statistical analysis is derived from the original input data, whereas with the backward type, quantization noise is involved in the analysis. Therefore, the forward type usually achieves more effective adaptation than the backward type. The latter, however, does not need to send quantizer setting parameters as side information to the receiver side, since the output values of quantization encoder (based on which the statistics are analyzed and parameters of the quantizer are adapted) are available in both the transmitter and receiver sides.

Switched quantization is another type of adaptive quantization. In this scheme, a bank of fixed quantizers is utilized, each quantizer having different input–output characteristics. A statistical analysis based on recent input decides which quantizer in the bank is suitable for the present input. The system then connects the input to this particular quantizer.

Nowadays, PCM is the most frequently used form of pulse modulation due to its robustness against noise. PCM consists of three stages: sampling, quantization, and encoding. First, the analog signals are sampled with a proper sampling frequency. Second, the sampled data are quantized using a uniform quantizer. Finally, the quantized values are encoded with NBC. It is the best established and most applied coding system. Despite its bit-consuming feature, it is utilized in almost all coding systems.

Exercises

1. Using your own words, define quantization and uniform quantization. What are the two features of uniform quantization?
2. What is optimum quantization? Why is uniform quantization sometimes desired, even when the input has a pdf different from uniform? How was this problem solved? Draw an input–output characteristic of an optimum uniform quantizer with an input obeying Gaussian pdf having zero mean, unit variance, and the number of reconstruction levels, N, equal to 8.

3. What are the conditions of optimum nonuniform quantization? From Table 2.2, what observations can you make?
4. Define variance mismatch and pdf mismatch. Discuss how you can resolve the mismatch problem.
5. What is the difference between forward and backward adaptive quantization? Comment on the merits and drawbacks for each.
6. What are PAM, PWM, PPM, and PCM? Why is PCM the most popular type of pulse modulation?

References

[fleischer 1964] P.E. Fleischer, Sufficient conditions for achieving minimum distortion in quantizer, *IEEE International Convention Records*, Part I, 12, 104–111, 1964.

[gersho 1977] A. Gersho, Quantization, *IEEE Communications Society Magazine*, 16–29, September 1977.

[gonzalez 1992] R.C. Gonzalez and R.E. Woods, *Digital Image Processing*, Addison-Wesley, Reading, MA, 1992.

[goodall 1951] W.M. Goodall, Television by pulse code modulation, *Bell System Technical Journal*, 30, 33–49, January 1951.

[huang 1965] T.S. Huang, PCM picture transmission, *IEEE Spectrum*, 2, 57–63, December 1965.

[jayant 1973] N.S. Jayant, Adaptive quantization with one word memory, *Bell System Technical Journal*, 52, 1119–1144, September 1973.

[jayant 1984] N.S. Jayant and P. Noll, *Digital Coding of Waveforms*, Prentice-Hall, Englewood Cliffs, NJ, 1984.

[li 1995] W. Li and Y.-Q. Zhang, Vector-based signal processing and quantization for image and video compression, *Proceedings of the IEEE*, 83, 2, 317–335, February 1995.

[lloyd 1957] S.P. Lloyd, *Least Squares Quantization in PCM*, Institute of Mathematical Statistics Meeting, Atlantic City, NJ, September 1957.

[lloyd 1982] S.P. Lloyd, Least squares quantization in PCM, *IEEE Transactions on Information Theory*, IT-28, 129–137, March 1982.

[max 1960] J. Max, Quantizing for minimum distortion, *IRE Transactions on Information Theory*, IT-6, 7–12, 1960.

[musmann 1979] H.G. Musmann, Predictive image coding, in *Image Transmission Techniques*, W.K. Pratt (Ed.), Academic Press, NY, 1979.

[paez 1972] M.D. Paez and T.H. Glisson, Minimum mean-squared-error quantization in speech PCM and DPCM Systems, *IEEE Transactions on Communications*, COM-20, 225–230, April 1972.

[panter 1951] P.F. Panter and W. Dite, Quantization distortion in pulse count modulation with nonuniform spacing of levels, *Proceedings of the IRE*, 39, 44–48, January 1951.

[sayood 1996] K. Sayood, *Introduction to Data Compression*, Morgan Kaufmann, San Francisco, CA, 1996.

[sklar 1988] B. Sklar, *Digital Communications: Fundamentals and Applications*, PTR Prentice-Hall, Englewood Cliffs, NJ, 1988.

[smith 1957] B. Smith, Instantaneous companding of quantized signals, *Bell System Technical Journal*, 36, 653–709, May 1957.

3
Differential Coding

Instead of encoding a signal directly, the differential coding technique codes the difference between the signal itself and its prediction. Therefore, it is also known as predictive coding. By utilizing spatial and temporal interpixel correlation, differential coding is an efficient and yet computationally simple coding technique. In this chapter, we will first describe the differential technique in general. And then its two components: prediction and quantization. There is an emphasis on (optimum) prediction, since quantization was already discussed in Chapter 2. When the difference signal (also known as prediction error) is quantized, the differential coding is called differential pulse code modulation (DPCM). Some issues in DPCM are discussed, after which delta modulation (DM) as a special case of DPCM is covered. The idea of differential coding involving image sequences is briefly discussed in this chapter. A more detailed coverage is presented in Parts III and IV, starting from Chapter 10. If quantization is not included, the differential coding is referred to as information-preserving differential coding and discussed at the end of the chapter.

3.1 Introduction to DPCM

As depicted in Figure 2.3, a source encoder consists of the following three components: Transformation, quantization, and code word assignment. The transformation converts input into a format for quantization followed by code word assignment. In other words, the component of transformation decides which format of input to be encoded. As mentioned in Chapter 2, input itself is not necessarily the most suitable format for encoding.

Consider the case of monochrome image encoding. The input is usually a 2-D array of gray level values of an image obtained via PCM coding. The concept of spatial redundancy, discussed in Section 1.2.1.1, tells us that neighboring pixels of an image are usually highly correlated. Therefore, it is more efficient to encode the gray difference between two neighboring pixels instead of encoding the gray level values of each pixel. At the receiver, the decoded difference is added back to reconstruct the gray level value of the pixel. As neighboring pixels are highly correlated, their gray level values bear a great similarity. Hence, we expect that the variance of the difference signal will be smaller than that of the original signal. Assume uniform quantization and natural binary coding for the sake of simplicity. Then we see that for the same bit rate (bits per sample) the quantization error will be smaller, i.e., a higher quality of reconstructed signal can be achieved. Or, for the same quality of reconstructed signal, we need a lower bit rate.

Assume a bit rate of 8 bits/sample in the quantization. We can see that although the dynamic range of the difference signal is theoretically doubled, from 256 to 512, the variance of the difference signal is actually much smaller. This can be confirmed from the histograms of the boy and girl image (refer to Figure 1.2) and its difference image obtained by horizontal pixel-to-pixel differencing, shown in Figure 3.1a and 3.1b, respectively. Figure 3.1b and its close-up (Figure 3.1c) indicate that by a rate of 42.44% the difference values fall into the range of −1, 0, and +1. In other words, the histogram of the difference signal is much more narrowly concentrated than that of the original signal.

3.1.1 Simple Pixel-to-Pixel DPCM

Denote the gray level values of pixels along a row of an image as z_i, $i = 1, \ldots, M$, where M is the total number of pixels within the row. Using the immediately preceding pixel's gray level value, z_{i-1}, as a prediction of that of the present pixel, $\hat{z}_i$, i.e.,

$$\hat{z}_i = z_{i-1}, \quad (3.1)$$

we then have the difference signal

$$d_i = z_i - \hat{z}_i = z_i - z_{i-1}. \quad (3.2)$$

A block diagram of the scheme described above is shown in Figure 3.2. There z_i denotes the sequence of pixels along a row, d_i is the corresponding difference signal, and $\hat{d}_i$ is the quantized version of the difference, i.e.,

$$\hat{d}_i = Q(d_i) = d_i + e_q \quad (3.3)$$

where e_q represents quantization error. In the decoder, $\bar{z}_i$ represents the reconstructed pixel gray value, and we have

$$\bar{z}_i = \bar{z}_{i-1} + \hat{d}_i. \quad (3.4)$$

This simple scheme, however, suffers from an accumulated quantization error. We can see this clearly from the following derivation [sayood 1996], where we assume the initial value z_0 is available for both the encoder and the decoder:

$$\text{as } i = 1, \; d_1 = z_1 - z_0$$
$$\hat{d}_1 = d_1 + e_{q,1} \quad (3.5)$$
$$\bar{z}_1 = z_0 + \hat{d}_1 = z_0 + d_1 + e_{q,1} = z_1 + e_{q,1}.$$

Similarly, we can have

$$\text{as } i = 2, \; \bar{z}_2 = z_2 + e_{q,1} + e_{q,2} \quad (3.6)$$

and, in general,

$$\bar{z}_i = z_i + \sum_{j=1}^{i} e_{q,j}. \quad (3.7)$$

This problem can be remedied by the following scheme, shown in Figure 3.3. Now we see that in both the encoder and the decoder, the reconstructed signal is generated in the same way, i.e.,

Differential Coding

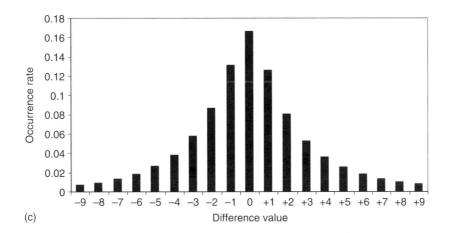

FIGURE 3.1
(a) Histogram of the original boy and girl image. (b) Histogram of the difference image obtained by using horizontal pixel-to-pixel differencing. (c) A close-up of the central portion of the histogram of the difference image.

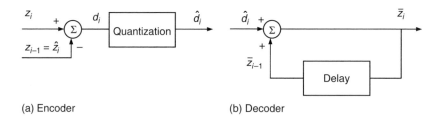

(a) Encoder (b) Decoder

FIGURE 3.2
Block diagram of a pixel-to-pixel differential coding system.

$$\bar{z}_i = \bar{z}_{i-1} + \hat{d}_i, \tag{3.8}$$

and in the encoder the difference signal changes to

$$d_i = z_i - \bar{z}_{i-1}. \tag{3.9}$$

That is, the previous reconstructed $\bar{z}_{i-1}$ is used as the prediction, $\hat{z}_i$, i.e.,

$$\hat{z}_i = \bar{z}_{i-1}. \tag{3.10}$$

In this way, we have

$$\begin{aligned} \text{as } i = 1, \quad & d_1 = z_1 - z_0 \\ & \hat{d}_1 = d_1 + e_{q,1} \\ & \bar{z}_1 = z_0 + \hat{d}_1 = z_0 + d_1 + e_{q,1} = z_1 + e_{q,1}. \end{aligned} \tag{3.11}$$

Similarly, we have

$$\begin{aligned} \text{as } i = 2, \quad & d_2 = z_2 - \bar{z}_1 \\ & \hat{d}_2 = d_2 + e_{q,2} \\ & \bar{z}_2 = \bar{z}_1 + \hat{d}_2 = z_2 + e_{q,2}. \end{aligned} \tag{3.12}$$

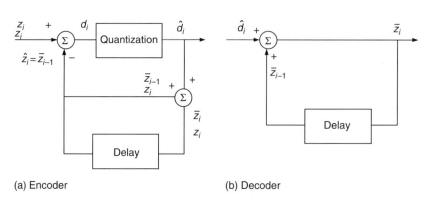

(a) Encoder (b) Decoder

FIGURE 3.3
Block diagram of a practical pixel-to-pixel differential coding system.

Differential Coding

In general,

$$\bar{z}_i = z_i + e_{q,i}. \quad (3.13)$$

Thus, we see that the problem of quantization error accumulation has been resolved by having both the encoder and the decoder work in the same fashion, as indicated in Figure 3.3, or in Equations 3.3, 3.9, and 3.10.

3.1.2 General DPCM Systems

In the above discussion, we can view the reconstructed neighboring pixel's gray value as a prediction of that of the pixel being coded. Now, we generalize this simple pixel-to-pixel DPCM. In a general DPCM system, a pixel's gray level value is first predicted from the preceding reconstructed pixels' gray level values. The difference between the pixel's gray level value and the predicted value is then quantized. Finally, the quantized difference is encoded and transmitted to the receiver. A block diagram of this general differential coding scheme is shown in Figure 3.4, where the code word assignment in the encoder and its counterpart in decoder are not included.

It is noted that, instead of using the previous reconstructed sample, $\bar{z}_{i-1}$, as a predictor, we now have the predicted version of z_i, $\hat{z}_i$, as a function of the n previous reconstructed samples, $\bar{z}_{i-1}, \bar{z}_{i-2}, \ldots, \bar{z}_{i-n}$. That is,

$$\hat{z}_i = f(\bar{z}_{i-1}, \bar{z}_{i-2}, \ldots, \bar{z}_{i-n}). \quad (3.14)$$

Linear prediction, i.e., the function f in Equation 3.14 is linear, is of particular interest and is widely used in differential coding. In linear prediction, we have

$$\hat{z}_i = \sum_{j=1}^{n} a_j \bar{z}_{i-j}, \quad (3.15)$$

where a_j are real parameters. Hence, we see that the simple pixel-to-pixel differential coding is a special case of general differential coding with linear prediction, i.e., $n=1$ and $a_1=1$.

In Figure 3.4, d_i is the difference signal and is equal to the difference between the original signal, z_i, and the prediction $\hat{z}_i$. That is,

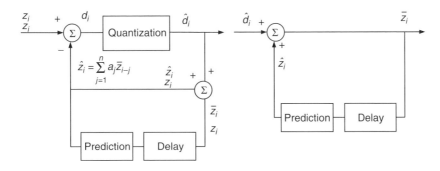

FIGURE 3.4
Block diagram of a general differential pulse code modulation (DPCM) system.

$$d_i = z_i - \hat{z}_i. \tag{3.16}$$

The quantized version of d_i is denoted by $\hat{d}_i$. The reconstructed version of z_i is represented by $\bar{z}_i$, and

$$\bar{z}_i = \hat{z}_i + \hat{d}_i. \tag{3.17}$$

Note that this is true for both the encoder and the decoder. Recall that the accumulation of the quantization error can be remedied by using this method.

The difference between the original input and the predicted input is called prediction error, which is denoted by e_p. That is,

$$e_p = z_i - \hat{z}_i, \tag{3.18}$$

where the e_p is understood as the prediction error associated with the index i. Quantization error, e_q, is equal to the reconstruction error or coding error, e_r, defined as the difference between the original signal, z_i, and the reconstructed signal, $\bar{z}_i$, when the transmission is error free:

$$\begin{aligned} e_q &= d_i - \hat{d}_i \\ &= (z_i - \hat{z}_i) - (\bar{z}_i - \hat{z}_i) \\ &= z_i - \bar{z}_i = e_r. \end{aligned} \tag{3.19}$$

This indicates that quantization error is the only source of information loss with an error free transmission channel.

The DPCM system depicted Figure 3.4 is also called closed-loop DPCM with feedback around the quantizer [jayant 1984]. This term reflects the feature in DPCM structure.

Before we leave this section, let us take a look at the history of the development of differential image coding. According to an excellent early article on differential image coding [musmann 1979], the first theoretical and experimental approaches to image coding involving linear prediction began in 1952 at the Bell Telephone Laboratories [harrison 1952; kretzmer 1952; oliver 1952]. The concepts of DPCM and DM were also developed in 1952 [cutler 1952; dejager 1952]. Predictive coding capable of preserving information for a PCM signal was established at the Massachusetts Institution of Technology [elias 1955].

Differential coding technique has played an important role in image and video coding. In the international coding standard for still images, JPEG, which is covered in Chapter 7, we can see that the differential coding is used in lossless mode, and in DCT-based mode for coding DC coefficients. Motion compensated (MC) coding has been a major development in video coding since 1980s and has been adopted by all the international video coding standards, such as H.261 and H.263 (Chapter 19), MPEG 1 and MPEG 2 (Chapter 16). MC coding is essentially a predictive coding technique applied to video sequences involving displacement motion vectors.

3.2 Optimum Linear Prediction

Figure 3.4 indicates that a differential coding system consists of two major components: prediction and quantization. Quantization was discussed in the Chapter 2. Hence, in this

Differential Coding

chapter, we emphasize prediction. Below, we formulate the optimum linear prediction problem and then present a theoretical solution to the problem.

3.2.1 Formulation

Optimum linear prediction can be formulated as follows. Consider a discrete-time random process z. At a typical moment i, it is a random variable z_i. We have n previous observations $\bar{z}_{i-1}, \bar{z}_{i-2}, \cdots, \bar{z}_{i-n}$ available and would like to form a prediction of z_i, denoted by $\hat{z}_i$. The output of the predictor, $\hat{z}_i$, is a linear function of the n previous observations. That is,

$$\hat{z}_i = \sum_{j=1}^{n} a_j \bar{z}_{i-j}, \tag{3.20}$$

with $a_j, j = 1, 2, \ldots, n$ being a set of real coefficients. An illustration of a linear predictor is shown in Figure 3.5. As defined above, the prediction error, e_p, is

$$e_p = z_i - \hat{z}_i. \tag{3.21}$$

The mean square prediction error, MSE_p, is

$$\text{MSE}_p = E[(e_p)^2] = E[(z_i - \hat{z}_i)^2] \tag{3.22}$$

The optimum prediction then refers to the determination of a set of coefficients $a_j, j = 1, 2, \ldots, n$ such that the MSE_p is minimized.

This optimization problem turns out to be computationally intractable for most practical cases due to the feedback around the quantizer shown in Figure 3.4, and the nonlinear nature of the quantizer. Therefore, the optimization problem is solved in two separate stages. That is, the best linear predictor is first designed ignoring the quantizer. Then, the quantizer is optimized for the distribution of the difference signal [habibi 1971]. Although the predictor thus designed is suboptimal, ignoring the quantizer in the optimum predictor design allows us to substitute the reconstructed $\bar{z}_{i-j}$ by z_{i-j} for $j = 1, 2, \ldots, n$, according to Equation 3.19. Consequently, we can apply the theory of optimum linear prediction to handle the design of the optimum predictor as shown below.

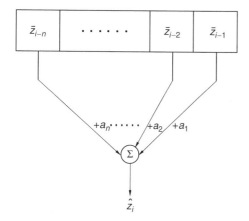

FIGURE 3.5
An illustration of a linear predictor.

3.2.2 Orthogonality Condition and Minimum Mean Square Error

By taking the differentiation of MSE_p with respect to coefficient a_j, one can derive the following necessary conditions, which are usually referred to as the orthogonality condition.

$$E(e_p \cdot z_{i-j}) = 0 \quad \text{for } j = 1, 2, \ldots, n. \tag{3.23}$$

The interpretation of Equation 3.23 is that the prediction error, e_p, must be orthogonal to all the observations, which are now the preceding samples: $z_{i-j}, j = 1, 2, \cdots, n$, according to our discussion made in Section 3.2.1. These are equivalent to

$$R_z(m) = \sum_{j=1}^{n} a_j R_z(m-j) \quad \text{for } m = 1, 2, \ldots, n, \tag{3.24}$$

where R_z represents the autocorrelation function of z. In a vector–matrix format, the above orthogonal conditions can be written as

$$\begin{bmatrix} R_z(1) \\ R_z(2) \\ \vdots \\ \vdots \\ R_z(n) \end{bmatrix} = \begin{bmatrix} R_z(0) & R_z(1) & \cdots & \cdots & R_z(n-1) \\ R_z(1) & R_z(2) & \cdots & \cdots & R_z(n-2) \\ \vdots & \vdots & \cdots & \cdots & \vdots \\ \vdots & \vdots & \cdots & \cdots & \vdots \\ R_z(n-1) & R_z(n) & \cdots & \cdots & R_z(0) \end{bmatrix} \cdot \begin{bmatrix} a_1 \\ a_2 \\ \vdots \\ \vdots \\ a_n \end{bmatrix} \tag{3.25}$$

Equations 3.24 and 3.25 are called Yule–Walker equations.

The minimum MSE_p is then found to be

$$\mathrm{MSE}_p = R_z(0) - \sum_{j=1}^{n} a_j R_z(j). \tag{3.26}$$

These results can be found in texts on random processes [leon-garcia 1994].

3.2.3 Solution to Yule–Walker Equations

Once autocorrelation data is available, the Yule–Walker equation can be solved by matrix inversion. A recursive procedure was developed by Levinson to solve the Yule–Walker equations [leon-garcia 1994]. When the number of previous samples used in the linear predictor is large, i.e., the dimension of the matrix is high, the Levinson recursive algorithm becomes more attractive. Note that in the field of image coding the autocorrelation function of various types of video frames is derived from measurements [o'neal 1966; habibi 1971].

3.3 Some Issues in the Implementation of DPCM

Several related issues in the implementation of DPCM are discussed in this section.

Differential Coding

3.3.1 Optimum DPCM System

As DPCM consists mainly of two parts, prediction and quantization, its optimization should not be carried out separately. The interaction between the two parts is quite complicated, however, and thus combined optimization of the whole DPCM system is difficult. Fortunately, with the mean square error criterion, the relation between quantization error and prediction error has been found as

$$\text{MSE}_q \approx \frac{9}{2N^2} \text{MSE}_p, \qquad (3.27)$$

where N is the total number of reconstruction levels in the quantizer [o'neal 1966; musmann 1979]. That is, the mean square error of quantization, MSE_q, is approximately proportional to the, MSE_p, mean square error of prediction. With this approximation, we can optimize two parts separately as mentioned in Section 3.2.1. While the optimization of quantization was addressed in Chapter 2, the optimum predictor was discussed in Section 3.2. A large amount of work has been done in this subject. For instance, the optimum predictor for color image coding was designed and tested in [pirsch 1977].

3.3.2 1-D, 2-D, and 3-D DPCM

In Section 3.1.2, we expressed linear prediction in Equation 3.15. However, so far we have not discussed how to predict a pixel's gray level value by using its neighboring pixels' coded gray level values.

Practical pixel-to-pixel differential coding system was discussed in Section 3.1.1. There, the reconstructed intensity of the immediately preceding pixel along the same scan line is used as a prediction of the pixel intensity being coded. This type of differential coding is referred to as 1-D DPCM. In general, 1-D DPCM may use the reconstructed gray level values of more than one preceding pixels within the same scan line to predict that of a pixel being coded. By far, however, the immediately preceding pixel in the same scan line is most frequently used in 1-D DPCM. That is, pixel A in Figure 3.6 is often used as a prediction of pixel Z, which is being DPCM coded.

Sometimes in DPCM image coding, both the decoded intensity values of adjacent pixels within the same scan line and the decoded intensity values of neighboring pixels in the different scan lines are involved in the prediction. This is called 2-D DPCM. A typical pixel arrangement in 2-D predictive coding is shown in Figure 3.6. Note that the pixels involved in the prediction are restricted to be either in the lines above the line where the pixel being coded, Z, is located or on the left-hand side of pixel Z if they are in the same line. Traditionally, a TV frame is scanned from top to bottom and from left to right. Hence,

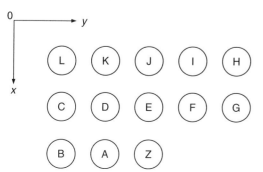

FIGURE 3.6
Pixel arrangement in 1-D and 2-D prediction.

the above restriction indicates that only those pixels, which have been coded, available in both the transmitter and the receiver, are used in the prediction. In 2-D system theory, this support is referred to as recursively computable [bose 1982]. An often used 2-D prediction involves pixels A, D, and E.

Obviously, 2-D predictive coding utilizes not only the spatial correlation existing within a scan line but also that existing in neighboring scan lines. In other words, the spatial correlation is utilized both horizontally and vertically. It was reported that 2-D predictive coding outperforms 1-D predictive coding by decreasing the prediction error by a factor of 2, or equivalently 3 dB in SNR. The improvement in subjective assessment is even larger [musmann 1979]. Furthermore, the transmission error in 2-D predictive image coding is much less severe than in 1-D predictive image coding. This is discussed in Section 3.6.

In the context of image sequences, neighboring pixels may be located not only in the same image frame but also in successive frames. That is, neighboring pixels along the time dimension are also involved. If the prediction of a DPCM system involves three types of neighboring pixels: those along the same scan line, those in the different scan lines of the same image frame, and those in the different frames, the DPCM is then called 3-D differential coding, discussed in Section 3.5.

3.3.3 Order of Predictor

The number of coefficients in the linear prediction, n, is referred to as the order of the predictor. The relation between the mean square prediction error, MSE_p, and the order of the predictor, n, has been studied. As shown in Figure 3.7, the MSE_p decreases as n increases quite effectively, but the performance improvement becomes negligible as $n > 3$ [habibi 1971].

3.3.4 Adaptive Prediction

Adaptive DPCM means adaptive prediction and adaptive quantization. As adaptive quantization was already discussed in Chapter 2, here we will discuss only adaptive prediction.

Similar to the discussion on adaptive quantization, adaptive prediction can be done in two different ways: forward adaptive and backward adaptive predictions. In the former, adaptation is based on the input of a DPCM system, while in the latter, adaptation is based on the output of the DPCM. Therefore, forward adaptive prediction is more sensitive to changes in local statistics. Prediction parameters (the coefficients of the predictor), however, need to be transmitted as side information to the decoder. On the other hand, quantization error is involved in backward adaptive prediction. Hence, the adaptation is less sensitive to local changing statistics. But, it does not need to transmit side information.

In either case, the data (either input or output) has to be buffered. Autocorrelation coefficients are analyzed, based on which the prediction parameters are determined.

3.3.5 Effect of Transmission Errors

Transmission error caused by channel noise may reverse the binary bit information from 0 to 1 or from 1 to 0 with what is known as bit error probability, or bit error rate. The effect of transmission error on reconstructed images varies depending on different coding techniques.

Differential Coding

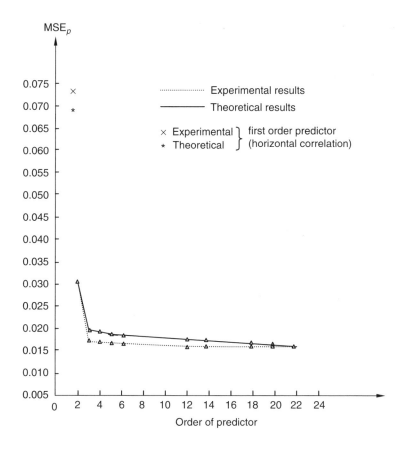

FIGURE 3.7
Mean square prediction error (MSE_p) versus order of predictor. (From Habibi, A., *IEEE Trans. Commun. Technol.*, COM-19, 948, 1971. With permission.)

In the case of the PCM-coding technique, each pixel is coded independently. Therefore, bit reversal in the transmission only affects the gray level value of the corresponding pixel in the reconstructed image. It does not affect other pixels in the reconstructed image.

In DPCM, however, the effect caused by transmission errors becomes more severe. Consider a bit reversal occurring in transmission. It causes error in the corresponding pixel. But, this is not the end of the effect. The affected pixel causes errors in reconstructing those pixels toward which the erroneous gray level value was used in the prediction. In this way, the transmission error propagates.

Interestingly, it is reported that the error propagation is more severe in 1-D differential image coding than in 2-D differential coding. This may be explained as follows: in 1-D differential coding, usually the immediate preceding pixel in the same scan line is involved in prediction. Therefore, an error will be propagated along the scan line until the beginning of the next line, where the pixel gray level value is reinitialized. In 2-D differential coding, the prediction of a pixel gray level value depends not only on the reconstructed gray level values of pixels along the same scan line but also on the reconstructed gray level values of the vertical neighbors. Hence, the effect caused by a bit reversal transmission error is less severe than in the 1-D differential coding.

For this reason, the bit error rate required by DPCM coding is lower than that required by PCM coding. For instance, while a bit error rate less than 5×10^{-6} is normally required for PCM to provide broadcast TV quality, for the same application a bit error rate less than 10^{-7} and 10^{-9} is required for DPCM coding with 2-D and 1-D predictions, respectively [musmann 1979].

Channel encoding with an error correction capability was applied to lower the bit error rate. For instance, to lower the bit error rate from the order of 10^{-6} to 10^{-9} for DPCM coding with 1-D prediction, an error correction code by adding 3% redundancy in channel coding has been used [bruders 1978].

3.4 Delta Modulation

Delta modulation is an important, simple, special case of DPCM, as discussed above. It was widely applied and is thus an important coding technique in and of itself.

The above discussion and characterization of DPCM systems are applicable to DM systems. This is because DM is essentially a special type of DPCM, with the following two features:

1. The linear predictor is of the first order, with the coefficient a_1 equal to 1.
2. The quantizer is a 1-bit quantizer. That is, depending on whether the difference signal is positive or negative, the output is either $+\Delta/2$ or $-\Delta/2$.

To perceive these two features, let us take a look at the block diagram of a DM system and the input–output characteristic of its 1-bit quantizer, shown in Figures 3.8 and 3.9, respectively. Due to the first feature listed above, we have

$$\hat{z}_i = \bar{z}_{i-1}. \tag{3.28}$$

Next, we see that there are only two reconstruction levels in quantization because of the second feature. That is,

$$\hat{d}_i = \begin{cases} +\Delta/2 & \text{if } z_i > \bar{z}_{i-1} \\ -\Delta/2 & \text{if } z_i < \bar{z}_{i-1} \end{cases}. \tag{3.29}$$

From the relation between the reconstructed value and the predicted value of DPCM discussed above and the fact that DM is a special case of DPCM, we have

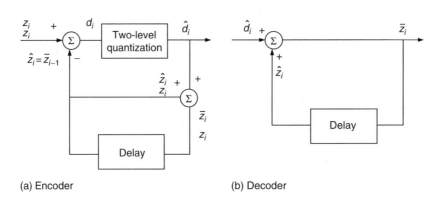

FIGURE 3.8
Block diagram of Delta modulation (DM) systems.

Differential Coding

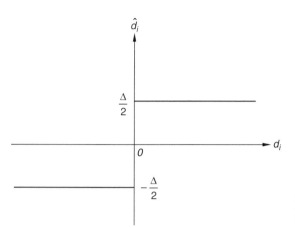

FIGURE 3.9
Input–output characteristic of two-level quantization in Delta modulation (DM).

$$\bar{z}_i = \hat{z}_i + \hat{d}_i. \quad (3.30)$$

Combining Equations 3.28 through 3.30, we have

$$\bar{z}_i = \begin{cases} \bar{z}_{i-1} + \Delta/2 & \text{if } z_i > \bar{z}_{i-1} \\ \bar{z}_{i-1} - \Delta/2 & \text{if } z_i < \bar{z}_{i-1} \end{cases}. \quad (3.31)$$

The above mathematical relationships are important in understanding DM systems. For instance, Equation 3.31 indicates that the step size Δ of DM is a crucial parameter. We note that a large step size compared with the magnitude of the difference signal causes granular error, as shown in Figure 3.10. Therefore, to reduce the granular error, we should choose a

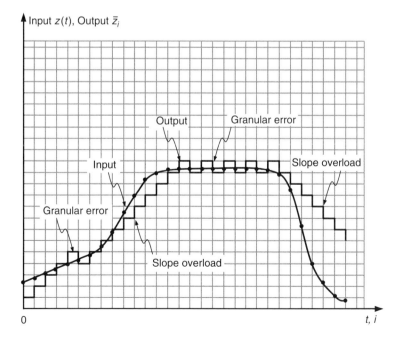

FIGURE 3.10
Delta modulation (DM) with fixed step size.

small step size. On the other hand, a small step size compared with the magnitude of the difference signal will lead to the overload error discussed in Chapter 2 for quantization. Since in DM systems it is the difference signal that is quantized, however, the overload error in DM becomes slope overload error, as shown in Figure 3.10. That is, it takes time (multiple steps) for the reconstructed samples to catch up with the sudden change in input. Therefore, the step size should be large to avoid the slope overload. Considering these two conflicting factors, a proper compromise in choosing the step size is common practice in DM.

To improve the performance of DM, an oversampling technique is often applied. That is, the input is oversampled before the application of DM. By oversampling, we mean that the sampling frequency is higher than the sampling frequency used in obtaining the original input signal. The increased sample density caused by oversampling decreases the magnitude of the difference signal. Consequently, a relatively small step size can be used so as to decrease the granular noise without increasing the slope overload error. At the last, the resolution of the DM-coded image is kept the same as that of the original input [Jayant 1984; Lim 1990].

To achieve better performance for changing inputs, an adaptive technique can be applied in DM. That is, either input (forward adaptation) or output (backward adaptation) data is buffered and the data variation is analyzed. The step size is then chosen accordingly. If it is forward adaptation, side information is required for transmission to the decoder. Figure 3.11 demonstrates step size adaptation. We see the same input as that shown in Figure 3.10. But, the step size is now not fixed. Instead, the step size is adapted according to the varying input. When the input changes with a large slope, the step size increases to avoid the slope overload error. On the other hand, when the input changes slowly, the step size decreases to reduce the granular error.

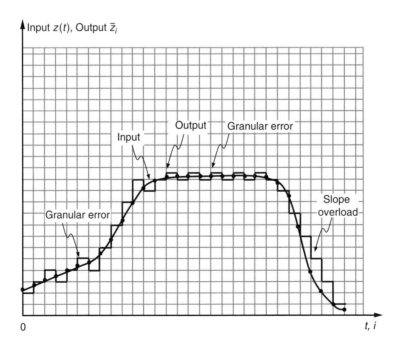

FIGURE 3.11
Adaptive Delta modulation (DM).

3.5 Interframe Differential Coding

As was mentioned in Section 3.3.2, 3-D differential coding involves an image sequence. Consider a sensor located in 3-D world space. For instances, in applications such as videophony and videoconferencing, the sensor is fixed in position for a while and it takes pictures. As time goes by, the images form a temporal image sequence. The coding of such an image sequence is referred to as interframe coding. The subject of image sequence and video coding is addressed in Parts III and IV. In this section, we will briefly discuss how differential coding is applied to interframe coding.

3.5.1 Conditional Replenishment

Recognizing the great similarity between consecutive TV frames, a conditional replenishment coding technique was proposed and developed [mounts 1969]. It was regarded one of the first real demonstrations of interframe coding exploiting interframe redundancy [netravali 1979].

In this scheme, the previous frame is used as a reference for the present frame. Consider a pair of pixels: one in the previous frame and the other in the present frame—both occupying the same spatial position in the frames. If the gray level difference between the pair of pixels exceeds a certain criterion, then the pixel is considered a changing pixel. The present pixel gray level value and its position information are transmitted to receiving side, where the pixel is replenished. Otherwise, the pixel is considered unchanged. At receiver its previous gray level is repeated. A block diagram of conditional replenishment is shown in Figure 3.12. There a frame memory unit in the transmitter is used to store frames. The differencing and thresholding of corresponding pixels in two consecutive frames can then be conducted there. A buffer in the transmitter is used to smooth the transmission data rate. This is necessary because the data rate varies from region-to-region

(a) Transmitter

(b) Receiver

FIGURE 3.12
Block diagram of conditional replenishment.

within an image frame and from frame-to-frame within an image sequence. A buffer in the receiver is needed for a similar consideration. In the frame memory unit, the replenishment is carried out for the changing pixels and the gray level values in the receiver are repeated for the unchanged pixels.

With conditional replenishment, a considerable savings in bit rate was achieved in applications such as videophony, videoconferencing, and TV broadcasting. Experiments in real time, using the head-and-shoulder view of a person in animated conversation as the video source, demonstrated an average bit rate of 1 bit/pixel with a quality of reconstructed video comparable with standard 8 bits/pixel PCM transmission [mount 1969]. Compared with pixel-to-pixel 1-D DPCM, the most popularly used coding technique at the time, conditional replenishment technique is more efficient due to the exploitation of high interframe redundancy. As pointed in [mount 1969] there is more correlation between television pixels along the frame-to-frame temporal dimension than there is between adjacent pixels within a signal frame. That is, the temporal redundancy is normally higher than spatial redundancy for TV signals.

Tremendous efforts have been made to improve the efficiency of this rudimentary technique. For an excellent review, readers are referred to [haskell 1972, 1979]. 3-D DPCM coding is among the improvements and is discussed next.

3.5.2 3-D DPCM

It is soon realized that it is more efficient to transmit the gray level difference than to transmit the gray level itself, resulting in interframe differential coding. Furthermore, instead of treating each pixel independently of its neighboring pixels, it is more efficient to utilize spatial redundancy as well as temporal redundancy, resulting in 3-D DPCM.

Consider two consecutive TV frames, each consisting of an odd and an even field. Figure 3.13 demonstrates the small neighborhood of a pixel, Z, in the context. As with the 1-D and 2-D DPCM discussed before, the prediction can only be based on the previously encoded pixels. If the pixel under consideration, Z, is located in the even field of the

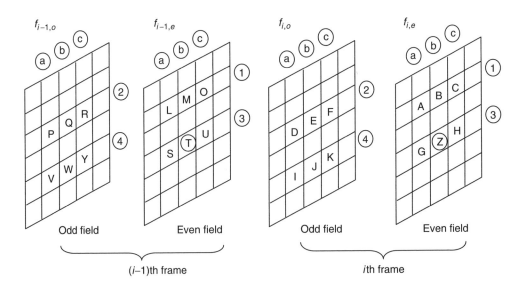

FIGURE 3.13
Pixel arrangement in two TV frames. (From Haskell, B.G., in *Image Transmission Techniques*, Academic Press, New York, 1979. With permission.)

Differential Coding

TABLE 3.1

Some Linear Prediction Schemes

	Original Signal (Z)	Prediction Signal (Ẑ)	Differential Signal (d_z)
Element difference	Z	G	$Z - G$
Field difference	Z	$\frac{E+J}{2}$	$Z - \frac{E+J}{2}$
Frame difference	Z	T	$Z - T$
Element difference of frame difference	Z	$T + G - S$	$(Z - G) - (T - S)$
Line difference of frame difference	Z	$T + B - M$	$(Z - B) - (T - M)$
Element difference of field difference	Z	$\left(T + \frac{E+J}{2}\right) - \left(\frac{Q+W}{2}\right)$	$\left(Z - \frac{E+J}{2}\right) - \left(T - \frac{Q+W}{2}\right)$

Source: From Haskell, B.G., in *Image Transmission Techniques*, Academic Press, New York, 1979. With permission.

present frame, then the odd field of the present frame and both odd and even fields of the previous frame are available. As mentioned in Section 3.3.2, it is assumed that in the even field of the present frame, only those pixels in the lines above the line where pixel Z lies and those pixels left of the Z in the line where Z lies are used for prediction.

Table 3.1 lists several utilized linear prediction scheme. It is recognized that the case of element difference is a 1-D predictor because the immediately preceding pixel is used as the predictor. The field difference is defined as the arithmetic average of two immediately vertical neighboring pixels in the previous odd field. As the odd field is generated first, followed by the even field, this predictor cannot be regarded as a pure 2-D predictor. Instead, it should be considered a 3-D predictor. The remaining cases belong to 3-D predictors. One thing is common in all the cases: the gray levels of pixels used in the prediction have already been coded and thus are available in both the transmitter and the receiver.

The prediction error of each changing pixel Z identified in thresholding process is then quantized and coded.

An analysis of the relationship between the entropy of moving areas (bits per changing pixel) and the motion speeds (pixels per frame interval) in the scenery containing a moving mannequin's head was studied with different linear predictions, listed in Table 3.1 [haskell 1979]. It was found that the element difference of field difference generally corresponds to the lowest entropy, meaning that this prediction is the most efficient. The frame difference and element difference correspond to higher entropy. It is recognized that, in the circumstances, transmission error will be propagated if the pixels in the previous line are used in prediction [connor 1973]. Hence, the linear predictor should use only pixels from the same line or the same line in the previous frame when bit reversal error in transmission needs to be considered. Combining these two factors, the element difference of frame difference prediction is preferred.

3.5.3 Motion Compensated Predictive Coding

When frames are taken densely enough, changes in successive frames can be attributed to the motion of objects during the interval between frames. Under this assumption, if we can analyze object motion from successive frames, then we should be able to predict objects in the next frame based on their positions in the previous frame and the estimated motion. The difference between the original frame and the predicted frame is thus generated and the motion vectors are then quantized and coded. If the motion estimation is accurate enough, the MC prediction error can be smaller than 3-D DPCM. In other words, the variance of the prediction error will be smaller, resulting in more efficient coding. Taking motion into consideration, this differential technique is called MC predictive coding. This technique was a major development in image sequence coding since the 1980s and was adopted by all international video coding standards. A more detailed discussion is provided in Chapter 10.

3.6 Information-Preserving Differential Coding

As emphasized in Chapter 2, quantization is not reversible in the sense that it causes information loss permanently. The DPCM technique, discussed above, includes quantization and hence is lossy coding. In applications such as those involving scientific measurements, information preserving is required. In this section, the following question is addressed: Under these circumstances, how should we apply differential coding to reduce bit rate while preserving information?

Figure 3.14 shows a block diagram of information-preserving differential coding. First, we see that there is no quantizer. Therefore, the irreversible information loss associated with quantization does not exist in this technique. Second, we observe that prediction and differencing are still used. That is, the differential (predictive) technique still applies. Hence it is expected that the variance of the difference signal is smaller than that of the original signal (Section 3.1). Consequently, the higher-peaked histograms make coding more efficient. Third, an efficient lossless coder is utilized. Since quantizers cannot be used here, PCM with natural binary coding is also not used here. As the histogram of the difference signal is narrowly concentrated about its mean, lossless coding techniques such as an efficient Huffman coder (discussed in Chapter 5) is naturally a suitable choice here.

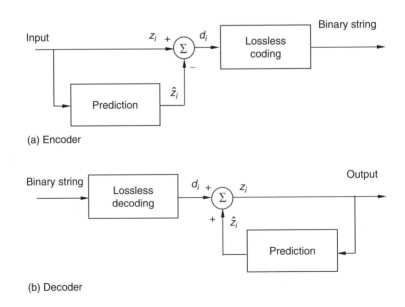

FIGURE 3.14
Block diagram of information-preserving differential coding.

Differential Coding

As mentioned earlier, input images are normally in a PCM coded format with a bit rate of 8 bits/pixel for monochrome pictures. The difference signal is therefore integer valued. Having no quantization and using an efficient lossless coder, the coding system depicted in Figure 3.14 is, therefore, an information-preserving differential coding technique.

3.7 Summary

Rather than coding the signal itself, differential coding, also known as predictive coding, encodes the difference between the signal and its prediction. Utilizing spatial and/or temporal correlation between pixels in the prediction, the variance of the difference signal can be much smaller than that of the original signal, thus making differential coding quite efficient.

Among differential coding methods, differential pulse code modulation (DPCM) is used most widely. In DPCM coding, the difference signal is quantized and code words are assigned to the quantized difference. Prediction and quantization are therefore the two major components in the DPCM systems. Since quantization was already addressed in Chapter 2, this chapter emphasizes prediction. The theory of optimum linear prediction is introduced. Here, optimum means minimization of mean square prediction error (MSE_p). The formulation of optimum linear prediction, the orthogonality condition, and the minimum MSE_p are presented. The orthogonality condition states that the prediction error must be orthogonal to each observation, i.e., the reconstructed sample intensity values are used in the linear prediction. By solving the Yule–Walker equation, the optimum prediction coefficients may be determined.

In addition, some fundamental issues in implementing the DPCM technique are discussed. One issue is the dimensionality of the predictor in DPCM. We discussed 1-D, 2-D, and 3-D predictors. DPCM with a 2-D predictor demonstrates better performance than that with 1-D predictor, because 2-D DPCM utilizes more spatial correlation, i.e., not only horizontally but also vertically. As a result, a 3 dB improvement in SNR was reported. 3-D prediction is encountered in what is known as interframe coding. There, temporal correlation exists and 3-D DPCM utilizes both spatial and temporal correlation between neighboring pixels in successive frames. Consequently, more redundancy can be removed. Motion compensated (MC) predictive coding as a very powerful technique in video coding belongs to differential coding. It uses a more advanced translational motion model in the prediction, however, and it is covered in Parts III and IV.

Another issue is the order of predictors and its effect on the performance of prediction in terms of MSE_p. Increasing prediction order can lower MSE_p effectively, but the performance improvement becomes not significant after the third order.

Adaptive prediction is another issue. Similar to adaptive quantization, discussed in Chapter 2, we can adapt the prediction coefficients in the linear predictor to varying local statistics.

The last issue is concerned with the effect of transmission error. Bit reversal in transmission causes a different effect on reconstructed images depending on what type of coding technique is used. PCM is known to be bit-consuming. (An acceptable PCM representation of monochrome images requires 6–8 bits/pixel.) But 1 bit reversal only affects an individual pixel. For the DPCM coding technique, however, a transmission error may propagate from one pixel to the other. In particular, DPCM with a 1-D predictor suffers from error propagation more severely than DPCM with a 2-D predictor.

Delta modulation is an important special case of DPCM, in which the predictor is of the first order. Specifically, the immediate preceding coded sample is used as a prediction of the present input sample. Furthermore, the quantizer has only two reconstruction levels.

Finally, an information-preserving differential coding technique is discussed. As mentioned in Chapter 2, quantization is an irreversible process: it causes information loss. To be able to preserve information, there is no quantizer in this type of system. To be efficient, lossless codes such as Huffman code or arithmetic code are used for difference signal encoding.

Exercises

1. Justify the necessity of the closed-loop DPCM with feedback around quantizers. That is, give a suitable reason why the quantization error will be accumulated if, instead of using the reconstructed preceding samples, we use the immediately preceding sample as the prediction of the sample being coded in DPCM.
2. Why does the overload error encountered in quantization appear to be the slope overload in DM?
3. What advantage does oversampling bring up in the DM technique?
4. What are the two features of DM that make it a subclass of DPCM?
5. Explain why DPCM with a 1-D predictor suffers from bit reversal transmission error more severely than DPCM with a 2-D predictor.
6. Explain why no quantizer can be used in information-preserving differential coding, and why the differential system can work without a quantizer.
7. Why do all the pixels involved in prediction of differential coding have to be in a recursively computable order from the point of view of the pixel being coded?
8. Discuss the similarity and dissimilarity between DPCM and MC predictive coding.

References

[bose 1982] N.K. Bose, *Applied Multidimensional System Theory*, Van Nostrand Reinhold, New York, 1982.
[bruders 1978] R. Bruders, T. Kummerow, P. Neuhold, and P. Stamnitz, Ein versuchssystem zur digitalen ubertragung von fernsehsignalen unter besonderer berucksichtigung von ubertragungsfehlern, Festschrift 50 Jahre Heinrich-Hertz-Institut, Berlin, 1978.
[connor 1973] D.J. Connor, *IEEE Transactions on Communications*, COM-21, 695–706, 1973.
[cutler 1952] C.C. Cutler, U.S. Patent 2,605,361, 1952.
[dejager 1952] F. DeJager, *Philips Research Report*, 7, 442–466, 1952.
[elias 1955] P. Elias, *IRE Transactions on Information Theory*, IT-1, 16–32, 1955.
[habibi 1971] A. Habibi, Comparison of nth-order DPCM encoder with linear transformations and block quantization techniques, *IEEE Transactions on Communication Technology*, COM-19, 6, 948–956, December 1971.
[harrison 1952] C.W. Harrison, *Bell System Technical Journal*, 31, 764–783, 1952.
[haskell 1972] B.G. Haskell, F.W. Mounts, and J.C. Candy, Interframe coding of videotelephone pictures, *Proceedings of the IEEE*, 60, 7, 792–800, July 1972.
[haskell 1979] B.G. Haskell, Frame replenishment coding of television, in *Image Transmission Techniques*, W.K. Pratt, (Ed.), Academic Press, New York, 1979.
[Jayant 1984] N.S. Jayant and P. Noll, *Digital Coding of Waveforms*, Prentice-Hall, Englewood Cliffs, NJ, 1984.
[kretzmer 1952] E.R. Kretzmer, Statistics of television signals, *Bell System Technical Journal*, 31, 751–763, July 1952.

[leon-garcia 1994] A. Leon-Garcia, *Probability and Random Processes for Electrical Engineering*, 2nd edn., Addison Wesley, Reading, MA, 1994.

[lim 1990] J.S. Lim, *Two-dimensional Signal and Image Processing*, Prentice-Hall, Englewood Cliffs, NJ, 1990.

[mounts 1969] F.W. Mounts, A video encoding system with conditional picture-element replenishment, *Bell System Technical Journal*, 48, 7, 2545–1554, September 1969.

[musmann 1979] H.G. Musmann, Predictive image coding, in *Image Transmission Techniques*, W.K. Pratt (Ed.), Academic Press, New York, 1979.

[netravali 1979] A.N. Netravali and J.D. Robbins, Motion compensated television coding: Part I, *The Bell System Technical Journal*, 58, 3, 631–670, March 1979.

[oliver 1952] B.M. Oliver, *Bell System Technical Journal*, 31, 724–750, 1952.

[o'neal 1966] J.B. O'Neal, *Bell System Technical Journal*, 45, 689–721, 1966.

[pirsch 1977] P. Pirsch and L. Stenger, *Acta Electronica*, 19, 277–287, 1977.

[sayood 1996] K. Sayood, *Introduction to Data Compression*, Morgan Kaufmann Publishers, San Francisco, CA, 1996.

4

Transform Coding

As introduced in Chapter 3, differential coding achieves high coding efficiency by utilizing the correlation between pixels existing in image frames. Transform coding (TC), which is the focus of this chapter, is another efficient coding scheme based on utilization of inter-pixel correlation. As we will see in Chapter 7, TC has become a fundamental technique recommended by the international still image coding standard Joint Photographic Experts Group coding (JPEG). In addition, TC was found to be efficient in coding prediction error in motion compensated (MC) predictive coding. As a result, it was also adopted by the international video coding standards such as H.261, H.263, and MPEG 1, 2, and 4. This will be discussed in Part IV.

4.1 Introduction

As shown in Figure 2.3 there are three components in a source encoder: transformation, quantization, and code word assignment. It is the transformation component that decides which format of input source is quantized and encoded. In differential pulse code modulation (DPCM), for instance, the difference between an original signal and a predicted version of the original signal is quantized and encoded. As long as the prediction error is small enough, i.e., the prediction resembles the original signal well (by using correlation between pixels), differential coding is efficient.

In TC, the main idea is that if the transformed version of a signal is less correlated compared with the original signal, then quantizing and encoding the transformed signal may lead to data compression. At the receiver, the encoded data are decoded and transformed back to reconstruct the signal. Therefore, in TC, the transformation component illustrated in Figure 2.3 is a transform. Quantization and code word assignment are carried out with respect to the transformed signal or the transformed domain.

We begin with the Hotelling transform, using it as an example of how a transform may decorrelate a signal in the transform domain.

4.1.1 Hotelling Transform

Consider an N-dimensional (N-D) vector $\vec{z}_s$. The ensemble of such vectors, $\{\vec{z}_s\}\ s \in I$, where I represents the set of all vector indexes, can be modeled by a random vector $\vec{z}$ with each of its component z_i, $i = 1, 2, \ldots, N$ as a random variable. That is,

$$\vec{z} = (z_1, z_2, \ldots, z_N)^T, \tag{4.1}$$

where T stands for the operator of matrix transposition. The mean vector of the population, $m_{\vec{z}}$, is defined as

$$m_{\vec{z}} = E[\vec{z}] = (m_1, m_2, \ldots, m_N)^T, \qquad (4.2)$$

where E stands for the expectation operator. Note that $m_{\vec{z}}$ is an N-D vector with the ith component, m_i, being the expectation value of the ith random variable component in $\vec{z}$.

$$m_i = E[z_i], \; i = 1, 2, \ldots, N. \qquad (4.3)$$

The covariance matrix of the population denoted by, $C_{\vec{z}}$, is equal to

$$C_{\vec{z}} = E[(\vec{z} - m_{\vec{z}})(\vec{z} - m_{\vec{z}})^T]. \qquad (4.4)$$

Note that the product inside the E operator is referred to as the outer product of the vector $(\vec{z} - m_{\vec{z}})$. Denote an entry at the ith row and jth column in the covariance matrix by $c_{i,j}$. From Equation 4.4, it can be seen that $c_{i,j}$ is the covariance between the ith and jth components of the random vector $\vec{z}$. That is,

$$c_{i,j} = E[(z_i - m_i)(z_j - m_j)] = \text{Cov}(z_i, z_j). \qquad (4.5)$$

On the main diagonal of the covariance matrix $C_{\vec{z}}$, the element $c_{i,j}$ is the variance of the ith component of $\vec{z}$, z_i.

Obviously, the covariance matrix $C_{\vec{z}}$ is a real and symmetric matrix. It is real because of the definition of random variables. It is symmetric because $\text{Cov}(z_i, z_j) = \text{Cov}(z_j, z_i)$. According to the theory of linear algebra, it is always possible to find a set of N orthonormal eigenvectors of the matrix $C_{\vec{z}}$, with which we can convert the real symmetric matrix $C_{\vec{z}}$ into a full-ranked diagonal matrix. This statement can be found in the texts of linear algebra [strang 1998].

Denote the set of N orthonormal eigenvectors and their corresponding eigenvalues of the covariance matrix $C_{\vec{z}}$ by $\vec{e}_i$ and λ_i, $i = 1, 2, \ldots, N$, respectively. Note that eigenvectors are column vectors. Form a matrix Φ such that its rows comprise the N eigenvectors. That is,

$$\Phi = (\vec{e}_1, \vec{e}_2, \ldots, \vec{e}_N)^T. \qquad (4.6)$$

Now, consider the following transformation.

$$\vec{y} = \Phi(\vec{z} - m_{\vec{z}}). \qquad (4.7)$$

It is easy to verify that the transformed random vector $\vec{y}$ has the following two characteristics:

1. The mean vector, $m_{\vec{y}}$, is a zero vector. That is,

$$m_{\vec{y}} = 0. \qquad (4.8)$$

2. The covariance matrix of the transformed random vector $C_{\vec{y}}$ is

$$C_{\vec{y}} = \Phi C_{\vec{z}} \Phi^T = \begin{bmatrix} \lambda_1 & & & & 0 \\ & \lambda_2 & & & \\ & & \ddots & & \\ & & & \ddots & \\ 0 & & & & \lambda_N \end{bmatrix}. \qquad (4.9)$$

This transform is called the Hotelling transform [hotelling 1933], or eigenvector transform [tasto 1971; wintz 1972].

The inverse Hotelling transform is defined as

$$\vec{z} = \Phi^{-1} \vec{y} + m_{\vec{z}}, \qquad (4.10)$$

where Φ^{-1} is the inverse matrix of Φ. It is easy to see from its formation discussed above that the matrix Φ is orthogonal. Therefore, we have $\Phi^T = \Phi^{-1}$. Hence the inverse Hotelling transform can be expressed as

$$\vec{z} = \Phi^T \vec{y} + m_{\vec{z}}. \qquad (4.11)$$

Note that in implementing the Hotelling transform, the mean vector $m_{\vec{z}}$ and the covariance matrix $C_{\vec{z}}$ can be calculated approximately by using a given set of K sample vectors [gonzalez 2001].

$$m_{\vec{z}} = \frac{1}{K} \sum_{s=1}^{K} \vec{z}_s \qquad (4.12)$$

$$C_{\vec{z}} = \frac{1}{K} \sum_{s=1}^{K} \vec{z}_s \vec{z}_s^T - m_{\vec{z}} m_{\vec{z}}^T \qquad (4.13)$$

The analogous transform for continuous data was devised by Karhunen and Loeve [karhunen 1947; loeve 1948]. Alternatively, the Hotelling transform can be viewed as the discrete version of the Karhunen–Loeve transform (KLT). We observe that the covariance matrix $C_{\vec{y}}$ is a diagonal matrix. The elements in the diagonal are the eigenvalues of the covariance matrix $C_{\vec{z}}$. That is, the two covariance matrices have the same eigenvalues and eigenvectors because the two matrices are similar. The fact that zero values are everywhere, except along the main diagonal in $C_{\vec{y}}$, indicates that the components of the transformed vector $\vec{y}$ are uncorrelated. That is, the correlation previously existing between the different components of the random vector $\vec{z}$ has been removed in the transformed domain. Therefore, if the input is split into blocks and the Hotelling transform is applied blockwise, the coding may be more efficient because the data in the transformed block are uncorrelated. At the receiver, we may produce a replica of the input with an inverse transform. This basic idea behind TC will be further illustrated. Note that TC is also referred to as block quantization [huang 1963].

4.1.2 Statistical Interpretation

Let us continue our discussion of the 1-D Hotelling transform, recalling that the covariance matrix of the transformed vector $\vec{y}$, $C_{\vec{y}}$, is a diagonal matrix. The elements in the main diagonal are eigenvalues of the covariance matrix $C_{\vec{y}}$. According to the definition of covariance matrix, these elements are the variances of the components of vector $\vec{y}$, denoted by $\sigma_{y,1}^2, \sigma_{y,2}^2, \ldots, \sigma_{y,N}^2$. Let us arrange the eigenvalues (variances) in a nonincreasing order, i.e., $\lambda_1 \geq \lambda_2 \geq, \ldots, \geq \lambda_N$. Choose an integer L, and $L < N$. Using the corresponding L eigenvectors, $\vec{e}_1, \vec{e}_2, \ldots, \vec{e}_L$, we form a matrix $\bar{\Phi}$ with these L eigenvectors (transposed) as its L rows. Obviously, the matrix $\bar{\Phi}$ is of $L \times N$. Hence, using the matrix $\bar{\Phi}$ in Equation 4.7 will have the transformed vector $\vec{y}$ of $L \times 1$. That is,

$$\vec{y} = \bar{\Phi}(\vec{z} - m_{\vec{z}}). \tag{4.14}$$

The inverse transform changes accordingly

$$\vec{z}\,' = \bar{\Phi}^T \vec{y} + m_{\vec{z}}. \tag{4.15}$$

Note that the reconstructed vector $\vec{z}$, denoted by $\vec{z}\,'$, is still an $N \times 1$ column vector. It can be shown [wintz 1972] that the mean square reconstruction error between the original vector $\vec{z}$ and the reconstructed vector $\vec{z}\,'$ is given by

$$\mathrm{MSE}_r = \sum_{i=L+1}^{N} \sigma_{y,i}^2. \tag{4.16}$$

Equation 4.16 indicates that the mean square reconstruction error equals the sum of variances of the discarded components. Note that although we discuss the reconstruction error here, we have not considered the quantization error and transmission error involved. Equation 4.15 implies that if, in the transformed vector $\vec{y}$, the first L components have their variances occupy a large percentage of the total variances, the mean square reconstruction error will not be large even though only the first L components are kept, i.e., the $(N-L)$ remaining components in the $\vec{y}$ are discarded. Quantizing and encoding only L components of vector $\vec{y}$ in the transform domain lead to higher coding efficiency, which is the basic idea behind TC.

4.1.3 Geometrical Interpretation

Transforming a set of statistically dependent data into another set of uncorrelated data, and then discarding the insignificant transform coefficients (having small variances) illustrated earlier using the Hotelling transform, can be viewed as a statistical interpretation of TC. Here, we give a geometrical interpretation of TC. For this purpose, we use 2-D vectors instead of N-D vectors.

Consider a binary image of a car in Figure 4.1a. Each pixel in the shaded object region corresponds to a 2-D vector with its two components being coordinates z_1 and z_2, respectively. Hence, the set of all pixels associated with the object forms a population of vectors. We can determine its mean vector and covariance matrix using Equations 4.12 and 4.13, respectively. We can then apply the Hotelling transform by using Equation 4.7. Figure 4.1b depicts the same object after the application of the Hotelling transform in the $y_1 - y_2$ coordinate system. We note that the origin of the new coordinate system is now located at the centroid of the binary object. Furthermore, the new coordinate system is aligned with the two eigenvectors of the covariance matrix $C_{\vec{z}}$.

As mentioned earlier, the elements along the main diagonal $C_{\vec{y}}$ (two eigenvalues of the $C_{\vec{y}}$ and $C_{\vec{z}}$) are the two variances of the two components of the $\vec{y}$ population. Since the covariance matrix $C_{\vec{y}}$ is a diagonal matrix, the two components are uncorrelated after

Transform Coding

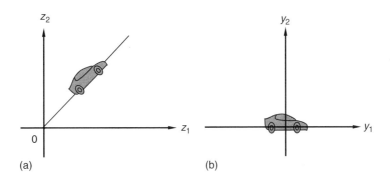

FIGURE 4.1
(a) A binary object in the z_1–z_2 coordinate system. (b) After the Hotelling transform, the object is aligned with its principal axes.

the transform. As one variance (along the y_1 direction) is larger than the other (along the y_2 direction), it is possible for us to achieve higher coding efficiency by ignoring the component associated with the smaller variance without too much sacrifice of the reconstructed image quality.

It is noted that the alignment of the object with the eigenvectors of the covariance matrix is of importance in pattern recognition [gonzalez 2001].

4.1.4 Basis Vector Interpretation

Basis vector expansion is another interpretation of TC. For simplicity, in this section we assume a zero mean vector. Under this assumption, the Hotelling transform and its inverse transform become

$$\vec{y} = \Phi \vec{z} \tag{4.17}$$

$$\vec{z} = \Phi^T \vec{y} \tag{4.18}$$

Recall that the row vectors in the matrix Φ are the transposed eigenvectors of the covariance matrix $C_{\vec{z}}$. Equation 4.18 can be written as

$$\vec{z} = \sum_{i=1}^{N} y_i \vec{e}_i. \tag{4.19}$$

In Equation 4.19, we can view vector $\vec{z}$ as a linear combination of basis vectors $\vec{e}_i$, $i = 1, 2, \ldots, N$. The components of the transformed vector $\vec{y}$, y_i, $i = 1, 2, \ldots, N$, serve as coefficients in the linear combination or as weights in the weighted sum of basis vectors. The coefficient y_i, $i = 1, 2, \ldots, N$, can be produced according to Equation 4.17:

$$y_i = \vec{e}_i^T \vec{z}. \tag{4.20}$$

That is, y_i is the inner product between vectors $\vec{e}_i$ and $\vec{z}$. Therefore, the coefficient y_i can be interpreted as the amount of correlation between the basis vector $\vec{e}_i$ and the original signal $\vec{z}$.

In the Hotelling transform, the coefficients y_i, $i = 1, 2, \ldots, N$, are uncorrelated. The variance of y_i can be arranged in a nonincreasing order. For $i > L$, the variance of the coefficient becomes insignificant. We can then discard these coefficients without introducing significant error in the linear combination of basis vectors and achieve higher coding efficiency.

In the above three interpretations of TC, we see that the linear unitary transform can provide the following two functions:

1. Decorrelate input data; i.e., transform coefficients are less correlated than the original data.
2. Have some transform coefficients more significant than others (with large variance, eigenvalue, or weight in basis vector expansion) such that transform coefficients can be treated differently: some can be discarded, some can be coarsely quantized, and some can be finely quantized.

Note that the definition of unitary transform is given shortly in Section 4.2.1.3.

4.1.5 Procedures of Transform Coding

In this section, we summarize the procedures of TC. There are three steps in TC as shown in Figure 4.2. First, the input data (frame) is divided into blocks (subimages). Each block is then linearly transformed. The transformed version is then truncated, quantized, and encoded. These last three functions, which are discussed in Section 4.4, can be grouped and termed as bit allocation. The output of encoder is a bit stream.

In the receiver, the bit stream is decoded and then inversely transformed to form reconstructed blocks. All the reconstructed blocks collectively produce a replica of the input image.

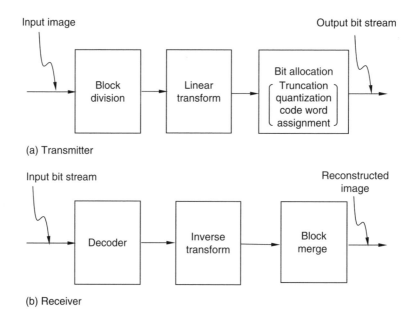

FIGURE 4.2
Block diagram of transform coding.

4.2 Linear Transforms

Here, we first discuss a general formulation of a linear unitary 2-D image transform. Then, a basis image interpretation of TC is given.

4.2.1 2-D Image Transformation Kernel

There are two different ways to handle image transformation. For example, in the first way, we convert a 2-D array representing a digital image into a 1-D array via row-by-row stacking. That is, from the second row on, the beginning of each row in the 2-D array is cascaded to the end of its previous row. Then we transform this 1-D array using a 1-D transform. After the transformation, we can convert the 1-D array back to a 2-D array. With the second way, a 2-D transform is directly applied to the 2-D array corresponding to an input image, resulting in a transformed 2-D array. These two ways are essentially the same. It can be straightforwardly shown that the difference between the two is simply a matter of notation [wintz 1972]. In this section, we use the second way to handle image transformation. That is, we work on 2-D image transformation.

Assume a digital image is represented by a 2-D array $g(x, y)$, where (x, y) is the coordinates of a pixel in the 2-D array, while g is the gray level value (also often called intensity or brightness) of the pixel. Denote the 2-D transform of $g(x, y)$ by $T(u, v)$, where (u, v) is the coordinates in the transformed domain. Assume that both $g(x, y)$ and $T(u, v)$ are a square 2-D array of $N \times N$, i.e., $0 \leq x, y, u, v \leq N-1$.

The 2-D forward and inverse transforms are defined as

$$T(u,v) = \sum_{x=0}^{N-1} \sum_{y=0}^{N-1} g(x,y) f(x,y,u,v) \qquad (4.21)$$

and

$$g(x,y) = \sum_{u=0}^{N-1} \sum_{v=0}^{N-1} T(u,v) i(x,y,u,v), \qquad (4.22)$$

where $f(x, y, u, v)$ and $i(x, y, u, v)$ are referred to as the forward and inverse transformation kernels, respectively.

A few characteristics of transforms are discussed below.

4.2.1.1 Separability

A transformation kernel is called separable (hence, the transform is said to be separable) if the following conditions are satisfied:

$$f(x,y,u,v) = f_1(x,u) f_2(y,v) \qquad (4.23)$$

and

$$i(x,y,u,v) = i_1(x,u) i_2(y,v). \qquad (4.24)$$

Note that a 2-D separable transform can be decomposed into two 1-D transforms. That is, a 2-D transform can be implemented by a 1-D transform rowwise followed by another 1-D transform columnwise. That is,

$$T_1(x,v) = \sum_{y=0}^{N-1} g(x,y) f_2(y,v), \qquad (4.25)$$

where $0 \leq x, v \leq N-1$, and

$$T(u,v) = \sum_{x=0}^{N-1} T_1(x,v) f_1(x,u), \qquad (4.26)$$

where $0 \leq u, v \leq N-1$. Of course, the 2-D transform can also be implemented in a reverse order with two 1-D transforms, i.e., columnwise first followed by rowwise. The counterparts of Equations 4.25 and 4.26 for the inverse transform can be derived similarly.

4.2.1.2 Symmetry

The transformation kernel is symmetric (hence, the transform is symmetric) if the kernel is separable and the following condition is satisfied:

$$f_1(y,v) = f_2(y,v). \qquad (4.27)$$

That is, f_1 is functionally equivalent to f_2.

4.2.1.3 Matrix Form

If a transformation kernel is symmetric (hence, separable) then the 2-D image transform discussed earlier can be expressed compactly in the following matrix form. Denote an image matrix by G and $G = \{g_{i,j}\} = \{g(i-1, j-1)\}$. That is, a typical element (at the ith row and jth column) in the matrix G is the pixel gray level value in the 2-D array $g(x, y)$ at the same geometrical position. Note that the subtraction of one in the notation $g(i-1, j-1)$ comes from Equations 4.21 and 4.22. That is, the indexes of a square 2-D image array are conventionally defined from 0 to $N-1$, while the indexes of a square matrix are from 1 to N. Denote the forward transform matrix by F and $F = \{f_{i,j}\} = \{f_1(i-1, j-1)\}$. We then have the following matrix form of a 2-D transform:

$$T = F^T G F, \qquad (4.28)$$

where T at the left-hand side of the equation denotes the matrix corresponding to the transformed 2-D array in the same fashion as that used in defining the G matrix. The inverse transform can be expressed as

$$G = I^T T I, \qquad (4.29)$$

where the matrix I is the inverse transform matrix and $I = \{i_{j,k}\} = \{i_1(j-1, k-1)\}$. The forward and inverse transform matrices have the following relation:

$$I = F^{-1}. \qquad (4.30)$$

Note that all of the matrices defined above, G, T, F, and I, are of $N \times N$.

It is known that the discrete Fourier transform (DFT) involves complex quantities. In this case, the counterparts of Equations 4.28 through 4.30 become Equations 4.31 through 4.33, respectively:

$$T = F^{*T} G F \qquad (4.31)$$

Transform Coding

$$G = I^{*T}TI \tag{4.32}$$

$$I = F^{-1} = F^{*T}, \tag{4.33}$$

where * indicates complex conjugation. Note that the transform matrices F and I contain complex quantities and satisfy Equation 4.33. They are called unitary matrices and the transform is referred to as a unitary transform.

4.2.1.4 Orthogonality

A transform is said to be orthogonal if the transform matrix is orthogonal. That is,

$$F^T = F^{-1}. \tag{4.34}$$

Note that an orthogonal matrix (orthogonal transform) is a special case of a unitary matrix (unitary transform), where only real quantities are involved. We will see that all the 2-D image transforms, presented in Section 4.3, are separable, symmetric, and unitary.

4.2.2 Basis Image Interpretation

Here we study the concept of basis images or basis matrices. Recall that we discussed basis vectors when we considered the 1-D transform. That is, the components of the transformed vector (also referred to as the transform coefficients) can be interpreted as the coefficients in the basis vector expansion of the input vector. Each coefficient is essentially the amount of correlation between the input vector and the corresponding basis vector. The concept of basis vectors can be extended to basis images in the context of 2-D image transforms.

The 2-D inverse transform introduced in Section 4.2.1 (Equation 4.22) is defined as

$$g(x,y) = \sum_{u=0}^{N-1} \sum_{v=0}^{N-1} T(u,v) i(x,y,u,v), \tag{4.35}$$

where $0 \leq x, y \leq N-1$. Equation 4.35 can be viewed as a component form of the inverse transform. As defined in Section 4.2.1.3, the whole image $\{g(x, y)\}$ is denoted by the image matrix G of $N \times N$. We now denote the image formed by the inverse transformation kernel $\{i(x, y, u, v), 0 \leq x, y \leq N - 1\}$ as a 2-D array $I_{u,v}$ of $N \times N$ for a specific pair of (u, v) with $0 \leq u, v \leq N - 1$. Recall that a digital image can be represented by a 2-D array of gray level values. In turn the 2-D array can be arranged into a matrix. Namely, we treat the following three: a digital image, a 2-D array (with proper resolution), and a matrix (with proper indexing), interchangeably. We then have

$$I_{u,v} = \begin{bmatrix} i(0,0,u,v) & i(0,1,u,v) & \cdots & \cdots & i(0,N-1,u,v) \\ i(1,0,u,v) & i(1,1,u,v) & \cdots & \cdots & i(1,N-1,u,v) \\ \vdots & \vdots & \cdots & \cdots & \vdots \\ \vdots & \vdots & \cdots & \cdots & \vdots \\ i(N-1,0,u,v) & i(N-1,1,u,v) & \cdots & \cdots & i(N-1,N-1,u,v) \end{bmatrix} \tag{4.36}$$

The 2-D array $I_{u,v}$ is referred to as a basis image. There are N^2 basis images in total because $0 \leq u, v \leq N-1$. The inverse transform expressed in Equation 4.35 can then be written in a collective form as

$$G = \sum_{u=0}^{N-1} \sum_{v=0}^{N-1} T(u,v) I_{u,v}. \qquad (4.37)$$

We can interpret Equation 4.37 as a series expansion of the original image G into a set of N^2 basis images $I_{u,v}$. The transform coefficients $T(u, v)$, $0 \leq u, v \leq N-1$, become the coefficients of the expansion. Alternatively, the image G is said to be a weighted sum of basis images. Note that, similar to the 1-D case, the coefficient or the weight $T(u, v)$ is a correlation measure between the image G and the basis image $I_{u,v}$ [wintz 1972].

Note that basis images have nothing to do with the input image. Instead, it is completely defined by the transform itself. That is, basis images are the attribute of 2-D image transforms. Different transforms have different sets of basis images.

The motivation behind TC is that with a proper transform, hence, a proper set of basis images, the transform coefficients are more independent than the gray scales of the original input image. In the ideal case, the transform coefficients are statistically independent. We can then optimally encode the coefficients independently, which can make coding more efficient and simple. As pointed out in [wintz 1972], however, this is generally impossible because of the following two reasons. First, it requires the joint probability density function (pdf) of the N^2 pixels, which have not been deduced from basic physical laws and cannot be measured. Second, even if the joint pdfs were known, the problem of devising a reversible transform that can generate independent coefficients is unsolved. The optimum linear transform we can have results in uncorrelated coefficients. When Gaussian distribution is involved, we can have independent transform coefficients. In addition to the uncorrelatedness of coefficients, the variance of the coefficients varies widely. Insignificant coefficients can be ignored without introducing significant distortion in the reconstructed image. Significant coefficients can be allocated more bits in encoding. The coding efficiency is thus enhanced.

As shown in Figure 4.3, TC can be viewed as expanding the input image into a set of basis images, then quantizing and encoding the coefficients associated with the basis images separately. At the receiver the coefficients are reconstructed to produce a replica of the input image. This strategy is similar to that of subband coding, which is discussed in Chapter 8. From this point of view, TC can be considered a special case of subband coding, though TC was devised much earlier than subband coding.

It is worth mentioning an alternative way to define basis images. That is, a basis image with indexes (u, v), $I_{u,v}$, of a transform can be constructed as the outer product of the uth basis vector, $\vec{b}_u$, and the vth basis vector, $\vec{b}_v$, of the transform. The basis vector, $\vec{b}_u$, is the uth column vector of the inverse transform matrix I [jayant 1984]. That is,

$$I_{u,v} = \vec{b}_u \vec{b}_v^T. \qquad (4.38)$$

4.2.3 Subimage Size Selection

The selection of subimage (block) size, N, is important. Normally, the larger the size the more decorrelation the TC can achieve. It has been shown, however, that the correlation between image pixels becomes insignificant when the distance between pixels becomes large, e.g., it exceeds 20 pixels [habibi 1971a]. On the other hand, a large size causes some

Transform Coding

(a) Transmitter

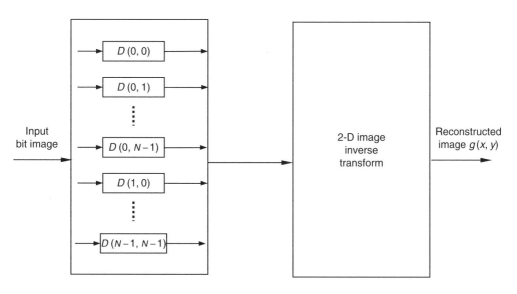

(b) Receiver

FIGURE 4.3
Basis image interpretation of TC (Q: quantizer, E: encoder, D: decoder).

problems. In adaptive TC, a large block cannot adapt to local statistics well. As will be discussed later in this chapter, a transmission error in TC affects the whole associated subimage. Hence a large size implies a possibly severe effect of transmission error on reconstructed images. As will be shown in video coding (Parts III and IV), TC is used together with motion compensated (MC) coding. Consider that large block size is not used in motion estimation; subimage sizes of 4, 8, and 16 are used most often. In particular, $N = 8$ is adopted by the international still image coding standard JPEG as well as video coding standards H.261, H.263, H. 264, MPEG 1, 2, and 4.

4.3 Transforms of Particular Interest

Several commonly used image transforms are discussed in this section. They include the DFT, the discrete Walsh transform (DWT), the discrete Hadamard transform (DHT), and the discrete cosine transform (DCT) and discrete sine transform. All of these transforms are symmetric (hence, separable as well), unitary, and reversible. For each transform, we define its transformation kernel and discuss its basis images.

4.3.1 Discrete Fourier Transform

The DFT is of great importance in the field of digital signal processing. Owing to the fast Fourier transform (FFT) based on the algorithm developed in [cooley 1965], the DFT is widely utilized for various tasks of digital signal processing. It has been discussed in many signal and image processing texts. Here we only define it using the transformation kernel just introduced above. The forward and inverse transformation kernels of the DFT are

$$f(x,y,u,v) = \frac{1}{N}\exp\{-j2\pi(xu+yv)/N\} \tag{4.39}$$

and

$$i(x,y,u,v) = \frac{1}{N}\exp\{j2\pi(xu+yv)/N\}. \tag{4.40}$$

Clearly, since complex quantities are involved in the DFT transformation kernels, the DFT is generally complex. Hence, we use the unitary matrix to handle the DFT (refer to Section 4.2.1.3). The basis vector of the DFT $\vec{b}_u$ is an $N \times 1$ column vector and is defined as

$$\vec{b}_u = \frac{1}{\sqrt{N}}\left[1, \exp\left(j2\pi\frac{u}{N}\right), \exp\left(j2\pi\frac{2u}{N}\right), \ldots, \exp\left(j2\pi\left(\frac{(N-1)u}{N}\right)\right)\right]^T. \tag{4.41}$$

As mentioned, the basis image with index (u,v), $I_{u,v}$, is equal to $\vec{b}_u \vec{b}_v^T$. A few basis images are listed below for $N=4$.

$$I_{0,0} = \frac{1}{4}\begin{pmatrix} 1 & 1 & 1 & 1 \\ 1 & 1 & 1 & 1 \\ 1 & 1 & 1 & 1 \\ 1 & 1 & 1 & 1 \end{pmatrix} \tag{4.42}$$

$$I_{0,1} = \frac{1}{4}\begin{pmatrix} 1 & j & -1 & -j \\ 1 & j & -1 & -j \\ 1 & j & -1 & -j \\ 1 & j & -1 & -j \end{pmatrix} \tag{4.43}$$

$$I_{1,2} = \frac{1}{4}\begin{pmatrix} 1 & 1 & 1 & -1 \\ j & -j & j & -j \\ -1 & 1 & -1 & 1 \\ -j & -j & -j & j \end{pmatrix} \tag{4.44}$$

$$I_{3,3} = \frac{1}{4}\begin{pmatrix} 1 & -j & -1 & j \\ -j & -1 & j & 1 \\ -1 & j & 1 & -j \\ j & 1 & -j & -1 \end{pmatrix} \tag{4.45}$$

4.3.2 Discrete Walsh Transform

The transformation kernels of the DWT [walsh 1923] are defined as

$$f(x,y,u,v) = \frac{1}{N} \prod_{i=0}^{n-1} [(-1)^{p_i(x)p_{n-1-i}(u)}(-1)^{p_i(y)p_{n-1-i}(v)}] \quad (4.46)$$

and

$$i(x,y,u,v) = f(x,y,u,v). \quad (4.47)$$

where $n = \log_2 N$, $p_i(\text{arg})$ represents the ith bit in the natural binary representation of the arg, the 0th bit corresponds to the least significant bit and the $(n-1)$th bit corresponds to the most significant bit. For instance, consider $N=16$, then $n=4$. The natural binary code of number 8 is 1000. Hence, $p_0(8) = p_1(8) = p_2(8) = 0$, and $p_3(8) = 1$. We see that if the factor $1/N$ is put aside then the forward transformation kernel is always an integer: either $+1$ or -1. In addition, the inverse transformation kernel is the same as the forward transformation kernel. Therefore, we conclude that the implementation of the DWT is simple.

When $N=4$, the 16 basis images of the DWT are shown in Figure 4.4. Each corresponds to a specific pair of (u, v) and is of resolution 4×4 in x–y coordinate system. They are binary images, where the bright represents $+1$, and the dark -1. The transform matrix of the DWT is shown below for $N=4$.

$$F = \frac{1}{2} \begin{pmatrix} 1 & 1 & 1 & 1 \\ 1 & 1 & -1 & -1 \\ 1 & -1 & 1 & -1 \\ 1 & -1 & -1 & 1 \end{pmatrix} \quad (4.48)$$

4.3.3 Discrete Hadamard Transform

The DHT [hadamard 1893] is closely related to the DWT. This can be seen from the following definition of the transformation kernels.

$$f(x,y,u,v) = \frac{1}{N} \prod_{i=0}^{n} [(-1)^{p_i(x)p_i(u)}(-1)^{p_i(y)p_i(v)}] \quad (4.49)$$

and

$$i(x,y,u,v) = f(x,y,u,v), \quad (4.50)$$

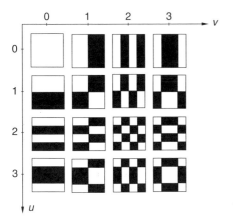

FIGURE 4.4
When $N=4$, a set of 16 basis images of DWT.

where the definition of n, i, and $p_i(\text{arg})$ are the same as in the DWT. For this reason, the term Walsh–Hadamard transform (DWHT) is frequently used to represent either of the two transforms.

When N is a power of 2, the transform matrices of the DWT and DHT have the same row (or column) vectors except that the order of row (or column) vectors in the matrices are different. This is the only difference between the DWT and DHT under the circumstance: $N = 2^n$. Because of this difference, while the DWT can be implemented by using the FFT algorithm with a straightforward modification, the DHT needs more work to use the FFT algorithm. On the other hand, the DHT possesses the following recursive feature, while the DWT does not:

$$F_2 = \begin{bmatrix} 1 & 1 \\ 1 & -1 \end{bmatrix} \quad (4.51)$$

and

$$F_{2N} = \begin{bmatrix} F_N & F_N \\ F_N & -F_N \end{bmatrix}, \quad (4.52)$$

where the subscripts indicate the size of the transform matrices. It is obvious that the transform matrix of the DHT can be easily derived by using the recursion.

Note that the number of sign changes between consecutive entries in a row (or a column) of a transform matrix (from positive to negative and from negative to positive) is known as sequency. It is observed that the sequency does not monotonically increase as the order number of rows (or columns) increases in the DHT. Since sequency bears some similarity to frequency in the Fourier transform, sequency is desired as an increasing function of the order number of rows (or columns). This is realized by the ordered Hadamard transform [gonzalez 2001].

The transformation kernel of the ordered Hadamard transform is defined as

$$f(x,y,u,v) = \frac{1}{N} \prod_{i=0}^{N-1} [(-1)^{p_i(x)d_i(u)}(-1)^{p_i(y)d_i(v)}], \quad (4.53)$$

where the definition of i, $p_i(\text{arg})$ are the same as defined above for the DWT and DHT. The $d_i(\text{arg})$ is defined as

$$\begin{aligned} d_0(\text{arg}) &= b_{n-1}(\text{arg}) \\ d_1(\text{arg}) &= b_{n-1}(\text{arg}) + b_{n-2}(\text{arg}) \\ &\vdots \\ d_{n-1}(\text{arg}) &= b_1(\text{arg}) + b_0(\text{arg}) \end{aligned} \quad (4.54)$$

The 16 basis images of the ordered Hadamard transform are shown in Figure 4.5 for $N = 4$. It is observed that the variation of the binary basis images becomes more frequent monotonically when u and v increase. We also see that the basis image expansion is similar to the frequency expansion of the Fourier transform in the sense that an image is decomposed into components with different variations. In TC, these components with different coefficients are treated differently.

Transform Coding

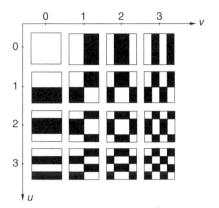

FIGURE 4.5
When $N=4$, a set of 16 basis images of the ordered DHT.

4.3.4 Discrete Cosine Transform

Discrete cosine transform is the most commonly used transform for image and video coding.

4.3.4.1 Background

The DCT, which plays an extremely important role in image and video coding, was established by Ahmed et al. [ahmed 1974]. There, it was shown that the basis member $\cos[(2x+1)u\pi/2N]$ is the uth Chebyshev polynomial $T_U(\xi)$ evaluated at the xth zero of $T_N(\xi)$. Recall that the Chebyshev polynomials are defined as

$$T_0(\xi) = 1/\sqrt{2} \tag{4.55}$$

$$T_K(\xi) = \cos[k \cos^{-1}(\xi)], \tag{4.56}$$

where $T_K(\xi)$ is the k order Chebyshev polynomial and it has k zeros, starting from the first zero to the kth zero. Furthermore, it was demonstrated that the basis vectors of 1-D DCT provide a good approximation to the eigenvectors of the class of Toeplitz matrices defined as

$$\begin{bmatrix} 1 & \rho & \rho^2 & \cdots & \rho^{N-1} \\ \rho & 1 & \rho & \cdots & \rho^{N-2} \\ \rho^2 & \rho & 1 & \cdots & \rho^{N-3} \\ \vdots & \vdots & \vdots & \cdots & \vdots \\ \rho^{N-1} & \rho^{N-2} & \rho^{N-3} & \cdots & 1 \end{bmatrix}, \tag{4.57}$$

where $0 < \rho < 1$.

4.3.4.2 Transformation Kernel

The transformation kernel of the 2-D DCT can be extended straightforwardly from that of 1-D DCT as follows:

$$f(x,y,u,v) = C(u)C(v) \cos\left(\frac{(2x+1)u\pi}{2N}\right) \cos\left(\frac{(2y+1)v\pi}{2N}\right), \tag{4.58}$$

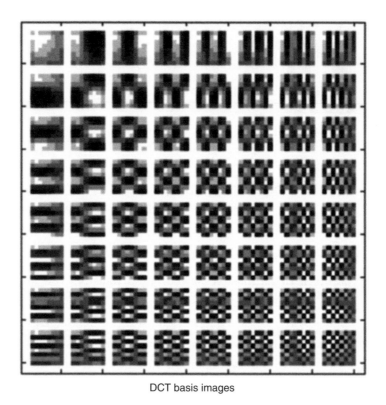

DCT basis images

FIGURE 4.6
When $N = 8$, a set of 64 basis images of the DCT.

where

$$C(u) = \begin{cases} \sqrt{\frac{1}{N}} & \text{for } u = 0 \\ \sqrt{\frac{2}{N}} & \text{for } u = 1, 2, \ldots, N-1 \end{cases} \quad (4.59)$$

$$i(x, y, u, v) = f(x, y, u, v). \quad (4.60)$$

Note that $C(v)$ is defined the same way as in Equation 4.59. The 64 basis images of the DCT are shown in Figure 4.6 for $N = 8$.

4.3.4.3 Relationship with DFT

The DCT is closely related to the DFT. This can be examined from an alternative method of defining the DCT. It is known that applying the DFT to an N-point sequence $g_N(n)$, $n = 0, 1, \ldots, N-1$, is equivalent to the following:

1. Repeating $g_N(n)$ every N points, form a periodic sequence, $\tilde{g}_N(n)$, with a fundamental period N, that is,

$$\tilde{g}_N(n) = \sum_{i=-\infty}^{\infty} g_N(n - iN). \quad (4.61)$$

Transform Coding

2. Determine the Fourier series expansion of the periodic sequence $\tilde{g}_N(n)$. That is, determine all the coefficients in the Fourier series, which are known to be periodic with the same fundamental period N.
3. Truncate the sequence of the Fourier series coefficients so as to have the same support as that of the given sequence $g_N(n)$. That is, only keep the N coefficients with indexes $0, 1, \ldots, N-1$, and set all the others to equal zero. These N Fourier series coefficients form the DFT of the given N-point sequence $g_N(n)$.

An N-point sequence $g_N(n)$ and the periodic sequence $\tilde{g}_N(n)$, generated from $g_N(n)$, are shown in Figure 4.7a and b, respectively. In summary, the DFT can be viewed as a correspondence between two periodic sequences. One is the periodic sequence $\tilde{g}_N(n)$, which is formed by periodically repeating $g_N(n)$. The other is the periodic sequence of Fourier series coefficients of $\tilde{g}_N(n)$.

(a) Original 1-D input sequence

(b) Formation of a periodic sequence with a fundamental period of N (DFT)

(c) Formation of a back-to-back 2N sequence

(d) Formation of a periodic sequence with a fundamental period of 2N (DCT)

FIGURE 4.7
An example to illustrate the differences and similarities between DFT and DCT.

The DCT of an N-point sequence is obtained through the following three steps:

1. Flip over the given sequence with respect to the end point of the sequence to form a 2N-point sequence, $g_{2N}(n)$, as shown in Figure 4.7c. Then form a periodic sequence $\tilde{g}_{2N}(n)$, shown in Figure 4.7d, according to

$$\tilde{g}_{2N}(n) = \sum_{i=-\infty}^{\infty} g_{2N}(n - 2iN) \qquad (4.62)$$

2. Find the Fourier series coefficients of the periodic sequences $\tilde{g}_{2N}(n)$.
3. Truncate the resultant periodic sequence of the Fourier series coefficients to have the support of the given finite sequence $g_N(n)$. That is, keeping only the N coefficients with indexes $0, 1, \ldots, N-1$, set all the others to equal zero. These N Fourier series coefficients form the DCT of the given N-point sequence $g_N(n)$.

A comparison between Figure 4.7b and d reveals that the periodic sequence $\tilde{g}_N(n)$ is not smooth. There exist discontinuities at the beginning and end of each period. These end-head discontinuities cause a high-frequency distribution in the corresponding DFT. On the contrary, the periodic sequence $\tilde{g}_{2N}(n)$ does not have this type of discontinuity due to flipping over the given finite sequence. As a result, there is no high-frequency component corresponding to the end-head discontinuities. Hence, the DCT possesses better energy compaction in the low frequencies than the DFT. By energy compaction, we mean more energy is compacted in a fraction of transform coefficients. For instance, it is known that the most energy of an image is contained in a small region of low frequency in the DFT domain. Vivid examples can be found in [gonzalez 2001]. In terms of energy compaction, when compared with the KLT (the Hotelling transform is its discrete version), which is known as the optimal, the DCT is the best among the DFT, DWT, DHT, and discrete Harr transform.

Besides this advantage, the DCT can be implemented using the FFT. This can be seen from the above discussion. There, It has been shown that the DCT of an N-point sequence, $g_N(n)$, can be obtained from the DFT of the 2N-point sequence $g_{2N}(n)$. Furthermore, the even symmetry in $\tilde{g}_{2N}(n)$ makes the computation required for the DCT of an N-point equal to that required for the DFT of the N-point sequence. Because of these two merits, the DCT is the most popular image transform used in image and video coding. No other transform has been proven to be better than the DCT from a practical standpoint [haskell 1996].

4.3.5 Performance Comparison

In this section, we compare the performance of a few commonly used transforms in terms of energy compaction, mean square reconstruction error, and computational complexity.

4.3.5.1 Energy Compaction

Since all the transforms we discussed are symmetric (hence separable) and unitary, the matrix form of the 2-D image transform can be expressed as $T = F^T G F$ as discussed in Section 4.2.1.3. In the 1-D case, the transform matrix F is the counterpart of the matrix Φ discussed in the Hotelling transform. Using F, one can transform a 1-D column vector $\vec{z}$ into another 1-D column vector $\vec{y}$. The components of the vector $\vec{y}$ are transform coefficients. The variances of these transform coefficients, and therefore the signal energy associated with the transform coefficients, can be arranged in a nondecreasing order.

Transform Coding

It can be shown that the total energy before and after the transform remains the same. Therefore, the more energy compacted in a fraction of total coefficients, the better energy compaction the transform has. One measure of energy compaction is the transform coding gain G_{TC}, which is defined as the ratio between the arithmetic mean and the geometric mean of the variances of all the components in the transformed vector [jayant 1984].

$$G_{TC} = \frac{\frac{1}{N}\sum_{i=0}^{N-1}\sigma_i^2}{\left(\prod_{i=0}^{N-1}\sigma_i^2\right)^{\frac{1}{N}}}. \quad (4.63)$$

A larger G_{TC} indicates higher energy compaction. The TC gains for a first-order autoregressive source with $\rho = 0.95$ achieved by using the DCT, DFT, and KLT were reported in [zelinski 1975; jayant 1984]. The TC gain afforded by the DCT compares very closely to that of the optimum KLT.

4.3.5.2 Mean Square Reconstruction Error

The performance of the transforms can be compared in terms of the mean square reconstruction error as well. This was mentioned in Section 4.1.2 when we provided a statistical interpretation for TC. That is, after arranging all the N transformed coefficients according to their variances in a nonincreasing order, if $L < N$ and we discard the last $N - L$ coefficients to reconstruct the original input signal $\vec{z}$ (similar to what we did with the Hotelling transform), then the mean square reconstruction error is

$$\text{MSE}_r = E\left[\|\vec{z} - \vec{z}')\|^2\right] = \sum_{i=L+1}^{N}\sigma_i^2, \quad (4.64)$$

where $\vec{z}'$ denotes the reconstructed vector. Note that in the above defined mean square reconstruction error, the quantization error and transmission error have not been included. Hence, it is sometimes referred to as the mean square approximation error.

Therefore it is desired to choose a transform so that the transformed coefficients are more independent and more energy is concentrated in the first L coefficients. Then it is possible to discard the remaining coefficients to save coding bits without causing significant distortion in input signal reconstruction.

In terms of the mean square reconstruction error, the performance of the DCT, KLT, DFT, DWT, and discrete Haar transform for the 1-D case was reported in [ahmed 1974]. The variances of the 16 transform coefficients are shown in Figure 4.8 when $N = 16$, $\rho = 0.95$. Note that N stands for the dimension of the 1-D vector, while the parameter ρ is shown in the Toeplitz matrix (refer to Equation 4.57). We can see that the DCT compares most closely to the KLT, which is known to be optimum.

Note that the unequal variance distribution among transform coefficients also found application in the field of pattern recognition. Similar results to those in [ahmed 1974] for the DFT, DWT, and Harr transform were reported in [andrews 1971].

A similar analysis can be carried out for the 2-D case [wintz 1972]. Recall that an image $g(x, y)$ can be expressed as a weighted sum of basis images $I_{u,v}$. That is,

$$G = \sum_{u=0}^{N-1}\sum_{v=0}^{N-1}T(u,v)I_{u,v}, \quad (4.65)$$

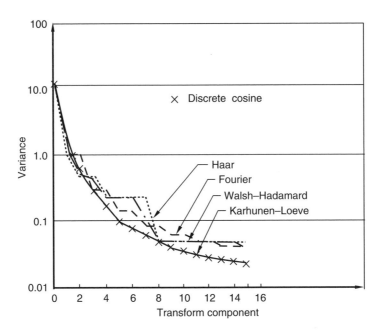

FIGURE 4.8
Transform coefficient variances when $N = 16$, $\rho = 0.95$. (From Ahmed, N., Nararajan, T., and Rao, K.R., *IEEE Trans. Comput.*, 90–93, 1974. With permission.)

where the weights are transform coefficients. We arrange the coefficients according to their variances in a nonincreasing order. For some choices of the transform (hence basis images), the coefficients become insignificant after the first L terms, and the image can be approximated well by truncating the coefficients after L. That is,

$$G = \sum_{u=0}^{N-1} \sum_{v=0}^{N-1} T(u,v) I_{u,v} \approx \sum_{u=0}^{L} \sum_{v=0}^{L} T(u,v) I_{u,v}. \tag{4.66}$$

The mean square reconstruction error is given by

$$\text{MSE}_r = \sum_{L}^{N-1} \sum_{L}^{N-1} \sigma_{u,v}^2 \tag{4.67}$$

A comparison among the KLT, DHT, and DFT in terms of the mean square reconstruction error for 2-D array of 16×16 (i.e., 256 transform coefficients) was reported in [Figure 5, wintz 1972]. Note that the discrete KLT is image dependent. In the comparison, the KLT is calculated with respect to an image named Cameraman. It shows that while the KLT achieves best performance, the other transforms perform closely.

In essence, the criteria of mean square reconstruction error and energy compaction are closely related. It has been shown that the discrete KLT, also known as the Hotelling transform, is the optimum in terms of energy compaction and mean square reconstruction error. The DWT, DHT, DFT, and DCT are close to the optimum [wintz 1972; ahmed 1974]; however, the DCT is the best among these several suboptimum transforms.

Note that the performance comparison among various transforms in terms of bit rate versus distortion in the reconstructed image was reported in [pearl 1972; ahmed 1974]. The

Transform Coding

same conclusion was drawn. That is, the KLT is optimum, while the DFT, DWT, DCT, and Harr transforms are close in performance. Among the suboptimum transforms, the DCT is the best.

4.3.5.3 Computational Complexity

Note that while the DWT, DHT, DFT, and DCT are input image independent, the discrete KLT (the Hotelling transform) is input dependent. More specifically, the row vectors of the Hotelling transform matrix are transposed eigenvectors of the covariance matrix of the input random vector. So far there is no fast transform algorithm available. This computational complexity prohibits the Hotelling transform from practical usage. It can be shown that the DWT, DFT, and DCT can be implemented using the FFT algorithm.

4.3.5.4 Summary

As pointed earlier, the DCT is the best among the suboptimum transforms in terms of energy compaction. Moreover, the DCT can be implemented using the FFT. Even though a $2N$-point sequence is involved, the even symmetry makes the computation involved in the N-point DCT equivalent to that of the N-point FFT. For these two reasons, the DCT finds the widest application in image and video coding.

4.4 Bit Allocation

As shown in Figure 4.2, in TC, an input image is first divided into blocks (subimages). Then a 2-D linear transform is applied to each block. The transformed blocks go through truncation, quantization, and code word assignment. The last three functions: truncation, quantization, and code word assignment are combined and called bit allocation.

From the previous section, it is known that the applied transform decorrelates subimages. Moreover, it redistributes image energy in the transform domain in such a way that most of the energy is compacted into a small fraction of coefficients. Therefore, it is possible to discard the majority of transform coefficients without introducing significant distortion.

As a result, we see that in TC there are mainly three types of error involved. One is due to truncation, i.e., the majority of coefficients are truncated to zero. The other comes from quantization. Transmission error is the third type of error. (Note that truncation can also be considered a special type of quantization.) The mean square reconstruction error discussed in Section 4.3.5.2 is in fact only related to truncation error. For this reason, it was referred to more precisely as mean square approximation error. In general, the reconstruction error, i.e., the error between the original image signal and the reconstructed image at the receiver, includes three types of error: truncation error, quantization error, and transmission error.

There are two different ways to truncate transform coefficients. One is called zonal coding, while the other is threshold coding. And they are discussed below.

4.4.1 Zonal Coding

In zonal coding, also known as zonal sampling, a zone in the transformed block is predefined according to a statistical average obtained from many blocks. All transform coefficients in the zone are retained, while all coefficients outside the zone are set to zero. As mentioned in Section 4.3.5.1, the total energy of the image remains the same after applying the transforms discussed there. Since it is known that the DC and low-frequency

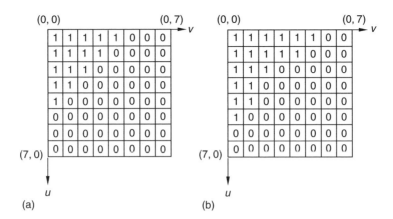

FIGURE 4.9
Two illustrations of zonal coding.

AC coefficients of the DCT occupy most of the energy, the zone is located in the top-left portion of the transformed block when the transform coordinate system is set conventionally. (Note that by DC we mean $u = v = 0$. By AC we mean u and v do not equal zero simultaneously.) That is, the origin is at the top-left corner of the transformed block. Two typical zones are shown in Figure 4.9. The simplest uniform quantization with natural binary coding can be used to quantize and encode the retained transform coefficients. With this simple technique, there is no overhead side information that needs to be sent to the receiver, since the structure of the zone, the scheme of the quantization, and encoding are known at both the transmitter and the receiver.

The coding efficiency, however, may not be very high. This is because the zone is predefined based on average statistics. Therefore some coefficients outside the zone might be large in magnitude, while some coefficients inside the zone may be small in quantity. Uniform quantization and natural binary encoding are simple, but they are known not to be efficient enough.

For further improvement of coding efficiency, an adaptive scheme has to be used. There, a two-pass procedure is applied. In the first pass, the variances of transform coefficients are measured or estimated. Based on the statistics, the quantization and encoding schemes are determined. In the second pass, quantization and encoding are carried out [habibi 1971a; chen 1977].

4.4.2 Threshold Coding

In threshold coding, also known as threshold sampling, there is not a predefined zone. Instead, each transform coefficient is compared with a threshold. If it is smaller than the threshold, then it is set to zero. If it is larger than the threshold, it will be retained for quantization and encoding. Compared with zonal coding, this scheme is adaptive in truncation in the sense that the coefficients with more energy are retained no matter where they are located. The address of these retained coefficients, however, has to be sent to the receiver as side information. Furthermore, the threshold is determined after an evaluation of all coefficients. Hence, it was usually a two-pass adaptive technique.

Chen and Pratt devised an efficient adaptive scheme to handle threshold coding [chen 1984]. It is a one-pass adaptive scheme, in contrast to two-pass adaptive schemes. Hence it is fast in implementation. With several effective techniques addressed here, it achieved excellent results in TC. Specifically, it demonstrated satisfied quality of reconstructed

Transform Coding

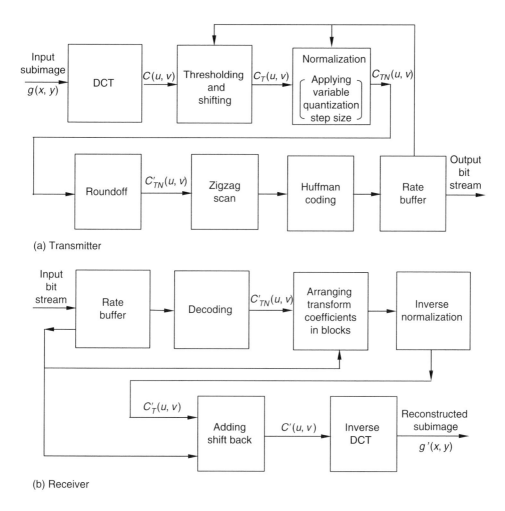

FIGURE 4.10
Block diagram of the algorithm proposed by Chen and Pratt. (From Chen, W.H. and Pratt, W.K., *IEEE Trans. Commun.*, COM-32, 225–232, 1984. With permission.)

frames at a bit rate of 0.4 bits/pixel for coding of color images, which corresponds to real-time color television transmission over a 1.5 Mbits/s channel. This scheme has been adopted by the international still coding standard JPEG. A block diagram of the threshold coding proposed by Chen and Pratt is shown in Figure 4.10. More details and modification made by JPEG will be described in Chapter 7.

4.4.2.1 Thresholding and Shifting

The DCT is used in the scheme because of its superiority, described in Section 4.3. Here we use $C(u,v)$ to denote the DCT coefficients. The DC coefficient, $C(0,0)$, is processed differently. As mentioned in Chapter 3, the DC coefficients are encoded with differential coding technique. For more detail, refer to Chapter 7. For all the AC coefficients, the following thresholding and shifting are carried out:

$$C_T(u,v) = \begin{cases} C(u,v) - T & \text{if } C(u,v) > T \\ 0 & \text{if } C(u,v) \leq T \end{cases} \quad (4.68)$$

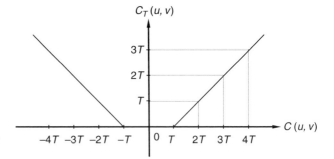

FIGURE 4.11
Input–output characteristic of thresholding and shifting.

where T on the right-hand side is the threshold. Note that Equation 4.68 also implies a shifting of transform coefficients by T. The input–output characteristic of the thresholding and shifting is shown in Figure 4.11.

Figure 4.12 demonstrates that more than 60% of the DCT coefficients normally fall below a threshold value as low as 5. This indicates that with a properly selected threshold value it is possible to set most of the DCT coefficients equal to zero. The threshold value is adjusted by the feedback from the rate buffer, or by the desired bit rate.

4.4.2.2 Normalization and Roundoff

The threshold subtracted transform coefficients $C_T(u,v)$ are normalized before roundoff. The normalization is implemented as follows:

$$C_{TN}(u,v) = \frac{C_T(u,v)}{\Gamma_{u,v}}, \qquad (4.69)$$

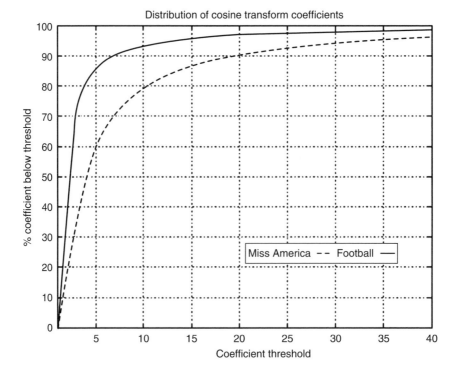

FIGURE 4.12
Amplitude distribution of the DCT coefficients.

Transform Coding

where the normalization factor $\Gamma_{u,v}$ is controlled by the rate buffer. The roundoff process converts floating point to integer as follows:

$$R[C_{TN}(u,v)] = C^*_{TN}(u,v) = \begin{cases} \lfloor C_{TN}(u,v) + 0.5 \rfloor & \text{if } C_{TN}(u,v) \geq 0 \\ \lceil C_{TN}(u,v) - 0.5 \rceil & \text{if } C_{TN}(u,v) < 0 \end{cases} \quad (4.70)$$

where the operator $\lfloor x \rfloor$ means the largest integer smaller than or equal to x, the operator $\lceil x \rceil$ means the smallest integer larger than or equal to x. The input–output characteristics of the normalization and roundoff are shown in Figure 4.13a and b, respectively.

From these input–output characteristics, we can see that the roundoff is a uniform midtread quantizer with a unit quantization step. The combination of normalization and roundoff is equivalent to a uniform midtread quantizer with the quantization step size equal to the normalization factor $\Gamma_{u,v}$. Normalization is a scaling process, which makes the resultant uniform midtread quantizer adapt to the dynamic range of the associated transform coefficient. It is therefore possible for one quantizer design to be applied to various coefficients with different ranges. Obviously, by adjusting the parameter $\Gamma_{u,v}$ (quantization step size) variable bit rate (VBR) and mean square quantization error can be achieved. The selection of the normalization factors for different transform coefficients can hence take

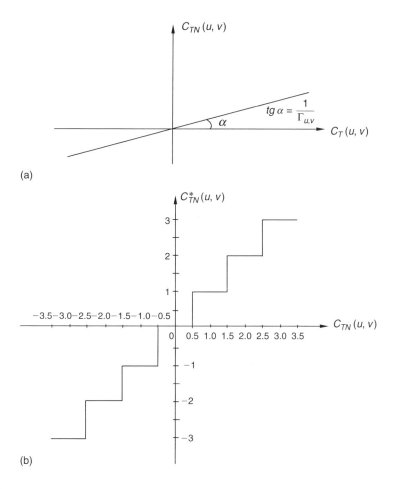

FIGURE 4.13
Input–output characteristic of (a) normalization and (b) roundoff.

FIGURE 4.14
Quantization tables. (a) Luminance quantization table. (b) Chrominance quantization table.

(a)

16	11	10	16	24	40	51	61
12	12	14	19	26	58	60	55
14	13	16	24	40	57	69	56
14	17	22	29	51	87	80	62
18	22	37	56	68	109	103	77
24	35	55	64	81	104	113	92
49	64	78	87	103	121	120	101
72	92	95	98	112	100	103	99

(b)

17	18	24	47	99	99	99	99
18	21	26	66	99	99	99	99
24	26	56	99	99	99	99	99
47	66	99	99	99	99	99	99
99	99	99	99	99	99	99	99
99	99	99	99	99	99	99	99
99	99	99	99	99	99	99	99
99	99	99	99	99	99	99	99

the statistical feature of the images and the characteristics of the HVS into consideration. In general, most image energy is contained in the DC and low-frequency AC transform coefficients. The HVS is more sensitive to a relatively uniform region than to a relatively detailed region, as discussed in Chapter 1. Chapter 1 also mentions that with regard to the color image, the HVS is more sensitive to the luminance component than to the chrominance components.

These have been taken into consideration in JPEG. A matrix consisting of all the normalization factors is called a quantization table in JPEG. A luminance quantization table and a chrominance quantization table used in JPEG are shown in Figure 4.14. We observe that in general in both tables the small normalization factors are assigned to the DC and low-frequency AC coefficients. The large Γ's are associated with the high-frequency transform coefficients. Compared with the luminance quantization table, the chrominance quantization table has larger quantization step sizes for the low and middle frequency coefficients and almost the same step sizes for the DC and high-frequency coefficients, indicating that the chrominance components are relatively coarsely quantized, compared with the luminance component.

4.4.2.3 Zigzag Scan

As mentioned at the beginning of this section, while threshold coding is adaptive to the local statistics and hence is more efficient in truncation, threshold coding needs to send the address of retained coefficients to the receiver as overhead side information. An efficient scheme, called zigzag scan, was proposed in [chen 1984] as shown in Figure 4.15. As shown in Figure 4.12, a great majority of transform coefficients has magnitude smaller than a threshold of 3. Consequently, most quantized coefficients are zero. Hence, in the 1-D sequence obtained by zigzag scanning, most of the numbers are zero. A code known as run-length code (RLC), discussed in Chapter 6, is very efficient under these circumstances to encode the address information of nonzero coefficients. Run-length of zero coefficients is understood as the number of consecutive zeros in the zigzag scan. Zigzag scanning minimizes the use of RLCs in the block, hence making codes most efficient.

4.4.2.4 Huffman Coding

Statistical studies of the magnitude of nonzero DCT coefficients and the run-length of zero DCT coefficients in zigzag scanning were conducted in [chen 1984]. The domination of the coefficients with small amplitude and the short run-lengths was found and is shown in Figures 4.16 and 4.17. This justifies the application of the Huffman coding to the magnitude of nonzero transform coefficients and run-lengths of zeros.

Transform Coding

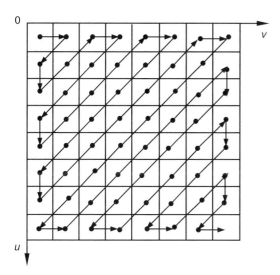

FIGURE 4.15
Zigzag scan of DCT coefficients within an 8 × 8 block.

4.4.2.5 Special Code Words

Two special code words were used in [chen 1984]. One is called end of block (EOB). Another is called run-length prefix. Once the last nonzero DCT coefficients in the zigzag are coded, EOB is appended, indicating the termination of coding the block. This further saves bits used in coding. Run-length prefix is used to discriminate the RLC words from the amplitude code words.

4.4.2.6 Rate Buffer Feedback and Equalization

As shown in Figure 4.10, a rate buffer accepts a variable-rate data input from the encoding process and provides a fixed-rate data output to the channel. The status of the rate buffer is monitored and fed back to control the threshold and the normalization factor. In this fashion a one-pass adaptation is achieved.

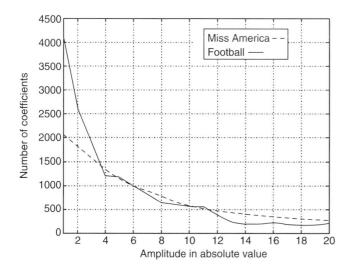

FIGURE 4.16
Histogram of DCT coefficients in absolute amplitude.

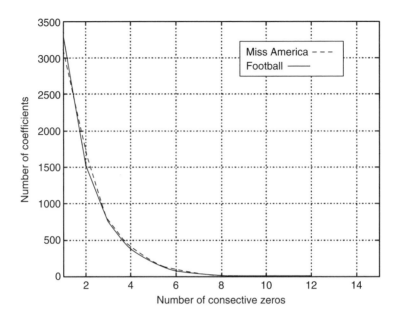

FIGURE 4.17
Histogram of zero run-length.

4.5 Some Issues

4.5.1 Effect of Transmission Error

In TC, each pixel in the reconstructed image relies on all transform coefficients in the subimage where the pixel is located. Hence, a bit reversal transmission error will be spread. That is, an error in a transform coefficient will lead to errors in all the pixels within the subimage. As discussed in Section 4.2.3, this is one of the reasons the selected subimage size cannot be very large. Depending on which coefficient is in error, the effect caused by a bit reversal error on the reconstructed image varies. For instance, an error in the DC or a low-frequency AC coefficient may be objectionable, while an error in the high-frequency coefficient may be less noticeable.

4.5.2 Reconstruction Error Sources

As discussed, three sources, truncation (discarding transform coefficients with small variances), quantization, and transmission, contribute to the reconstruction error. It is noted that in TC transform is applied block by block. Quantization and encoding of transform coefficients are also conducted blockwise. At the receiver, reconstructed blocks are put together to form the whole reconstructed image. In the process, block artifacts are produced. Sometimes, even though it may not severely affect an objective assessment of the reconstructed image quality, block artifacts can be annoying to the HVS, especially when the coding rate is low.

To alleviate the blocking effect, several techniques have been proposed. One is to overlap blocks in the source image. Another is to postfilter the reconstructed image along block boundaries. The selection of advanced transforms is an additional possible method [lim 1990].

Transform Coding

In the block overlapping method, when the blocks are finally organized to form the reconstructed image, each pixel in the overlapped regions takes an average value of all its reconstructed gray level values from multiple blocks. In this method, extra bits are used for those pixels involved in the overlapped regions. For this reason, the overlapped region is usually only one pixel wide.

Due to the sharp transition along block boundaries, block artifacts are of high frequency in nature. Low-pass filtering is hence normally used in the postfiltering method. To avoid the blurring effect caused by low-pass filtering on the nonboundary image area, low-pass postfiltering is only applied to block boundaries. Unlike the block overlapping method, the postfiltering method does not need extra bits. Moreover, it has been shown that the postfiltering method can achieve better results in combating block artifacts [reeve 1984; ramamurthi 1986]. For these two reasons, the postfiltering method has been adopted by the international coding standards.

4.5.3 Comparison between DPCM and TC

As mentioned at the beginning of the chapter, both differential coding and TC utilize interpixel correlation and are efficient coding techniques. Comparisons between these two techniques have been reported in [habibi 1971b]. Take a look at the techniques discussed in the Chapters 3 and 4. We can see that differential coding is simpler than TC. This is because the linear prediction and differencing involved in differential coding are simpler than the 2-D transform involved in TC. In terms of the memory requirement and processing delay, differential coding such as DPCM is superior to TC. That is, DPCM needs less memory and has less processing delay than TC. The design of the DPCM system, however, is sensitive to image-to-image variation, and so is its performance. That is, an optimum DPCM design is matched to the statistics of a certain image. When the statistics change, the performance of the DPCM will be affected. On the contrary, TC is less sensitive to the variation in the image statistics. In general, the optimum DPCM coding system with a third or higher-order predictor performs better than TC when the bit rate is about 2–3 bits/pixel for single images. When the bit rate is below 2–3 bits/pixel, TC is normally preferred. As a result, the international still image coding standard JPEG is based on TC, whereas, in JPEG, DPCM is used for coding the DC coefficients of DCT, and information-preserving differential coding is used for lossless still image coding.

4.5.4 Hybrid Coding

A method called hybrid transform/waveform coding, or simply hybrid coding, was devised to combine the merits of the two methods. By waveform coding, we mean coding techniques that code the waveform of a signal instead of the transformed signal. DPCM is a waveform coding technique. Hybrid coding combines TC and DPCM coding. That is, TC can be applied first rowwise followed by DPCM coding columnwise, or vice versa. In this way, the two techniques complement each other. That is, the hybrid coding technique simultaneously has TC's small sensitivity to variable image statistics and DPCM's simplicity in implementation.

It is worth mentioning a successful hybrid coding scheme in interframe coding: predictive coding along the temporal domain. Specifically, it uses MC predictive coding. That is, the motion analyzed from successive frames is used to more accurately predict a frame. The prediction error (in the 2-D spatial domain) is transform coded. This hybrid coding scheme has been very efficient and was adopted by the international video coding standards H.261, H.263, and MPEG 1, 2 and 4.

4.6 Summary

In TC, instead of the original image or some function of the original image in the spatial and temporal domain, the image in the transform domain is quantized and encoded. The main idea behind TC is that the transformed version of the image is less correlated. Moreover, the image energy is compacted into a small proper subset of transform coefficients.

The basis vector (1-D) and the basis image (2-D) provide a meaningful interpretation of TC. This type of interpretation considers the original image to be a weighted sum of basis vectors or basis images. The weights are the transform coefficients, each of which is essentially a correlation measure between the original image and the corresponding basis image. These weights are less correlated than the gray level values of pixels in the original image. Furthermore they have a great disparity in variance distribution. Some weights have large variances. They are retained and finely quantized. Some weights have small energy. They are retained and coarsely quantized. A vast majority of weights are insignificant and discarded. In this way, a high coding efficiency is achieved in TC. Because the quantized nonzero coefficients have a very nonuniform probability distribution, they can be encoded by using efficient variable-length codes. In summary, three factors, truncation (discarding a great majority of transform coefficients), adaptive quantization, and variable-length coding, contribute mainly to a high coding efficiency of TC.

Several linear, reversible, unitary transforms have been studied and utilized in TC. They include the discrete KLT (the Hotelling transform), the DFT, the Walsh transform, the Hadamard transform, and the discrete cosine transform. It is shown that the KLT is the optimum. The transform coefficients of the KLT are uncorrelated. The KLT can compact the most energy in the smallest fraction of transform coefficients. However, the KLT is image dependent. There is no fast algorithm to implement it. This prohibits the KLT from practical use in TC. While the rest of the transforms perform closely, the DCT appears to be the best. Its energy compaction is very close to the optimum KLT and it can be implemented using the FFT. The DCT has been found to be efficient not only for still images coding but also for coding residual images (predictive error) in MC interframe predictive coding. These features make the DCT the most widely used transform in the image and video coding.

There are two ways to truncate transform coefficients: zonal coding and threshold coding. In zonal coding, a zone is predefined based on average statistics. The transform coefficients within the zone are retained, while those outside the zone are discarded. In threshold coding, each transform coefficient is compared with a threshold. Those coefficients larger than the threshold are retained, while those smaller are discarded. Threshold coding is adaptive to local statistics. A two-pass procedure was usually taken. That is, the local statistics are measured or estimated in the first pass. The truncation takes place in the second pass. The addresses of the retained coefficients need to be sent to the reviver as overhead side information.

A one-step adaptive framework of TC has evolved as a result of the tremendous research efforts in image coding. It became a base of the international still image coding standard JPEG. Its fundamental components include the DCT transform, thresholding and adaptive quantization of transform coefficients, zigzag scan, Huffman coding of magnitude of the nonzero DCT coefficients and run-length of zeros in the zigzag scan, the code word of EOB, and rate buffer feedback control.

Threshold and normalization factor are controlled by rate buffer feedback. Since the threshold decides how many transform coefficients are retained and the normalization factor is actually the quantization step size, the rate buffer has direct impact on the bit rate of the TC system. Selection of quantization steps takes the energy compaction of the DCT and the characteristics of the HVS into consideration. That is, it uses not only statistical redundancy but also psychovisual redundancy to enhance coding efficiency.

After thresholding, normalization, and roundoff are applied to the DCT transform coefficients in a block, a great majority of transform coefficients are set to zero. Zigzag scan can convert the 2-D array of transform coefficients into a 1-D sequence. The number of consecutive zero-valued coefficients in the 1-D sequence is referred to as run-length of zeros and is used to provide address information of nonzero DCT coefficients. Both magnitude of nonzero coefficients and run-length information need to be coded. The statistical analysis has demonstrated that small magnitude and short run-length are dominant. Therefore, efficient lossless entropy coding methods such as Huffman coding and arithmetic coding (the focus of the next chapter) can be applied to magnitude and run-length.

In a reconstructed subimage, there are three types of error involved: truncation error (some transform coefficients have been set to zero), quantization error, and transmission error. In a broad sense, the truncation can be viewed as a part of the quantization. That is, these truncated coefficients are quantized to zero. The transmission error in terms of bit reversal will affect the whole reconstructed subimage. This is because, in the inverse transform (such as the inverse DCT), each transform coefficient makes a contribution.

In reconstructing the original image all the subimages are organized to form the whole image. Therefore the independent processing of individual subimages causes block artifacts. Though they may not severely affect the objective assessment of reconstructed image quality, block artifacts can be annoying, especially in low bit rate image coding. Block overlapping and postfiltering are the two effective ways to alleviate block artifacts. In the former, neighboring blocks are purposely overlapped by one pixel. In reconstructing the image, those pixels that have been coded more than once take an average of the multiple decoded values. Extra bits are used. In the latter technique, a low-pass filter is applied along boundaries of blocks. No extra bits are required in the process and the effect of combating block artifacts is better than with the former technique.

The selection of subimage size is an important issue in the implementation of TC. In general, the large size will remove more interpixel redundancy. But it has been shown that the pixel correlation becomes insignificant when the distance of pixels exceeds 20. On the other hand, large size is not suitable for adaptation to local statistics, while adaptation is required in handling nonstationary images. Large size also makes the effect of transmission error spread more widely. For these reasons, subimage size should not be large. In MC predictive interframe coding, motion estimation is normally carried out in sizes of 16×16 or 8×8. To be compatible, the subimage size in TC is frequently chosen as 8×8.

Both predictive coding, say, DPCM, and TC utilize interpixel correlation and are efficient coding schemes. Compared with TC, DPCM is simpler in computation. It needs less storage and has less processing delay. But it is more sensitive to image-to-image variation. On the other hand, TC provides higher adaptation to statistical variation. TC is capable of removing more interpixel correlation, thus providing higher coding efficiency. Traditionally, people consider that predictive coding is preferred if bit rate is in the range of 2–3 bites/pixel, while TC is preferred when bit rate is below 2–3 bits/pixel. However, the situation changes. TC becomes the core technology in image and video coding. Many special VLSI chips are designed and manufactured for reducing computational complexity. The complexity becomes less important. Consequently, predictive coding such as DPCM is only used in some very simple applications.

In the context of interframe coding, 3-D (two spatial dimensions and one temporal dimension) TC has not found wide application in practice due to the complexity in computation and storage. Hybrid transform/waveform coding has proven to be very efficient in interframe coding. There, the MC predictive coding is used along temporal dimension, while TC is used to code the prediction error in two spatial dimensions.

Exercises

1. Consider the following eight points in a 3-D coordinate system: $(0,0,0)^T$, $(1,0,0)^T$, $(0,1,0)^T$, $(0,0,1)^T$, $(0,1,1)^T$, $(1,0,1)^T$, $(1,1,0)^T$, $(1,1,1)^T$. Find the mean vector and covariance matrix using the Equations 4.12 and 4.13.
2. For $N=4$, find the basis images of the DFT, $I_{u,v}$ when (a) $u=0$, $v=0$, (b) $u=1$, $v=0$, (c) $u=2$, $v=2$, (d) $u=3$, $v=2$. Use both methods discussed in the text; i.e., the method with basis image and the method with basis vectors.
3. For $N=4$, find the basis images of the ordered DHT when (a) $u=0$, $v=2$, (b) $u=1$, $v=3$, (c) $u=2$, $v=3$, (d) $u=3$, $v=3$. Verify your results by comparing them with Figure 4.5.
4. Repeat the previous problem for the DWT, and verify your results by comparing them with Figure 4.4.
5. Repeat problem 3 for the DCT and $N=4$.
6. When $N=8$, draw the transform matrix F for the DWT, DHT, the order DHT, DFT, and DCT.
7. The matrix form of forward and inverse 2-D symmetric image transforms are expressed in texts such as [Jayant 1984] as $T=FGF^T$ and $G=ITI^T$, which are different from Equations 4.28 and 4.29. Can you explain this discrepancy?
8. Derive Equation 4.64. [NB: Use the concept of basis vectors and the orthogonality of basis vectors.]
9. Justify that the normalization factor is the quantization step.
10. The transform used in TC has two functions: decorrelation and energy compaction. Does decorrelation automatically lead to energy compaction? Comment.
11. Using your own words, explain the main idea behind TC.
12. Read the techniques by Chen and Pratt presented in Section 4.4.2. Compare them with JPEG discussed in Chapter 7. Comment on the similarity and dissimilarity between them.
13. How is the one-pass adaptation to local statistics in the algorithm of [chen and pratt] achieved?
14. Using your own words, explain why the DCT is superior to the DFT in terms of energy compaction.
15. Why is the subimage size of 8×8 widely used?

References

[ahmed 1974] N. Ahmed, T. Nararajan, and K.R. Rao, Discrete cosine transform, *IEEE Transactions on Computers*, 90–93, January 1974.
[andrews 1971] H.C. Andrews, Multidimensional rotations in feature selection, *IEEE Transactions on Computers*, c-20, 1045–1051, September 1971.
[chen 1977] W.H. Chen and C.H. Smith, Adaptive coding of monochrome and color images, *IEEE Transactions on Communications*, COM-25, 1285–1292, November 1977.

[chen 1984] W.H. Chen and W.K. Pratt, Scene adaptive coder, *IEEE Transactions on Communications*, COM-32, 225–232, March, 1984.

[cooley 1965] J.W. Cooley and J.W. Tukey, An algorithm for the machine calculation of complex Fourier series, *Mathematics of Computation*, 19, 297–301, 1965.

[gonzalez 2001] R.C. Gonzalez and R.E. Woods, *Digital Image Processing*, 2nd edition, Prentice Hall, Upper Saddle River, NJ, 2001.

[habibi 1971a] A. Habibi and P.A. Wintz, Image coding by linear transformations and block quantization, *IEEE Transactions on Communication Technology*, COM-19, 50–60, February 1971.

[habibi 1971b] A. Habibi, Comparison of nth-order DPCM encoder with linear transformations and block quantization techniques, *IEEE Transactions on Communication Technology*, COM-19, 6, 948–956, December 1971.

[hadamard 1893] J. Hadamard, Resolution d'une question relative aux determinants, *Bulletin des Sciences Mathematiques Series* 2, 17, Part I, 240–246, 1893.

[haskell 1996] B.G. Haskell, A. Puri, and A.N. Netravali, *Digital Video: An Introduction to MPEG-2*, Chapman & Hall, 1996.

[hotelling 1933] H. Hotelling. Analysis of a complex of statistical variables into principal components, *Journal of Educational Psychology*, 24, 417–441, 498–520, 1933.

[huang 1963] J.-Y. Huang and P.M. Schultheiss, Block quantization of correlated Gaussian random variables, *IEEE Transactions on Communication Systems*, cs-11, 289–296, September 1963.

[jayant 1984] N.S. Jayant and P. Noll, *Digital Coding of Waveforms*, Prentice-Hall, Englewood Cliffs, NJ, 1984.

[karhunen 1947] H. Karhunen, Über lineare Methoden in der Wahrscheinlich-Keitsrechnung, *Ann. Acad. Sci. Fenn.*, Ser. A. I. 37, Helsinki, 1947. (An English translation is available as "On linear methods in Probability theory" (I. Selin transl.), The RAND Corp., Dec. T-131, Aug. 11, 1960.)

[lim 1990] J.S. Lim, *Two-Dimensional Signal and Image Processing*, Prentice-Hall, Englewood Cliffs, NJ, 1990.

[loeve 1948] M. Loéve, Fonctions aleatoires de seconde ordre, in P. Levy, *Processus Stochastiques et Mouvement Brownien*, Hermann, Paris, France, 1948.

[pearl 1972] J. Pearl, H.C. Andrews, and W.K. Pratt, Performance measures for transform data coding, *IEEE Transactions on Communication Technology*, col. com-20, 411–415, June 1972.

[ramamurthi 1986] B. Ramamurthi and A. Gersho, Nonlinear space-variant postprocessing of block coded images, *IEEE Transactions on Acoustics, Speech and Signal Processing*, 34, 1258–1267, October 1986.

[reeve 1984] H.C. Reeve III and J.S. Lim, Reduction of blocking effects in image coding, *Journal of Optical Engineering*, 23, 34–37, January/February 1984.

[strang 1998] G. Strang, *Introduction to Linear Algebra*, Wellesley-Cambridge Press, Cambridge, MA, June 1998.

[tasto 1971] M. Tasto and P.A. Wintz, Image coding by adaptive block quantization, *IEEE Transactions on Communication Technology*, COM-19, 6, 957–972, December 1971.

[walsh 1923] J.L. Walsh, A closed set of normal orthogonal functions, *American Journal of Mathematics*, 45, 1, 5–24, 1923.

[wintz 1972] P.A. Wintz, Transform picture coding, *Proceedings of The IEEE*, 60, 7, 809–820, July 1972.

[zelinski 1975] R. Zelinski and P. Noll, Adaptive block quantization of speech signals (in German), Technical Report no. 181, Heinrich Hertz Institut, Berlin, 1975.

5

Variable-Length Coding: Information Theory Results (II)

There are three stages that take place in an encoder: transformation, quantization, and code word assignment (Figure 2.3). Quantization was discussed in Chapter 2. Differential coding and transform coding using two different transformation components were covered in Chapters 3 and 4, respectively. In differential coding, it is the difference signal that is quantized and encoded, whereas in transform coding it is the transformed signal that is quantized and encoded. In this chapter and the next chapter, we discuss several code word assignment (encoding) techniques. In this chapter, we cover two types of variable-length coding (VLC): Huffman coding and arithmetic coding.

First we introduce some fundamental concepts of encoding. Then, the rules that must be obeyed by all optimum and instantaneous codes are discussed. On the basis of these rules, the Huffman coding algorithm is presented. A modified version of the Huffman coding algorithm is introduced as an efficient way to dramatically reduce codebook memory while keeping almost the same optimality.

The promising arithmetic coding algorithm, which is quite different from Huffman coding, is another focus of the chapter. While Huffman coding is a block-oriented coding technique, arithmetic coding is a stream-oriented coding technique. With improvements in implementation, arithmetic coding has gained increasing popularity. Both Huffman and arithmetic codings are included in the international still image coding standard JPEG (Joint Photographic (image) Experts Group coding). The adaptive arithmetic coding algorithms are adopted by the international bi-level image coding standard JBIG (Joint Bi-level Image Experts Group coding). Note that the material presented in this chapter can be viewed as a continuation of the information theory results presented in Chapter 1.

5.1 Some Fundamental Results

Before presenting Huffman coding and arithmetic coding, we first provide some fundamental concepts and results as necessary background.

5.1.1 Coding an Information Source

Consider an information source, represented by a source alphabet S.

$$S = \{s_1, s_2, \ldots, s_m\}, \tag{5.1}$$

where s_i, $i = 1, 2, \ldots, m$, are source symbols. Note that the terms source symbol and information message are used interchangeably in the literature. In this book, however, we would like to distinguish them. That is, an information message can be a source symbol, or a combination of source symbols. We denote code alphabet by A and

$$A = \{a_1, a_2, \ldots, a_r\}, \tag{5.2}$$

where a_j, $j = 1, 2, \ldots, r$, are code symbols. A message code is a sequence of code symbols that represents a given information message. In the simplest case, a message consists of only a source symbol. Encoding is then a procedure to assign a code word to the source symbol. Namely,

$$s_i \rightarrow A_i = (a_{i1}, a_{i2}, \ldots, a_{ik}), \tag{5.3}$$

where the code word A_i is a string of k code symbols assigned to the source symbol s_i. The term message ensemble is defined as the entire set of messages. A code, also known as an ensemble code, is defined as a mapping of all the possible sequences of symbols of S (message ensemble) into the sequences of symbols in A.

Note that in binary coding, the number of code symbols r is equal to 2, since there are only two code symbols available: the binary digits "0" and "1". Two examples are given below to illustrate the above concepts.

Example 5.1

Consider an English article and the ASCII code. Refer to Table 5.1. In this context, the source alphabet consists of all the English letters in both lower and upper cases and all the punctuation marks. The code alphabet consists of the binary 1 and 0. There are a total of 128 7-bit binary code words. From Table 5.1, we see that the code word assigned to the capital letter A is 1000001. That is, A is a source symbol, while 1000001 is its code word.

Example 5.2

Table 5.2 lists what is known as the (5,2) code. It is a linear block code. In this example, the source alphabet consists of the four (2^2) source symbols listed in the left column of the table: 00, 01, 10, and 11. The code alphabet consists of the binary 1 and 0. There are four code words listed in the right column of the table. From this table, we see that the code assigns a 5-bit code word to each source symbol. Specifically, the code word of the source symbol 00 is 00000. The source symbol 01 is encoded as 10100. The code word assigned to 10 is 01111. The symbol 11 is mapped to 11011.

5.1.2 Some Desired Characteristics

To be practical in use, codes need to have some desired characteristics [abramson 1963]. Some of the characteristics are addressed in this subsection.

5.1.2.1 Block Code

A code is said to be a block code if it maps each source symbol in S into a fixed code word in A. Hence, the codes listed in the above two examples are block codes.

5.1.2.2 Uniquely Decodable Code

A code is uniquely decodable if it can be unambiguously decoded. Obviously, a code has to be uniquely decodable if it is to be in use.

TABLE 5.1

Seven-bit American Standard Code for Information Interchange

Bits				5 6	0 0	0 0	1 0	0 1	1 1	0 0	1 0	0 1	1 1
1	2	3	4	7	0	0	0	0	1	1	1	1
0	0	0	0	NUL	DLE	SP	0	@	P	'	p
1	0	0	0	SOH	DC1	!	1	A	Q	a	q
0	1	0	0	STX	DC2	"	2	B	R	b	r
1	1	0	0	ETX	DC3	#	3	C	S	c	s
0	0	1	0	EOT	DC4	$	4	D	T	d	t
1	0	1	0	ENQ	NAK	%	5	E	U	e	u
0	1	1	0	ACK	SYN	&	6	F	V	f	v
1	1	1	0	BEL	ETB	'	7	G	W	g	w
0	0	0	1	BS	CAN	(	8	H	X	h	x
1	0	0	1	HT	EM	)	9	I	Y	i	y
0	1	0	1	LF	SUB	*	:	J	Z	j	z
1	1	0	1	VT	ESC	+	;	K	[	k	{
0	0	1	1	FF	FS	,	<	L	\	l	|
1	0	1	1	CR	GS	-	=	M	]	m	}
0	1	1	1	SO	RS	.	>	N	^	n	~
1	1	1	1	SI	US	/	?	O	—	o	DEL

NUL	Null, or all zeros	DC1	Device control 1
SOH	Start of heading	DC2	Device control 2
STX	Start of text	DC3	Device control 3
ETX	End of text	DC4	Device control 4
EOT	End of transmission	NAK	Negative acknowledgment
ENQ	Enquiry	SYN	Synchronous idle
ACK	Acknowledge	ETB	End of transmission block
BEL	Bell, or alarm	CAN	Cancel
BS	Backspace	EM	End of medium
HT	Horizontal tabulation	SUB	Substitution
LF	Line feed	ESC	Escape
VT	Vertical tabulation	FS	File separator
FF	Form feed	GS	Group separator
CR	Carriage return	RS	Record separator
SO	Shift out	US	Unit separator
SI	Shift in	SP	Space
DLE	Data link escape	DEL	Delete

Example 5.3

Table 5.3 specifies a code. Obviously it is not uniquely decodable since if a binary string "00" is received we do not know which of the following two source symbols has been sent out: s_1 or s_3.

TABLE 5.2

(5,2) Linear Block Code

Source Symbol	Code Word
s_1 (0 0)	00000
s_2 (0 1)	10100
s_3 (1 0)	01111
s_4 (1 1)	11011

TABLE 5.3

Not Uniquely Decodable Code

Source Symbol	Code Word
s_1	00
s_2	10
s_3	00
s_4	11

Nonsingular Code
A block code is nonsingular if all the code words are distinct.

Example 5.4
Table 5.4 gives a nonsingular code since all four code words are distinct. If a code is not a nonsingular code, i.e., at least two code words are identical then the code is not uniquely decodable. Notice that, however, a nonsingular code does not guarantee unique decodability. The code shown in Table 5.4 is such an example in that it is nonsingular while it is not uniquely decodable. It is not uniquely decodable because once the binary string "11" is received, we do not know if the source symbols transmitted are s_1 followed by s_1 or simply s_2.

The nth Extension of a Block Code
The nth extension of a block code, which maps the source symbol s_i into the code word A_i, is a block code that maps the sequences of source symbols $s_{i1} s_{i2} \cdots s_{in}$ into the sequences of code words $A_{i1} A_{i2} \cdots A_{in}$.

A Necessary and Sufficient Condition of Block Codes' Unique Decodability
A block code is uniquely decodable if and only if the nth extension of the code is nonsingular for every finite n.

Example 5.5
The second extension of the nonsingular block code shown in Example 5.4 is listed in Table 5.5. Clearly, this second extension of the code is not a nonsingular code, since the entries $s_1 s_2$ and $s_2 s_1$ are the same. This confirms the nonunique decodability of the nonsingular code in Example 5.4.

5.1.2.3 Instantaneous Codes

5.1.2.3.1 Definition of Instantaneous Codes

A uniquely decodable code is said to be instantaneous if it is possible to decode each code word in a code symbol sequence without knowing the succeeding code words.

TABLE 5.4

Nonsingular Code

Source Symbol	Code Word
s_1	1
s_2	11
s_3	00
s_4	01

TABLE 5.5

Second Extension of the Nonsingular Block Code Shown in Example 5.4

Source Symbol	Code Word	Source Symbol	Code Word
$s_1 s_1$	11	$s_3 s_1$	001
$s_1 s_2$	111	$s_3 s_2$	0011
$s_1 s_3$	100	$s_3 s_3$	0000
$s_1 s_4$	101	$s_3 s_4$	0001
$s_2 s_1$	111	$s_4 s_1$	011
$s_2 s_2$	1111	$s_4 s_2$	0111
$s_2 s_3$	1100	$s_4 s_3$	0100
$s_2 s_4$	1101	$s_4 s_4$	0101

Example 5.6

Table 5.6 lists three uniquely decodable codes. The first one is in fact a 2-bit natural binary code. In decoding, we can immediately tell which source symbols are transmitted since each code word has the same length. In the second code, code symbol "1" functions like a comma. Whenever we see a "1," we know it is the end of the code word. The third code is different from the earlier two codes in that if we see a "10" string we are not sure if it corresponds to s_2 until we see a succeeding "1." Specifically, if the next code symbol is "0," we still cannot tell if it is s_3 since the next one may be "0" (hence s_4) or "1" (hence s_3). In this example, the next "1" belongs to the succeeding code word. Therefore, we see that code 3 is uniquely decodable. However, it is not instantaneous.

Definition of the jth Prefix

Assume a code word $A_i = a_{i1} a_{i2} \cdots a_{ik}$. Then the sequences of code symbols $a_{i1} a_{i2} \cdots a_{ij}$ with $1 \leq j \leq k$ is the jth-order prefix of the code word A_i.

Example 5.7

If a code word is 11001, it has the following five prefixes: 11001, 1100, 110, 11, 1. The first-order prefix is 1, while the fifth-order prefix is 11001.

Necessary and Sufficient Condition of Being Instantaneous Codes

A code is instantaneous if and only if no code word is a prefix of some other code word. This condition is often referred to as the prefix condition. Hence, the instantaneous code is also called the prefix condition code or sometimes simply the prefix code. In many applications, we need a block code that is nonsingular, uniquely decodable, and instantaneous.

TABLE 5.6

Three Uniquely Decodable Codes

Source Symbol	Code 1	Code 2	Code 3
s_1	00	1	1
s_2	01	01	10
s_3	10	001	100
s_4	11	0001	1000

5.1.2.4 Compact Code

A uniquely decodable code is said to be compact if its average length is the minimum among all other uniquely decodable codes based on the same source alphabet S and code alphabet A. A compact code is also referred to as a minimum redundancy code, or an optimum code.

Note that the average length of a code was defined in Chapter 1 and is restated below.

5.1.3 Discrete Memoryless Sources

This is the simplest model of an information source. In this model, the symbols generated by the source are independent of each other. That is, the source is memoryless or it has a zero-memory.

Consider the information source expressed in Equation 5.1 as a discrete memoryless source. The occurrence probabilities of the source symbols can be denoted by $p(s_1)$, $p(s_2), \ldots, p(s_m)$. The lengths of the code words can be denoted by $l_1, l_2, \ldots, l_m$. The average length of the code is then equal to

$$L_{\text{avg}} = \sum_{i=1}^{m} l_i p(s_i). \tag{5.4}$$

Recall Shanon's first theorem, i.e., the noiseless coding theorem, described in Chapter 1. The average length of the code is bounded below by the entropy of the information source. The entropy of the source S is defined as $H(S)$ and

$$H(S) = -\sum_{i=1}^{m} p(s_i) \log_2 p(s_i). \tag{5.5}$$

Recall that entropy is the average amount of information contained in a source symbol. In Chapter 1, the efficiency of a code, η, is defined as the ratio between the entropy and the code. That is, $\eta = H(S)/L_{\text{avg}}$. The redundancy of the code, ζ, is defined as $\zeta = 1 - \eta$.

5.1.4 Extensions of a Discrete Memoryless Source

Instead of coding each source symbol in a discrete source alphabet, it is often useful to code blocks of symbols. It is, therefore, necessary to define the nth extension of a discrete memoryless source.

5.1.4.1 Definition

Consider the zero-memory source alphabet S defined in Equation 5.1. That is, $S = \{s_1, s_2, \ldots, s_m\}$. If n symbols are grouped into a block, then there are a total of m^n blocks. Each block is considered as a new source symbol. These m^n blocks thus form an information source alphabet, called the nth extension of the source S, which is denoted by S^n.

5.1.4.2 Entropy

Let each block be denoted by β_i and

$$\beta_i = (s_{i1}, s_{i2}, \ldots, s_{in}). \tag{5.6}$$

Variable-Length Coding: Information Theory Results (II)

TABLE 5.7

Discrete Memoryless Source Alphabet

Source Symbol	Occurrence Probability
s_1	0.6
s_2	0.4

Then we have the following relation due to the memoryless assumption.

$$p(\beta_i) = \prod_{j=1}^{n} p(s_{ij}). \tag{5.7}$$

Hence the relationship between the entropy of the source S and the entropy of its nth extension is as follows:

$$H(S^n) = n \cdot H(S). \tag{5.8}$$

Example 5.8

Table 5.7 lists a source alphabet. Its second extension is listed in Table 5.8.

The entropy of the source and its second extension are calculated.

$$H(S) = -0.6 \cdot \log_2(0.6) - 0.4 \cdot \log_2(0.4) \approx 0.97,$$
$$H(S^2) = -0.36 \cdot \log_2(0.36) - 2(0.24) \cdot \log_2(0.24) - 0.16 \cdot \log_2(0.16) \approx 1.94.$$

It is seen that $H(S^2) = 2H(S)$.

5.1.4.3 Noiseless Source Coding Theorem

The noiseless source coding theorem, also known as Shanon's first theorem, was presented in Chapter 1, but without a mathematical expression. Here, we provide some mathematical expressions to give more insight about the theorem.

For a discrete zero-memory information source S, the noiseless coding theorem can be expressed as

$$H(S) \leq L_{\text{avg}} < H(S) + 1, \tag{5.9}$$

TABLE 5.8

Second Extension of the Source Alphabet Shown in Table 5.7

Source Symbol	Occurrence Probability
$s_1 s_1$	0.36
$s_2 s_2$	0.24
$s_2 s_1$	0.24
$s_2 s_2$	0.16

that is, there exists a VLC whose average length is bounded below by the entropy of the source (that is encoded) and bounded above by the entropy plus 1. Since the nth extension of the source alphabet, S^n, is itself a discrete memoryless source, we can apply the above result to it. That is,

$$H(S^n) \leq L_{\text{avg}}^n < H(S^n) + 1, \qquad (5.10)$$

where L_{avg}^n is the average code word length of a VLC for the S^n. Since $H(S^n) = nH(S)$ and $L_{\text{avg}}^n = nL_{\text{avg}}$, we have

$$H(S) \leq L_{\text{avg}} < H(S) + \frac{1}{n}. \qquad (5.11)$$

Therefore, when coding blocks of n source symbols, the noiseless source coding theory states that for an arbitrary positive number ε, there is a VLC, which satisfies the following:

$$H(S) \leq L_{\text{avg}} < H(S) + \varepsilon \qquad (5.12)$$

as n is large enough. That is, the average number of bits used in coding per source symbol is bounded below by the entropy of the source and is bounded above by the sum of the entropy and an arbitrary positive number. To make ε arbitrarily small, i.e., to make the average length of the code arbitrarily close to the entropy, we have to make the block size n large enough. This version of the noiseless coding theorem suggests a way to make the average length of a VLC approach the source entropy. It is known, however, that the high coding complexity that occurs when n approaches infinity makes implementation of the code impractical.

5.2 Huffman Codes

Consider the source alphabet defined in Equation 5.1. The method of encoding source symbols according to their probabilities suggested in [shannon 1948; fano 1949] is not optimum. It approaches the optimum, however, when the block size n approaches infinity. This results in a large storage requirement and high computational complexity. In many cases, we need a direct encoding method that is optimum and instantaneous (hence uniquely decodable) for an information source with finite source symbols in source alphabet S. Huffman code is the first such optimum code [huffman 1952], and is the technique most frequently used at present. It can be used for r-ary encoding as $r > 2$. For the notational brevity, however, we discuss only the Huffman coding used in the binary case presented here.

5.2.1 Required Rules for Optimum Instantaneous Codes

Let us rewrite Equation 5.1 as follows:

$$S = (s_1, s_2, \ldots, s_m). \qquad (5.13)$$

Without loss of generality, assume the occurrence probabilities of the source symbols are as follows:

$$p(s_1) \geq p(s_2) \geq \cdots \geq p(s_{m-1}) \geq p(s_m). \qquad (5.14)$$

As we are seeking the optimum code for S, the lengths of code words assigned to the source symbols should be

$$l_1 \leq l_2 \leq \cdots \leq l_{m-1} \leq l_m. \quad (5.15)$$

On the basis of the requirements of the optimum and instantaneous code, Huffman derived the following rules (restrictions):

1. $l_1 \leq l_2 \leq \cdots \leq l_{m-1} = l_m.$ (5.16)

Equations 5.14 and 5.16 imply that when the source symbol occurrence probabilities are arranged in a nonincreasing order, the length of the corresponding code words should be in a nondecreasing order. In other words, the code word length of a more probable source symbol should not be longer than that of a less probable source symbol. Furthermore, the length of the code words assigned to the two least probable source symbols should be the same.

2. The code words of the two least probable source symbols should be the same except for their last bits.
3. Each possible sequence of length l_{m-1} bits must be used either as a code word or must have one of its prefixes used as a code word.

Rule 1 can be justified as follows. If the first part of the rule, i.e., $l_1 \leq l_2 \leq \cdots \leq l_{m-1}$ is violated, say, $l_1 > l_2$, then we can exchange the two code words to shorten the average length of the code. This means the code is not optimum, which contradicts the assumption that the code is optimum. Hence it is impossible. That is, the first part of rule 1 has to be the case. Now assume that the second part of the rule is violated, i.e., $l_{m-1} < l_m$. (Note that $l_{m-1} > l_m$ can be shown to be impossible by using the same reasoning we just used to prove the first part of the rule.) Since the code is instantaneous, code word A_{m-1} is not a prefix of code word A_m. This implies that the last bit in the code word A_m is redundant. It can be removed to reduce the average length of the code, implying that the code is not optimum. This contradicts the assumption, thus proving rule 1.

Rule 2 can be justified as follows. As in the above, A_{m-1} and A_m are the code words of the two least probable source symbols. Assume that they do not have the identical prefix of the order l_{m-1}. Since the code is optimum and instantaneous, code words A_{m-1} and A_m cannot have prefixes of any order that are identical to other code words. This implies that we can drop the last bits of A_{m-1} and A_m to achieve a lower average length. This contradicts the optimum code assumption. It proves that rule 2 has to be the case.

Rule 3 can be justified using a similar strategy to that used above. If a possible sequence of length l_{m-1} has not been used as a code word and any of its prefixes have not been used as code words, then it can be used in place of the code word of the mth source symbol, resulting in a reduction of the average length L_{avg}. This is a contradiction to the optimum code assumption and it justifies the rule.

5.2.2 Huffman Coding Algorithm

On the basis of these three rules, we see that the two least probable source symbols have equal-length code words. These two code words are identical except for the last bits, the binary 0 and 1, respectively. Therefore, these two source symbols can be combined to form a single new symbol. Its occurrence probability is the sum of two source symbols,

TABLE 5.9

Source Alphabet and Huffman Codes in Example 5.9

Source Symbol	Occurrence Probability	Code Word Assigned	Length of Code Word
s_1	0.3	00	2
s_2	0.1	101	3
s_3	0.2	11	2
s_4	0.05	1001	4
s_5	0.1	1000	4
s_6	0.25	01	2

i.e., $p(s_{m-1}) + p(s_m)$. Its code word is the common prefix of order l_{m-1} of the two code words assigned to s_m and s_{m-1}, respectively. The new set of source symbols thus generated is referred to as the first auxiliary source alphabet, which is one source symbol less than the original source alphabet. In the first auxiliary source alphabet, we can rearrange the source symbols according to a nonincreasing order of their occurrence probabilities. The same procedure can be applied to this newly created source alphabet. A binary 0 and a binary 1 are, respectively, assigned to the last bits of the two least probable source symbols in the alphabet. The second auxiliary source alphabet will again have one source symbol less than the first auxiliary source alphabet. The procedure continues. In some step, the resultant source alphabet will have only two source symbols. At this time, we combine them to form a single source symbol with a probability of 1. Then the coding is complete.

Let's go through the following example to illustrate the above Huffman algorithm.

Example 5.9

Consider a source alphabet whose six source symbols and their occurrence probabilities are listed in Table 5.9. Figure 5.1 demonstrates the Huffman coding procedure applied. In the example, among the two least probable source symbols encountered at each step, we assign binary 0 to the top symbol and binary 1 to the bottom symbol.

5.2.2.1 Procedures

In summary, the Huffman coding algorithm consists of the following steps:

FIGURE 5.1
Huffman coding procedure in Example 5.9.

Variable-Length Coding: Information Theory Results (II)

1. Arrange all source symbols in such a way that their occurrence probabilities are in a nonincreasing order.
2. Combine the two least probable source symbols:
   - Form a new source symbol with a probability equal to the sum of the probabilities of the two least probable symbols.
   - Assign a binary 0 and a binary 1 to the two least probable symbols.
3. Repeat until the newly created auxiliary source alphabet contains only one source symbol.
4. Start from the source symbol in the last auxiliary source alphabet and trace back to each source symbol in the original source alphabet to find the corresponding code words.

5.2.2.2 Comments

First, it is noted that the assignment of the binary 0 and 1 to the two least probable source symbols in the original source alphabet and each of the first $(u - 1)$ auxiliary source alphabets can be implemented in two different ways. Here u denotes the total number of the auxiliary source symbols in the procedure. Hence, there is a total of 2^u possible Huffman codes. In Example 5.9, there are five auxiliary source alphabets, hence a total of $2^5 = 32$ different codes. Note that each is optimum: that is, each has the same average length.

Second, in sorting the source symbols, there may be more than one symbol having equal probabilities. This results in multiple arrangements of symbols, hence multiple Huffman codes. While all of these Huffman codes are optimum, they may have some other different properties. For instance, some Huffman codes result in the minimum code word length variance [sayood 1996]. This property is desired for applications in which a constant bit rate is required.

Third, Huffman coding can be applied to r-ary encoding with $r > 2$. That is, code symbols are r-ary with $r > 2$.

5.2.2.3 Applications

As a systematic procedure to encode a finite discrete memoryless source, the Huffman code has found wide application in image and video coding. Recall that it has been used in differential coding and transform coding. In transform coding, as introduced in Chapter 4, the magnitude of the quantized transform coefficients and the run length of zeros in the zigzag scan are encoded by using the Huffman code.

5.3 Modified Huffman Codes

5.3.1 Motivation

As a result of Huffman coding, a set of all the code words, called a codebook, are created. It is an agreement between the transmitter and the receiver. Consider the case where the occurrence probabilities are skewed, i.e., some are large, whereas some are small. Under these circumstances, the improbable source symbols take a disproportionately large amount of memory space in by the codebook. The size of the codebook will be very large if the number of the improbable source symbols is large. A large size codebook requires a large memory space and increases the computational complexity. A modified Huffman (MH) procedure was therefore devised to reduce the memory requirement while keeping almost the same optimality [hankamer 1979].

Example 5.10
Consider a source alphabet consisting of 16 symbols, each being a 4-bit binary sequence. That is, $S = \{s_i, i = 1, 2, \ldots, 16\}$. The occurrence probabilities are

$$p(s_1) = p(s_2) = 1/4,$$
$$p(s_3) = p(s_4) = \cdots = p(s_{16}) = 1/28.$$

The source entropy can be calculated as follows:

$$H(S) = 2\left(-\frac{1}{4} \log_2 \frac{1}{4}\right) + 14\left(-\frac{1}{28} \log_2 \frac{1}{28}\right) \approx 3.404 \text{ bits/symbol}.$$

Applying the Huffman coding algorithm, we find that the code word lengths associated with the symbols are $l_1 = l_2 = 2$, $l_3 = 4$, and $l_4 = l_5 = \cdots = l_{16} = 5$, where l_i denotes the length of the ith code word. The average length of Huffman code is

$$L_{\text{avg}} = \sum_{i=1}^{16} p(s_i) l_i = 3.464 \text{ bits/symbol}.$$

We see that the average length of Huffman code is quite close to the lower entropy bound. It is noted, however, that the required codebook memory, M (defined as the sum of the code word lengths), is quite large:

$$M = \sum_{i=1}^{16} l_i = 73 \text{ bits}.$$

This number is obviously larger than the average code word length multiplied by the number of code words. This should not come as a surprise since the average here is in the statistical sense instead of in the arithmetic sense. When the total number of improbable symbols increases, the required codebook memory space will increase dramatically, resulting in a great demand on memory space.

5.3.2 Algorithm

Consider a source alphabet S that consists of 2^v binary sequences, each of length v. In other words, each source symbol is a v-bit code word in the natural binary code. The occurrence probabilities are highly skewed and there is a large number of improbable symbols in S. The MH coding algorithm is based on the following idea: lumping all the improbable source symbols into a category named ELSE [weaver 1978]. The algorithm is described below.

Variable-Length Coding: Information Theory Results (II)

1. Categorize the source alphabet S into two disjoint groups, S_1 and S_2, such that

$$S_1 = \left\{ s_i | p(s_i) > \frac{1}{2^v} \right\} \quad (5.17)$$

and

$$S_2 = \left\{ s_i | p(s_i) \leq \frac{1}{2^v} \right\}. \quad (5.18)$$

2. Establish a source symbol ELSE with its occurrence probability equal to $p(S_2)$.
3. Apply the Huffman coding algorithm to the source alphabet S_3 with $S_3 = S_1 \cup \text{ELSE}$.
4. Convert the codebook of S_3 to that of S as follows:
   - Keep the same code words for those symbols in S_1.
   - Use the code word assigned to ELSE as a prefix for those symbols in S_2.

5.3.3 Codebook Memory Requirement

Codebook memory M is the sum of the code word lengths. The M required by Huffman coding with respect to the original source alphabet S is

$$M = \sum_{i \in S} l_i = \sum_{i \in S_1} l_i + \sum_{i \in S_2} l_i, \quad (5.19)$$

where l_i denotes the length of the ith code word, as defined earlier. In the case of the MH coding algorithm, the memory required M_{mH} is

$$M_{mH} = \sum_{i \in S_3} l_i = \sum_{i \in S_1} l_i + l_{\text{ELSE}}, \quad (5.20)$$

where l_{ELSE} is the length of the code word assigned to ELSE. The above equation reveals the big savings in memory requirement when the probability is skewed. The following example is used to illustrate the MH coding algorithm and the resulting dramatic memory savings.

Example 5.11
In this example, we apply the MH coding algorithm to the source alphabet presented in Example 5.10. We first lump the 14 symbols having the least occurrence probabilities together to form a new symbol ELSE. The probability of ELSE is the sum of the 14 probabilities. That is,

$$p(\text{ELSE}) = \frac{1}{28} \cdot 14 = \frac{1}{2}.$$

Apply Huffman coding to the new source alphabet, $S_3 = \{s_1, s_2, \text{ELSE}\}$, as shown in Figure 5.2.

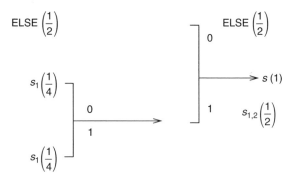

FIGURE 5.2
The modified Huffman (MH) coding procedure in Example 5.11.

From Figure 5.2, it is seen that the code words assigned to symbols s_1, s_2, and ELSE are, respectively, 10, 11, and 0. Hence, for every source symbol lumped into ELSE, its code word is 0 followed by the original 4-bit binary sequence. Therefore,

$$M_{mH} = 2 + 2 + 1 = 5 \text{ bits},$$

i.e., the required codebook memory is only 5 bits. Compared with 73 bits required by Huffman coding (refer to Example 5.10), there is a savings of 68 bits in codebook memory space. Similar to the comment made in Example 5.10, the memory savings will be even larger if the probability distribution is skewed more severely and the number of improbable symbols is larger. The average length of the MH algorithm is

$$L_{\text{avg},mH} = \frac{1}{4} \cdot 2 \cdot 2 + \frac{1}{28} \cdot 5 \cdot 14 = 3.5 \text{ bits/symbol}.$$

This demonstrates that MH coding retains almost the same coding efficiency as that achieved by Huffman coding.

5.3.4 Bounds on Average Code Word Length

It has been shown that the average length of the MH codes satisfies the following condition:

$$H(S) \leq L_{\text{avg}} < H(S) + 1 - p \log_2 p, \quad (5.21)$$

where $p = \sum_{s_i \in S_2} p(S_i)$. It is seen that, compared with the noiseless source coding theorem, the upper bound of the code average length is increased by a quantity of $-p \log_2 p$. In Example 5.11 it is seen that the average length of the MH code is close to that achieved by the Huffman code. Hence the MH code is almost optimum.

5.4 Arithmetic Codes

Arithmetic coding, which is quite different from Huffman coding, is gaining increasing popularity. In this section, we first analyze the limitations of Huffman coding. Then the principle of arithmetic coding is been introduced. Finally some implementation issues are discussed briefly.

5.4.1 Limitations of Huffman Coding

As seen in Section 5.2, Huffman coding is a systematic procedure for encoding a source alphabet, with each source symbol having an occurrence probability. Under these circumstances, Huffman coding is optimum in the sense that it produces a minimum coding redundancy. It has been shown that the average code word length achieved by Huffman coding satisfies the following inequality [gallagher 1978].

$$H(S) \leq L_{\text{avg}} < H(S) + p_{\text{max}} + 0.086, \quad (5.22)$$

where $H(S)$ is the entropy of the source alphabet, and p_{max} denotes the maximum occurrence probability in the set of the source symbols. This inequality implies that the upper bound of the average code word length of Huffman code is determined by the entropy and the maximum occurrence probability of the source symbols being encoded.

In the case where the probability distribution among source symbols is skewed (some probabilities are small, while some are quite large), the upper bound may be large, implying that the coding redundancy may not be small. Imagine the following extreme situation. There are only two source symbols. One has a very small probability, while the other has a very large probability (very close to 1). The entropy of the source alphabet in this case is close to 0 since the uncertainty is very small. Using Huffman coding, however, we need 2 bits: one for each. That is, the average code word length is 1, which means that the redundancy is very close to 1. This agrees with Equation 5.22. This inefficiency is due to the fact that Huffman coding always encodes a source symbol with an integer number of bits.

The noiseless coding theorem (reviewed in Section 5.1) indicates that the average code word length of a block code can approach the source alphabet entropy when the block size approaches infinity. As the block size approaches infinity, the storage required, the codebook size, and the coding delay will approach infinity, however, and the complexity of the coding will be out of control. Fortunately, it is often the case that when the block size is large enough in practice the average code word length of a block code has been rather close to the source alphabet entropy.

The fundamental idea behind Huffman coding and Shannon–Fano coding (devised a little earlier than Huffman coding [bell 1990]) is block coding. That is, some code word having an integral number of bits is assigned to a source symbol. A message may be encoded by cascading the relevant code words. It is the block-based approach that is responsible for the limitations of Huffman codes.

Another limitation is that when encoding a message that consists of a sequence of source symbols the nth extension Huffman coding needs to enumerate all possible sequences of source symbols having the same length, as discussed in coding the nth extended source alphabet. This is not computationally efficient.

Quite different from Huffman coding, arithmetic coding is stream based. It overcomes the drawbacks of Huffman coding. A string of source symbols are encoded as a string of code symbols. It is hence free of the integral-bits-per-source-symbol restriction and is more efficient. Arithmetic coding may reach the theoretical bound to coding efficiency specified in the noiseless source coding theorem for any information source. Below, we introduce the principle of arithmetic coding, from which we can see the stream-based nature of arithmetic coding.

5.4.2 The Principle of Arithmetic Coding

To understand the different natures of Huffman coding and arithmetic coding, let us look at Example 5.12, where we use the same source alphabet and the associated occurrence probabilities used in Example 5.9. In this example, however, a string of source symbols $s_1 s_2 s_3 s_4 s_5 s_6$ are encoded. Note that we consider the terms string and stream to be slightly different. By stream, we mean a message or possibly several messages, which may correspond to quite a long sequence of source symbols. Moreover, stream gives a dynamic flavor. Later we see that arithmetic coding is implemented in an incremental manner. Hence stream is a suitable term to use for arithmetic coding. In this example, however, only six source symbols are involved. Hence we consider the term string to be suitable, aiming at distinguishing it from the term block.

Example 5.12
The set of six source symbols and their occurrence probabilities are listed in Table 5.10. In this example, the string to be encoded using arithmetic coding is $s_1 s_2 s_3 s_4 s_5 s_6$. In the following four sections, we will use this example to illustrate the principle of arithmetic coding and decoding.

5.4.2.1 Dividing Interval [0, 1) into Subintervals

As pointed out by Elias, it is not necessary to sort out source symbols according to their occurrence probabilities. Therefore, in Figure 5.3a, the six symbols are arranged in their natural order from symbols $s_1, s_2, \ldots, s_6$. The real interval between 0 and 1 is divided into six subintervals, each having a length of $p(s_i)$, $i = 1, 2, \ldots, 6$. Specifically, the interval denoted by [0, 1)—where 0 is included in (the left end is closed) and 1 is excluded from (the right end is open) the interval—is divided into six subintervals. The first subinterval [0, 0.3) corresponds to s_1 and has a length of $p(s_1)$, i.e., 0.3. Similarly, the subinterval [0, 0.3) is said to be closed on the left and open on the right. The remaining five subintervals are similarly constructed. All six subintervals thus formed are disjoint and their union is equal to the interval [0, 1). This is because the sum of all the probabilities is equal to 1.

We list the sum of the preceding probabilities, known as cumulative probability (CP) [langdon 1984], in the right most column of Table 5.10 as well. Note that the concept of CP is slightly different from that of cumulative distribution function (CDF) in probability theory. Recall that in the case of discrete random variables the CDF is defined as follows:

$$\mathrm{CDF}(s_i) = \sum_{j=1}^{i} p(s_j). \tag{5.23}$$

TABLE 5.10
Source Alphabet and Cumulative Probabilities in Example 5.12

Source Symbol	Occurrence Probability	Associated Subintervals	Cumulative Probability
s_1	0.3	[0, 0.3)	0
s_2	0.1	[0.3, 0.4)	0.3
s_3	0.2	[0.4, 0.6)	0.4
s_4	0.05	[0.6, 0.65)	0.6
s_5	0.1	[0.65, 0.75)	0.65
s_6	0.25	[0.75, 1.0)	0.75

Variable-Length Coding: Information Theory Results (II)

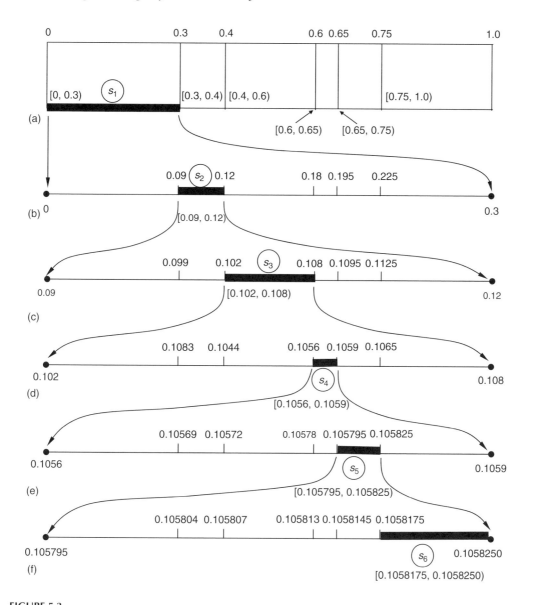

FIGURE 5.3
Arithmetic coding working on the same source alphabet as that given in Example 5.9. The encoded symbol string is $s_1 s_2 s_3 s_4 s_5 s_6$.

The CP is defined as

$$\text{CP}(s_i) = \sum_{j=1}^{i-1} p(s_j), \quad (5.24)$$

where $\text{CP}(s_1) = 0$ is defined. Now we see each subinterval has its lower end point located at $\text{CP}(s_i)$. The width of each subinterval is equal to the probability of the corresponding source symbol. A subinterval can be completely defined by its lower end point and its width. Alternatively, it is determined by its two end points: the lower and upper end points (sometimes also called the left and right end points).

Now consider encoding the string of source symbols $s_1 s_2 s_3 s_4 s_5 s_6$ with the arithmetic coding method.

5.4.2.2 Encoding

5.4.2.2.1 Encoding the First Source Symbol

As the first symbol is s_1, we pick up its subinterval [0, 0.3). Picking up the subinterval [0, 0.3) means that any real number in the subinterval, i.e., any real number equal to or greater than 0 and smaller than 0.3, can be a pointer to the subinterval, thus representing the source symbol s_1. This can be justified by considering that all the six subintervals are disjoint (see Figure 5.3a).

5.4.2.2.2 Encoding the Second Source Symbol

We use the same procedure as used in Figure 5.3a to divide the interval [0, 0.3) into six subintervals (Figure 5.3b). Since the second symbol to be encoded is s_2, we pick up its subinterval [0.09, 0.12).

Notice that the subintervals are recursively generated from Figure 5.3a to b. It is known that an interval may be completely specified by its lower end point and width. Hence, the subinterval recursion in the arithmetic coding procedure is equivalent to the following two recursions: end point recursion and width recursion.

From interval [0, 0.3) derived in Figure 5.3a to interval [0.09, 0.12) obtained in Figure 5.3b, we can conclude the following lower end point recursion:

$$L_{new} = L_{current} + W_{current} \cdot CP_{new}, \qquad (5.25)$$

where L_{new} and $L_{current}$ represent, respectively, the lower end points of the new and current recursions, and the $W_{current}$ and the CP_{new} denote the width of the interval in the current recursion and the CP in the new recursion, respectively. The width recursion is

$$W_{new} = W_{current} \cdot p(s_i), \qquad (5.26)$$

where W_{new} and $p(s_i)$ are, respectively, the width of the new subinterval and the probability of the source symbol s_i that is being encoded. These two recursions, also called double recursion [langdon 1984], play a central role in arithmetic coding.

5.4.2.2.3 Encoding the Third Source Symbol

When the third source symbol is encoded, the subinterval generated above in Part (b) is similarly divided into six subintervals. Since the third symbol to encode is s_3, its subinterval [0.102, 0.108) is picked up (see Figure 5.3c).

5.4.2.2.4 Encoding the Fourth, Fifth, and Sixth Source Symbols

The subinterval division is carried out according to Equations 5.25 and 5.26. The symbols $s_4, s_5,$ and s_6 are encoded. The final subinterval generated is [0.1058175, 0.1058250) (see Figure 5.3d through f).

That is, the resulting subinterval [0.1058175, 0.1058250) can represent the source symbol string $s_1 s_2 s_3 s_4 s_5 s_6$. Note that in this example decimal digits instead of binary digits are used. In binary arithmetic coding, the binary digits 0 and 1 are used.

5.4.2.3 Decoding

As seen in this example, for the encoder of arithmetic coding, the input is a source symbol string, and the output is a subinterval. Let us call this the final subinterval or the resultant subinterval. Theoretically, any real numbers in the interval can be the code string for the input symbol string since all subintervals are disjoint. Often, however, the lower end of the final subinterval is used as the code string. Now let us examine how the decoding process is carried out with the lower end of the final subinterval.

Decoding sort of reverses what encoding has done. The decoder knows the encoding procedure and therefore has the information contained in Figure 5.3a. It compares the lower end point of the final subinterval 0.1058175 with all the end points. It is determined that

$$0 < 0.1058175 < 0.3.$$

That is, the lower end falls into the subinterval associated with the symbol s_1. Therefore, the symbol s_1 is first decoded.

Once the first symbol is decoded, the decoder may know the partition of subintervals shown in Figure 5.3b. It is then determined that

$$0.09 < 0.1058175 < 0.12.$$

That is, the lower end is contained in the subinterval corresponding to the symbol s_2. As a result, s_2 is the second decoded symbol.

The procedure repeats itself until all six symbols are decoded. That is, based on Figure 5.3c, it is found that

$$0.102 < 0.1058175 < 0.108.$$

The symbol s_3 is decoded. Then, the symbols s_4, s_5, s_6 are subsequently decoded because the following inequalities are determined:

$$0.1056 < 0.1058175 < 0.1059$$
$$0.105795 < 0.1058175 < 0.1058250$$
$$0.1058145 < 0.1058175 < 0.1058250$$

Note that a terminal symbol is necessary to inform the decoder to stop decoding.

The above procedure gives us an idea of how decoding works. The decoding process, however, does not need to construct Figure 5.3b through f. Instead, the decoder only needs the information contained in Figure 5.3a. Decoding can be split into the following three steps: comparison, readjustment (subtraction), and scaling [langdon 1984].

As described above, through comparison we decode the first symbol s_1. From the way Figure 5.3b is constructed, we know the decoding of s_2 can be accomplished as follows. We subtract the lower end of the subinterval associated with s_1 in Figure 5.3a, i.e., 0 in this example from the lower end of the final subinterval 0.1058175, resulting in 0.1058175. Then we divide this number by the width of the subinterval associated with s_1, i.e., the probability of s_1, 0.3, resulting in 0.352725. From Figure 5.3a, it is found that

$$0.3 < 0.352725 < 0.4$$

That is, the number is within the subinterval corresponding to s_2. Therefore, the second decoded symbol is s_2. Note that these three decoding steps exactly undo what encoding has done.

To decode the third symbol, we subtract the lower end of the subinterval with s_2, 0.3 from 0.352725, obtaining 0.052725. This number is divided by the probability of s_2, 0.1, resulting in 0.52725. The comparison of 0.52725 with end points in Figure 5.3a reveals that the third decoded symbol is s_3.

In decoding the fourth symbol, we first subtract the lower end of the s_3's subinterval in Figure 5.3a, 0.4 from 0.52725, getting 0.12725. Dividing 0.12725 by the probability of s_3, 0.2, results in 0.63625. Referring to Figure 5.3a, we decode the fourth symbol as s_4 by comparison.

Subtraction of the lower end of the subinterval of s_4 in Figure 5.3a, 0.6, from 0.63625 leads to 0.03625. Division of 0.03625 by the probability of s_4, 0.05, produces 0.725. The comparison between 0.725 and the end points decodes the fifth symbol as s_5.

Subtracting 0.725 by the lower end of the subinterval associated with s_5, 0.65, gives 0.075. Dividing 0.075 by the probability of s_5, 0.1, generates 0.75. The comparison indicates that the sixth decoded symbol is s_6.

In summary, considering the way in which Figure 5.3b through f is constructed, we see that the three steps discussed in the decoding process: comparison, readjustment, and scaling exactly undo what the encoding procedure has done.

5.4.2.4 Observations

Both encoding and decoding involve only arithmetic operations (addition and multiplication in encoding, subtraction and division in decoding). This explains the name arithmetic coding.

We see that an input source symbol string $s_1 s_2 s_3 s_4 s_5 s_6$, via encoding, corresponds to a subinterval [0.1058175, 0.1058250). Any number in this interval can be used to denote the string of the source symbols.

We also observe that arithmetic coding can be carried out in an incremental manner. That is, source symbols are fed into the encoder one by one and the final subinterval is refined continually, i.e., the code string is generated continually. Furthermore, it is done in a manner called first in first out (FIFO). That is, the source symbol encoded first is decoded first. This manner is superior to that of last in first out (LIFO). This is because FIFO is suitable for adaptation to the statistics of the symbol string.

Obviously, the width of the final subinterval becomes smaller and smaller when the length of the source symbol string becomes larger and larger. This causes what is known as the precision problem. It is this problem that prohibited arithmetic coding from practical usage for quite a long period of time. Only after this problem was solved in the late 1970s, did arithmetic coding become an increasingly important coding technique.

It is necessary to have a termination symbol at the end of an input source symbol string. In this way, an arithmetic coding system is able to know when to terminate decoding.

Compared with Huffman coding, arithmetic coding is quite different. Basically, Huffman coding converts each source symbol into a fixed code word with an integral number of bits, whereas arithmetic coding converts a source symbol string to a code symbol string. To encode the same source symbol string, Huffman coding can be implemented in two different ways. One way is shown in Example 5.9. We construct a fixed code word for each source symbol. Since Huffman coding is instantaneous, we can cascade the corresponding code words to form the output, a 17-bit code string 00.101.11.1001.1000.01, where, for easy reading, the five periods are used to indicate different code words. As we see that for the same source symbol string, the final subinterval obtained by using arithmetic coding is [0.1058175, 0.1058250). It is noted that a decimal in binary number system, 0.000110111111111, which is of 15 bits, is equal to the decimal in decimal number system, 0.1058211962, which falls into the final subinterval representing the string $s_1 s_2 s_3 s_4 s_5 s_6$. This indicates that, for this example, arithmetic coding is more efficient than Huffamn coding.

Another way is to form a 6th extension of the source alphabet as discussed in Section 5.1.4: treat each group of six source symbols as a new source symbol; calculate its occurrence probability by multiplying the related six probabilities; then apply the Huffman coding algorithm to the 6th extension of the discrete memoryless source. This is called the 6th extension of Huffman block code (refer to Section 5.1.2.2). In other words, to encode the source string $s_1 s_2 s_3 s_4 s_5 s_6$, (the 6th extension of) Huffman coding encodes all of the $6^6 = 46656$ code words in the 6th extension of the source alphabet. This implies a high complexity in implementation and a large codebook. It is therefore not efficient.

Note that we use the decimal fraction in this section. In binary arithmetic coding, we use the binary fraction. In [langdon 1984] both binary source and code alphabets are used in binary arithmetic coding.

Similar to the case of Huffman coding, arithmetic coding is also applicable to r-ary encoding with $r > 2$.

5.4.3 Implementation Issues

As mentioned, the final subinterval resulting from arithmetic encoding of a source symbol stream becomes smaller and smaller as the length of the source symbol string increases. That is, the lower and upper bounds of the final subinterval become closer and closer. This causes a growing precision problem. It is this problem that prohibited arithmetic coding from practical usage for a long period. This problem has been resolved and the finite precision arithmetic is now used in arithmetic coding. This advance is due to the incremental implementation of arithmetic coding.

5.4.3.1 Incremental Implementation

Recall in Example 5.12 as source symbols come in one by one, the lower and upper ends of the final subinterval get closer and closer. In Figure 5.3, these lower and upper ends in Example 5.12 are listed. We observe that after the third symbol, s_3, is encoded, the resultant subinterval is [0.102, 0.108). That is, the two most significant decimal digits are the same and they remain the same in the encoding process. Hence, we can transmit these two digits without affecting the final code string. After the fourth symbol s_4 is encoded, the resultant subinterval is [0.1056, 0.1059). That is, one more digit, 5, can be transmitted. Or we say the cumulative output is now 0.105. After the sixth symbol is encoded, the final subinterval is [0.1058175, 0.1058250). The cumulative output is 0.1058. Refer to Table 5.11. This important observation reveals that we are able to incrementally transmit output (the code symbols) and receive input (the source symbols that need to be encoded).

TABLE 5.11

Final Subintervals and Cumulative Output in Example 5.12

Source Symbol	Final Subinterval		Cumulative Output
	Lower End	Upper End	
s_1	0	0.3	—
s_2	0.09	0.12	—
s_3	0.102	0.108	0.10
s_4	0.1056	0.1059	0.105
s_5	0.105795	0.105825	0.105
s_6	0.1058175	0.1058250	0.1058

5.4.3.2 Finite Precision

With the incremental manner of transmission of encoded digits and reception of input source symbols, it is possible to use finite precision to represent the lower and upper bounds of the resultant subinterval, which gets closer and closer as the length of the source symbol string becomes long.

Instead of floating-point math, integer math is used. The potential problems namely the underflow and overflow, however, need to be carefully monitored and controlled [bell 1990].

5.4.3.3 Other Issues

There are some other problems that need to be handled in implementation of binary arithmetic coding. Two of them are listed below [langdon 1981].

5.4.3.3.1 Eliminating Multiplication

The multiplication in the recursive division of subintervals is expensive in hardware as well as software. It can be avoided in binary arithmetic coding so as to simplify the implementation of binary arithmetic coding. The idea is to approximate the lower end of the interval by the closest binary fraction 2^{-Q}, where Q is an integer. Consequently, the multiplication by 2^{-Q} becomes a right shift by Q bits. A simpler approximation to eliminate multiplication is used in the Skew Coder [langdon 1982] and the Q-Coder [pennebaker 1988].

5.4.3.3.2 Carry-Over Problem

Carry-over takes place in the addition required in the recursion updating the lower end of the resultant subintervals. A carry may propagate over q bits. If the q is larger than the number of bits in the fixed-length (FL) register utilized in finite precision arithmetic, the carry-over problem occurs. To block the carry-over problem, a technique known as bit stuffing is used, in which an additional buffer register is utilized.

For detailed discussion on the various issues involved, readers are referred to [langdon 1981, 1982, 1984; pennebaker 1988, 1992]. Some computer programs of arithmetic coding in C language can be found in [bell 1990; nelson 1996].

5.4.4 History

The idea of encoding by using cumulative probability in some ordering, and decoding by comparison of magnitude of binary fraction was introduced in Shannon's celebrated paper [shannon 1948]. The recursive implementation of arithmetic coding was devised by Elias. This unpublished result was first introduced by Abramson as a note in his book on information theory and coding [abramson 1963]. The result was further developed by Jelinek in his book on information theory [jelinek 1968]. The growing precision problem prevented arithmetic coding from practical usage, however. The proposal of using finite precision arithmetic was made independently by Pasco [pasco 1976] and Rissanen [rissanen 1976]. Practical arithmetic coding was developed by several independent groups [rissanen 1979; rubin 1979; guazzo 1980]. A well-known tutorial paper on arithmetic coding appeared in [langdon 1984]. The tremendous efforts made in IBM lead to a new form of adaptive binary arithmetic coding known as the Q-coder [pennebaker 1988]. On the basis of the Q-coder, the activities of the international still image coding standards JPEG and JBIG combined the best features of the various existing arithmetic coders and developed the binary arithmetic coding procedure known as the QM-coder [pennebaker 1992].

5.4.5 Applications

Arithmetic coding is becoming popular. Note that in text and bi-level image applications there are only two source symbols (black and while), and the occurrence probability is skewed. Therefore, binary arithmetic coding achieves high coding efficiency. It has been successfully applied to bi-level image coding [langdon 1981] and adopted by the international standards for bi-level image compression JBIG. It has also been adopted by the international still image coding standard JPEG. More in this regard is covered in the next chapter when we introduce JBIG.

5.5 Summary

So far in this chapter, not much has been explicitly discussed regarding the term variable-length codes (VLC). It is known that if source symbols in a source alphabet are equally probable, i.e., their occurrence probabilities are the same, then fixed-length codes (FLC) such as the natural binary code are a reasonable choice. When the occurrence probabilities are, however, unequal, VLCs should be used to achieve high coding efficiency. This is one of the restrictions on the minimum redundancy codes imposed by Huffman. That is, the length of the code word assigned to a probable source symbol should not be larger than that associated with a less probable source symbol. If the occurrence probabilities happen to be the integral powers of $1/2$, then choosing the code word length equal to $-\log_2 p(s_i)$ for a source symbol s_i having the occurrence probability $p(s_i)$ results in minimum redundancy coding. In fact, the average length of the code thus generated is equal to the source entropy.

Huffman devised a systematic procedure to encode a source alphabet consisting of finitely many source symbols, each having an occurrence probability. It is based on some restrictions imposed on the optimum, instantaneous codes. By assigning code words with variable lengths according to variable probabilities of source symbols, Huffman coding results in minimum redundancy codes, or optimum codes for short. These have found wide applications in image and video coding and have been adopted in the international still image coding standard JPEG and video coding standards H.261, H.263, MPEG 1, 2.

When some source symbols have small probabilities and their number is large, the size of the codebook of Huffman codes will require a large memory space. The modified Huffman (MH) coding technique employs a special symbol to lump all the symbols with small probabilities together. As a result, it can reduce the codebook memory space drastically while retaining almost the same coding efficiency as that achieved by the conventional Huffman coding technique.

On the one hand, Huffman coding is optimum as a block code for a fixed-source alphabet. On the other hand, compared with the source entropy (the lower bound of the average code word length) it is not efficient when the probabilities of a source alphabet are skewed with the maximum probability being large. This is caused by the restriction that Huffman coding can only assign an integral number of bits to each code word.

Another limitation of Huffman coding is that it has to enumerate and encode all the possible groups of n source symbols in the nth extension Huffman code, even though there may be only one such group that needs to be encoded.

Arithmetic coding can overcome the limitations of Huffman coding because it is stream oriented rather than block-oriented. It translates a stream of source symbols into a stream of code symbols. It can work in an incremental manner. That is, the source symbols are fed into the coding system one by one and the code symbols are output continually. In this stream-oriented way, arithmetic coding is more efficient. It can approach the lower coding bounds set by the noiseless source coding theorem for various sources.

The recursive subinterval division (equivalently, the double recursion: the lower end recursion and width recursion) is the heart of arithmetic coding. Several measures have been taken in the implementation of arithmetic coding. They include the incremental manner, finite precision, and the elimination of multiplication. As a result of its merits, binary arithmetic coding has been adopted by the international bi-level image coding standard JBIG and still image coding standard JPEG. It is becoming an increasingly important coding technique.

Exercises

1. What does the noiseless source coding theorem state (using your own words)? Under what condition does the average code length approach the source entropy? Comment on the method suggested by the noiseless source coding theorem.
2. What characterizes a block code? Consider another definition of block code in [blahut 1986]: a block code breaks the input data stream into blocks of fixed length (FL) n and encodes each block into a code word of FL m. Are these two definitions (the one above and the one in Section 5.1, which comes from [abramson 1963]) essentially the same? Explain.
3. Is a uniquely decodable code necessarily a prefix condition code?
4. For text encoding, there are only two source symbols for black and white. It is said that Huffman coding is not efficient in this application. But it is known as the optimum code. Is there a contradiction? Explain.
5. A set of source symbols and their occurrence probabilities is listed in Table 5.12. Apply the Huffman coding algorithm to encode the alphabet.
6. Find the Huffman code for the source alphabet shown in Example 5.10.
7. Consider a source alphabet $S = \{s_i, i=1,2,\ldots,32\}$ with $p(s_1)=1/4$, $p(s_i)=3/124$, $i=2,3,\ldots,32$. Determine the source entropy, the average length of Huffman code if applied to the source alphabet. Then apply the MH coding algorithm. Calculate the average length of the MH code. Compare the codebook memory required by Huffman code and the MH code.
8. A source alphabet consists of the following four source symbols: $s_1, s_2, s_3,$ and s_4 with their occurrence probability equal to 0.25, 0.375, 0.125, and 0.25, respectively. Applying arithmetic coding as shown in Example 5.12 to the source symbol string $s_2 s_1 s_3 s_4$, determine the lower end of the final subinterval.

TABLE 5.12

Source Alphabet in Problem 5

Source Symbol	Occurrence Probability	Code Word Assigned
s_1	0.20	
s_2	0.18	
s_3	0.10	
s_4	0.10	
s_5	0.10	
s_6	0.06	
s_7	0.06	
s_8	0.04	
s_9	0.04	
s_{10}	0.04	
s_{11}	0.04	
s_{12}	0.04	

9. For the above problem, show step-by-step how we can decode the original source string from the lower end of the final subinterval.
10. In Problem 8, find the code word of the symbol string $s_2 s_1 s_3 s_4$ by using the 4th extension of Huffman code. Compare the two methods, arithmetic coding and Huffman coding.
11. Discuss how modern arithmetic coding overcame the growing precision problem.

References

[abramson 1963] N. Abramson, *Information Theory and Coding*, McGraw-Hill, New York, 1963.

[bell 1990] T.C. Bell, J.G. Cleary, and I.H. Witten, *Text Compression*, Prentice-Hall, Englewood Cliffs, NJ, 1990.

[blahut 1986] R.E. Blahut, *Principles and Practice of Information Theory*, Addison-Wesley, Reading, MA, 1986.

[fano 1949] R.M. Fano, The transmission of information, *Technical Report* 65, Research Laboratory of Electronics, MIT, Cambridge, MA, 1949.

[gallagher 1978] R.G. Gallagher, Variations on a theme by Huffman, *IEEE Transactions on Information Theory*, IT-24, 6, 668–674, November 1978.

[guazzo 1980] M. Guazzo, A general minimum-redundacy source-coding algorithm, *IEEE Transactions on Information Theory*, IT-26, 1, 15–25, January 1980.

[hankamer 1979] M. Hankamer, A modified Huffman procedure with reduced memory requirement, *IEEE Transactions on Communications*, COM-27, 6, 930–932, June 1979.

[huffman 1952] D.A. Huffman, A method for the construction of minimum-redundancy codes, *Proceedings of the IRE*, 40, 1098–1101, September 1952.

[jelinek 1968] F. Jelinek, *Probabilistic Information Theory*, McGraw-Hill, New York, 1968.

[langdon 1981] G.G. Langdon, Jr. and J. Rissanen, Compression of black-white images with arithmetic coding, *IEEE Transactions on Communications*, COM-29, 6, 858–867, June 1981.

[langdon 1982] G.G. Langdon, Jr. and J. Rissanen, A simple general binary source code, *IEEE Transactions on Information Theory*, IT-28, 800 (1982).

[langdon 1984] G.G. Langdon, Jr., An introduction to arithmetic coding, *IBM Journal of Research and Development*, 28, 2, 135–149, March 1984.

[nelson 1996] M. Nelson and J. Gailly, *The Data Compression Book*, 2nd edn., M&T Books, New York, 1996.

[pasco 1976] R. Pasco, Source coding algorithms for fast data compression, Ph.D. dissertation, Stanford University, Palo Alto, CA, 1976.

[pennebaker 1988] W.B. Pennebaker, J.L. Mitchell, G.G. Langdon, Jr., and R.B. Arps, An overview of the basic principles of the Q-coder adaptive binary arithmetic Coder, *IBM Journal of Research and Development*, 32, 6, 717–726, November 1988.

[pennebaker 1992] W.B. Pennebaker and J.L. Mitchell, *JPEG: Still Image Data Compression Standard*, Van Nostrand Reinhold, New York, 1992.

[rissanen 1976] J.J. Rissanen, Generalized Kraft inequality and arithmetic coding, *IBM Journal of Research and Development*, 20, 198–203, May 1976.

[rissanen 1979] J.J. Rissanen and G.G. Landon, Arithmetic coding, *IBM Journal of Research and Development*, 23, 2, 149–162, March 1979.

[rubin 1979] F. Rubin, Arithmetic stream coding using fixed precision registers, *IEEE Transactions on Information Theory*, IT-25, 6, 672–675, November 1979.

[sayood 1996] K. Sayood, *Introduction to Data Compression*, Morgan Kaufmann Publishers, San Francisco, CA, 1996.

[shannon 1948] C.E. Shannon, A mathematical theory of communication, *Bell System Technical Journal*, 27, 379–423 (Part I), July 1948, pp. 623–656 (Part II), October 1948.

[weaver 1978] C.S. Weaver, Digital ECG data compression, *Digital Encoding of Electrocardiograms*, H.K. Wolf (Ed.), Springer-Verlag, Berlin, New York, 1979.

6

Run-Length and Dictionary Coding: Information Theory Results (III)

As mentioned at the beginning of Chapter 5, we study some code word assignment (encoding) techniques in Chapters 5 and 6. In this chapter, we focus on run-length coding (RLC) and dictionary-based coding techniques. We first introduce Markov models as a type of dependent source model in contrast to memoryless source model discussed in Chapter 5. Based on Markov model, RLC is suitable for facsimile encoding. Its principle and application to facsimile encoding are discussed, followed by an introduction to dictionary-based coding, which is quite different from Huffman and arithmetic coding techniques discussed in Chapter 5. Two types of adaptive dictionary coding techniques, the LZ77 and LZ78 algorithms, are presented. Finally, a summary of and a performance comparison between international standard algorithms for lossless still image coding are presented.

Since the Markov source model, RLC and dictionary-based coding are the core of this chapter we consider this chapter as a third part of information theory results presented in the book. It is noted that, however, the emphasis is placed on their applications to image and video compression.

6.1 Markov Source Model

In Chapter 5, we discussed the discrete memoryless source model, in which source symbols are assumed to be independent of each other. In other words, the source has zero memory, i.e., the previous status does not affect the present one at all. In reality, however, many sources are dependent in nature. Namely, the source has memory in the sense that the previous status has an influence on the present status. For instance, as mentioned in Chapter 1, there is an interpixel correlation in digital images. That is, pixels in a digital image are not independent of each other. As discussed in this chapter, there is some dependence between characters in text. For instance, the letter u often follows the letter q in English. Therefore, it is necessary to introduce models that can reflect this type of dependence. A Markov source model is often in this regard.

6.1.1 Discrete Markov Source

Here, as in Chapter 5, we denote a source alphabet by $S = \{s_1, s_2, \ldots, s_m\}$, the occurrence probability by p. An lth-order Markov source is characterized by the following equation of conditional probabilities:

$$p(s_j|s_{i1},s_{i2},\ldots,s_{il},\ldots) = p(s_j|s_{i1},s_{i2},\ldots,s_{il}), \quad (6.1)$$

where $j, i1, i2, \ldots, il, \ldots \in \{1, 2, \ldots, m\}$, i.e., the symbols $s_j, s_{i1}, s_{i2}, \ldots, s_{il}, \ldots$ are chosen from the source alphabet S. This equation states that the source symbols are not independent of each other. The occurrence probability of a source symbol is determined by some of its previous symbols. Specifically, the probability of s_j given its history being $s_{i1}, s_{i2}, \ldots, s_{il}, \ldots$ (also called the transition probability) is determined completely by the immediately previous l symbols $s_{i1}, \ldots, s_{il}$. That is, the knowledge of the entire sequence of previous symbols is equivalent to that of the l symbols immediately preceding the current symbol s_j.

An lth-order Markov source can be described by what is called a state diagram. A state is a sequence of $(s_{i1}, s_{i2}, \ldots, s_{il})$ with $i1, i2, \ldots, il \in \{1, 2, \ldots, m\}$. That is, any group of l symbols from the m symbols in the source alphabet S forms a state. When $l = 1$, it is called a first-order Markov source. The state diagrams of the first-order Markov sources, with their source alphabets having two and three symbols, are shown in Figure 6.1a and b, respectively. Obviously, an lth-order Markov source with m symbols in the source alphabet has a total of m^l different states. Therefore, we conclude that a state diagram consists of all the m^l states. In the diagram, all the transition probabilities together with appropriate arrows are used to indicate the state transitions.

The source entropy at a state $(s_{i1}, s_{i2}, \ldots, s_{il})$ is defined as

$$H(S|s_{i1},s_{i2},\ldots,s_{il}) = -\sum_{j=1}^{m} p(s_j|s_{i1},s_{i2},\ldots,s_{il}) \log_2 p(s_j|s_{i1},s_{i2},\ldots,s_{il}), \quad (6.2)$$

The source entropy is defined as the statistical average of the entropy at all the states. That is

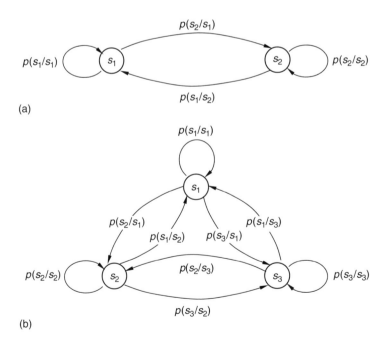

FIGURE 6.1
State diagrams of the first-order Markov sources with their source alphabets having (a) two symbols and (b) three symbols.

$$H(S) = \sum_{(s_{i1},s_{i2},\ldots,s_{il})\in S^l} p(s_{i1}, s_{i2}, \ldots, s_{il})H(S|s_{i1}, s_{i2}, \ldots, s_{il}), \quad (6.3)$$

where, as defined in Chapter 5, S^l denotes the lth extension of the source alphabet S. That is, the summation is carried out with respect to all l-tuples taking over the S^l. Extensions of a Markov source are defined below.

6.1.2 Extensions of a Discrete Markov Source

An extension of a Markov source can be defined in a similar way to that of an extension of a meomoryless source in Chapter 5. The definition of extensions of a Markov source and the relation between the entropy of the original Markov source and the entropy of the nth extension of the Markov source are presented below without derivation. For the derivation, readers are referred to [abramson 1963].

6.1.2.1 Definition

Consider an lth-order Markov source $S = \{s_1, s_2, \ldots, s_m\}$ and a set of conditional probabilities $p(s_j|s_{i1}, s_{i2}, \ldots, s_{il})$, where $j, i1, i2, \ldots, il \in \{1, 2, \ldots, m\}$. Similar to the memoryless source discussed in Chapter 5, if n symbols are grouped into a block, then there are a total of m^n blocks. Each block can be viewed as a new source symbol. Hence, these m^n blocks form a new information source alphabet, called the nth extension of the source S, and denoted by S^n. The nth extension of the lth-order Markov source is a kth-order Markov source, where k is the smallest integer greater than or equal to the ratio between l and n. That is,

$$k = \left\lceil \frac{l}{n} \right\rceil, \quad (6.4)$$

where the notation $\lceil a \rceil$ represents the operation of taking the smallest integer greater than or equal to the quantity a.

6.1.2.2 Entropy

Denote the entropy of the lth-order Markov source S by $H(S)$, and the entropy of the nth extension of the lth-order Markov source, S^n, by $H(S^n)$, respectively. The relation between the two entropies can be shown as

$$H(S^n) = nH(S) \quad (6.5)$$

6.1.3 Autoregressive Model

The Markov source discussed earlier represents a kind of dependence between source symbols in terms of the transition probability. Concretely, in determining the transition probability of a present source symbol given all the previous symbols, only the set of finitely many immediately preceding symbols matters. The autoregressive (AR) model is another kind of dependent source model that has been used often in image coding. It is defined as

$$s_j = \sum_{k=1}^{l} a_k s_{ik} + x_j, \quad (6.6)$$

where
  s_j represents the currently observed source symbol, while s_{ik} with $k = 1, 2, \ldots, l$ denote the l preceding observed symbols
  a_k's represent coefficients
  x_j represents the current input to the model

If $l = 1$, the model defined in Equation 6.6 is referred to as the first-order AR model. Clearly, in this case, the current source symbol is a linear function of its preceding symbol.

6.2 Run-Length Coding

The term "run" is used to indicate the repetition of a symbol, while the term "run-length" is used to represent the number of repeated symbols, in other words, the number of consecutive symbols of the same value. Instead of encoding the consecutive symbols, it is obvious that encoding the run-length and the value that these consecutive symbols commonly share may be more efficient. According to an excellent early review on binary image compression [arps 1979], RLC has been in use since the earliest days of information theory [shannon 1949; laemmel 1951].

From the discussion of the Joint Photographic (image) Experts Group coding (JPEG) in Chapter 4 (with more detail in Chapter 7), it is seen that most of the DCT coefficients within a block of 8×8 are zero after certain manipulations. The DCT coefficients are zigzag scanned. The nonzero DCT coefficients and their address in the 8×8 block need to be encoded and transmitted to the receiver side. There, the nonzero DCT values are referred to as labels. The position information about the nonzero DCT coefficients is represented by the run-length of zeros between the nonzero DCT coefficients in the zigzag scan. The labels and the run-length of zeros are then Huffman coded.

Many documents such as letters, forms, and drawings can be transmitted using facsimile machines over the general switched telephone network (GSTN). In digital facsimile techniques, these documents are quantized into binary levels: black and white. The resolution of these binary tone images is usually very high. In each scan line, there are many consecutive white and black pixels, i.e., many alternate white runs and black runs. Therefore, it is not surprising to see that RLC has proven to be efficient in binary document transmission. RLC has been adopted in the international standards for facsimile coding: the CCITT Recommendations T.4 and T.6.

RLC using only the horizontal correlation between pixels on the same scan line is referred to as 1-D RLC. It is noted that the first-order Markov source model with two symbols in the source alphabet depicted in Figure 6.1a can be used to characterize 1-D RLC. To achieve higher coding efficiency, 2-D RLC utilizes both horizontal and vertical correlation between pixels. Both 1-D and 2-D RLC algorithms are introduced below.

6.2.1 1-D Run-Length Coding

In this technique, each scan line is encoded independently. Each scan line can be considered as a sequence of alternating, independent white and black runs. As an agreement between encoder and decoder, the first run in each scan line is assumed to be a white run. If the first actual pixel is black, then the run-length of the first white run is set to be zero. At the end of each scan line, there is a special code word called end-of-line (EOL). The decoder knows the end of a scan line when it encounters an EOL code word.

Denote run-length by r, which is integer-valued. All of the possible run-lengths construct a source alphabet R, which is a random variable. That is,

$$R = \{r : r \in 0, 1, 2, \ldots\}. \tag{6.7}$$

Measurements on typical binary documents have shown that the maximum compression ratio, ζ_{max}, which is defined below, is about 25% higher when the white and black runs are encoded separately [hunter 1980]. The average white run-length, $\bar{r}_W$, can be expressed as

$$\bar{r}_W = \sum_{r=0}^{m} r \cdot P_W(r), \tag{6.8}$$

where
 m is the maximum value of the run-length
 $P_W(r)$ denotes the occurrence probability of a white run with length r.

The entropy of the white runs, H_W, is

$$H_W = -\sum_{r=0}^{m} P_W(r) \log_2 P_W(r). \tag{6.9}$$

For the black runs, the average run-length $\bar{r}_B$ and the entropy H_B can be defined similarly. The maximum theoretical compression factor ζ_{max} is

$$\zeta_{max} = \frac{\bar{r}_W + \bar{r}_B}{H_W + H_B}. \tag{6.10}$$

Huffman coding is then applied to two source alphabets. According to CCITT Recommendation T.4, A4 size (210 × 297 mm) documents should be accepted by facsimile machines. In each scan line, there are 1728 pixels. This means that the maximum run-length for both white and black runs is 1728, i.e., $m = 1728$. Two source alphabets of such a large size imply the requirement of two large codebooks, hence the requirement of large storage space. Therefore, some modification was made, resulting in the modified Huffman (MH) code.

In the MH code, if the run-length is larger than 63, then the run-length is represented as

$$r = M \times 64 + T \quad \text{as } r > 63, \tag{6.11}$$

where M takes integer values from 1 to 27 and $M \times 64$ is referred to as the makeup run-length; T takes integer values from 0 to 63, and is called the terminating run-length. That is, if $r \leq 63$, the run-length is represented by a terminating code word only. Otherwise, if $r > 63$, the run-length is represented by a makeup code word and a terminating code word. A portion of the MH code table [hunter 1980] is shown in Table 6.1. In this way, the requirement of large storage space is alleviated. The idea is similar to that behind MH coding, discussed in Chapter 5.

6.2.2 2-D Run-Length Coding

The 1-D RLC discussed above only utilizes correlation between pixels within a scan line. In order to utilize correlation between pixels in neighboring scan lines and to achieve higher coding efficiency, 2-D RLC was developed. In Recommendation T.4, the modified relative

TABLE 6.1

Modified Huffman Code Table

Run-Length	White Runs	Black Runs
	Terminating code words	
0	00110101	0000110111
1	000111	010
2	0111	11
3	1000	10
4	1011	011
5	1100	0011
6	1110	0010
7	1111	00011
8	10011	000101
⋮	⋮	⋮
60	01001011	000000101100
61	00110010	000001011010
62	00110011	000001100110
63	00110100	000001100111
	Make-up code words	
64	11011	0000001111
128	10010	000011001000
192	010111	000011001001
256	0110111	000001011011
⋮	⋮	⋮
1536	010011001	0000001011010
1600	010011010	0000001011011
1664	011000	0000001100100
1728	010011011	0000001100101
EOL	000000000001	000000000001

Source: From Hunter, R. and Robinson, A.H., *Proc. IEEE*, 68, 7, 854–867, 1980. With permission.

element address designate (READ) code, also known as the modified READ code or simply the MR code, is adopted.

The modified READ code operates in a line-by-line manner. In Figure 6.2 two lines are shown. The top line is called the reference line, which has been coded, while the bottom line is referred to as the coding line, which is being coded. There are a group of five changing pixels, a_0, a_1, a_2, b_1, b_2, in the two lines. Their relative positions decide which of the three coding modes is used. The starting changing pixel a_0 (hence, five changing points) moves from left to right and from top to bottom as 2-D RLC proceeds. The five changing pixels and the three coding modes are defined below.

6.2.2.1 Five Changing Pixels

By a changing pixel, we mean the first pixel encountered in white or black runs when we scan an image line-by-line, from left to right, and from top to bottom. The five changing pixels are defined below.

> a_0: The reference changing pixel in the coding line. Its position is defined in the previous coding mode, whose meaning will be explained shortly. At the beginning of a coding line, a_0 is an imaginary white changing pixel located before the first actual pixel in the coding line.
> a_1: The next changing pixel in the coding line. Because of the above-mentioned left-to-right and top-to-bottom scanning order, it is at the right-hand side of a_0. Since it is a changing pixel, it has an opposite "color" to that of a_0.

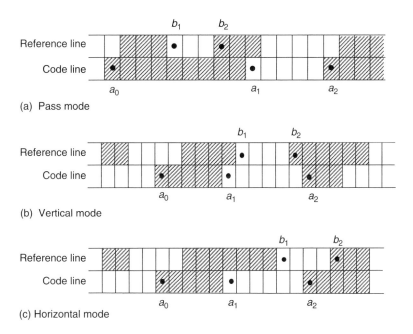

FIGURE 6.2
2-D run-length coding.

a_2: The next changing pixel after a_1 in the coding line. It is to the right of a_1 and has the same color as that of a_0.
b_1: The changing pixel in the reference line that is closest to a_0 from the right and has the same color as a_1.
b_2: The next changing pixel in the reference line after b_1.

6.2.2.2 Three Coding Modes

6.2.2.2.1 Pass Coding Mode

If the changing pixel b_2 is located to the left of the changing pixel a_1, it means that the run in the reference line starting from b_1 is not adjacent to the run in the coding line starting from a_1. Note that these two runs have the same color. This is called pass coding mode. A special code word, "0001," is sent out from transmitter. The receiver then knows that the run starting from a_0 in the coding line does not end at the pixel below b_2. This pixel (below b_2 in the coding line) is identified as the reference changing pixel a_0 of the new set of five changing pixels for the next coding mode.

6.2.2.2.2 Vertical Coding Mode

If the relative distance along the horizontal direction between the changing pixels a_1 and b_1 is not larger than three pixels, the coding is conducted in vertical coding mode. That is, the position of a_1 is coded with reference to the position of b_1. Seven different code words are assigned to seven different cases: the distance between a_1 and b_1 equals 0, ±1, ±2, ±3, where + means a_1 is to the right of b_1, while − means a_1 is to the left of b_1. The a_1 then becomes the reference changing pixel a_0 of the new set of five changing pixels for the next coding mode.

6.2.2.2.3 Horizontal Coding Mode

If the relative distance between the changing pixels a_1 and b_1 is larger than three pixels, the coding is conducted in horizontal coding mode. Here, 1-D RLC is applied. Specifically,

TABLE 6.2

2-D Run-Length Coding Table, $|x_iy_j|$: Distance between x_i and y_j, $x_iy_j > 0$: x_i is Right to y_j, $x_iy_j < 0$: x_i is Left to y_j. (x_iy_j): Code Word of the Run Denoted by x_iy_j Taken from the Modified Huffman Code

Mode	Conditions	Output Code Word	Position of New a_0
Pass coding mode	$b_2a_1 < 0$	0001	Under b_2 in Coding Line
Vertical coding mode	$a_1b_1 = 0$	1	a_1
	$a_1b_1 = 1$	011	
	$a_1b_1 = 2$	000011	
	$a_1b_1 = 3$	0000011	
	$a_1b_1 = -1$	010	
	$a_1b_1 = -2$	000010	
	$a_1b_1 = -3$	0000010	
Horizontal coding mode	$|a_1b_1| > 3$	001 + $(a_0a_1) + (a_1a_2)$	a_2

Source: From Hunter, R. and Robinson, A.H., *Proc. IEEE*, 68, 7, 854–867, 1980.

the transmitter sends out a code word consisting the following three parts: a flag "001"; a 1-D RLC word for the run from a_0 to a_1; a 1-D RLC word for the run from a_1 to a_2. The a_2 then becomes the reference changing pixel a_0 of the new set of five changing pixels for the next coding mode.

Table 6.2 contains three coding modes and the corresponding output code words. There, (a_0a_1) and (a_1a_2) represent 1-D run-length code words of run-length a_0a_1 and a_1a_2, respectively.

6.2.3 Effect of Transmission Error and Uncompressed Mode

In this section, effect of transmission error in the 1-D and 2-D RLC cases and uncompressed mode is discussed.

6.2.3.1 Error Effect in the 1-D RLC Case

As introduced above, the special code word EOL is used to indicate the end of each scan line. With the EOL, 1-D RLC encodes each scan line independently. If a transmission error occurs in a scan line, there are two possibilities that the effect caused by the error is limited within the scan line. One possibility is that resynchronization is established after a few runs. One example is shown in Figure 6.3. There the transmission error takes place in the

FIGURE 6.3
Establishment of resynchronization after a few runs.

second run from the left. Resynchronization is established in the fifth run in this example. Another possibility lies in the EOL, which forces resynchronization.

In summary, it is seen that the 1-D RLC will not propagate transmission error between scan lines. In other words, a transmission error will be restricted within a scan line. Although error detection and retransmission of data through an automatic repeat request (ARQ) system are supposed to be able to effectively handle the error susceptibility issue effectively, the ARQ technique was not included into Recommendation T.4 due to the computational complexity and extra transmission time required.

Once the number of decoded pixels between two consecutive EOL code words is not equal to 1728 (for an A4 size document), an error has been identified. Some error concealment techniques can be used to reconstruct the scan line [hunter 1980]. For instance, we can repeat the previous line, or replace the damaged line by a white line, or use a correlation technique to recover the line as much as possible.

6.2.3.2 Error Effect in the 2-D RLC Case

From the above discussion, we realize that 2-D RLC is more efficient than 1-D RLC on the one hand. On the other hand 2-D RLC is more susceptible to transmission errors than the 1-D RLC. To prevent error propagation, there is a parameter used in 2-D RLC, known as the K-factor, which specifies the number of scan lines that are 2-D RLC coded.

Recommendation T.4 defined that no more than $K-1$ consecutive scan lines be 2-D RLC coded after a 1-D RLC coded line. For binary documents scanned at normal resolution, $K=2$. For documents scanned at high resolution, $K=4$.

According to Arps [arps 1979], there are two different types of algorithms in binary image coding: raster algorithms and area algorithms. Raster algorithms operate only on data within one or two raster scan lines. They are hence mainly 1-D in nature. Area algorithms are truly 2-D in nature. They require that all, or a substantial portion, of the image is in random access memory. From our discussion above, we see that both 1-D and 2-D RLC defined in T.4 belong to the category of raster algorithms. Area algorithms require large memory space and are susceptible to transmission noise.

6.2.3.3 Uncompressed Mode

For some detailed binary document images, both 1-D and 2-D RLC may result in data expansion instead of data compression. Under these circumstances the number of coding bits is larger than the number of bilevel pixels. An uncompressed mode is created as an alternative way to avoid data expansion. Special code words are assigned for the uncompressed mode.

For the performances of 1-D and 2-D RLC applied to eight CCITT test document images, and issues such as fill bits and minimum scan line time (MSLT), to only name a few, readers are referred to [hunter 1980].

6.3 Digital Facsimile Coding Standards

Facsimile transmission, an important means of communication in modern society, is often used as an example to demonstrate the mutual interaction between widely used applications and standardization activities. Active facsimile applications and the market brought on the necessity for international standardization to facilitate interoperability between facsimile machines worldwide. Successful international standardization, in turn,

TABLE 6.3

Facsimile Coding Standards

Group of Facsimile Apparatuses	Speed Requirement for A4 Size Document	Analog or Digital Scheme	CCITT Recommendation	Compression Technique		
				Model	Basic Coder	Algorithm Acronym
G_1	6 min	Analog	T.2	—	—	—
G_2	3 min	Analog	T.3	—	—	—
G_3	1 min	Digital	T.4	1-D RLC	Modified Huffman	MH
				2-D RLC (optional)		MR
G_4	1 min	Digital	T.6	2-D RLC	Modified Huffman	MMR

has stimulated wider use of facsimile transmission and, hence, a more demanding market. Facsimile has also been considered as a major application for binary image compression.

So far facsimile machines are classified in four different groups. Facsimile apparatuses in groups 1 and 2 use analog techniques. They can transmit an A4 size (210 × 297 mm) document scanned at 3.85 lines/mm in 6 and 3 minutes, respectively, over the GSTN. International standards for these two groups of facsimile apparatuses are CCITT (now ITU) Recommendations T.2 and T.3, respectively. Group 3 facsimile machines use digital techniques and hence achieve high coding efficiency. They can transmit the A4 size binary document scanned at a resolution of 3.85 lines/mm and sampled at 1728 pixels/line in about 1 minute at a rate of 4800 bits/s over the GSTN. The corresponding international standard is CCITT Recommendation T.4. Group 4 facsimile apparatuses have the same transmission speed requirement as that for group 3 machines, but the coding technique is different. Specifically, the coding technique used for group 4 machines is based on 2-D RLC, discussed above, but modified to achieve higher coding efficiency. Hence it is referred to as the modified modified READ (MMR) coding. The corresponding standard is CCITT Recommendation T.6. Table 6.3 summarizes the above descriptions.

6.4 Dictionary Coding

Dictionary coding, the focus of this section, is different from Huffman and arithmetic coding techniques, discussed in Chapter 5. Both Huffman and arithmetic coding techniques are based on a statistical model, and the occurrence probabilities play a particularly important role. Recall that in the Huffman coding the shorter code words are assigned to more frequently occurring source symbols. In dictionary-based data compression techniques a symbol or a string of symbols generated from a source alphabet is represented by an index to a dictionary constructed from the source alphabet. A dictionary is a list of symbols and strings of symbols. There are many examples of this in our daily lives. For instance, the string "September" is sometimes represented by an index "9," while a social security number represents a person in the United States.

Dictionary coding is widely used in text coding. Consider English text coding. The source alphabet includes 26 English letters in both upper and lower cases, numbers, various punctuation marks, and the space bar. Huffman or arithmetic coding treats each symbol based on its occurrence probability. That is, the source is modeled as a memoryless source. It is well known, however, that this is not true in many applications. In text coding, structure or context plays a significant role. As mentioned earlier, it is very likely that the letter u appears after the letter q. Similarly, it is likely that the word "concerned" will appear after "As far as the weather is." The strategy of the dictionary coding is to build a dictionary that contains frequently occurring symbols and string of symbols. When a symbol or a string is encountered and it is contained in the dictionary, it is encoded with an index to the dictionary. Otherwise, if not in the dictionary, the symbol or the string of symbols is encoded in a less efficient manner.

6.4.1 Formulation of Dictionary Coding

To facilitate further discussion, we define dictionary coding in a precise manner [bell 1990]. We denote a source alphabet by S. A dictionary consisting of two elements is defined as $D = (P, C)$, where P is a finite set of phrases generated from the S, and C is a coding function mapping P onto a set of code words.

The set P is said to be complete if any input string can be represented by a series of phrases chosen from the P. The coding function C is said to obey the prefix property if there is no code word that is a prefix of any other code word. For practical usage, i.e., for reversible compression of any input text, the phrase set P must be complete and the coding function C must satisfy the prefix property.

6.4.2 Categorization of Dictionary-Based Coding Techniques

The heart of dictionary coding is the formulation of the dictionary. A successfully built dictionary results in data compression; the opposite case may lead to data expansion. According to the ways in which dictionaries are constructed, dictionary coding techniques can be classified as static or adaptive.

6.4.2.1 Static Dictionary Coding

In some particular applications, the knowledge about the source alphabet and the related strings of symbols, also known as phrases, is sufficient for a fixed dictionary to be produced before the coding process. The dictionary is used at both the transmitting and the receiving ends. This is referred to as static dictionary coding. The merit of the static approach is its simplicity. Its drawback lies on its relatively lower coding efficiency and less flexibility compared with adaptive dictionary techniques. By less flexibility, we mean that a dictionary built for a specific application is not normally suitable for utilization in other applications.

An example of static algorithms occurs is digram coding. In this simple and fast coding technique, the dictionary contains all source symbols and some frequently used pairs of symbols. In encoding, two symbols are checked at once to see if they are in the dictionary. If so, they are replaced by the index of the two symbols in the dictionary, and the next pair of symbols is encoded in the next step. If not, then the index of the first symbol is used to encode the first symbol. The second symbol is combined with the third symbol to form a new pair, which is encoded in the next step.

The digram can be straightforwardly extended to *n-gram*. In the extension, the size of the dictionary increases and so does its coding efficiency.

6.4.2.2 Adaptive Dictionary Coding

As opposed to the static approach, with the adaptive approach a completely defined dictionary does not exist before the encoding process and the dictionary is not fixed.

At the beginning of coding, only an initial dictionary exists. It adapts itself to the input during the coding process. All the adaptive dictionary coding algorithms can be traced back to two different original works by Ziv and Lempel [ziv 1977, 1978]. The algorithms based on [ziv 1977] are referred to as the LZ77 algorithms, while those based on [ziv 1978] are referred to as the LZ78 algorithms. Before introducing the two landmark works, we will discuss the parsing strategy.

6.4.3 Parsing Strategy

Once we have a dictionary, we need to examine the input text and find a string of symbols that matches an item in the dictionary. Then the index of the item to the dictionary is encoded. This process of segmenting the input text into disjoint strings (whose union equals the input text) for coding is referred to as parsing. Obviously, the way to segment the input text into strings is not unique.

In terms of the highest coding efficiency, optimal parsing is essentially a shortest-path problem [bell 1990]. In practice, however, a method called greedy parsing is used most often. In fact, it is used in all the LZ77 and LZ78 algorithms [nelson 1995]. With greedy parsing, the encoder searches for the longest string of symbols in the input that matches an item in the dictionary at each coding step. Greedy parsing may not be optimal, but it is simple in implementation.

Example 6.1

Consider a dictionary, D, whose phrase set is $P = \{a, b, ab, ba, bb, aab, bbb\}$. The code words assigned to these strings are $C(a) = 10$, $C(b) = 011$, $C(ab) = 010$, $C(ba) = 0101$, $C(bb) = 01$, $C(aab) = 11$, and $C(bbb) = 0110$. Now the input text is *abbaab*.

Using greedy parsing, we then encode the text as $C(ab).C(ba).C(ab)$, which is a 10-bit string: 010.0101.010. In the above representations, the periods are used to indicate the division of segments in the parsing. This, however, is not an optimum solution. Obviously, the following parsing will be more efficient, i.e., $C(a).C(bb).C(aab)$, which is a 6-bit string: 10.01.11.

6.4.4 Sliding Window (LZ77) Algorithms

As mentioned earlier, LZ77 algorithms are a group of adaptive dictionary coding algorithms rooted in the pioneering work in [ziv 1977]. Since they are adaptive, there is no complete and fixed dictionary before coding. Instead, the dictionary changes as the input text changes.

6.4.4.1 Introduction

In the LZ77 algorithms [bell 1990; nelson 1995], the dictionary used is actually a portion of the input text, which has been recently encoded. The text that needs to be encoded is compared with the strings of symbols in the dictionary. The longest matched string in the dictionary is characterized by a pointer (sometimes called a token), which is represented by a triple of data items. Note that this triple functions as an index to the dictionary, as mentioned earlier. In this way, a variable-length string of symbols is mapped to a fixed-length (FL) pointer.

There is a sliding window in the LZ77 algorithms. The window consists of two parts: a search buffer and a look-ahead buffer. The search buffer contains the portion of the text stream that has recently been encoded which, as mentioned, is the dictionary; while the look-ahead buffer contains the text to be encoded next. The window slides through the input text stream from beginning to end during the entire encoding process.

This explains the term sliding window. The size of the search buffer is much larger than that of the look-ahead buffer. This is expected because what is contained in the search buffer is in fact the adaptive dictionary. The sliding window is usually on the order of a few thousand symbols, whereas the look-ahead buffer is on the order of several tens to one hundred symbols.

6.4.4.2 Encoding and Decoding

Below we present more detail about the sliding window dictionary coding technique, i.e., the LZ77 approach, via a simple illustrative example.

Example 6.2

Figure 6.4 shows a sliding window. The input text stream is

$$ikaccbadaccbaccbaccgikmoabcc$$

In Figure 6.4a, a search buffer of nine symbols and a look-ahead buffer of six symbols are shown. All the symbols in the search buffer, *accbadacc*, have just been encoded. All the symbols in the look-ahead buffer, *baccba*, are to be encoded. (It is understood that the symbols before the search buffer have been encoded and the symbols after the look-ahead buffer are to be encoded.) The strings of symbols, *ik* and *ccgikmoabcc*, are not covered by the sliding window at the moment.

At the moment, or in other words, in the first step of encoding, the symbol (symbols) to be encoded begins (begin) with the symbol *b*. The pointer starts searching for the symbol *b* from the last symbol in the search buffer, *c*, which is immediately to the left of the first symbol *b* in the look-ahead buffer. It finds a match at the sixth position from *b*. It further determines that the longest string of the match is *ba*. That is, the maximum matching length is two. The pointer is then represented by a triple, $<i, j, k>$. The first item, i, represents the distance between the first symbol in the look-ahead buffer and the position of the pointer (the position of the first symbol of the matched string). This distance is called offset. In this step, the offset is six. The second item in the triple, j, indicates the length of the matched string. Here, the length of the matched string *ba* is two. The third item, k, is the code word assigned to the symbol immediately following the matched string in the look-ahead buffer. In this step, the third item is C(c), where C is used to represent a function to map symbol(s)

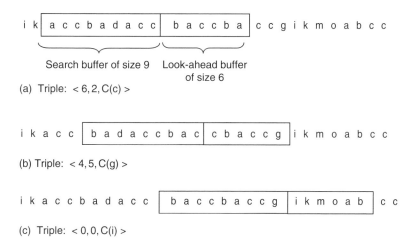

FIGURE 6.4
An encoding example using LZ77.

to a code word, as defined in Section 6.4.1. That is, the resulting triple after the first step is $<6, 2, C(c)>$.

The reason to include the third item k into the triple is as follows. In the case where there is no match in the search buffer, both i and j will be zero. The third item at this moment is the code word of the first symbol in the look-ahead buffer itself. This means that even in the case where we cannot find a match string, the sliding window still works. In the third step of the encoding process described below, we will see that the resulting triple is $<0, 0, C(i)>$. The decoder hence understands that there is no matching, and the single symbol i is decoded.

The second step of the encoding is illustrated in Figure 6.4b. The sliding window has been shifted to the right by three positions. The first symbol to be encoded now is c, which is the leftmost symbol in the look-ahead buffer. The search pointer moves towards the left from the symbol c. It first finds a match in the first position with a length of one. It then finds another match in the fourth position from the first symbol in the look-ahead buffer. Interestingly, the maximum matching can exceed the boundary between the search and the look-ahead buffers and can enter the look-ahead buffer. Why this is possible will be explained shortly, when we discuss the decoding process. In this manner, it is found that the maximum length of matching is five. The last match is found at the fifth position. The length of the matched string is, however, only one. As greedy parsing is used, the match with a length five is chosen. That is, the offset is four and the maximum match length is five. Consequently, the triple resulting from the second step is $<4, 5, C(g)>$.

The sliding window is then shifted to the right by six positions. The third step of the encoding is depicted in Figure 6.4c. Obviously, there is no matching of i in the search buffer. The resulting triple is hence $<0, 0, C(i)>$.

The encoding process can continue in this way. The possible cases we may encounter in the encoding, however, are described in the above-mentioned three steps. Hence, we end our discussion of the encoding process and start discussing the decoding process. Compared with the encoding, the decoding is simpler because there is no need for matching, which involves many comparisons between the symbols in the look-ahead buffer and the symbols in the search buffer. The decoding process is illustrated in Figure 6.5.

In the above-mentioned three steps, the resulting triples are: $<6, 2, C(c)>$, $<4, 5, C(g)>$, and $<0, 0, C(i)>$. Now let us see how the decoder works. That is, how the decoder recovers the string *baccbaccgi* from these three triples.

In Figure 6.5a, the search buffer is the same as that in Figure 6.4a. That is, the string *accbadacc* stored in the search window is what was just decoded.

Once the first triple $<6, 2, C(c)>$ is received, the decoder will move the decoding pointer from the first position in the look-ahead buffer to the left by six positions. That is, the pointer will point to the symbol b. The decoder then copies the two symbols starting from b, i.e., *ba*, into the look-ahead buffer. The symbol c will be copied right to *ba*. This is shown in Figure 6.5b. The window is then shifted to the right by three positions, as shown in Figure 6.5c.

After the second triple $<4, 5, C(g)>$ is received, the decoder move the decoding pointer from the first position of the look-ahead buffer to the left by four positions. The pointer points the symbol c. The decoder then copies five successive symbols starting from the symbol c pointed by the pointer. We see that at the beginning of this copying process there are only four symbols available for copying. Once the first symbol is copied, however, all five symbols are available. After copying, the symbol g is added to the end of the five copied symbols in the look-ahead buffer. The results are shown in Figure 6.5c. Figure 6.5d then shows the window shifting to the right by six positions.

After receiving the triple $<0, 0, C(i)>$, the decoder knows that there is no matching and a single symbol i is encoded. Hence, the decoder adds the symbol i following the symbol g. This is shown in Figure 6.5f.

FIGURE 6.5
A decoding example using LZ77.

In Figure 6.5, for each part, the last encoded symbol c before receiving the three triples is shaded. From Figure 6.5f, we see that the string added after the symbol c due to the three triples is *baccbaccgi*. This agrees with the sequence mentioned at the beginning of our discussion about the decoding process. We thus conclude that the decoding process has correctly decoded the encoded sequence from the last encoded symbol and the received triples.

6.4.4.3 Summary of the LZ77 Approach

The sliding window consists of two parts: the search buffer and the look-ahead buffer. The most recently encoded portion of the input text stream is contained in the search buffer, while the portion of the text that needs to be encoded immediately is in the look-ahead buffer. The first symbol in the look-ahead buffer, located to the right of the boundary between the two buffers, is the symbol or the beginning of a string of symbols to be encoded at the moment. Let us call it the symbol s. The size of the search buffer is usually much larger than that of the look-ahead buffer.

In encoding, the search pointer moves to the left, away from the symbol s, to find a match of the symbol s in the search buffer. Once a match is found, the encoding process will further determine the length of the matched string. When there are multiple matches, the match that produces the longest matched string is chosen. The match is denoted by a triple $<i, j, k>$. The first item in the triple, i, is the offset, which is the distance between the pointer pointing to the symbol giving the maximum match and the symbol s. The second item, j, is the length of the matched string. The third item, k, is the code word of the symbol following the matched string in the look-ahead buffer. The sliding window is then shifted to the right by $j + 1$ position, before the next coding step takes place.

When there is no matching in the search buffer, the triple is represented by $< 0, 0, C(s) >$, where C(s) is the code word assigned to the symbol s. The sliding window is then shifted to the right by one position.

The sliding window is shifted along the input text stream during the encoding process. The symbol s moves from the beginning symbol to the ending symbol of the input text stream.

At the very beginning, the content of the search buffer can be arbitrarily selected. For instance, the symbols in the search buffer may all be the space symbol.

Let us denote the size of the search buffer by SB, the size of the look-ahead buffer by L, and the size of the source alphabet by A. Assume that the natural binary code (NBC) is used. Then we see that the LZ77 approach encodes variable-length strings of symbols with fixed-length code words. Specifically, the offset i is of coding length $\lceil \log_2 (SB) \rceil$, the length of matched string j is of coding length $\lceil \log_2 (SB + L) \rceil$, and the code word k is of coding length $\lceil \log_2 (A) \rceil$, where the sign $\lceil a \rceil$ denotes the smallest integer larger than a.

The length of the matched string is equal to $\lceil \log_2 (SB + L) \rceil$ because the search for the maximum matching can enter into the look-ahead buffer as shown in Example 6.2.

The decoding process is simpler than the encoding process since there is no comparison involved in the decoding.

The most recently encoded symbols in the search buffer serve as the dictionary used in the LZ77 approach. The merit of doing so is that the dictionary is well adapted to the input text. The limitation of the approach is that if the distance between the repeated patterns in the input text stream is larger than the size of the search buffer, then the approach cannot utilize the structure to compress the text. A vivid example can be found in [sayood 1996].

A window with a moderate size, say, $SB + L \leq 8192$, can compress a variety of texts well. Several reasons have been analyzed in [bell 1990].

Many variations have been made to improve coding efficiency of the LZ77 approach. The LZ77 produces a triple in each encoding step; i.e., the offset (position of the matched string), the length of the matched string, and the code word of the symbol following the matched string. The transmission of the third item in each coding step is not efficient. This is true especially at the beginning of coding. A variant of the LZ77, referred to as the LZSS algorithm [bell 1986], improves this inefficiency.

6.4.5 LZ78 Algorithms

6.4.5.1 Introduction

As mentioned earlier, the LZ77 algorithms use a sliding window of fixed size, and both the search buffer and the look-ahead buffer have a fixed size. This means that if the distance between two repeated patterns is larger than the size of the search buffer, the LZ77 algorithms cannot work efficiently. The fixed size of the both buffers implies that the matched string cannot be longer than the sum of the sizes of the two buffers, meaning another limitation on coding efficiency. Increasing the sizes of the search buffer and the look-ahead buffer will seemingly resolve the problems. A close look, however, reveals that it also leads to increases in the number of bits required to encode the offset and matched string length as well as an increase in processing complexity.

The LZ78 algorithms [ziv 1978; bell 1990; nelson 1995] eliminate the use of the sliding window. Instead these algorithms use the encoded text as a dictionary which, potentially, does not have a fixed size. Each time a pointer (token) is issued, the encoded string is included in the dictionary. Theoretically the LZ78 algorithms reach optimal performance as the encoded text stream approaches infinity. In practice, however, as mentioned above with respect to the LZ77, a very large dictionary will affect coding efficiency negatively. Therefore, once a preset limit to the dictionary size has been reached, either the dictionary is fixed for the future (if the coding efficiency is good), or it is reset to zero, i.e., it must be restarted.

Instead of the triples used in the LZ77, only pairs are used in the LZ78. Specifically, only the position of the pointer to the matched string and the symbol following the matched string need to be encoded. The length of the matched string does not need to be encoded because both the encoder and the decoder have exactly the same dictionary, i.e., the decoder knows the length of the matched string.

6.4.5.2 Encoding and Decoding

Like the discussion of the LZ77 algorithms, we will go though an example to describe the LZ78 algorithms.

Example 6.3

Consider the text stream: *baccbaccacbcabccbbacc*. Table 6.4 shows the coding process. We see that for the first three symbols there is no match between the individual input symbols and the entries in the dictionary. Therefore, the doubles are $<0, C(b)>$, $<0, C(a)>$, and $<0, C(c)>$, respectively, where 0 means no match, and C(b), C(a), and C(c) represent the code words of *b*, *a*, and *c*, respectively. After symbols b, a, c, comes *c*, which finds a match in the dictionary (the third entry). Therefore, the next symbol *b* is combined to be considered. Since the string *cb* did not appear before, it is encoded as a double and it is appended as a new entry into the dictionary. The first item in the double is the index of the matched entry *c*, 3, the second item is the index/code word of the symbol following the match *b*, 1. That is, the double is $<3, 1>$. The following input symbol is *a*, which appeared in the dictionary. Hence the next symbol *c* is taken into consideration. Since the string *ac* is not an entry of the dictionary, it is encoded with a double. The first item in the double is the index of symbol *a*, 2, the second item is the index of symbol *c*, 3, i.e., $<2, 3>$. The encoding proceeds in this way. In Table 6.4, as the encoding proceeds the entries in the dictionary become longer and longer. First, entries with single symbols come out, later more and more entries with two symbols show up. After that more and more entries with three symbols appear. This means that coding efficiency is increasing.

Now consider the decoding process. Since the decoder knows the rule applied in the encoding, it can reconstruct the dictionary and decode the input text stream from the received doubles. When the first double $<0, C(b)>$ is received, the decoder knows that there is no match. Hence, the first entry in the dictionary is *b*. So is the first decoded symbol. From the second double $<0, C(a)>$, symbol *a* is known as the second entry in the

TABLE 6.4

An Encoding Example Using the LZ78 Algorithm

Index	Doubles	Encoded Symbols
1	$<0, C(b)>$	b
2	$<0, C(a)>$	a
3	$<0, C(c)>$	c
4	$<3, 1>$	cb
5	$<2, 3>$	ac
6	$<3, 2>$	ca
7	$<4, 3>$	cbc
8	$<2, 1>$	ab
9	$<3, 3>$	cc
10	$<1, 1>$	bb
11	$<5, 3>$	acc

dictionary as well as the second decoded symbol. Similarly, the next entry in the dictionary and the next decoded symbol are known as *c*. When the following double $< 3, 1 >$ is received. The decoder knows from two items, 3 and 1, that the next two symbols are the third and the first entries in the dictionary. This indicates that the symbols *c* and *b* are decoded, and the string *cb* becomes the fourth entry in the dictionary.

We omit the next two doubles and take a look at the double $< 4, 3 >$, which is associated with index 7 in Table 6.4. Since the first item in the double is 4, it means that the maximum matched string is *cb*, which is associated with index 4 in Table 6.4. The second item in the double, 3, implies that the symbol following the match is the third entry *c*. Therefore, the decoder decodes a string *cbc*. Also the string *cbc* becomes the seventh entry in the reconstructed dictionary. In this way, the decoder can reconstruct the exact same dictionary as that established by the encoder and decode the input text stream from the received doubles.

6.4.5.3 LZW Algorithm

Both the LZ77 and LZ78 approaches, when published in 1977 and 1978, respectively, were theory oriented. The effective and practical improvement over the LZ78 in [welch 1984] brought much attention to the LZ dictionary coding techniques. The resulting algorithm is referred to as the LZW algorithm [bell 1990; nelson 1995]. It removed the second item in the double (the index of the symbol following the longest matched string) and hence, it enhanced coding efficiency. In other words, the LZW only sends the indexes of the dictionary to the decoder. For the purpose, the LZW first forms an initial dictionary, which consists of all the individual source symbols contained in the source alphabet. Then, the encoder examines the input symbol. Since the input symbol matches to an entry in the dictionary, its succeeding symbol is cascaded to form a string. The cascaded string does not find a match in the initial dictionary. Hence the index of the matched symbol is encoded and the enlarged string (the matched symbol followed by the cascaded symbol) is listed as a new entry in the dictionary. The encoding process continues in this manner.

For the encoding and decoding processes, let us go through an example to see how the LZW algorithm can encode only the indexes and the decoder can still decode the input text string.

Example 6.4

Consider the following input text stream: *accbadaccbaccbacc*. We see that the source alphabet is $S = \{a,b,c,d,\}$. The top portion of Table 6.5 (with indexes 1, 2, 3, 4) gives a possible initial dictionary used in the LZW. When the first symbol *a* is input, the encoder finds that it has a match in the dictionary. Therefore, the next symbol *c* is taken to form a string *ac*. As the string *ac* is not in the dictionary, it is listed as a new entry in the dictionary and is given an index, 5. The index of the matched symbol *a*, 1, is encoded. When the second symbol, *c*, is input, the encoder takes the following symbol *c* into consideration because there is a match to the second input symbol *c* in the dictionary. Since the string *cc* does not match any existing entry, it becomes a new entry in the dictionary with an index, 6. The index of the matched symbol (the second input symbol), *c*, is encoded. Now consider the third input symbol *c*, which appeared in the dictionary. Hence, the following symbol *b* is cascaded to form a string *cb*. Since the string *cb* is not in the dictionary, it becomes a new entry in the dictionary and is given an index, 7. The index of matched symbol *c*, 3, is encoded. The process proceeds in this fashion. Take a look at entry 11 in the dictionary shown in Table 6.5. The input symbol at this point is *a*. Since it has a match in the previous entries, its next symbol *c* is considered. Since the string *ac* appeared in entry 5, the succeeding symbol *c* is combined. Now the new enlarged string becomes *acc* and it does not have a match in

TABLE 6.5

An Example of the Dictionary Coding Using the LZW Algorithm

Index	Entry	Input Symbols	Encoded Index
1	a	} Initial dictionary	
2	b		
3	c		
4	d		
5	ac	a	1
6	cc	c	3
7	cb	c	3
8	ba	b	2
9	ad	a	1
10	da	d	4
11	acc	a, c	5
12	cba	c, b	7
13	accb	a, c, c	11
14	bac	b, a	8
15	cc…	c, c, …	

the previous entries. It is thus added to the dictionary. And a new index, 11, is given to the string *acc*. The index of the matched string *ac*, 5, is encoded and transmitted. The final sequence of encoded indexes is 1, 3, 3, 2, 1, 4, 5, 7, 11, 8. Like the LZ78, the entries in the dictionary become longer and longer in the LZW algorithm. This implies high coding efficiency since long strings can be represented by indexes.

Now let us take a look at the decoding process to see how the decoder can decode the input text stream from the received index. Initially, the decoder has the same dictionary (the top four rows in Table 6.5) as that in the encoder. Once the first index 1 comes, the decoder decodes a symbol *a*. The second index is 3, which indicates that the next symbol is *c*. From the rule applied in encoding, the decoder knows further that a new entry *ac* has been added to the dictionary with an index 5. The next index is 3. It is known that the next symbol is also *c*. It is also known that the string *cc* has been added into the dictionary as the sixth entry. In this way, the decoder reconstructs the dictionary and decodes the input text stream.

6.4.5.4 Summary

The LZW algorithm, as a representative of the LZ78 approach, is summarized below.

The initial dictionary contains the indexes for all the individual source symbols. At the beginning of encoding, when a symbol is input, since it has a match in the initial dictionary, the next symbol is cascaded to form a two-symbol string. Since the two-symbol string cannot find a match in the initial dictionary, the index of the first symbol is encoded and transmitted, and the two-symbol string is added to the dictionary with a new, incremented index. The next encoding step starts with the second symbol among the two symbols.

In the middle, the encoding process starts with the last symbol of the latest added dictionary entry. Since it has a match in the previous entries, its succeeding symbol is cascaded after the symbol to form a string. If this string as appeared before in the dictionary (i.e., the string finds a match), the next symbol is cascaded as well. This process continues until such an enlarged string cannot find a match in the dictionary. At this moment, the index of the last matched string (the longest match) is encoded and transmitted, and the enlarged and unmatched string is added into the dictionary as a new entry with a new, incremented index.

Decoding is a process of transforming the index string back to the corresponding symbol string. To do so, however, the dictionary must be reconstructed exactly the same as that established in the encoding process. That is, the initial dictionary is constructed first in the same way as that in the encoding. When decoding the index string, the decoder reconstructs the same dictionary as that in the encoder according to the rule used in the encoding.

Specifically, at the beginning of the decoding, after receiving an index, a corresponding single symbol can be decoded. Through the next received index, another symbol can be decoded. From the rule used in the encoding, the decoder knows that the two symbols should be cascaded to form a new entry added into the dictionary with an incremented index. The next step in the decoding will start from the second symbol among the two symbols.

Now consider the middle of the decoding process. The presently received index is used to decode a corresponding string of input symbols according to the reconstructed dictionary at the moment. (Note that this string is said to be with the present index.) It is known from the encoding rule that the symbols in the string associated with the next index should be considered. (Note that this string is said to be with the next index). That is, the first symbol in the string with the next index should be appended to the last symbol in the string with the present index. The resultant combination, i.e., the string with the present index followed by the first symbol in the string with the next index, cannot find a match to an entry in the dictionary. Therefore, the combination should be added to the dictionary with an incremented index. At this moment, the next index becomes the new present index, and the index following the next index becomes the new next index. The decoding process then proceeds in the same fashion in a new decoding step.

Compared with the LZ78 algorithm, the LZW algorithm eliminates the necessity of having the second item in the double, an index/code word of the symbol following a matched string. That is, the encoder only needs to encode and transmit the first item in the double. This greatly enhances the coding efficiency and reduces the complexity of the LZ algorithm.

6.4.5.5 Applications

The CCITT Recommendation V.42 bis is a data compression standard used in modems that connect computers with remote users via the GSTN. In the compressed mode, the LZW algorithm is recommended for data compression.

In image compression, the LZW finds its application as well. Specifically, it is utilized in the graphic interchange format (GIF) that was created to encode graphical images. GIF is now also used to encode natural images, though it is not very efficient in this regard. For more information, readers are referred to [sayood 1996]. The LZW algorithm is also used in the Unix Compress command.

6.5 International Standards for Lossless Still Image Compression

In Chapter 5, we studied Huffman and arithmetic coding techniques. We also briefly discussed the international standard for bilevel image compression, known as the JBIG. In this chapter, so far we have discussed another two coding techniques: the RLC and dictionary coding techniques. We have also introduced the international standards for facsimile compression, in which the techniques known as the MH, MR, and MMR were recommended. All of these techniques involve lossless compression. In Chapter 7, the international still image coding standard JPEG will be introduced. As we will see, the JPEG has four different modes. They can be divided into two compression categories: lossy and lossless. Hence, we can discuss about the lossless JPEG. Before leaving this chapter, however, we will briefly discuss, compare, and summarize various techniques used in the international standards for lossless still image compression. For more detail, readers are referred to an excellent survey paper [arps 1994].

6.5.1 Lossless Bilevel Still Image Compression

6.5.1.1 *Algorithms*

As mentioned earlier, there are four different international standard algorithms falling into this category.

MH (modified Huffman coding): This algorithm defined in CCITT Recommendation T.4 for facsimile coding uses the 1-D RLC technique followed by the MH coding technique.

MR (modified READ (relative element address designate) coding): This algorithm defined in CCITT Recommendation T.4 for facsimile coding uses the 2-D RLC technique followed by the MH coding technique.

MMR (modified modified READ coding): This algorithm defined in CCITT Recommendation T.6 is based on MR, but is modified to maximize compression.

JBIG (Joint Bilevel Image experts Group coding): This algorithm defined in CCITT Recommendation T.82 uses an adaptive 2-D coding model, followed by an adaptive arithmetic coding technique.

6.5.1.2 *Performance Comparison*

The JBIG test image set was used to compare the performance of the above-mentioned algorithms. The set contains scanned business documents with different densities, graphic images, digital halftones, and mixed (document and halftone) images.

Note that digital halftones, also named (digital) halftone images, are generated by using only binary devices. Some small black units are imposed on a white background. The units may assume different shapes: circle, square, and so on. The denser the black units in a spot of an image, the darker the spot appears. The digital halftoning method has been used for printing gray-level images in newspapers and books. Digital halftoning through character overstriking, used to generate digital images in the early days for the experimental work associated with courses on digital image processing, is described in [gonzalez 1992].

The following two observations on the performance comparison were made after the application of several techniques to the JBIG test image set.

For bilevel images excluding digital halftones, the compression ratio achieved by these techniques ranges from 3 to 100. The compression ratio increases monotonically in the order of the following standard algorithms: MH, MR, MMR, and JBIG.

For digital halftones, MH, MR, and MMR result in data expansion, while JBIG achieves compression ratios in the range of 5–20. This demonstrates that among the techniques, JBIG is the only one suitable for the compression of digital halftones.

6.5.2 Lossless Multilevel Still Image Compression

6.5.2.1 *Algorithms*

There are two international standards for multilevel still image compression:

JBIG (Joint Bilevel Image experts Group coding): Defined in CCITT Recommendation T. 82 uses an adaptive arithmetic coding technique. To encode multilevel images, the JBIG decomposes multilevel images into bit-planes, then compresses these bit-planes using its bilevel image compression technique. To further enhance compression ratio, it uses Gary coding to represent pixel amplitudes instead of weighted binary coding.

JPEG (Joint Photographic (image) Experts Group coding): Defined in CCITT Recommendation T. 81. For lossless coding, it uses the differential coding technique. The predictive error is encoded using either Huffman coding or adaptive arithmetic coding techniques.

6.5.2.2 Performance Comparison

A set of color test images from the JPEG standards committee was used for performance comparison. The luminance component (Y) is of resolution 720×576 pixels, while the chrominance components (U and V) are of 360×576 pixels. The compression ratios calculated are the combined results for all the three components. The following observations have been reported:

When quantized in 8 bits/pixel, the compression ratios vary much less for multilevel images than for bilevel images, and are roughly equal to 2.

When quantized with 5 bits/pixel down to 2 bits/pixel, compared with the lossless JPEG, the JBIG achieves an increasingly higher compression ratio, up to a maximum of 29%.

When quantized with 6 bits/pixel, JBIG and lossless JPEG achieve similar compression ratios.

When quantized with 7–8 bits/pixel, the lossless JPEG achieves a 2.4%–2.6% higher compression ratio than JBIG.

6.6 Summary

Both Huffman coding and arithmetic coding, discussed in Chapter 5, are referred to as variable-length coding techniques, because the lengths of code words assigned to different entries in a source alphabet are different. In general, a code word of a shorter length is assigned to an entry with higher occurrence probabilities. They are also classified as fixed-length-to-variable-length coding techniques [arps 1979], since the entries in a source alphabet have the same fixed length. Run-length coding (RLC) and dictionary coding, which are the focus of this chapter, are opposite and are referred to as variable-length-to-fixed-length coding techniques. This is because the runs in the RLC and the string in the dictionary coding are variable and are encoded with code words of the same fixed length.

Based on RLC, the international standard algorithms for facsimile coding, MH, MR, and MMR have worked successfully except for dealing with digital halftones. That is, these algorithms result in data expansion when applied to digital halftones. The JBIG, based on an adaptive arithmetic coding technique not only achieves a higher coding efficiency than MH, MR, and MMR for facsimile coding, but also compresses the digital halftones effectively.

Note that 1-D RLC utilizes the correlation between pixels within a scan line, whereas 2-D RLC utilizes the correlation between pixels within a few scan lines. As a result, 2-D RLC can obtain higher coding efficiency than 1-D RLC on the one hand. On the other hand, 2-D RLC is more susceptible to transmission errors than 1-D RLC.

In text compression, the dictionary-based techniques have proven to be efficient. All the adaptive dictionary-based algorithms can be classified into two groups. One is based on a pioneering work by Ziv and Lempel in 1977, and another is based on their pioneering work in 1978. They are called the LZ77 and LZ78 algorithms, respectively. With the LZ77 algorithms, a fixed size window slides through the input text stream. The sliding window consists of two parts: the search buffer and the look-ahead buffer. The search buffer contains the most recently encoded portion of the input text, while the look-ahead buffer contains the portion of the input text to be encoded immediately. For the symbols to be encoded, the LZ77 algorithms search for the longest match in the search buffer. The information about the match: the distance between the matched string in the search buffer and that in the look-ahead buffer, the length of the matched string, and the code word of the symbol following the matched string in the look-ahead buffer are encoded. Many improvements have been made in the LZ77 algorithms.

The performance of the LZ77 algorithms is limited by the sizes of the search buffer and the look-ahead buffer. With a finite size for the search buffer, the LZ77 algorithms will not work well in the case where repeated patterns are apart from each other by a distance longer than the size of the search buffer. With a finite size for the sliding window, the LZ77 algorithms will not work well in the case where matching strings are longer than the window. In order to be efficient, however, these sizes cannot be very large.

In order to overcome the problem, the LZ78 algorithms work in a different way. They do not use the sliding window at all. Instead of using the most recently encoded portion of the input text as a dictionary, the LZ78 algorithms use the index of the longest matched string as an entry of the dictionary. That is, each matched string cascaded with its immediate next symbol is compared with the existing entries of the dictionary. If this combination (a new string) does not find a match in the dictionary constructed at the moment, the combination will be included as an entry in the dictionary. Otherwise, the next symbol in the input text will be appended to the combination and the enlarged new combination will be checked with the dictionary. The process continues until the new combination cannot find a match in the dictionary. Among various variants of the LZ78 algorithms, the LZW algorithm is perhaps the most important one. It only needs to encode the indexes of the longest matched strings to the dictionary. It can be shown that the decoder can decode the input text stream from the given index stream. In doing so, the same dictionary as that established in the encoder needs to be reconstructed at the decoder, and this can be implemented since the same rule used in the encoding is known in the decoder.

The size of the dictionary cannot be infinitely large because, as mentioned above, the coding efficiency will not be high. The common practice of the LZ78 algorithms is to keep the dictionary fixed once a certain size has been reached and the performance of the encoding is satisfactory. Otherwise, the dictionary will be set to empty and will be reconstructed from scratch.

Considering the fact that there are several international standards concerning still image coding (for both bilevel and multilevel images), a brief summary and a performance comparison are presented at the end of this chapter. At the beginning of this chapter, a description of the discrete Markov source and its nth extensions are provided. The Markov source and the auto regressive (AR) model serve as important models for the dependent information sources.

Exercises

1. Draw the state diagram of a second-order Markov source with two symbols in the source alphabet. That is, $S = \{s_1, s_2\}$. It is assumed that the conditional probabilities are

$$p(s_1|s_1s_1) = p(s_2|s_2s_2) = 0.7,$$
$$p(s_2|s_1s_1) = p(s_1|s_2s_2) = 0.3, \text{ and}$$
$$p(s_1|s_1s_2) = p(s_1|s_2s_1) = p(s_2|s_1s_2) = p(s_2|s_2s_1) = 0.5.$$

2. What are the definitions of raster algorithm and area algorithm in binary image coding? Which category does 1-D RLC belong to? Which category does 2-D RLC belong to?
3. What effect does a transmission error have on 1-D RLC and 2-D RLC, respectively? What is the function of the code word EOL?
4. Make a convincing argument that the MH algorithm reduces the requirement of large storage space.
5. Which three different modes does 2-D RLC have? How do you view the vertical mode?
6. Using your own words, describe the encoding and decoding processes of the LZ77 algorithms. Go through Example 6.2.
7. Using your own words, describe the encoding and decoding processes of the LZW algorithm. Go through Example 6.3.
8. Read the reference paper [arps 1994], which is an excellent survey on the international standards for lossless still image compression. Pay particular attention to all the figures and Table 6.1.

References

[**abramson 1963**] N. Abramson, *Information Theory and Coding*, McGraw-Hill, New York, 1963.
[**arps 1979**] R.B. Arps, Binary image compression, in *Image Transmission Techniques*, W.K. Pratt (Ed.), Academic Press, New York, 1979.
[**arps 1994**] R.B. Arps and T.K. Truong, Comparison of international standards for lossless still image compression, *Proceedings of the IEEE*, 82, 6, 889–899, June 1994.
[**bell 1986**] T.C. Bell, Better OPM/L text compression, *IEEE Transactions on Communications*, COM-34, 1176–1182, December 1986.
[**bell 1990**] T.C. Bell, J.G. Cleary, and I.H. Witten, *Text Compression*, Prentice-Hall, Englewood Cliffs, NJ, 1990.
[**gonzalez 1992**] R.C. Gonzalez and R.E. Woods, *Digital Image Processing*, Addison Wesley, Reading, MA, 1992.
[**hunter 1980**] R. Hunter and A.H. Robinson, International digital facsimile coding standards, *Proceedings of the IEEE*, 68, 7, 854–867, 1980.
[**laemmel 1951**] A.E. Laemmel, Coding processes for bandwidth reduction in picture transmission, Rep. R-246-51, PIB-187, Microwave Research Institute, Polytechnic Institute of Brooklyn, New York.
[**nelson 1995**] M. Nelson and J.-L. Gailly, *The Data Compression Book*, 2nd edn., M&T Books, New York, 1995.
[**sayood 1996**] K. Sayood, *Introduction to Data Compression*, Morgan Kaufmann Publishers, San Francisco, CA, 1996.
[**shannon 1949**] C.E. Shannon and W. Weaver, *The Mathematical Theory of Communication*, University of Illinois Press, Urbana, IL, 1949.
[**welch 1984**] T. Welch, A technique for high-performance data compression, *IEEE Computer*, 17, 6, 8–19, June 1984.
[**ziv 1977**] J. Ziv and A. Lempel, A universal algorithm for sequential data compression, *IEEE Transactions on Information Theory*, 23, 3, 337–343, May 1977.
[**ziv 1978**] J. Ziv and A. Lempel, Compression of individual sequences via variable-rate coding, *IEEE Transactions on Information Theory*, 24, 5, 530–536, September 1978.

Part II

Still Image Compression

7
Still Image Coding: Standard JPEG

In this chapter, the JPEG standard is introduced. This standard allows for lossy and lossless encoding of still images, and four distinct modes of operation are supported: sequential DCT-based mode, progressive DCT-based mode, lossless mode, and hierarchical mode.

7.1 Introduction

Still image coding is an important application of data compression. When an analog image or picture is digitized, each pixel is represented by a fixed number of bits, which correspond to a certain number of gray levels. In this uncompressed format, the digitized image requires a large number of bits to be stored or transmitted. As a result, compression becomes necessary due to the limited communication bandwidth or storage size. Since the mid-1980s, the ITU and ISO have been working together to develop a joint international standard for the compression of still images. Officially, JPEG [jpeg 1992] is the ISO/IEC international standard 10918-1: digital compression and coding of continuous-tone still images, or the ITU-T Recommendation T.81. JPEG became an international standard in 1992. The JPEG standard allows for both lossy and lossless encoding of still images. The algorithm for lossy coding is a discrete cosine transform (DCT)-based coding scheme. This is the baseline of JPEG and is sufficient for many applications. However, to meet the needs of applications that cannot tolerate loss, e.g., compression of medical images, a lossless coding scheme is also provided and is based on a predictive coding scheme. From the algorithmic point of view, JPEG includes four distinct modes of operation: sequential DCT-based mode, progressive DCT-based mode, lossless mode, and hierarchical mode. In the following sections, an overview of these modes is provided. Further technical details can be found in the books by Pennelbaker and Symes [pennelbaker 1992, symes 1998].

In the sequential DCT-based mode, an image is first partitioned into blocks of 8×8 pixels. The blocks are processed from left to right and top to bottom. The 8×8 two-dimensional (2-D) forward DCT is applied to each block and the 8×8 DCT coefficients are quantized. Finally, the quantized DCT coefficients are entropy encoded and output as part of the compressed image data.

In the progressive DCT-based mode, the process of block partitioning and forward DCT transform is the same as in the sequential DCT-based mode. However, in the progressive mode, the quantized DCT coefficients are first stored in a buffer before the encoding is performed. The DCT coefficients in the buffer are then encoded by a multiple scanning process. In each scan, the quantized DCT coefficients are partially encoded by either spectral selection or successive approximation. In the method of spectral selection, the quantized DCT coefficients are divided into multiple spectral bands according to a zigzag (ZZ) order. In each scan, a specified band is encoded. In the method of successive

FIGURE 7.1
(a) Sequential coding and (b) progressive coding.

approximation, a specified number of most significant bits of the quantized coefficients are first encoded, followed by the least significant bits in later scans.

The difference between sequential coding and progressive coding is shown in Figure 7.1. In the sequential coding an image is encoded part-by-part according to the scanning order while in the progressive coding, the image is encoded by multiscanning process and in each scan the full image is encoded to a certain quality level.

As mentioned earlier, lossless coding is achieved by a predictive coding scheme. In this scheme, three neighboring pixels are used to predict the current pixel to be coded. The prediction difference is entropy coded using either Huffman or arithmetic coding. Because the prediction is not quantized, the coding is lossless.

Finally, in the hierarchical mode, an image is first spatially down-sampled to a multi-layered pyramid, resulting in a sequence of frames as shown in Figure 7.2. This sequence of frames is encoded by a predictive coding scheme. Except for the first frame, the predictive coding process is applied to the differential frames, i.e., the differences between the

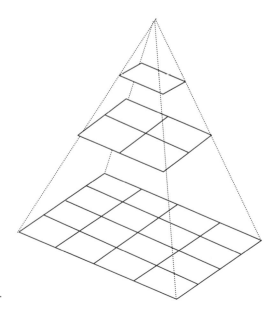

FIGURE 7.2
Hierarchical multiresolution encoding.

Still Image Coding: Standard JPEG

frame to be coded and the predictive reference frame. It is important to note that the reference frame is equivalent to the earlier frame that would be reconstructed in the decoder. The coding method for the difference frame may either use the DCT-based coding method, the lossless coding method, or the DCT-based processes with a final lossless process. Down-sampling and up-sampling filters are used in the hierarchical mode. The hierarchical coding mode provides a progressive presentation similar to progressive DCT-based mode, but is also useful in the applications that have multiresolution requirements. The hierarchical coding mode also provides the capability of progressive coding to a final lossless stage.

7.2 Sequential DCT-Based Encoding Algorithm

The sequential DCT-based coding algorithm is the baseline algorithm of the JPEG coding standard. The block diagram of encoding process is shown in Figure 7.3. As shown in Figure 7.4, the digitized image data is first partitioned into blocks of 8×8 pixels. The 2-D forward DCT is applied to each 8×8 block. The 2-D forward and inverse DCT of 8×8 block are defined as follows:

FDCT

$$S_{uv} = \frac{1}{4} C_u C_v \sum_{i=0}^{7} \sum_{j=0}^{7} s_{ij} \cos\frac{(2i+1)u\pi}{16} \cos\frac{(2j+1)v\pi}{16}$$

IDCT

$$s_{ij} = \frac{1}{4} \sum_{u=0}^{7} \sum_{v=0}^{7} C_u C_v S_{uv} \cos\frac{(2i+1)u\pi}{16} \cos\frac{(2j+1)v\pi}{16}$$

$$C_u C_v = \begin{cases} \frac{1}{\sqrt{2}} & \text{for } u, v = 0 \\ 1 & \text{otherwise} \end{cases}$$

(7.1)

where
  s_{ij} is the value of the pixel at position (i,j) in the block
  S_{uv} is the transformed (u,v) DCT coefficient

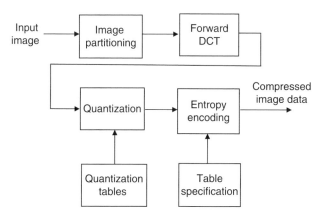

FIGURE 7.3
Block diagram of sequential discrete cosine transform (DCT)-based encoding process.

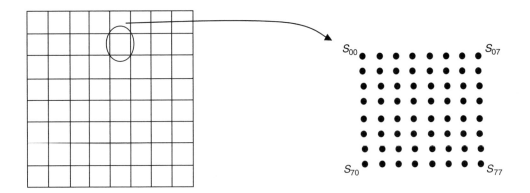

FIGURE 7.4
Partitioning to 8 × 8 blocks.

After the forward DCT, quantization of the transformed DCT coefficients is performed. Each of the 64 DCT coefficients is quantized by a uniform quantizer:

$$S_{quv} = \text{round}\left(\frac{S_{uv}}{Q_{uv}}\right) \qquad (7.2)$$

where

S_{quv} is quantized value of the DCT coefficient S_{uv}, and Q_{uv} is the quantization step obtained from the quantization table

There are four quantization tables that may be used by the encoder, but there is no default quantization table specified by the standard. Two particular quantization tables are shown in Table 7.1.

At the decoder, the dequantization is performed as follows:

$$R_{quv} = S_{quv} \times Q_{uv} \qquad (7.3)$$

where R_{quv} is the value of the dequantized DCT coefficient. After quantization, the DC coefficient, S_{q00}, is treated separately from the other 63 AC coefficients. The DC coefficients are encoded by a predictive coding scheme. The encoded value is the difference (DIFF)

TABLE 7.1

Two Examples of Quantization Tables Used by JPEG

Luminance Quantization Table							Chrominance Quantization Table						
16	11	10	16	24	40	51	61	17	18	24	47	99	99	99	99
12	12	14	19	26	58	60	55	18	21	26	66	99	99	99	99
14	13	16	24	40	57	69	56	24	26	56	99	99	99	99	99
14	17	22	29	51	87	80	62	47	66	99	99	99	99	99	99
18	22	37	56	68	109	103	77	99	99	99	99	99	99	99	99
24	35	55	64	81	104	113	92	99	99	99	99	99	99	99	99
49	64	78	87	103	121	120	101	99	99	99	99	99	99	99	99
72	92	95	98	112	100	103	99	99	99	99	99	99	99	99	99

TABLE 7.2

Huffman Coding of DC Coefficients

SSSS	Difference (DIFF) Values	Additional Bits
0	0	—
1	−1, 1	0, 1
2	−3, −2, 2, 3	00, 01, 10, 11
3	−7, ..., −4, 4, ..., 7	000, ..., 011, 100, ..., 111
4	−15, ..., −8, 8, ..., 15	0000, ..., 0111, 1000, ..., 1111
5	−31, ..., −16, 16, ..., 31	00000, ..., 01111, 10000, ..., 11111
6	−63, ..., −32, 32, ..., 63	..., ...
7	−127, ..., −64, 64, ..., 127	..., ...
8	−255, ..., −128, 128, ..., 255	..., ...
9	−511, ..., −256, 256, ..., 511	..., ...
10	−1023, ..., −512, 512, ..., 1023	..., ...
11	−2047, ..., −1024, 1024, ..., 2047	..., ...

between the quantized DC coefficient of the current block (S_{q00}) and that of the earlier block of the same component (PRED):

$$\text{DIFF} = S_{q00} - \text{PRED} \tag{7.4}$$

The value of DIFF is entropy coded with Huffman tables. More specifically, the two's complement of the possible DIFF magnitudes are grouped into 12 categories, "SSSS." The Huffman codes for these 12 difference categories and additional bits are shown in Table 7.2.

For each nonzero category, additional bits are added to the code word to uniquely identify which difference within the category actually occurred. The number of additional bits is defined by "SSSS" and the additional bits are appended to the least significant bit of the Huffman code (most significant bit first) according the following rule. If the difference value is positive, the "SSSS" low-order bits of DIFF are appended; if the difference value is negative, then the "SSSS" low-order bits of DIFF-1 are appended. As an example, the Huffman tables used for coding the luminance and chrominance DC coefficients are shown in Tables 7.3 and 7.4, respectively. These two tables have been developed from the average statistics of a large set of images with 8-bit precision.

In contrast to the coding of DC coefficients, the quantized AC coefficients are arranged to a zigzag order before being entropy coded. This scan order is shown in Figure 7.5.

TABLE 7.3

Huffman Table for Luminance DC Coefficient Differences

Category	Code Length	Code Word
0	2	00
1	3	010
2	3	011
3	3	100
4	3	101
5	3	110
6	4	1110
7	5	11110
8	6	111110
9	7	1111110
10	8	11111110
11	9	111111110

TABLE 7.4

Huffman Table for Chrominance DC Coefficient Differences

Category	Code Length	Code Word
0	2	00
1	2	01
2	2	10
3	3	110
4	4	1110
5	5	11110
6	6	111110
7	7	1111110
8	8	11111110
9	9	111111110
10	10	1111111110
11	11	11111111110

According to the zigzag scanning order, the quantized coefficients can be represented as

$$ZZ(0) = S_{q00},\ ZZ(1) = S_{q01},\ ZZ(2) = S_{q10},\ \ldots,\ ZZ(63) = S_{q77} \quad (7.5)$$

When many of the quantized AC coefficients become zero, they can be very efficiently encoded by exploiting the run length of zeros. The run length of zeros are identified by the nonzero coefficients. An 8-bit code 'RRRRSSSS' is used to represent the nonzero coefficient. The four least significant bits, 'SSSS,' define a category for the value of the next nonzero coefficient in the zigzag sequence, which ends the zero-run. The four most significant bits, 'RRRR,' define the run length of zeros in the zigzag sequence or the position of the nonzero coefficient in the zigzag sequence. The composite value, RRRRSSSS, is shown in Figure 7.6. The value 'RRRRSSSS' = '11110000' is defined as ZRL, "RRRR" = "1111" represents a run length of 16 zeros and "SSSS" = "0000" represents a zero-amplitude. Therefore, ZRL is used to represent a run length of 16 zero coefficients followed by a zero-amplitude coefficient, it is not an abbreviation. In the case of a run length of zero coefficients that exceeds 15, multiple symbols will be used. A special value 'RRRRSSSS' = '00000000' is used to code the end-of-block (EOB). An EOB occurs when the remaining coefficients in the block are zero. The entries marked N/A are undefined.

The composite value, RRRRSSSS, is then Huffman coded. SSSS is actually the number to indicate category in the Huffman code table. The coefficient values for each category are shown in Table 7.5.

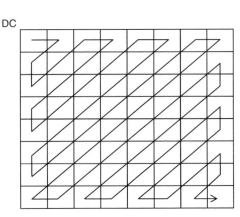

FIGURE 7.5
Zigzag scanning order of DCT coefficients.

Still Image Coding: Standard JPEG

		SSSS				
		0	1	2	9	10
RRRR	0	EOB	N/A			
	.	N/A		Composite values		
	.	N/A				
	.	N/A				
	15	ZRL				

FIGURE 7.6
Two-dimensional (2-D) value array for Huffman coding.

Each Huffman code is followed by additional bits that specify the sign and exact amplitude of the coefficients. As with the DC code tables, the AC code tables have also been developed from the average statistics of a large set of images with 8-bit precision. Each composite value is represented by a Huffman code in the AC code table. The format for the additional bits is the same as in the coding of DC coefficients. The value of SSSS gives the number of additional bits required to specify the sign and precise amplitude of the coefficient. The additional bits are either the low-order SSSS bits of $ZZ(k)$ when $ZZ(k)$ is positive or the low-order SSSS bits of $ZZ(k)-1$ when $ZZ(k)$ is negative. Here, $ZZ(k)$ is the kth coefficient in the zigzag scanning order of coefficients being coded. The Huffman tables for AC coefficients can be found in Annex K of the JPEG standard [jpeg 1992] and are not listed here due to space limitations.

As described above, Huffman coding is used as the means of entropy coding. However, an adaptive arithmetic coding procedure can also be used. As with the Huffman coding technique, the binary arithmetic coding technique is also lossless. It is possible to transcode between two systems without either the FDCT or IDCT processes. Moreover, this transcoding is a lossless process; it does not affect the picture quality of the reconstructed image. The arithmetic encoder encodes a series of binary symbols, zeros or ones, where each symbol represents the possible result of a binary decision. The binary decisions include the choice between positive and negative signs, a magnitude being zero or nonzero, or a particular bit in a sequence of binary digits being zero or one. There are four steps in the

TABLE 7.5

Huffman Coding for AC Coefficients

Category (SSSS)	AC Coefficient Range
1	$-1, 1$
2	$-3, -2, 2, 3$
3	$-7, \ldots, -4, 4, \ldots, 7$
4	$-15, \ldots, -8, 8, \ldots, 15$
5	$-31, \ldots, -16, 16, \ldots, 31$
6	$-63, \ldots, -32, 32, \ldots, 63$
7	$-127, \ldots, -64, 64, \ldots, 127$
8	$-255, \ldots, -128, 128, \ldots, 255$
9	$-511, \ldots, -256, 256, \ldots, 511$
10	$-1023, \ldots, -512, 512, \ldots, 1023$
11	$-2047, \ldots, -1024, 1024, \ldots, 2047$

arithmetic coding: initializing the statistical area, initializing the encoder, terminating the code string, and adding restart markers.

7.3 Progressive DCT-Based Encoding Algorithm

In progressive DCT-based coding, the input image is first partitioned to blocks of 8×8 pixels. The 2-D 8×8 DCT is then applied to each block. The transformed DCT-coefficient data is then encoded with multiple scans. In each scan, a portion of the transformed DCT coefficient data is encoded. This partial encoded data can be reconstructed to obtain a full image size with picture of lower quality. The coded data of each additional scan will enhance the reconstructed image quality until the full quality has been achieved at the completion of all scans. Two methods have been used in JPEG standard to perform the DCT-based progressive coding. These include spectral selection and successive approximation.

In the method of spectral selection, the transformed DCT coefficients are first reordered as zigzag sequence and then divided into several bands. A frequency band is defined in the scan header by specifying the starting and ending indices in the zigzag sequence. The band containing DC coefficient is encoded at the first scan. In the following scan, it is not necessary for the coding procedure to follow the zigzag ordering. In the method of the successive approximation, the DCT coefficients are reduced in precision by the point transform. The point transform of the DCT coefficients is an arithmetic-shift-right by a specified number of bits, or divided by a power of 2 (near zero, there is slight difference in truncation of precision between arithmetic shift and divide by 2, see annex K10 of [JPEG]). This specified number is the successive approximation of bit position. To encode using succesive approximations, the significant bits of DCT coefficient are encoded in the first scan, and each successive scan that follows progressively improves the precision of the coefficients by 1 bit. This continues until full precision is reached.

The principles of spectral selection and successive approximation are shown in Figure 7.7. For both methods, the quantized coefficients are coded with either Huffman or arithmetic codes at each scan. In spectral selection and the first scan of successive approximation for an image, the AC coefficient coding model is similar to that used by in the sequential DCT-based coding mode. However, the Huffman code tables are extended to include coding of runs of end-of-bands (EOBs). For distinguishing the end-of-band and end-of-block, a number, n, which is used to indicate the range of run length, is added to the end-of-band (EOBn). The EOBn code sequence is defined as follows. Each EOBn is followed by an extension field, which has the minimum number of bits required to specify the run length. The EOBn run structure allows efficient coding of blocks, which have only zero coefficients. For example, an EOB run of length 5 means that the current block and the next four blocks have an EOBn with no intervening nonzero coefficients. The Huffman coding structure of the subsequent scans of successive approximation for a given image is similar to the coding structure of the first scan of that image. Each nonzero quantized coefficient is described by a composite 8-bit run length–magnitude value of the form: RRRRSSSS. The four most significant bits, RRRR, indicate the number of zero coefficients between the current coefficient and the previously coded coefficient. The four least significant bits, SSSS, give the magnitude category of the nonzero coefficient. The run length–magnitude composite value is Huffman coded. Each Huffman code is followed by additional bits: 1 bit is used to code the sign of the nonzero coefficient and another one bit is used to code the correction, where 0 means no correction and 1 means add one to the

Still Image Coding: Standard JPEG

FIGURE 7.7
Progressive coding with spectral selection and successive approximation.

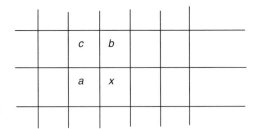

FIGURE 7.8
Spatial relation between the pixels to be coded and three decoded neighbors.

decoded magnitude of the coefficient. Although the above technique has been described using Huffman coding, it should be noted that arithmetic encoding can also be used in its place.

7.4 Lossless Coding Mode

In the lossless coding mode, the coding method is spatial-based coding instead of DCT-based coding. However, the coding method is extended from the method for coding the DC coefficients in the sequential DCT-based coding mode. Each pixel is coded with a predictive coding method, where the predicted value is obtained from one of three one-dimensional (1-D) or one of four 2-D predictors which are shown in Figure 7.8.

In Figure 7.8, the pixel to be coded is denoted by x, and the three causal neighbors are denoted by a, b, and c. The predictive value of x, Px, is obtained from three neighbors, a, b, and c in one of seven ways as listed in Table 7.6.

In Table 7.6, the selection value 0 is only used for differential coding in hierarchical coding mode. Selections 1, 2, and 3 are 1-D predictions and 4, 5, 6, and 7 are 2-D predictions. Each prediction is performed with full integer precision, and without clamping of either underflow or overflow beyond the input bounds. To achieve lossless coding, the prediction differences are coded with either Huffman coding or arithmetic coding. The prediction difference values can be from 0 to 2^{16} for 8-bit pixels. The Huffman tables developed for coding DC coefficients in the sequential DCT-based coding mode are used with one additional entry to code the prediction differences. For arithmetic coding, the statistical model defined for the DC coefficients in sequential DCT-based coding mode is

TABLE 7.6

Predictors for Lossless Coding

Selection-Value	Prediction
0	No prediction (hierarchical mode)
1	P$x = a$
2	P$x = b$
3	P$x = c$
4	P$x = a + b - c$
5	P$x = a + [(b - c)/2]$*
6	P$x = b + [(a - c)/2]$*
7	P$x = (a + b)/2$

Note: *Represents the shift right arithmetic operation.

generalized to a 2-D form in which differences are conditioned on the pixel to the left and the line above.

7.5 Hierarchical Coding Mode

The hierarchical coding mode provides a progressive coding similar to the progressive DCT-based coding mode, but it offers more functionality. This functionality addresses applications with multiresolution requirements. In hierarchical coding mode, an input image frame is first decomposed to a sequence of frames, such as the pyramid shown in Figure 7.2. Each frame is obtained through a down-sampling process, i.e., low-pass filtering followed by subsampling. The first frame (the lowest resolution) is encoded as a nondifferential frame. The following frames are encoded as differential frames, where the differential is with repect to the earlier coded frame. Note that an up-sampled version that would be reconstructed in the decoder is used. The first frame can be encoded by the methods of sequential DCT-based coding, spectral selection method of progressive coding, or lossless coding with either Huffman code or arithmetic code. However, within an image, the differential frames are either coded by the DCT-based coding method, the lossless coding method, or the DCT-based process with a final lossless coding. All frames within the image must use the same entropy coding, either Huffman or arithmetic, with the exception that nondifferential frame coded with the baseline coding may occur in the same image with frames coded with arithmetic coding methods. The differential frames are coded with the same method used for the nondifferential frame except the final frame. The final differential frame for each image may use differential lossless coding method. In the hierarchical coding mode, the resolution changes of frames may occur. These resolution changes occur if down-sampling filters are used to reduce the spatial resolution of some or all frames of an image. When the resolution of a reference frame does not match the resolution of the frame to be coded, an up-sampling filter is used to increase the resolution of reference frame. The block diagram of coding a differential frame is shown in Figure 7.9.

The up-sampling filter increases the spatial resolution by a factor of two in both horizontal and vertical directions by using bilinear interpolation of two neighboring pixels. The up-sampling with bilinear interpolation is consistent with the down-sampling filter that is used for the generation of down-sampled frames. It should be noted that the hierarchical coding mode allows one to improve the quality of the reconstructed frames at a given spatial resolution.

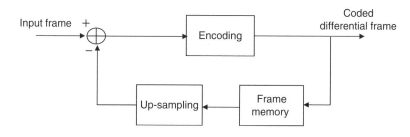

FIGURE 7.9
Coding of differential frame in hierarchical coding.

7.6 Summary

In this chapter, the still-image coding standard, JPEG, has been introduced. The JPEG coding standard includes four coding modes: sequential DCT-based coding mode, progressive DCT-based coding mode, lossless coding mode, and hierarchical coding mode. The DCT-based coding method is probably the one that most are familiar with; however, the lossless coding modes in JPEG which use a spatial domain predictive coding process have many interesting applications as well. For each coding mode, the entropy coding can be implemented with either Huffman coding or arithmetic coding. JPEG has been widely adopted for many applications.

Exercises

1. What is the difference between sequential coding and progressive coding in JPEG? Conduct a project to encode an image with sequence coding and progressive coding, respectively.
2. Use JPEG lossless mode to code several images and explain why different bit rates are obtained.
3. Generate a Huffman code table using a set of images with 8-bit precision (approximately $2 \sim 3$) using the method presented in Annex C of the JPEG specification. This set of images is called the training set. Use this table to code an image within the training set and an image, which is not in the training set, and explain the results.
4. Design a three-layer progressive JPEG coder using (a) spectral selection and (b) progressive approximation (0.3 bits/pixel at the first layer, 0.2 bits/pixel at the second layer, and 0.1 bits/pixel at the third layer).

References

[jpeg 1992] Digital compression and coding of continuous-tone still images: Requirements and Guidelines, ISO-/IEC International Standard 10918-1, CCITT T.81, September, 1992.

[pennelbaker 1992] W.B. Pennelbaker and J.L. Mitchell, *JPEG: Still Image Data Compression Standard*, Van Nostrand Reinhold, New York, September 1992.

[symes 1998] P. Symes, *Video Compression: Fundamental Compression Techniques and an Overview of the JPEG and MPEG Compression Systems*, McGraw-Hill, New York, April 1998.

8

Wavelet Transform for Image Coding: JPEG2000

Since the mid-1980s, a number of signal processing applications have emerged using wavelet theory. Among those applications, the most widespread developments have occurred in the area of data compression. Wavelet techniques have demonstrated the ability to provide not only high coding efficiency but also spatial and quality scalability features. In this chapter, we focus on the utility of the wavelet transform for image data compression applications.

We first introduce wavelet transform theory by starting with the short-time Fourier transform (STFT). Then discrete wavelet transform (DWT) is presented. Finally, the lifting scheme, known as the second generation wavelet transform, is described.

We then discuss the basic concept of image wavelet transform coding with an emphasis on embedded image wavelet transform coding algorithms, which is the base of JPEG2000.

Finally, JPEG2000, the newest still image coding standard, is described in this chapter with emphasis on its functionality and current status.

8.1 A Review of Wavelet Transform

8.1.1 Definition and Comparison with Short-Time Fourier Transform

The wavelet transform, as a specialized research field, started over more than two decades ago [grossman 1984]. It is known that the wavelet transform is rather different from the Fourier transform. The former is suitable to study transitional property of signals, while the latter is not suitable to study signals in the time–frequency space. The so-called STFT was designed to overcome the drawback of the Fourier transform. For a better understanding, we first give a very short review of the STFT because there are some similarities between the STFT and the wavelet transform. As we know, the STFT uses sinusoidal waves as its orthogonal basis and it is defined as

$$F(\omega,\tau) = \int_{-\infty}^{+\infty} f(t)w(t-\tau)e^{-j\omega t}\,dt \tag{8.1}$$

where $w(t)$ is a time domain windowing function, the simplest of which is a rectangular window that has a unit value over a time interval and has zero elsewhere. The value τ is the starting position of the window. Thus, the STFT maps a function $f(t)$ into a 2-D plane (ω, τ), where ω and τ stand for frequency and time moment, respectively. The STFT is also referred to as Gabor transform [cohen 1989]. Similar to the STFT, the wavelet transform also maps a time or spatial function into a 2-D function of a and τ, where a and τ denote dilation and translation in time, respectively. The wavelet transform is defined as follows. Let $f(t)$ be any square integrable function, i.e., it satisfies

$$\int_{-\infty}^{+\infty} |f(t)|^2 dt < \infty \qquad (8.2)$$

The continuous-time wavelet transform of $f(t)$ with respect to a wavelet $\psi(t)$ is defined as

$$W(a,\tau) = \int_{-\infty}^{+\infty} f(t)|a|^{-1/2} \psi^*\left[(t-\tau)a^{-1}\right] dt \qquad (8.3)$$

where a and τ are real variables and * denotes complex conjugation. The wavelet, denoted by $\psi_{a\tau}(t)$, is expressed as

$$\psi_{a\tau}(t) = |a|^{-1/2} \psi\left[(t-\tau)a^{-1}\right] \qquad (8.4)$$

Equation 8.4 represents a set of functions that are generated from a single function, $\psi(t)$, by dilations and translations. The variable τ represents the time shift and the variable a corresponds to the amount of time scaling or dilation. If $a > 1$, there is an expansion of $\psi(t)$, while if $0 < a < 1$, there is a contraction of $\psi(t)$. For negative values of a, the wavelet experiences a time reversal in combination with a dilation. The function, $\psi(t)$, is referred to as the mother wavelet and it must satisfy two conditions:

1. The function $\psi(t)$ integrates to zero:

$$\int_{-\infty}^{+\infty} \psi(t) dt = 0 \qquad (8.5)$$

2. The function is square integrable, or has finite energy:

$$\int_{-\infty}^{+\infty} |\psi(t)|^2 dt < \infty \qquad (8.6)$$

The continuous-time wavelet transform can now be rewritten as

$$W(a,\tau) = \int_{-\infty}^{+\infty} f(t) \psi_{a\tau}^*(t) dt \qquad (8.7)$$

In the following, we give two well-known examples of $\psi(t)$ and their Fourier transform. The first example is the Morlet (modulated Gaussian) wavelet [daubechies 1992],

$$\begin{aligned} \psi(t) &= e^{-\frac{1}{2}t^2} \cdot e^{j\omega_0 t} \\ \Psi(\omega) &= (2\pi)^{\frac{1}{2}} \exp\left[(\omega - \omega_0)^2/2\right] \end{aligned} \qquad (8.8)$$

Wavelet Transform for Image Coding: JPEG2000

and the second example is the Haar wavelet:

$$\psi(t) = \begin{cases} 1 & 0 \leq t \leq 1/2 \\ -1 & 1/2 \leq t \leq 1 \\ 0 & \text{otherwise} \end{cases} \quad (8.9)$$

$$\Psi(\omega) = je^{-j\omega/2} \frac{\sin^2(\omega/4)}{\omega/4}$$

From the above definitions and examples, we can find that the wavelets have zero DC value. This is clear from Equation 8.5. To have good time localization, the wavelets are usually bandpass signals and they decay rapidly toward zero with time. We can also find several other important properties of the wavelet transform and several differences between the STFT and the wavelet transform.

The STFT uses a sinusoidal wave as its basis functions, which keep the same frequency over the entire time interval. In contrast, the wavelet transform uses a particular wavelet as its basis function. Hence, wavelets vary in both position and frequency over the time interval. Examples of two basis functions for the sinusoidal wave and wavelet are shown in Figure 8.1a and b, respectively, where the vertical axes stand for magnitude and the horizontal axes for time.

The STFT uses a single analysis window. In contrast, the wavelet transform uses a short-time window at high frequencies and a long-time window at low frequencies. This is referred to as constant Q-factor filtering or relative constant bandwidth frequency analysis. A comparison of constant bandwidth analysis of the STFT and relative constant bandwidth wavelet transform is shown in Figure 8.2a and b, respectively.

This feature can be further explained based on the concept of a time–frequency plane (Figure 8.3). It is known from the Heisenberg inequality [rioul 1991] that the product of time resolution and frequency resolution has been lower bounded as follows:

$$\Delta t \cdot \Delta f \geq 1/(4\pi)$$

From the above expression, we see that these two resolutions cannot be arbitrarily small. As shown in Figure 8.3, the window size of the STFT in the time domain is always chosen

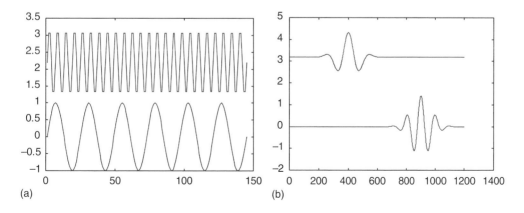

FIGURE 8.1
Wave versus wavelet: (a) two sinusoidal waves and (b) two wavelets, where vertical axes stand for magnitude and horizontal axes for time. (From Castleman, K.R., *Digital Image Processing*, Prentice-Hall, Englewood Cliffs, NJ, 1996. With permission.)

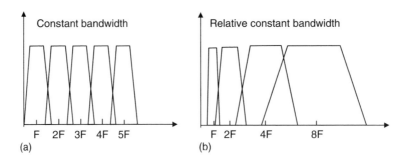

FIGURE 8.2
(a) Constant bandwidth analysis (for Fourier transform) and (b) relative constant bandwidth analysis (for wavelet transform). (From Rioul, O. and Vetterli, M., *IEEE Signal Process. Mag.*, 8, 14, 1991. With permission.)

to be constant. The corresponding frequency bandwidth is also constant. In the wavelet transform, the window size in time domain varies with the frequency. A longer time window is used for lower frequency and shorter time window is used for higher frequency. This property is very important for image data compression. For image data, the concept of time–frequency plane becomes spatial-frequency plane. The spatial resolution of a digital image is measured with pixels, as described in Chapter 15. To overcome the limitations of discrete cosine transform (DCT)-based coding, the wavelet transform allows the spatial resolution and frequency bandwidth to vary in the spatial-frequency plane. With this variation, better bit allocation for active and smooth areas can be achieved.

The continuous-time wavelet transform can be considered as a correlation. For fixed a, it is clear from Equation 8.3 that $W(a,\tau)$ is the cross-correlation of functions $f(t)$ with related wavelet conjugate dilated to scale factor a at time lag τ. This is an important property of the wavelet transform for multiresolution analysis of image data. While the convolution can be seen as a filtering operation, the integral wavelet transform can be seen as a bank of linear filters acting upon $f(t)$. This implies that the image data can be decomposed by a bank of filters defined by the wavelet transform.

The continuous-time wavelet transform can be seen as an operator. First, it has the property of linearity. If we rewrite $W(a,\tau)$ as $W_{a\tau}[f(t)]$, then we have

$$W_{a\tau}[\alpha f(t) + \beta g(t)] = \alpha W_{a\tau}[f(t)] + \beta W_{a\tau}[g(t)] \tag{8.10}$$

where α and β are constant scalars.

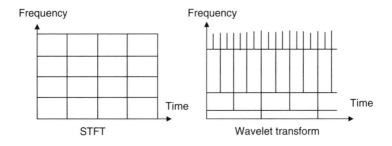

FIGURE 8.3
Comparison of the short-time Fourier transform (STFT) and the wavelet transform in the time–frequency plane. (From Rioul, O. and Vetterli, M., *IEEE Signal Process. Mag.*, 8, 14, 1991. With permission.)

Second, it has the property of translation:

$$W_{a\tau}[f(t-\lambda)] = W(a, \tau - \lambda) \tag{8.11}$$

where λ is a time lag.

Finally, it has the property of scaling:

$$W_{a\tau}[f(t/\alpha)] = W(a/\alpha, \tau/\alpha) \tag{8.12}$$

8.1.2 Discrete Wavelet Transform

In the continuous-time wavelet transform, the function $f(t)$ is transformed to a function $W(a, \tau)$ using the wavelet $\psi(t)$ as a basis function. Recall that the two variables, a and τ, are the dilation and translation in time, respectively. Now let us find a means of obtaining the inverse transform, i.e., given $W(a, \tau)$, find $f(t)$. If we know how to get the inverse transform, we can then represent any arbitrary function $f(t)$ as a summation of wavelets such as in the Fourier transform and discrete cosine transform that provide a set of coefficients for reconstructing the original function using sine and cosine as the basis functions. In fact, this is possible if the mother wavelet satisfies the admissibility condition:

$$C = \int_{-\infty}^{+\infty} |\Psi(\omega)|^2 |\omega|^{-1} d\omega \tag{8.13}$$

where
  C is a finite constant
  $\Psi(\omega)$ is the Fourier transform of the mother wavelet function $\psi(t)$

Then, the inverse wavelet transform is

$$f(t) = \frac{1}{C} \int_{-\infty}^{+\infty} \int_{-\infty}^{+\infty} |a|^{-2} W(a, \tau) \psi_{a\tau}(t) da\, d\tau \tag{8.14}$$

The above results can be extended for 2-D signals. If $f(x,y)$ is a 2-D function, its continuous-time wavelet transform is defined as

$$W(a, \tau_x, \tau_y) = \int_{-\infty}^{+\infty} \int_{-\infty}^{+\infty} f(x,y) \psi^*_{a\tau_x\tau_y}(x,y) dx\, dy \tag{8.15}$$

where τ_x and τ_y specify the transform in 2-D. The inverse 2-D continuous-time wavelet transform is then defined as

$$f(x,y) = \frac{1}{C} \int_{-\infty}^{+\infty} \int_{-\infty}^{+\infty} \int_{-\infty}^{+\infty} |a|^{-3} W(a, \tau_x, \tau_y) \psi_{a\tau_x\tau_y}(x,y) da\, d\tau_x\, d\tau_y \tag{8.16}$$

where C is defined as in Equation 8.13 and $\psi(x,y)$ is a 2-D wavelet

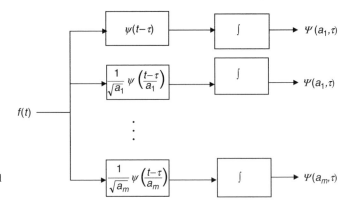

FIGURE 8.4
The wavelet transform implemented with a bank of filters.

$$\psi_{a\tau_x\tau_y}(x,y) = |a|^{-3}\psi\left(\frac{x-\tau_x}{a}, \frac{y-\tau_y}{a}\right) \qquad (8.17)$$

For image coding, the wavelet transform is used to decompose the image data into wavelets. As indicated in the third property of the wavelet transform, the wavelet transform can be viewed as the cross-correlation of the function $f(t)$ and the wavelets $\psi_{a\tau}(t)$. Therefore, the wavelet transform is equivalent to finding the output of a bank of bandpass filters specified by the wavelets of $\psi_{a\tau}(t)$ as shown in Figure 8.4. This process decomposes the input signal into several subbands. As each subband can be further partitioned, the filter bank implementation of the wavelet transform can be used for multiresolution analysis (MRA). Intuitively, when the analysis is viewed as a filter bank, the time resolution must increase with the central frequency of the analysis filters. This can be exactly obtained by the scaling property of the wavelet transform, where the center frequencies of the bandpass filters increase as the bandwidth widens. Again, the bandwidth becomes wider as the dilation parameter a reduces. It should be noted that such an MRA is consistent with the constant Q-factor property of the wavelet transform. Furthermore, the resolution limitation of the STFT does not exist in the wavelet transform because the time–frequency resolutions in the wavelet transform vary, as shown in Figure 8.2b.

For digital image compression, it is preferred to represent $f(t)$ as a discrete superposition sum rather than an integral. With this move to the discrete space, the dilation parameter a in Equation 8.10 takes the values $a = 2^k$ and the translation parameter τ takes the values $\tau = 2^k l$, where both k and l are integers. From Equation 8.4, the discrete version of $\psi_{a\tau}(t)$ becomes

$$\psi_{kl}(t) = 2^{-k/2}\psi(2^{-k}t - l) \qquad (8.18)$$

Its corresponding wavelet transform can be rewritten as

$$W(k,l) = \int_{-\infty}^{+\infty} f(t)\psi_{kl}^*(t)dt \qquad (8.19)$$

and the inverse transform becomes

$$f(t) = \sum_{k=-\infty}^{+\infty}\sum_{l=-\infty}^{+\infty} d(k,l)2^{-k/2}\psi(2^{-k}t - l) \qquad (8.20)$$

The values of the wavelet transform at those a and τ are represented by $d(k,l)$:

$$d(k,l) = W(k,l)/C \qquad (8.21)$$

The $d(k,l)$ coefficients are referred to as the DWT of the function $f(t)$ [daubechies 1992; vetterli 1995]. It is noted that the discretization so far is only applied to the parameters a and τ, $d(k,l)$ is still a continuous-time function. If the discretization is further applied to the time domain by letting $t = mT$, where m is an integer and T is the sampling interval (without loss of generality, we assume $T = 1$), then the discrete-time wavelet transform is defined as

$$W_d(k,l) = \sum_{m=-\infty}^{+\infty} f(m) \psi_{kl}^*(m) \qquad (8.22)$$

Of course, the sampling interval has to be chosen according to the Nyquist sampling theorem so that no information has been lost in the process of sampling. The inverse discrete-time wavelet transform is then

$$f(m) = \sum_{k=-\infty}^{+\infty} \sum_{l=-\infty}^{+\infty} d(k,l) 2^{-k/2} \psi(2^{-k}m - l) \qquad (8.23)$$

8.1.3 Lifting Scheme

An alternative implementation of the DWT has been proposed recently, known as the *lifting scheme* [sweldens 1995; daubechies 1998]. The implementation of the lifting scheme is very efficient and similar to the fast Fourier transform (FFT) implementation for Fourier transform. In this subsection, we first introduce how the lifting scheme works. Then, we comment on its features, merits, and its application in JPEG2000.

8.1.3.1 Three Steps in Forward Wavelet Transform

Similar to the discrete Fourier transform (DFT), the lifting scheme is conducted from one resolution level to the next lower resolution level iteratively. To facilitate representation, let us consider here only the case of 1-D data sequence. A 2-D data extension should be understandable straightforwardly afterward. One iteration of lifting scheme is described as follows. Denote the 1-D data sequence by X_i, and $X_i = \{x_j\}$. After the iteration of lifting scheme, the date sequence X_i becomes two data sequences X_{i+1} and Y_{i+1}, say, the former is the low-pass component and the latter is the high-pass component. The lifting scheme iterative consists of the following three steps: splitting, prediction, and updated.

Splitting (often referred to as **lazy wavelet transform**)
   The data sequence X_i is split into two parts: X_{i+1} and Y_{i+1}.
Prediction (often referred to as **dual lifting**)
   In this step, the data sequence Y_{i+1} is predicted with the data sequence X_{i+1}, and then Y_{i+1} is replaced by the prediction error

$$Y_{i+1} \leftarrow Y_{i+1} - P(X_{i+1})$$

Updated (often referred to as **primary lifting**)
   In this step, the data sequence X_{i+1} is updated with the data sequence Y_{i+1}, and then X_{i+1} is replaced as follows.

$$X_{i+1} \leftarrow X_{i+1} + U(Y_{i+1})$$

After this iteration, a new iteration will apply the same three steps to X_{i+1} to generate X_{i+2} and Y_{i+2}.

It is observed that the 1-D DWT via lifting scheme is now very simple and efficient if the prediction and update operators are simple. In the example that follows, we can see this is the case.

8.1.3.2 Inverse Transform

The inverse wavelet transform via lifting scheme is exactly the reverse process of the above forward transform in the sense that plus and minus are reverses, splitting and merging are reversed, and the order to the three steps is reversed. The corresponding three steps in inverse transform are shown below.

Inverse update

$$X_{i+1} \leftarrow X_{i+1} - U(Y_{i+1})$$

Inverse prediction

$$Y_{i+1} \leftarrow Y_{i+1} + P(X_{i+1})$$

Merging
The data sequence X_i is formed by union of X_{i+1} and Y_{i+1}.

8.1.3.3 Lifting Version of CDF (2,2)

In this section, a specific lifting scheme is introduced, which essentially implements the well-known CDF(2,2) wavelet transform, named after its three inventors: Cohen, Daubechies, and Feauveau.

Forward Transform:
Splitting
The dataset $X_i = \{x_j\}$ is split into two parts: even data samples and odd data samples. This is similar to the fast Fourier transform (FFT) technique.
That is,

$$s_j \leftarrow x_{2j}$$
$$d_j \leftarrow x_{2j+1}$$

Note that we use the notations $\{s_j\}$ and $\{d_j\}$ to denote the even and odd data sequence of $\{x_j\}$ in order to avoid confusion in notation of subscripts.

Prediction
Predict the odd data samples with the even data samples and replace the odd data samples by the prediction error as follows:

$$d_j \leftarrow d_j - \frac{1}{2}(s_j + s_{j+1})$$

Update
Update the even data samples with the odd data samples to preserve the mean of data samples.

$$s_j \leftarrow s_j + \frac{1}{4}(d_{j-1} + d_j)$$

It is observed that both prediction and update are linear in this example. The inverse transform runs just opposite to the forward transform as shown below.

Inverse Transformation:

Inverse update

$$S_j \leftarrow S_j - \frac{1}{4}(d_{j-1} + d_j)$$

Inverse prediction

$$d_j \leftarrow d_j + \frac{1}{2}(S_j + S_{j+1})$$

Merging

The data sequence X_i is formed by union of S_j and d_j.

8.1.3.4 A Demonstration Example

We present a simple numerical example for illustration purpose. In this example, we only consider a four-point data sequence, i.e.,

$$X_i = \{1, 2, 3, 4\}$$

After splitting, we have

$$\{s_j\} = \{2, 4\} \quad \text{and} \quad \{d_j\} = \{1, 3\}$$

After prediction, we have

$$d_1 = d_1 - \frac{1}{2}(s_1 + s_2) = -2 \quad \text{and} \quad d_2 = d_2 - \frac{1}{2}(s_2 + s_3) = 1$$

Note that the value of s_3 is not available and is hence treated as $s_3 = 0$ for the sake of simplicity. For a discussion on boundary treatment, readers are referred to literature, e.g., [uytterhoeven 1999].

After update, we have

$$s_1 = s_1 + \frac{1}{4}(d_0 + d_1) = 1.5 \quad \text{and} \quad s_2 = s_2 + \frac{1}{4}(d_1 + d_2) = 3.75$$

Similarly we do not have d_0 and hence treat $d_0 = 0$. Hence, after CDF (2,2) via lifting scheme we have $X_{i+1} = \{1.5, 3.75\}$ and $Y_{i+1} = \{-2, 1\}$. The former is the low-pass frequency part of the data sequence $X_i = \{1,2,3,4\}$, and the latter is the high-pass frequency part of X_i.

For inverse transform, after inverse update, we have

$$s_1 = s_1 - \frac{1}{4}(d_0 + d_1) = 2 \quad \text{and} \quad s_2 = s_2 - \frac{1}{4}(d_1 + d_2) = 4$$

After inverse prediction, we have

$$d_1 = d_1 + \frac{1}{2}(s_1 + s_2) = 1 \quad \text{and} \quad d_2 = d_2 + \frac{1}{2}(s_2 + s_3) = 3$$

Hence, inverse transform via lifting scheme produces the original sequence $X_i = \{1,2,3,4\}$.

8.1.3.5 (5,3) Integer Wavelet Transform

An additional advantage which lifting scheme possesses, has something to do with integer wavelet transform (IWT). That is, IWT can map integer to integer. Image grayscale values are known as integers. Hence, IWT can be used for reversible transformation, which finds applications in lossless image compression. Furthermore, both the forward IWT and the inverse IWT can easily be conducted via using the lifting scheme. In this subsection, we present the integer version of CDF (2,2), which is referred to as (5,3) IWT. Note that the (5,3) IWT been adopted in JPEG2000 [rabbani 2001; skodras 2001]. In addition, the IWT has also been used for reversible image data embedding [xuan 2002].

Forward (5,3) IWT

Splitting

$$s_j \leftarrow x_{2j} \quad \text{and} \quad d_j \leftarrow x_{2j+1}$$

Prediction

$$d_j \leftarrow d_j - \left\lfloor \frac{1}{2}(s_j + s_{j+1}) + \frac{1}{2} \right\rfloor$$

Update

$$s_j \leftarrow s_j + \left\lfloor \frac{1}{4}(d_{j-1} + d_j) + \frac{1}{2} \right\rfloor$$

where the notation $\lfloor z \rfloor$ indicates the largest integer not larger than z.

Inverse (5,3) IWT

Inverse primary lifting

$$s_j \leftarrow s_j - \left\lfloor \frac{1}{4}(d_{j-1} + d_j) + \frac{1}{2} \right\rfloor$$

Inverse dual lifting

$$d_j \leftarrow d_j + \left\lfloor \frac{1}{2}(s_j + s_{j+1}) + \frac{1}{2} \right\rfloor$$

Merging

$$x_{2j} \leftarrow s_j \quad \text{and} \quad x_{2j+1} \leftarrow d_j$$

8.1.3.6 A Demonstration Example of (5,3) IWT

Here we work on the same 1-D data sequence as used in the example shown in Section 8.1.3.4. That is,

$$X_i = \{1, 2, 3, 4\}$$

After splitting, we have

$$\{s_j\} = \{2, 4\} \quad \text{and} \quad \{d_j\} = \{1, 3\}$$

After prediction, we have

Wavelet Transform for Image Coding: JPEG2000

$$d_1 = d_1 - \left\lfloor \frac{1}{2}(s_1 + s_2) + \frac{1}{2} \right\rfloor = -2 \quad \text{and} \quad d_2 = d_2 - \left\lfloor \frac{1}{2}(s_2 + s_3) + \frac{1}{2} \right\rfloor = 1$$

After update, we have

$$s_1 = s_1 + \left\lfloor \frac{1}{4}(d_0 + d_1) + \frac{1}{2} \right\rfloor = 2 \quad \text{and} \quad s_2 = s_2 + \left\lfloor \frac{1}{4}(d_1 + d_2) + \frac{1}{2} \right\rfloor = 4$$

Because the values of s_3 and d_0 are not available we treat $s_3 = 0$ and $d_0 = 0$ for the sake of simplicity. For a discussion on boundary treatment, readers are referred to literature, e.g., [uytterhoeven 1999]. We then see that after (5,3) IWT with lifting scheme we have $X_{i+1} = \{2,4\}$ and $Y_{i+1} = \{-2,1\}$. The former is the low-pass frequency part of the data sequence $X_i = \{1,2,3,4\}$, and the latter is the high-pass frequency part of X_i. It is observed that both parts are integers. That is, IWT maps integer to integer.

Now we show that the inverse (5,3) IWT gives back exactly the same original data sequence $X_i = \{1,2,3,4\}$ as follows:

For inverse transform, after inverse primary lifting, we have

$$s_1 = s_1 - \left\lfloor \frac{1}{4}(d_0 + d_1) + \frac{1}{2} \right\rfloor = 2 \quad \text{and} \quad s_2 = s_2 - \left\lfloor \frac{1}{4}(d_1 + d_2) + \frac{1}{2} \right\rfloor = 4$$

After inverse dual lifting, we have

$$d_1 = d_1 + \left\lfloor \frac{1}{2}(s_1 + s_2) + \frac{1}{2} \right\rfloor = 1 \quad \text{and} \quad d_2 = d_2 + \left\lfloor \frac{1}{2}(s_2 + s_3) + \frac{1}{2} \right\rfloor = 3$$

After merging, we have $X_i = \{1,2,3,4\}$.

8.1.3.7 Summary

In this section, we summarize the merits possessed by lifting scheme. Because of these merits, lifting scheme has been adopted by JPEG2000 [rabbani 2001]

1. Lifting scheme provides another different way to illustrate wavelet transform. It is noted that most of wavelet transform theory starts from Fourier transform theory. Lifting scheme, however, does provide one way to view wavelet transform without using Fourier transform.
2. Lifting scheme is simple and hence efficient in implementation.
3. In addition, lifting scheme can reduce memory requirement significantly. It can provide so-called in-place computation of wavelet coefficients. That is, it can overwrite the memory used to store input data with wavelet coefficients. This bears similarity to the fast Fourier transform.
4. Lifting scheme lends itself easily to integer wavelet transform computation.

8.2 Digital Wavelet Transform for Image Compression

8.2.1 Basic Concept of Image Wavelet Transform Coding

From the last section, we have learned that the wavelet transform has several features that are different from traditional transforms. It is noted from Figure 8.2 that each transform

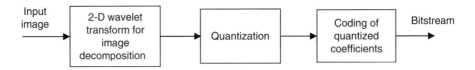

FIGURE 8.5
Block diagram of the image coding with the wavelet transform coding.

coefficient in the STFT represents a constant interval of time regardless of which band the coefficient belongs to, whereas for the wavelet transform, the coefficients at the course level represent a larger time interval but a narrower band of frequencies. This feature of the wavelet transform is very important for image coding. In traditional image transform coding, which make use of the Fourier transform or the discrete cosine transform, one difficult problem is to choose the block size or window width so that statistics computed within that block provide good models of the image signal behavior. The choice of the block size has to be compromised so that it can handle both active and smooth areas. In the active areas, the image data is more localized in the spatial domain, while in the smooth areas, the image data is more localized in the frequency domain. With traditional transform coding, it is very hard to reach a good compromise. The main contribution of wavelet transform theory is that it provides an elegant framework in which both statistical behaviors of image data can be analyzed with equal importance. This is because that wavelets can provide a signal representation in which some of the coefficients represent long data lags corresponding to a narrow band or low frequency range, and some of the coefficients represent short data lags corresponding to a wide band or high frequency range. Therefore, it is possible to obtain a good trade-off between spatial and frequency domain with the wavelet representation of image data.

To use the wavelet transform for image coding applications, an encoding process is needed, which includes three major steps: image data decomposition, quantization of the transformed coefficients, and coding of the quantized transformed coefficients. A simplified block diagram of this process is shown in Figure 8.5. The image decomposition is usually a lossless process, which converts the image data from the spatial domain to frequency domain, where the transformed coefficients are decorrelated. The information loss happens in the quantization step and the compression is achieved in the coding step. To begin the decomposition, the image data is first partitioned into four subbands labeled as LL_1, HL_1, LH_1, and HH_1, as shown in Figure 8.6a. Each coefficient represents a spatial area corresponding to the one-quarter of the original image size. The low frequencies represent a bandwidth corresponding to $0<|\omega|<\pi/2$, while the high frequencies represent the band $\pi/2<|\omega|<\pi$. To obtain the next level of decomposition, the LL_1 subband is further decomposed into the next level of four subbands, as shown in Figure 8.6b. The low frequencies of the second level decomposition correspond to $0<|\omega|<\pi/4$, while the high frequencies at the second level correspond to $\pi/4<|\omega|<\pi/2$. This decomposition can be

FIGURE 8.6
A 2-D wavelet transform, (a) first-level decomposition and (b) second-level decomposition (L denotes low band, H denotes high band, and the subscript denotes number of level; for example LL_1 denotes the low-low band at level 1).

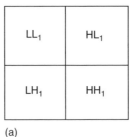

(a) (b)

continued to as many levels as needed. The filters used to compute the DWT are generally the symmetric quadrature mirror filters (QMFs), which is described in [woods 1986]. A QMF-pyramid subband decomposition is illustrated in Figure 8.6b.

During quantization, each subband is quantized differently depending on its importance, which is usually based on its energy or variance [jayant 1984]. To reach the predetermined bit rate or compression ratio, the coarse quantizers or large quantization steps would be used to quantize the low-energy subbands while the finer quantizers or small quantization steps would be used to quantize the high-energy subbands. This results in fewer bits allocated to those low-energy subbands and more bits for high-energy subbands.

8.2.2 Embedded Image Wavelet Transform Coding Algorithms

In this section, the wavelet transform based image coding methods, which form the basic framework of JPEG2000, are discussed. We first comment on the drawbacks of early wavelet transform based image coding methods. Then, we introduce the concept of modern wavelet transform based image coding methods, i.e., embedded image wavelet transform coding algorithm, in particular, the embedded image coding using zerotrees of wavelet coefficients (EZW).

8.2.2.1 Early Wavelet Image Coding Algorithms and Their Drawbacks

As with other transform coding schemes, most wavelet coefficients in the high-frequency bands have very low energy. After quantization, many of these high-frequency wavelet coefficients are quantized to zero. Based on the statistical property of the quantized wavelet coefficients, Huffman coding tables can be designed. Generally, most of the energy in an image is contained in the low-frequency bands. The data structure of the wavelet transformed coefficients is suitable to exploit this statistical property. Consider a multilevel decomposition of an image with the DWT, where the lowest levels of decomposition would correspond to the highest frequency subbands and the finest spatial resolution and the highest level of decomposition would correspond to the lowest frequency subband and the coarsest spatial resolution. Arranging the subbands from lowest to highest frequency, we expect a decrease in energy. Also, we expect that if the wavelet transformed coefficients at a particular level have lower energy, then coefficients at the lower levels or high frequency subbands, which correspond to the same spatial location, would have smaller energy. Another feature of the wavelet coefficient data structure is spatial self-similarity across subbands, which is shown in Figure 8.7. The early wavelet image coding methods [vetterli 1984; woods 1986; antonini 1992] utilizing the above features of wavelet transform are referred to as early or conventional wavelet image coding methods.

The drawbacks of these conventional wavelet image coding methods lie on [usevitch 2001] the following:

- Quantizer can only be optimal as coding rate larger than 1 bpp (bits per pixel).
- Optimal bit allocation changes as overall bit rate changes, requiring coding process repeated entirely for each new target bit rate desired.
- It is difficult to code an input to give an exact target bit rate (or predefined output size).

8.2.2.2 Modern Wavelet Image Coding

Embedded image coding is the basic concept of modern wavelet image coding algorithm. By embedding code, it means that the code arranges its bits in the order of their

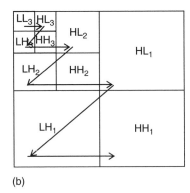

FIGURE 8.7
(a) Parent–children dependencies of subbands, the arrow points from the subband of the parents to the subband of children. The top left is the lowest frequency band. (b) The scanning order of the subbands for encoding a significance map. (From Shapiro, J., *IEEE Trans. Signal Process.*, 41, 3445, 1993. With permission.)

importance. In other words, the bits corresponding a lower bit rate coding will be arranged in the beginning portion of the embedded code. Because the embedded code have all lower rate codes arranged at the beginning portion of the bit stream, the embedded codes can be truncated to fit a targeted bit rate by simply truncating the bit stream accordingly. Embedded wavelet image coding began with the embedded zerotree wavelet algorithm (EZW) by Shapiro in the early 1990s [shapiro 1993]. This revolutionary breakthrough marks the beginning of the modern wavelet image coding age. The typical algorithms including the EZW, set partitioning in hierarchical trees (SPIHT) by Said and Pearlman [said 1996], and embedded block coding with optimized truncation of the embedded bit streams (EBCOT) [taubman 2000]. Consequently, the modern wavelet coding has been adopted by JPEG2000 for still image coding. A given image can be coded once and it can then be easily and optimally truncated according to any given bit rate.

Several algorithms have been developed to exploit this and the above-mentioned properties for image coding. Among them, one of the first was proposed by Shapiro [shapiro 1993] and used an embedded image coding using zerotrees of wavelet coefficients technique referred to as EZW. Another algorithm is the so-called SPIHT developed by Said and Pearlman [said 1996]. This algorithm also produces an embedded bitstream. The advantage of the embedded coding schemes allows an encoding process to terminate at any point so that a target bit rate or distortion metric can be met exactly. Intuitively, for a given bit rate or distortion requirement, a non embedded code should be more efficient than an embedded code. Since, it has no constraints imposed by embedding requirements. However, embedded wavelet transform coding algorithms are the best currently. The additional constraints do not seem to have deleterious effect. In the following, we introduce the two embedded coding algorithms: the zerotree coding and the set partitioning in hierarchical tree coding with an emphasis on the former.

8.2.2.3 Embedded Zerotree Wavelet Coding

As with DCT-based coding, an important aspect of wavelet-based coding is to code the positions of those coefficients that will be transmitted as nonzero values. After quantization, the probability of the zero symbol must be extremely high for the very low bit rate case. It is well known that the most of energy of an image is contained in the low-low (LL) subband at the highest scale level and the distribution of wavelet coefficients of high-frequency subbands follows a generalized Laplacian density. That is, it has a high peak around zero and long tail toward two sides, which means large number of coefficients having zero and small magnitude and yet still some small number of coefficients with large magnitudes. The statistics just mentioned implies that a large portion of the bit budget will then be spent on encoding the significance map, or the binary decision map that indicates whether a transformed coefficient has a zero- or nonzero-quantized value. Therefore, the ability to efficiently encode the significance map becomes a key issue for coding images at very low bit rates. A new data structure, the zerotree, has been proposed for this purpose [shapiro 1993]. To describe zerotree, we must first define insignificance. A wavelet coefficient is insignificant with respect to a given threshold value if the absolute value of this coefficient is smaller than this threshold. From the nature of the wavelet transform we can assume that every wavelet transform coefficients at a given scale can be strong related to a set of coefficients at the next finer scale of similar orientation. More specifically, we can further assume that if a wavelet coefficient at a coarse scale is insignificant with respect to the preset threshold, then all wavelet coefficients at finer scales are likely to be insignificant with respect to this threshold. Therefore, we can build a tree with these parent–child relationships, such that, coefficients at a coarse scale are called parents, and all coefficients corresponding to the same spatial location at the next finer scale of similar orientation are called children. Furthermore, for a parent, the set of all coefficients at all finer scales of similar orientation corresponding to the same spatial location are called descendants. For a QMF-pyramid decomposition the parent–children dependencies are shown in Figure 8.7a. For a multiscale wavelet transform, the scan of the coefficients begins at the lowest frequency subband and then takes the order of LL, HL, LH, and HH from the coarser scale to the next finer scale as shown in Figure 8.7b.

The zerotree is defined such that if a coefficient itself and all of its descendants are insignificant with respect to a threshold, then this coefficient is considered as an element of a zerotree. An element of a zerotree is considered as a zerotree root if this element is not the descendant of a previous zerotree root with respect to the same threshold value. The significance map can then be efficiently represented by a string with three symbols: zerotree root, isolated zero, and significant. The isolated zero means that the coefficient is insignificant, but it has some significant descendant. At the finest scale, only two symbols are needed because all coefficients have no children; thus the symbol for zerotree root is not used. The symbol string is then entropy encoded. Zerotree coding efficiently reduces the cost for encoding the significance map by using self-similarity of the coefficients at different scales. Additionally, it is different from the traditional run-length coding (RLC) that is used in DCT-based coding schemes. Each symbol in a zerotree is a single terminating symbol, which can be applied to all depth of the zerotree, similar to the end-of-block (EOB) symbol in the JPEG and MPEG video coding standards. The difference between the zerotree and EOB is that the zerotree represents the insignificance information at a given orientation across different scale layers. Therefore, the zerotree can efficiently exploit the self-similarity of the coefficients at the different scales corresponding to the same spatial location. The EOB only represents the insignificance information over the spatial area at the same scale. In summary, the zerotree coding scheme tries to reduce the number of bits to encode the significance map, which is used to encode the insignificant coefficients. Therefore, more bits can be allocated to encode the important significant coefficients. It should be emphasized that this zerotree coding scheme of wavelet coefficients is an embedded coder, which means that an encoder can terminate the encoding at any point according to a given target bit rate or target distortion metric. Similarly, a decoder, which receives this embedded stream, can terminate at any point to reconstruct an image that has been scaled in quality.

In summary, the statistics of wavelet transform coefficients indicates that how to code coefficients' magnitude and position is a key issue in wavelet image coding. The EZW coding has proposed to code the positions of the zero coefficients using wavelet transform's self-similarity characteristics instead of coding positions of significant coefficients directly. This is referred to as significance map coding using zerotrees [usevitch 2001]. In addition to this key point, the EZW has developed a successive approximation quantization, which generates large number of zero coefficients and led to embedded coding [usevitch 2001]. In either [shapiro 1993] or [usevitch 2001], there is one example of 8×8 image with three-level wavelet transform. In the examples, step by step, the EZW coding scheme is implemented. Readers are encouraged to go through the examples to get first-hand experience about the EZW algorithm, to see how the above-mentioned two key techniques enhance the coding efficiency, and see how embedded coding is realized. A problem, similar to these two examples is provided in Exercises.

8.2.2.4 Set Partitioning in Hierarchical Trees Coding

Another embedded wavelet coding method is the SPIHT-based algorithm [said 1996]. This algorithm includes two major core techniques: the set partitioning sorting algorithm and the spatial orientation tree. The set partitioning sorting algorithm is the algorithm that hierarchically divides coefficients into significant and insignificant from the most significant bit to the least significant bit by decreasing the threshold value at each hierarchical step for constructing a significance map. At each threshold value, the coding process consists of two passes: the sorting pass and the refinement pass, except for the first threshold that has only the sorting pass. Let $c(i,j)$ represent the wavelet transformed coefficients and m is an integer. The sorting pass involves selecting the coefficients such that $2^m \leq |c(i,j)| \leq 2^{m+1}$, with m being decreased at each pass. This process divides the coefficients into subsets and then tests each of these subsets for significant coefficients. The significance map constructed in the procedure is tree encoded. The significant information is stored in three ordered lists: list of insignificant pixels (LIP), list of significant pixels (LSP), and list of insignificant sets (LIS). At the end of each sorting pass, the LSP contains the coordinates of all significant coefficients with respect to the threshold at that step. The entries in the LIS can be one of two types: type A represents all its descendants and type B represents all its descendants from its grandchildren onward. The refinement pass involves transmitting the mth most significant bit of all the coefficients with respect to the threshold, 2^{m+1}.

The idea of a spatial orientation tree is based on the following observation. Normally, among the transformed coefficients, most of the energy is concentrated in the low frequencies. For the wavelet transform, when we move from the highest to the lowest levels of the subband pyramid, the energy usually decreases. It is also observed that there exists strong

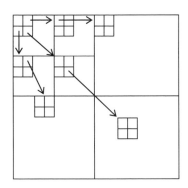

FIGURE 8.8
Relationship between pixels in the spatial orientation tree.

spatial self-similarity between subbands in the same spatial location such as in the zerotree case. Therefore, a spatial orientation tree structure has been proposed for the SPIHT algorithm. The spatial orientation tree naturally defines the spatial relationship on the hierarchical pyramid as shown in Figure 8.8.

During the coding, the wavelet transformed coefficients are first organized into spatial orientation trees as in Figure 8.8. In the spatial orientation tree, each pixel (i,j) from the former set of subbands is seen as a root for the pixels $(2i, 2j)$, $(2i + 1, 2j)$, $(2i, 2j + 1)$, and $(2i + 1, 2j + 1)$ in the subbands of the current scale. For a given n-level decomposition, this structure is used to link pixels of the adjacent subbands from level n until level 1. In the highest level n, the pixels in the low-pass subband are linked to the pixels in the three high-pass subbands at the same level. In the subsequent levels, all the pixels of a subband are involved in the tree-forming process. Each pixel is linked to the pixels of the adjacent subband at the next lower level. The tree stops at the lowest level.

The implementation of the SPIHT algorithm consists of four steps: initialization, sorting pass, refinement pass, and quantization scale update. In the initialization step, we find an integer $m = \lfloor \log_2 (\max_{(i,j)} \{|c(i,j)|\}) \rfloor$. Here $\lfloor \ \rfloor$ represent an operation of obtaining the largest integer less than $|c(i,j)|$. The value of m is used for testing the significance of coefficients and constructing the significance map. The LIP is set as an empty list. The LIS is initialized to contain all the coefficients in the low-pass subbands that have descendants. These coefficients can be used as roots of spatial trees. All these coefficients are assigned to be of type A. The LIP is initialized to contain all the coefficients in the low-pass subbands.

In the sorting pass, each entry of the LIP is tested for significance with respect to the threshold value 2^m. The significance map is transmitted in the following way. If it is significant, a "1" is transmitted, a sign bit of the coefficient is transmitted, and the coefficient coordinates are moved to the LSP. Otherwise, a "0" is transmitted. Then, each entry of the LIS is tested for finding the significant descendants. If there are none, a "0" is transmitted. If the entry has at least one significant descendant, then a "1" is transmitted and each of the immediate descendants are tested for significance. The significance map for the immediate descendants is transmitted in such a way that if it is significant, a "1" plus a sign bit are transmitted and the coefficient coordinates are appended to the LSP. If it is not significant, a "0" is transmitted and the coefficient coordinates are appended to the LIP. If the coefficient has more descendants then it is moved to the end of the LIS as an entry of type B. If an entry in the LIS is of type B then its descendants are tested for significance. If at least one of them is significant then this entry is removed from the list, and its immediate descendants are appended to the end of the list of type A. For the refinement pass, the mth most significant bit of the magnitude of each entry of the LSP is transmitted except those in the current sorting pass. For the quantization scale update step, m is decreased by 1 and the procedure is repeated from the sorting pass.

8.3 Wavelet Transform for JPEG2000

8.3.1 Introduction of JPEG2000

Most image coding standards had exploited the DCT as their core technology for image decomposition for a while. However, this has been changed later. The wavelet transform has been adopted by MPEG-4 for still image coding [mpeg4]. Also, JPEG2000 has used the wavelet transform as its core technology for the next generation of the still image coding standard [jpeg2000 vm]. This is because the wavelet transform can provide not only excellent coding efficiency but also good spatial and quality scalable functionality.

JPEG2000 is a new type of image compression system under development by Joint Photographic Experts Group for still image coding. This standard is intended to meet a need for image compression with great flexibility and efficient interchangeability. JPEG2000 is also intended to offer unprecedented access into the image while still in compressed domain. Thus, images can be accessed, manipulated, edited, transmitted, and stored in a compressed form.

8.3.1.1 Requirements of JPEG2000

As a new coding standard, the detailed requirements of JPEG2000 include:

Low bit rate compression performance: JPEG2000 is required to offer excellent coding performance at bit rates lower than 0.25 bits/pixel for highly detailed gray level images as the current JPEG (10918-1) cannot provide satisfactory results at this range of bit rates. This is the primary feature of JPEG2000.

Lossless and lossy compression: it is desired to provide lossless compression naturally in the course of progressive decoding. This feature is especially important for medical image coding where the loss is not always allowed. Also, other applications, such as high-quality image archival systems, and network applications desire to have functionality of lossless reconstruction.

Large images: currently, the JPEG image compression algorithm does not allow for images greater than 64K by 64K without tiling.

Single decomposition architecture: the current JPEG standard has 44 modes, many of these modes are for specific applications and not used by the majority of JPEG decoders. It is desired to have a single decomposition architecture that can encompass the interchange between applications.

Transmission in noisy environments: it is desirable to consider error robustness while designing the coding algorithm. This is important for the application of wireless communication. The current JPEG has provision for restart intervals, but image quality suffers dramatically when bit errors are encountered.

Computer generated imagery: the current JPEG is optimized for natural imagery and does not perform well on computer generated imagery or computer graphics.

Compound documents: the new coding standard is desired to be capable of compressing both continuous-tone and bi-level images. The coding scheme can compress and decompress images from 1 to 16-bit for each color component. The current JPEG standard does not work well for bi-level images.

Progressive transmission by pixel accuracy and resolution: progressive transmission that allows images to be transmitted with increasing pixel accuracy or spatial resolution is important for many applications. The image can be reconstructed with different resolutions and pixel accuracy as needed for different target devices such as in applications of World Wide Web and image archiving.

Real-time encoding and decoding: for real-time applications, the coding scheme should be capable of compressing and decompressing with a single sequential pass. Of course, the optimal performance cannot be guaranteed in this case.

Fixed-rate, fixed-size, and limited workspace memory: the requirement of fixed bit rate allows decoder to run in real time through channels with limited bandwidth. The limited memory space is required by the hardware implementation of decoding.

There are also some other requirements, such as backwards compatibility with JPEG, open architecture for optimizing the system for different image types and applications, interface with MPEG-4, and so on. All these requirements have been seriously considered during the development of JPEG2000. It is no doubt that the basic requirement on the coding performance at very low bit rate for still image coding have been achieved by using

8.3.1.2 Parts of JPEG2000

JPEG2000 consists of several parts. Some parts have become International Standards (ISs), while some are still in the developing stage. We first present in this subsection some completed parts, then introduce some parts that are still in their evolution stages.

Part 1 of JPEG2000 is entitled as JPEG2000 Image Coding System: Core Coding System. This is the counterpart of the JPEG baseline system and was issued as an IS in December 2000. It is royalty free.

Part 2 is entitled as JPEG2000 Image Coding Systems: Extension. It involves more advanced technologies, higher computational complexity, and provides enhanced performance, compared with those in Part 1.

Part 3 is Motion JPEG2000 (MJP2). It encodes motion pictures one frame by one frame using JPEG2000 technologies. As a result, it offers random access to any frame in the motion pictures and is much less complicated than MPEG coding schemes. It has been adopted by Hollywood as a format for digital cinema. It is also used for digital cameras.

Part 4 is Conformance Testing.

Part 5 is about Reference Software for Part 1. Two available implementations are as follows [rabbani 2001]. One is a C implementation by the Image Power and University of British Columbia [adams 2000] and another is a Java implementation by the JJ2000 group [JJ2000].

Part 6 is Compound Image File Format for document scanning and fax applications [rabbani 2001].

While the above-mentioned six parts have become ISs before 2004, there are some parts that are still in the working stage. For instance, Part 8, Secure JPEG2000, abbreviated as JPSEC, had its IS published in April 2007. Part 11 is about Wireless JPEG2000 (JPWL) and is close to the completion of IS. JPEG2000 is still going on with some new issues and new parts at this writing.

8.3.2 Verification Model of JPEG2000

As in other standards such as MPEG-2 and MPEG-4, the verification model (VM) plays an important role during the development of standards. This is because the VM or TM (test model for MPEG-2) is a platform for verifying and testing the new techniques before they are adopted by the standards. The VM is updated by completing a set of core experiments from one meeting to another. Experience has shown that the decoding part of the final version of VM is very close to the final standard. Therefore, to give an overview of the related wavelet transform parts of the JPEG2000, we start to introduce the newest version of JPEG2000 VM [jpeg2000 vm]. The VM of JPEG2000 describes the encoding process, decoding process, and the bitstream syntax, which eventually completely defines the functionality of the existing JPEG2000 compression system.

The newest version of JPEG2000 verification model, currently VM 4.0, was revised on April 22, 1999. In this VM, the final convergence has not been reached, but several candidates have been introduced. These techniques include a DCT-based coding mode, which is currently the baseline JPEG, and a wavelet-based coding mode. In the wavelet-based coding mode, several algorithms have been proposed: overlapped spatial segmented wavelet transform (SSWT), nonoverlapped SSWT, and the embedded block-based coding with optimized truncation (EBCOT). Among these techniques, and according to the consensus, the EBCOT had been included into the final JPEG2000 standard.

FIGURE 8.9
Example of subblock partitioning for a block of 64 × 64.

B_i^1	B_i^2	B_i^3	B_i^4
B_i^5	B_i^6	B_i^7	B_i^8
B_i^9	B_i^{10}	B_i^{11}	B_i^{12}
B_i^{13}	B_i^{14}	B_i^{15}	B_i^{16}

The basic idea of EBCOT is the combination of block coding with wavelet transform. First the image is decomposed into subbands using the wavelet transform. The wavelet transform is not restricted to any particular decomposition. However, the Mallat wavelet provides the best compression performance on average for natural images; therefore, the current bitstream syntax is restricted to the standard Mallat wavelet transform in VM 4.0. After decomposition, each subband is divided into 64 × 64 blocks except at image boundaries where some blocks may have smaller sizes. Every block is then coded independently. For each block, a separate bitstream is generated without utilizing any information from other blocks. The key techniques used for coding include embedded quad-tree algorithm and fractional bit-plane coding.

The idea of embedded quad-tree algorithm is that it uses a single bit to represent whether or not each leading bit-plane contains any significant samples. The quad-tree is formed in the following way. The subband is partitioned into a basic block. The basic block size is 64 × 64. Each basic block is further partitioned into 16 × 16 subblocks, as shown in Figure 8.9. Let $\sigma^j(B_i^k)$ denote the significance of subblock, B_i^k (k is the kth subblock as shown in Figure 8.9), in jth bit-plane of ith block. If one or more samples in the subblock have the magnitude greater than 2^j, then $\sigma^j(B_i^k) = 1$; otherwise, $\sigma^j(B_i^k) = 0$. For each bit-plane, the information concerning the significant subblocks is first encoded. All other subblocks can then be bypassed in the remaining coding procedure for that bit-plane. To specify the exact coding sequence, we define a two-level quad-tree for the block size of 64 × 64 and subblock size of 16 × 16. The level-1 quads, $Q_i^1[k]$, consist of four subblocks, B_i^1, B_i^2, B_i^3, and B_i^4 from Figure 8.9. In the same way, we define level-2 quads, $Q_i^2[k]$, to be 2 × 2 groupings of level-1 quads. Let $\sigma^j(Q_i^1[k])$ denote the significance of the level-1 quad, $Q_i^1[k]$, in jth bit-plane. If at least one member subblock is significant in the jth bit-plane then $\sigma^j(Q_i^1[k]) = 1$; otherwise, $\sigma^j(Q_i^1[k]) = 0$. At each bit-plane, the quad-tree coder visits the level-2 quad first and followed by level-1 quads. When visiting a particular quad, $Q_i^L[k]$ ($L = 1$ or 2, is the number of level), the coder sends the significance of each of the four child quads, $\sigma^j(Q_i^L[k])$, or subblocks, $\sigma^j(B_i^k)$, as appropriate, except if the significance value can be deduced from the decoder. Under following three cases, the significance may be deduced by the decoder: (1) the relevant quad or subblock was significant in the previous bit-plane; (2) the entire subblock is insignificant; or (3) this is the last child or subblock visited in $Q_i^L[k]$ and all earlier quads or subblocks are insignificant.

The idea of bit-plane coding is to code the most significant bit first for all samples in the subblocks with entropy coding and to send the resulting bits. Then, the next most significant bit will be coded and sent; this process will be continued until all bit-planes have been coded and sent. This kind of bitstream structure can be used for robust transmission. If the bitstream is truncated due to transmission error or some other reason, then some or all the samples in the block may lose one or more least significant bits. This will be equivalent to having used a coarser quantizer for the relevant samples and we can still obtain a reduced quality reconstructed image. The idea of fractional bit-plane coding is to code each bit-plane with four passes: forward significance propagation pass, backward significance propagation pass, magnitude refinement pass, and normalization pass. For the technical detail of fractional bit-plane coding, the interested readers can refer to the VM of JPEG2000 [jpeg2000 vm].

Finally, we briefly describe the optimization issue of EBCOT. The encoding optimization algorithm is not a part of standard, as the decoder does not need to know how the encoder generates the bitstream. From the viewpoint of the standard, the only requirement from the decoder to the encoder is that the bitstream must be compliant with the syntax of standard. However, from the other side, the bitstream syntax could always be defined to favor certain coding algorithms for generating optimized bitstreams. The optimization algorithm described here is justified only if the distortion measure adopted for the code blocks is additive. That is, the final distortion, D, of the whole reconstructed image should satisfy

$$D = \sum D_i^{T_i} \quad (8.24)$$

where
  D_i is the distortion for block for B_i
  T_i is the truncation point for B_i

Let R be the total number of bits for coding all blocks of the image for a set of truncation point T_i, then

$$R = \sum R_i^{T_i} \quad (8.25)$$

where $R_i^{T_i}$ are the bits for coding block B_i. The optimization process wishes to find the suitable set of T_i values, which minimizes D subject to the constraint $R \leq R_{max}$. R_{max} is the maximum number of bits assigned for coding the image. The solution is obtained by the method of Langrage multipliers:

$$L = \sum (R_i^{T_i} - \lambda D_i^{T_i}) \quad (8.26)$$

where the value λ must be adjusted until the rate obtained by the truncation points, which minimize the value of L, satisfy $R = R_{max}$. From Equation 8.26, we have a separate trivial optimization problem for each individual block. Specially, for each block, B_i, we find the truncation point, T_i, which minimizes the value $(R_i^{T_i} - \lambda D_i^{T_i})$. This can be achieved by finding the slope turning point of rate distortion curves. In the VM, the set of truncation points and the slopes of rate distortion curves are computed immediately after each block is coded, and we only store enough information to later determine the truncation points which correspond to the slope turning points of rate distortion curves. This information is generally much smaller than the bitstream itself which is stored for the block. Also, the search for the optimal λ is extremely fast and occupies a negligible proportion of the overall computation time.

8.3.3 An Example of Performance Comparison between JPEG and JPEG2000

Before ending this chapter, we present an example to compare the performance of JPEG and JPEG2000 in low bit rate compression. We apply both JPEG and JPEG2000 algorithm to one of the JPEG2000 test images, named Woman or N1A, with 0.1 bpp (bits per pixel). At this rather low bit rate, the advantage of JPEG2000 over JPEG is rather obvious. At the 0.1 bpp, the PSNR of the JPEG2000 compressed Woman image is 25.50 dB, while the JPEG compressed Woman image is 23.91 dB. From the human visual system point of view, the superior visual quality of the JPEG2000 compressed image is rather obvious because it can be clearly observed that there is severe distortion called false-countering (as discussed in Section 1.2.2.2) in the JPEG compressed Woman image (see Figure 8.10).

(a)   (b)

FIGURE 8.10
Performance comparison between the JPEG compressed and JPEG2000 compressed Woman image at 0.1 bits/pixel (bits per pixel). (a) JPEG compressed with PSNR 23.91 dB (b) JPEG2000 compressed with PSNR 25.50 dB.

8.4 Summary

In this chapter, image coding using wavelet transform has been introduced. First, an overview of wavelet theory was given, and second, the principles of image coding using wavelet transform have been presented. Additionally, two particular embedded image coding algorithms have been explained, namely the embedded zero tree (EZW) and set partitioning in hierarchical trees (SPIHT). Finally, the new standard for still image coding, JPEG2000, which adopts the wavelet transform as its core technique, has been described.

Exercises

1. For a given function, the Mexican hat wavelet,

$$f(t) = \begin{cases} 1, & \text{for } |t| \leq 1 \\ 0, & \text{otherwise} \end{cases},$$

   use Equations 8.3 and 8.4 to derive a closed-form expression for the continuous wavelet transform, $\psi_{ab}(t)$.

2. Consider the dilation equation

$$\varphi(t) = \sqrt{2} \sum_k h(k)\, \varphi(2t - k)$$

How does $\varphi(t)$ change if $h(k)$ is shifted? Specifically, let

$$g(k) = h(n - l)$$
$$u(t) = 2^{1/2} \sum_k g(k) u(2t - k)$$

How does $u(t)$ relate to $\varphi(t)$?

3. Let $\varphi_a(t)$ and $\varphi_b(t)$ be two scaling functions generated by the two scaling filters $h_a(k)$ and $h_b(k)$. Show that the convolution $\varphi_a(t) * \varphi_b(t)$ satisfies a dilation equation with $h_a(k) * h_b(k)/\sqrt{2}$.

4. In the applications of denoising and image enhancement, how can the wavelet transform improve the results?

5. For a given function

$$f(t) = \begin{cases} 0 & t < 0 \\ t & 0 \leq t < 1 \\ 1 & t \geq 1 \end{cases}$$

show that the wavelet transform of $f(t)$ will be

$$W(a, b) = \text{sgn}\left\{|a|^{-\frac{1}{2}}\left[2f\left(b + \frac{a}{2}\right) - f(b) - f(b + a)\right]\right\}$$

where $\text{sgn}(x)$ is the signum function defined as

$$\text{sgn}(x) = \begin{cases} -1 & t < 0 \\ 1 & t > 0 \\ 0 & t = 0 \end{cases}$$

6. Given an 8×8 block of pixels from the central portion (255:262, 255:262) of Barbara image, whose 8×8 pixel values are shown below.

144	163	194	210	195	151	136	191
170	194	209	200	162	136	178	226
185	200	205	183	152	159	207	218
186	201	188	167	172	197	200	176
204	194	166	171	203	199	162	167
208	167	163	203	212	164	155	215
180	148	186	219	181	141	191	234
141	170	217	198	146	166	231	194

a. Apply three-level 9/7, or 5/3, or Haar wavelet transform to this 8×8 image block. If you have difficulty on this, the result of 5/3 is listed below.
Three-level integer 5/3 discrete wavelet transform (DWT)

144	45	26	-49	-6	15	-15	55
56	-63	-15	1	8	3	-21	7
9	-16	56	111	9	-14	18	6
-32	-2	62	61	-30	32	-41	32
5	11	-15	6	5	4	-16	15
-9	2	-5	10	7	-8	13	-32
16	-13	20	-14	-6	4	-5	36
-39	33	-34	26	26	-19	22	-80

b. Then, following the examples shown in [shapiro 1993] or the example shown in [usevitch 2001], apply the EZW method to this 8×8 wavelet coefficient array completely. It means that you come up with a set of four different types of symbols (positive significant coefficient, negative significant coefficient, zerotree root, and isolated zero), which represent this 8×8 three-level wavelet coefficient 2-D array. Provide your coding results in terms of the four types of symbols in tables.

c. Comment on the efficiency and the nature embedding code of the EZW.

References

[adams 2000] M.D. Adams and F. Kossentini, JasPer: A software-based JPEG-2000 codec implementation, Proceedings of the IEEE International Conference on Image Processing, Vancouver, CA, September 2000. Also, refer to the JasPer home page at http://www.ece.ubc.ca/~mdadams/jasper/.

[antonini 1992] M. Antonini, M. Barlaud, P. Mathieu, and I. Daubechies, Image coding using wavelet transform, *IEEE Transactions on Image Processing*, 1, 205–220, April 1992.

[castleman 1996] K.R. Castleman, *Digital Image Processing*, Prentice-Hall, Englewood Cliffs, NJ, 1996.

[cohen 1989] L. Cohen, Time–frequency distributions—A review, *Proceedings of the IEEE*, 77, 7, 941–981, July 1989.

[daubechies 1992] I. Daubechies, *Ten Lectures on Wavelets*, CBMS-NSF series in Applied Mathematics, Philadelphia, SIAM, 1992.

[daubechies 1998] I. Daubechies, W. Sweldens, Factoring Wavelet Transform into Lifting Steps, *Journal of Fourier Analysis*, 4, 3, 247–269, 1998.

[grossman 1984] A. Grossman and J. Morlet, Decompositions of hardy functions into square integrable wavelets of constant shape, *SIAM Journal of Mathematical Analysis*, 15, 4, 723–736, July 1984.

[jayant 1984] N.S. Jayant and P. Noll, *Digital Coding of Waveforms*, Prentice-Hall, Englewood Cliffs, NJ, 1984.

[jj2000] JJ2000: An implementation of JPEG2000 in JAVA®, available at http://jj2000.epfl.ch

[jpeg2000 vm] JPEG2000 Verification Model 4.0 (Technical description), sc29wg01 N1282, April 22, 1999.

[mpeg4] ISO/IEC 14496-2, Coding of audio-visual objects, November 1998.

[rabbani 2001] M. Rabbani and R. Joshi, An overview of the JPEG 2000 still image compression standard, ISO/IEC JTC 1/SC 29/WG1 N2233, July 2001.

[rioul 1991] O. Rioul and M. Vetterli, Wavelets and signal processing, *IEEE Signal Processing Magazine*, 8, 4, 14–38, October 1991.

[said 1996] A. Said and W.A. Pearlman, A new fast and efficient image codec based on set partitioning in hierarchical trees, *IEEE Transactions on Circuits and Systems for Video Technology*, 6, 243–250, June 1996.

[shapiro 1993] J. Shapiro, Embedded image coding using zerotrees of wavelet coefficients, *IEEE Transactions on Signal Processing*, 41, 12, 3445–3462, December 1993.

[skodras 2001] A. Skodras, C. Christopoulos, and T. Ebrahimi, The JPEG2000 Still Image Compression Standard, *IEEE Signal Processing Magazine*, 18, 5, 36–58, September 2001.

[sweldens 1995] W. Sweldens, The lifting scheme: A new philosophy in biorthogonal wavelet constructions, *Proceedings of SPIE*, 2569, 68–79, 1995.

[taubman 2000] D. Taubman, High performance scalable image compression with EBCOT, *IEEE Transaction on Image Processing*, 8, 7, 1158–1170, July 2000.

[usevitch 2001] B.E. Usevitch, A tutorial on modern lossy wavelet image compression: Foundations of JPEG2000, *IEEE Signal Processing Magazine*, 18, 5, 22–35, September 2001.

[uytterhoeven 1999] G. Uytterhoeven, Wavelets: Software and applications, doctoral dissertation, Department of Computer Science. K.U. Leuven, Belgium, April 1999.

[vetterli 1984] M. Vetterli, Multidimensional subbands coding: Some theory and algorithms, *Signal Processing*, 6, 97–112, February 1984.

[vetterli 1995] M. Vetterli and J. Kovacevic, *Wavelets and Subband Coding*, Prentice-Hall, Englewood Cliffs, NJ, 1995.

[woods 1986] J. Woods and S. O'Neill, Subband coding of images, *IEEE Transactions Acoustics Speech and Signal Processing*, 34, 1278–1288, October 1986.

[xuan 2002] G. Xuan, J. Zhu, J. Chen, Y.Q. Shi, Z. Ni, and W. Su, Distortionless data hiding based on integer wavelet transform, *IEEE Electronics Letters*, 38, 25, 1646–1648, December 2002.

9
Nonstandard Still Image Coding

In this chapter, we introduce three nonstandard image coding techniques: vector quantization (VQ) [nasrabadi 1988] fractal coding [barnsley 1993; jacquin 1993; fisher 1994], and model-based coding [li 1994].

9.1 Introduction

The VQ, fractal coding, and model-based coding techniques have not been adopted by any image coding standard. However, due to their unique features these techniques may find some special applications. VQ is an effective technique for performing data compression. Theoretically, VQ is always better than scalar quantization because it fully exploits the correlation between components within the vector. The optimal coding performance will be obtained when the dimension of vector approaches to infinity, and then the correlation between all components is exploited for compression. Another very attractive feature of image VQ is that its decoding procedure is very simple since it only consists of table lookups. However, there are two major problems with image VQ techniques. The first is that the complexity of VQ exponentially increases with the increasing dimensionality of vectors. Therefore, for VQ, it is important to solve the problem of how to design a practical coding system which can provide a reasonable performance under a given complexity constraint. The second major problem of image VQ is the need of a codebook which causes several problems in practical application, such as generating a universal codebook for a large number of images, scaling the codebook to fit the bit rate requirement and so on. Recently, the lattice VQ schemes have been proposed to address these problems [li 1997].

Fractal theory has a long history. Fractal-based techniques have been used in several areas of digital image processing, such as image segmentation, image synthesis, and computer graphics. But only recently it has been extended to the applications of image compression [jacquin 1993]. A fractal is a geometric form, which has the unique feature of having extremely high visual self-similar irregular details while containing very low information content. Several methods for image compression have been developed based on different characteristics of fractals. One method is based on iterated function system (IFS) proposed in [barnsley 1988]. This method uses the self-similar and self-affine property of fractals. Such a system consists of sets of transformations, including translation, rotation, and scaling. In the encoder side of fractal image coding system, a set of fractals is generated from the input image. These fractals can be used to reconstruct the image at the decoder side. Because these fractals are represented by very compact fractal transformations, they require a very small amount of data to be expressed and stored as formulas. Therefore, the information needed to be transmitted is very small. The second fractal image coding method is based on the fractal dimension [jang 1990; lu 1993]. Fractal dimension is a good representation of roughness of image surfaces. In this method, the image is first segmented using the fractal dimension and then the resulted uniform segments can be efficiently coded using the properties of human visual system (HVS). Another fractal image coding scheme is based on fractal geometry, which is used to measure the length of a curve with a yardstick [walach 1986]. The details of these coding methods are discussed in Section 9.3.

The basic idea of model-based coding is to reconstruct an image with a set of model parameters. The model parameters are then encoded and transmitted to the decoder. At the decoder, the decoded model parameters are used to reconstruct the image with the same model used at the encoder. Therefore, the key techniques in the model-based coding are image modeling, image analysis, and image synthesis.

9.2 Vector Quantization

9.2.1 Basic Principle of Vector Quantization

An N-level vector quantizer, Q, is a mapping from a K-dimensional vector set $\{V\}$, into a finite codebook, $W = \{w_1, w_2, \ldots, w_N\}$:

$$Q: V \to W \qquad (9.1)$$

In other words, it assigns an input vector, v, to a representative vector (code word), w, from a codebook, W. The vector quantizer, Q, is completely described by the codebook, $W = \{w_1, w_2, \ldots, w_N\}$, together with the disjoint partition, $R = \{r_1, r_2, \ldots, r_N\}$, where

$$r_i = \{v: Q(v) = w_i\} \qquad (9.2)$$

and w and v are K-dimensional vectors. The partition should identically minimize the quantization error [gersho 1982]. A block diagram of the various steps involved in image VQ is depicted in Figure 9.1.

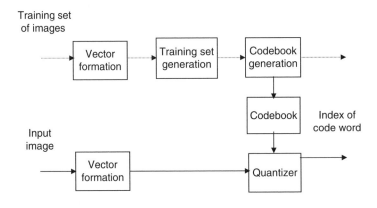

FIGURE 9.1
Principle of image vector quantization (VQ). The dashed lines correspond to the training set generation, codebook generation, and transmission (if it is necessary).

The first step of image VQ is the image formation. The image data is first partitioned into a set of vectors. A large number of vectors from various images are then used to form a training set. The training set is used to generate a codebook, normally using an iterative clustering algorithm. The quantization or coding step involves searching, for each input vector, the closest code word in the codebook. Then the corresponding index of the selected code word is coded and transmitted to the decoder. At the decoder, the index is decoded and converted to the corresponding vector with the same codebook as at the encoder by look-up table. Thus, the design decisions in implementing image VQ include (1) vector formation, (2) training set generation, (3) codebook generation, and (4) quantization.

9.2.1.1 Vector Formation

The first step of VQ is vector formation; that is, the decomposition of the images into a set of vectors. Many different decompositions have been proposed; examples include the intensity values of a spatially contiguous block of pixels [gersho/ramamuthi 1982; baker 1983]; these same intensity values but now normalized by the mean and variance of the block [murakami 1982]; the transformed coefficients of the block pixels [li 1995]; and the adaptive linear predictive coding coefficients for a block of pixels [sun 1984]. Basically, the approaches of vector formation can be classified into two categories: direct spatial or temporal and feature extraction. Direct spatial or temporal is a simple approach to forming vectors from the intensity values of a spatial or temporal contiguous block of pixels in an image or an image sequence. A number of image VQ schemes have been investigated with this method. The other kind of methods is feature extraction. An image feature is a distinguishing primitive characteristic. Some features are natural in the sense that such features are defined by the visual appearance of an image, while the other so-called artificial features result from specific manipulations or measurements of images or image sequences. In vector formation, it is well known that the image data in spatial domain can be converted to a different domain so that subsequent quantization and joint entropy encoding can be more efficient. For this purpose, some features of image data, such as transformed coefficients, block means, can be extracted and vector quantized. The practical significance of feature extraction is that it can result in the reduction of vector size, and consequently, reduce the complexity of coding procedure.

9.2.1.2 Training Set Generation

An optimal vector quantizer should ideally match the statistics of the input vector source. However, if the statistics of an input vector source is unknown, a training set, representative of the expected input vector source, can be used to design the vector quantizer. If the expected vector source has a large variance, then a large training set is needed. To alleviate the implementation complexity caused by large training set, the input vector source can be divided to the subsets. For example in [gersho 1982], the single input source is divided into edge and shade vectors, and then the separate training sets are used to generate the separate codebooks, respectively. Those separate codebooks are then concatenated into a final codebook. In other methods, small local input sources corresponding to portions of the image are used as the training sets; thus the codebook can better match the local statistics. However, the codebook needs to be updated to track the changes in local statistics of the input sources. This may increase the complexity and reduce the coding efficiency. Practically, in most coding systems a set of typical images is selected as the training set and used to generate the codebook. The coding performance can then be insured for the images with the training set or those not in training set but with statistics similar to those in the training set.

9.2.1.3 Codebook Generation

The key step of conventional image VQ is the development of a good codebook. The optimal codebook, using the mean squared error (MSE) criterion, must satisfy two necessary conditions [gersho 1982]. First, the input vector source is partitioned into a pre-decided number of regions with the minimum distance rule. The number of regions is decided by the requirement of the bit rate, or compression ratio and coding performance. Second, the code word or the representative vector of this region is the mean value, or the statistical center, of the vectors within the region. Under these two conditions, a generalized Lloyd clustering algorithm proposed by Linde, Buzo, and Gray (the so-called LBG algorithm [linde 1980]) has been extensively used to generate the codebook. The clustering algorithm is an iterative process, minimizing a performance index calculated from the distances between the sample vectors and their cluster centers. The LBG clustering algorithm can only generate a codebook with a local optimum, which depends on the initial cluster seeds. Two basic procedures have been used to obtain the initial codebook or cluster seeds. In the first approach, the starting points involve finding a small codebook with only two code words, and then recursively splitting the codebook until the required number of code words is obtained. This approach is referred to as binary splitting. The second starts with initial seeds for the required number of code words, these seeds being generated by preprocessing the training sets. To address the problem of local optimum, Equitz [equitz 1989] proposed a new clustering algorithm, the pairwise nearest neighbor (PNN) algorithm. The PNN algorithm begins with a separate cluster for each vector in the training set and merges two clusters at a time until the desired codebook size is obtained. At the beginning of the clustering process, each cluster contains only one vector. In the following process, two closest vectors in the training set are merged to their statistical mean value, in such a way the error incurred by replacing these two vectors with a single code word is minimized. The PNN algorithm significantly reduces computation complexity without sacrificing performance. This algorithm can also be used as an initial codebook generation for LBG algorithm.

9.2.1.4 Quantization

Quantization in the context of a VQ involves selecting a code word in the codebook for each input vector. The optimal quantization, in turn, implies that for each input vector, v, the closest code word, w_i, is found as shown in Figure 9.2. The measure criterion could be either MSE, absolute error, or other distortion measures.

A full search quantization is an exhaustive search process over the entire codebook for finding the closest code word as shown in Figure 9.3a. It is optimal for the given codebook, but the computation is more expensive. An alternative approach is a tree-search quantization, where the search is carried out based on a hierarchical partition. A binary tree-search is shown in Figure 9.3b. Tree-search is much faster than full search, but it is clear that the tree-search is suboptimal for the given codebook and requires more memory for the codebook.

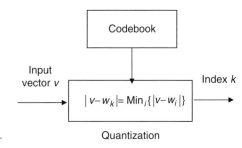

FIGURE 9.2
Principle of vector quantization (VQ).

Nonstandard Still Image Coding

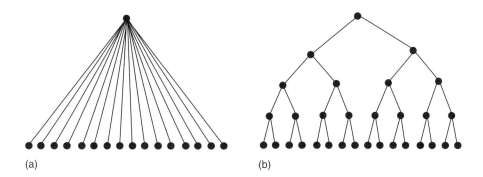

FIGURE 9.3
(a) Full search quantization and (b) binary tree-search quantization.

9.2.2 Several Image Coding Schemes with Vector Quantization

In this section, we present several image coding schemes using VQ that include residual VQ, classified VQ, transform domain VQ, predictive VQ, and block truncation coding (BTC), which can be seen as a binary VQ.

9.2.2.1 Residual VQ

In the conventional image VQ, the vectors are formed by spatially partitioning the image data into blocks of 8×8 or 4×4 pixels. In the original spatial domain the statistics of vectors may be widely spread in the multidimensional vector space. This causes the difficulty for generating the codebook with a finite size and limits the coding performance. Residual VQ is proposed to alleviate this problem. In residual VQ, the mean of the block is extracted and coded separately. The vectors are formed by subtracting the block mean from the original pixel values. This scheme can be further modified by considering the variance of the blocks. The original blocks are converted to the vectors with zero mean and unit standard deviation with the following conversion formula [murakami 1982]:

$$m_i = \frac{1}{K} \sum_{j=0}^{K-1} s_j \tag{9.3}$$

$$x_j = \frac{(s_j - m_i)}{\sigma_i} \tag{9.4}$$

$$\sigma_i = \left[\frac{1}{K} \sum_{j=0}^{K-1} (s_j - m_i)^2 \right]^{\frac{1}{2}} \tag{9.5}$$

where
  m_i is the mean value of ith block
  σ_i is the variance of ith block
  s_j is the pixel value of pixel j ($j = 0, \ldots, K-1$) in the ith block
  K is the total number of pixels in the block
  x_j is the normalized value of pixel j

The new vector X_i is now formed by x_j ($j = 0, 1, \ldots, K-1$):

$$X_i = [x_0, x_1, \ldots, x_K]_i \tag{9.6}$$

With the above normalization, the probability function $P(X)$ of input vector X is approximately similar for image data from different scenes. Therefore, it is easy to generate a codebook for the new vector set. The problem of this method is that the mean and variance values of blocks have to be coded separately. This increases the overhead and limits the coding efficiency. Several methods have been proposed to improve the coding efficiency. One of these methods is to use predictive coding to code the block mean values. The mean value of the current block can be predicted by the one of previously coded neighbors. In such a way, the coding efficiency increases as the use of inter-block correlation.

9.2.2.2 Classified VQ

In image VQ, the codebook is usually generated using training set under constraint of minimizing the MSE. This implies that the code word is the statistical mean of the region. During the quantization, each input vector is replaced by its closest code word. Therefore, the coded images usually suffer from edge distortion at very low bit rates since edges are smoothed by the operation of averaging with the small-sized codebook. To overcome this problem, we can classify the training vector as edge vectors and shade vectors [gersho 1982]. Two separate codebooks can then be generated with the two types of training sets, respectively. Each input vector can be coded by the appropriate code word in the codebook. However, the edge vectors can be further classified into many types according to their location and angular orientation. The classified VQ can be extended into a system, which contains many sub-codebooks; each represents a type of edges. However, this would increase the complexity of the system and would be hard to implement in practical applications.

9.2.2.3 Transform Domain VQ

The VQ can be performed in the transform domain. A spatial block of 4×4 or 8×8 pixels is first transformed to the 4×4 or 8×8 transformed coefficients. There are several methods to form vectors with transformed coefficients. In the first method, a number of high-order coefficients can be discarded because most of the energy is usually contained in the low-order coefficients for most of the blocks. This reduces the VQ computational complexity at the expense of a small increase of distortion. However, for some active blocks, the edge information is contained in the high frequencies, or high-order coefficients. It will cause serious subjective distortion by discarding high frequencies. In the second method, the transformed coefficients are divided into several bands and each band is used to form its corresponding vector set. This method is equivalent to the classified VQ in spatial domain. An adaptive scheme is then developed by using two kinds of vector formation methods. The first method is used for the blocks containing the moderate intensity variation and the second method is used for the blocks with high spatial activities. However, the complexity increases, as more codebooks are needed in such kinds of adaptive coding systems.

9.2.2.4 Predictive VQ

The vectors are usually formed by the spatially consecutive blocks. The consecutive vectors are then highly statistically dependent. Therefore, better coding performance can be achieved if the correlation between vectors is exploited. Several predictive VQ schemes have been proposed to address this problem. One kind of predictive VQ is finite state VQ [dunham 1985; foster 1985]. The finite-state VQ is similar to a trellis coder. In finite state VQ, the codebook consists of a set of sub-codebooks. A state variable, used to specify which sub-codebook should be selected for coding the input vector. The information about state variable must be inferred from the received sequence of state symbols and initial state such as in a trellis coder [stewart 1982]. Therefore, there is no side information or no overhead is needed to be transmitted to the decoder. The new encoder state is a function of previous encoder state and the selected sub-codebook. This permits the decoder to track the encoder state if the initial condition is known. The finite state VQ needs additional memory to store the previous state, but it takes advantage of correlation between successive input vectors by choosing the appropriate codebook for the given history. It should be noted that the minimum distortion selection rule of conventional VQ is not necessarily optimum for finite state VQ for a given decoder because a low distortion code word may lead to a bad state and hence to poor long-term behavior. Therefore, the key design issue of finite state VQ is to find a good next-state function.

Another predictive VQ was proposed in [hang 1985]. In this system, the input vector is formed in such a way that the current pixel is as the first element of the vector and the previous inputs as the remaining elements in the vector. The system is like a mapping or a recursive filter which is used to predict the next pixel. The mapping is implemented by a vector quantizer look-up table and provides the predictive errors.

9.2.2.5 Block Truncation Coding

In the block truncation coding (BTC) [delp 1979], an image is first divided into 4×4 blocks. Each block is then coded individually. The pixels in each block are first converted into two level signals by using the first two moments of the block:

$$a = m + \sigma\sqrt{\frac{N-q}{q}}$$
$$b = m - \sigma\sqrt{\frac{q}{N-q}}$$
(9.7)

where
m is the mean value of the block
σ is the standard deviation of the block
N is the number of total pixels in the block
q is the number of pixels which are greater in value than m

Therefore, each block can be described by the values of block mean, variance, and a binary-bit plane, which indicates whether the pixels have the values above or below the block mean. The binary-bit plane can be seen as a binary vector quantizer. If the mean and variance of the block are quantized to 8 bits, then 2 bits/pixel is achieved for the blocks of 4×4 pixels. The conventional BTC scheme can be modified to increase the coding efficiency. For example, the block mean can be coded by DPCM coder that exploits the inter-block correlation. The bit plane can be coded with an entropy coder on the patterns [udpikar 1987].

9.2.3 Lattice VQ for Image Coding

In the conventional image VQ schemes, there are several issues, which cause some difficulties for the practical applications of image VQ. The first problem is the limitation of vector dimension. As it is indicated that the coding performance of VQ increases as vector dimension the coding complexity exponentially increases at the same time as increasing vector dimension. Therefore, in practice, only a small size of vector dimension is possible under the complexity constraint. Another important issue in VQ is the need for a codebook. Much research effort has gone into finding out how to generate a codebook. However, in practical applications, there is another problem of how to scale the codebook for various rate distortion requirements. The codebook generated by LBG-like algorithms with a training set is usually only suitable for a specified bit rate and it does not have the flexibility of codebook scalability. For example, a codebook generated for an image with small resolution may not be suitable for the images with high resolution. Even for the same spatial resolution, different bit rates would require different codebooks. Additionally, the VQ needs a table to specify the codebook and consequently, the complexity of storing and searching the table is too high to have a very large table. This further limits the coding performance of image VQ. These problems become major obstacles of image VQ for implementation. Recently, an algorithm of lattice VQ has been proposed to address these problems [li 1997]. Lattice VQ does not have the above problems. The codebook for lattice VQ is simply a collection of lattice points uniformly distributed over the vector space. Scalability can be achieved by scaling the cell size associated with every lattice points just like in the scalar quantizer by scaling the quantization step. The basic concept of lattice can be found in [conway 1991]. A Typical Lattice VQ scheme is shown in Figure 9.4. There are two steps involved in the image lattice VQ. The first step is to find the closest lattice point for the input vector. The second step is to label the lattice point, i.e., mapping a lattice point to an index. Since lattice VQ does need a codebook, the index assignment is based on lattice labeling algorithm instead of look-up table such as in the conventional VQ. Therefore, the key issue of lattice VQ is to develop an efficient lattice-labeling algorithm. Using this algorithm, the closest lattice point and its corresponding index within a finite boundary can be obtained by calculation at the encoder for each input vector.

At the decoder, the index is converted to the lattice point by the same labeling algorithm. The vector is then reconstructed with the lattice point. The efficiency of a labeling algorithm for lattice VQ is measured by how many bits needed to represent the indices of the lattice points within a finite boundary. We use a two-dimensional (2-D) lattice to explain the lattice labeling efficiency. A 2-D lattice is shown in Figure 9.5.

In Figure 9.5, there are seven lattice points. One method to label these seven 2-D lattice points is to use their coordinates (x,y) to label each point. If we label x and y separately, we need 2 bits to label three values of x and 3 bits to label five possible values of y and thus, we need a total of 5 bits. It is clear that 3 bits are sufficient to label seven lattice points. Therefore, different labeling algorithms may have different labeling efficiency. Several algorithms have been developed for multidimensional lattice labeling. In [conway 1983],

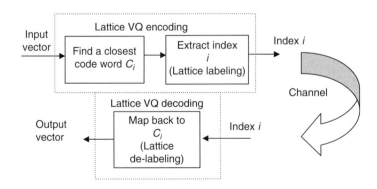

FIGURE 9.4
Block diagram of lattice vector quantization (VQ).

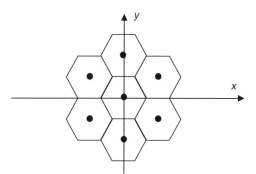

FIGURE 9.5
Labeling a two-dimensional (2-D) lattice.

the labeling method assigns an index to every lattice point within a Voronoi boundary where the shape of the boundary is the same as the shape of Voronoi cells. Apparently, for different dimensions, the boundaries have different shapes. In the algorithm proposed in [laroia 1993], the same method is used to assign an index to each lattice point. However, the boundaries are defined by the labeling algorithm; this algorithm might not achieve a 100% labeling efficiency for a prespecified boundary such as a pyramid boundary. The algorithm proposed in [fischer 1986] can assign an index to every lattice point within a prespecified pyramid boundary and achieves a 100% labeling efficiency, but this algorithm can only be used for the Z^n lattice. In the recent proposed algorithm [wang 1998], the technical breakthrough has been obtained. In this algorithm, a labeling method has been developed for Construction-A and Construction-B lattices [conway 1983], which are very useful for VQ with proper vector dimension such as 16 and achieve 100% efficiency. Additionally, these algorithms are used for labeling lattice points with dimension 16 and provide the minimum distortion. These algorithms are developed based on the relations between lattices and linear block codes. Construction-A and Construction-B are the two simplest ways to construct a lattice from a binary linear block code $C = (n, k, d)$, where n, k, and d are the length, the dimension, and the minimum distance of the code, respectively.

A construct-A lattice is defined as

$$\Lambda_n = C + 2Z^n \tag{9.8}$$

where Z^n is the n-dimensional cubic lattice and C is a binary linear block code. There are two steps involved for labeling a Construct-A lattice. First is to order the lattice points according to the binary linear block code C, and then to order the lattice points associated with a particular nonzero binary code word. For the lattice points associated with nonzero binary code word, two sub-lattices are considered separately. One sub-lattice consists of all the dimensions that have 0 component in the binary code word and the other consists of all the dimensions that have 1 component in the binary code word. The first sub-lattice is considered as a 2Z lattice, whereas the second is considered as a translated 2Z lattice. Therefore, the labeling problem is reduced to label the Z lattice at the final stage.

A Construction-B lattice is defined as

$$\Lambda_n = C + 2D_n \tag{9.9}$$

where D_n is an n-dimensional Construction-A lattice with the definition as

$$D_n = (n, n-1, 2) + 2Z^n \tag{9.10}$$

and C is a binary doubly even linear block code. When n is equal to 16, the binary even linear block code associated with Λ_{16} is $C = (16, 5, 8)$. The method for labeling a Construction-B lattice is similar to the method for labeling a Construction-A lattice with two minor differences. The first difference is that for any vector $y = C + 2x$, $x \in Z^n$, if y is a Construction-A lattice point; and $x \in D_n$, if y is a Construction-B lattice point. The second difference is that C is a binary doubly even linear block code for Construction-B lattices while it is not necessarily doubly even for Construction-A lattices. In the implementation of these lattice point labeling algorithms, the encoding and decoding functions for lattice VQ have been developed in [li 1997]. For a given input vector, an index representing the closest lattice point will be found by the encoding function and for an input index, the reconstructed vector will be generated by the decoding function. In summary, the idea of lattice VQ for image coding is an important achievement for eliminating the need of codebook for image VQ. The development of efficient algorithms for lattice point labeling makes lattice VQ feasible for image coding.

9.3 Fractal Image Coding

9.3.1 Mathematical Foundation

A fractal is a geometric form whose irregular details can be represented by some objects with different scale and angle, which can be described by a set of transformations such as affine transformations. Additionally, the objects used to represent the image irregular details have some form of self-similarity and these objects can be used to represent an image with simple recursive way. An example of fractals is Von Koch curve as shown in Figure 9.6. The fractals can be used to generate an image. The fractal image coding that is based on IFS is the inverse process of image generation with fractals; therefore, the key technology of fractal image coding is the generation of fractals with an IFS.

To explain what an IFS is, we start from the contractive affine transformation. A 2-D affine transformation A is defined as follows:

$$A\begin{bmatrix} x \\ y \end{bmatrix} = \begin{bmatrix} a & b \\ c & d \end{bmatrix}\begin{bmatrix} x \\ y \end{bmatrix} + \begin{bmatrix} e \\ f \end{bmatrix} \tag{9.11}$$

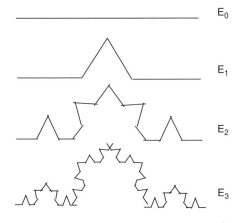

FIGURE 9.6
Construction of the Von Koch curve.

This is a transformation, which consists of a linear transformation followed by a shift or translation and maps points in the Euclidean plane into new points in the another Euclidean plane. We define that a transformation is contractive if the distance of two points P_1 and P_2 in the new plane is smaller than their distance in the original plane, i.e.,

$$d(A(P_1), A(P_2)) < s\, d(P_1, P_2) \tag{9.12}$$

where s is a constant and $0 < s < 1$. The contractive transformations have the property that when the contractive transformations are repeatedly applied to the points in a plane, these points will converge to a fixed point. An IFS is defined as a collection of contractive affine transformations. A well-known example of the IFS contains four following transformations:

$$A_i \begin{bmatrix} x \\ y \end{bmatrix} = \begin{bmatrix} a & b \\ c & d \end{bmatrix} \begin{bmatrix} x \\ y \end{bmatrix} + \begin{bmatrix} e \\ f \end{bmatrix} i = 1, 2, 3, 4 \tag{9.13}$$

This is the IFS of a fern leaf of which parameters are shown in Table 9.1.

The transformation A_1, A_2, A_3, and A_4 are used to generate the stalk, right leaf, left leaf, and main fern, respectively. A fundamental theorem of fractal geometry is that each IFS defines a unique fractal image. This image is referred to as the attractor of the IFS. In other words, an image corresponds to the attractor of an IFS. Now let us explain how to generate the image using the IFS. Let us suppose that an IFS contains N affine transformations, $A_1, A_2, \ldots, A_N$; each transformation has an associated probability, $p_1, p_2, \ldots, p_N$, respectively. Suppose that this is a complete set and the sum of the probability equals to 1, i.e.,

$$p_1 + p_2 + \ldots + p_N = 1 \quad \text{and} \quad p_i > 0 \quad \text{for} \quad i = 0, 1, \ldots, N. \tag{9.14}$$

The procedure of generating an attractor is as follows. For any given point (x_0, y_0) in Euclidean plane, one transformation in the IFS according to its probability is selected and applied to this point to generate a new point (x_1, y_1). Then another transformation is selected according to its probability and applied to the point (x_1, y_1) to obtain a new point (x_2, y_2). This process is repeated over and over again to obtain a long sequence of points: $(x_0, y_0), (x_1, y_1), \ldots, (x_n, y_n), \ldots$. According to the theory of iterated function systems, these points will be converged to an image that is the attractor of the given IFS. The above described procedure is shown in the flowchart of Figure 9.7. With the above algorithm and the parameters in Table 9.1, initially the point can be anywhere within the large square, but after several iterations it will converge onto the fern. The 2-D affine transformations are extended to three-dimensional (3-D) transformations, which can be used to create fractal surfaces with the iterated function systems. This fractal surface can be considered as the gray level or brightness of a 2-D image.

TABLE 9.1
The Parameters of the Iterated Function System (IFS) of a Fern Leaf

	a	b	c	d	e	f
A_1	0	0	0	0.16	0	0.2
A_2	0.2	−0.26	0.23	0.22	0	0.2
A_3	−0.15	0.28	0.26	0.24	0	0.2
A_4	0.85	0.04	−0.04	0.85	0	0.2

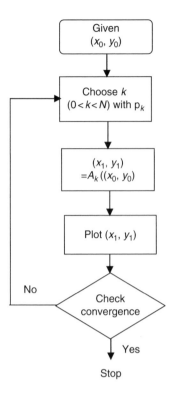

FIGURE 9.7
Flowchart of generating an image with an iterated function system (IFS).

9.3.2 IFS-Based Fractal Image Coding

As it is described in the last section, an IFS can be used to generate a unique image, which is referred to as an attractor of the IFS. In other words, an image is the attractor of an IFS; this image can be simply represented by the parameters of the IFS. Therefore, if we can use an inverse procedure to generate a set of transformations, i.e., an IFS, from an image, then these transformations or the IFS can be used to represent the approximation of the image. The image coding system can use the parameters of the transformations in the IFS instead of the original image data for storage or transmission. As the IFS contains only very limited data such as transformation parameters, this image coding method may result in a very high compression ratio. For example, the fern image is represented by 24 integers or 192 bits (if each integer is represented by 8 bits). This number is much smaller than the number needed to represent the fern image in the way of pixel by pixel. Now the key issue of the IFS-based fractal image coding is to generate the IFS for the given input image. Three methods have been proposed to obtain the IFS [lu 1993]. The first method is the direct method, which directly finds a set of contractive affine transformations from the image based on the self-similarity of the image. The second method partitions an image into the smaller objects whose IFSs are known. These IFSs are used to form a library. The encoding procedure is to look for an IFS from the library for each small object. The third method is called partitioned IFS (PIFS). In this method, the image is first divided into the smaller blocks and then the IFS for each block is found by mapping a larger block into a small block.

In the first direct approach, the image is first partitioned into nonoverlapped blocks in such a way that each block is similar to the whole image and a transformation can map the whole image to the block. The transformation for each individual block may be different. The combination of these transformations can be taken as the IFS of the given image.

Then much less data is required to represent the IFS or the transformations than to transmit or store the given image in the pixel-by-pixel way. For the second approach, the key issue is how to partition the given image into objects whose IFSs are known. The image processing techniques, such as color separation, edge detection, spectrum analysis, and texture-variation analysis can be used for the image partitioning. However, for natural images or arbitrary images, it may be impossible or very difficult to find an IFS whose attractor perfectly covers the original image. Therefore, for most natural images the PIFS method has been proposed [lu 1993]. In this method, the transformations do not map the whole image into small blocks. For encoding an image, the whole image is first partitioned into a number of larger blocks that are referred to as domain blocks. The domain blocks can be overlapped. Then the image is partitioned into a number of smaller blocks, called range blocks. The range blocks do not overlap and the sum of total range blocks covers the whole image. In the third step, a set of contractive transformations is chosen. Each range block is mapped into a domain block with a searching method and a matching criterion. The combination of the transformations is used to form a PIFS. The parameters of PIFS are transmitted to the decoder. It is noted that no domain blocks are transmitted. The decoding starts with a flat background. The iterated process is then applied with the set of transformations. The reconstructed image is then obtained after the process converges. From the above discussion, it is found that there are three main design issues involved in the block fractal image coding system. First is partitioning techniques which include the range block partitioning and domain block partitioning. As mentioned earlier, the domain block is larger than range block. Dividing the image into square blocks is the simplest partitioning approach. The second issue is the choice of distortion measure and searching method. The common distortion measure in the block fractal image coding is the root mean square (RMS) error. The closest matching between range block and transformed domain block is found by the RMS distortion measure. The third is the selection of a set of contractive transformations defined consistently with a partition.

It is noted that the PIFS-based fractal image coding has several similar features with image VQ. Both coding schemes are block-based coding schemes and need a codebook for encoding. For PIFS-based fractal image coding, the domain blocks can be seen as forming a virtual codebook. One difference is that the fractal image coding does not need to transmit the codebook data (domain blocks) to the decoder while VQ needs. The second difference is the block size. For VQ, block size for code vector and input vector is the same while in PIFS fractal coding the size of the domain block is different from the size of the range blocks. Another difference is that in fractal image coding the image itself serves the codebook while this is not true for VQ image coding.

9.3.3 Other Fractal Image Coding Methods

Apart from the IFS-based fractal image coding, there are several other fractal image coding methods. One is the segmentation-based coding scheme using fractal dimension. In this method, the image is segmented into regions based on the properties of the HVS. The image is segmented into the regions; each of these regions is homogeneous in the sense of having similar features in visual perception. This is different from the traditional image segmentation techniques that try to segment an image into regions of constant intensity. For complicated image, good representation of an image needs a large number of small segmentations. However, to obtain high compression ratio, the number of segmentations is limited. The trade-off between image quality and bit rate has to be considered. A parameter, fractal dimension, is used as a measure to control the trade-off. Fractal dimension is a characteristic of a fractal. It is related to a metric property such as the length of a curve and the area of a surface. The fractal dimension can provide a good measure of perceptual roughness of the curve and surface. For example, if we use many segments of straight lines to approximate a curve, with increasing length of straight line, the perceptual rougher curves are represented.

9.4 Model-Based Coding

9.4.1 Basic Concept

In model-based coding, an image model that can be 2-D model for still images or 3-D model for video sequence is first constructed. At the encoder, the model is used to analyze the input image. The model parameters are then transmitted to the decoder. At the decoder, the reconstructed image is synthesized by the model parameters with the same image model used at the encoder. This basic idea of model-based coding is shown in the Figure 9.8. Therefore, the basic techniques in model-based coding are the image modeling, image analysis, and image synthesis techniques. Both image analysis and synthesis are based on the image model. The image modeling techniques used for image coding can normally be divided into two classes: structure modeling and motion modeling. The motion modeling is usually used for video sequences and moving pictures, whereas the structure modeling is usually used for still image coding. The structure model is used for reconstruction of 2-D or 3-D scene model.

9.4.2 Image Modeling

The geometric model is usually used for image structure description. The geometric model can be classified into surface-based description and volume-based description. The major advantage of surface description is that such description is easily converted into surface representation that can be encoded and transmitted. In these models, the surface is approximated by planar polygonal patches such as triangle patches. The surface shape is represented by a set of points that represent the vertices of these triangle meshes. The size of these triangle patches can be adjusted according to the surface complexity. In other words, for more complicated area, more triangle meshes are needed to approximate the surface, whereas for smoothing area, the mesh sizes can be larger or fewer vertices of the triangle meshes are needed to represent the surface. The volume-based description is a natural approach for modeling most of solid world objects. Most existing research work on

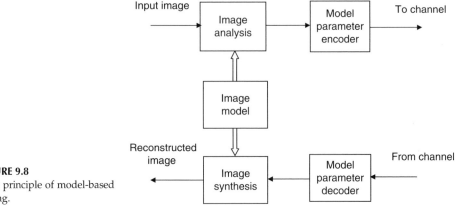

FIGURE 9.8
Basic principle of model-based coding.

volume-based description focuses on the parametric volume description. The volume-based description is, of course, used for 3-D objects or video sequences.

However, model-based coding is successfully applicable only to certain kinds of images since it is very hard to find general image models suitable for most natural scenes. The few successful examples of image models include the human face, head, and body. These models are developed for analysis and synthesis of moving images. The face animation has been adopted by the MPEG-4 visual coding. The body animation is under consideration for the version 2 of MPEG-4 visual coding.

9.5 Summary

In this chapter, three kinds of image coding techniques, VQ, fractal image coding, and model-based coding, which are not used in the current standards, have been presented. All three techniques have several important features such as very high compression ratio for certain kinds of images and very simple decoding procedure (special for VQ). However, due to some limitations these techniques have not been adopted by industry standards. It should be noted that recently the facial model, face animation technique, has been adopted by MPEG-4 visual standard [mpeg4 visual].

Exercises

1. In the modified residual VQ described in Equation 9.5, with 4×4 block size, and 8 bit for each pixel of original image, if we use 8 bits for coding block mean and block variance. We want to obtain the final bit rate is 2 bits/pixel, what codebook size we have to use for coding residual, assuming that we use fixed-length coding (FLC) to code vector indices?
2. In the block truncation coding (BTC) described in Equation 9.7, what is the bit rate for a block size of 4×4 if the mean and variance are both encoded with 8 bits? Do you have any suggestions for reducing the bit rate without seriously affecting the reconstruction quality?
3. Is the codebook generated with the LBG algorithm local optimum? List the several important factors that will affect the quality of codebook generation.
4. In image coding using VQ, what kind of problems will be caused by using codebook in the practical applications (NB: changing bit rate).
5. What is the most important improvement of the lattice VQ over traditional VQ in the practical application? What is the key issue for lattice VQ for image coding application?
6. Write a subroutine to generate a fern leaf (using C).

References

[baker 1983] R.L. Baker and R.M. Gray, Image compression using an-adaptive spatial vector quantization, International Symposium on Circuits and Systems (ISCAS'83), 1983, pp. 55–61.

[barnsley 1988] Michael F. Barnsley and A.E. Jacquin, Application of recurrent iterated function systems, SPIE, 1001, *Visual Communications and Image Processing*, 122–131, 1988.

[barnsley 1993] M.F. Barnsley and L.P. Hurd, *Fractal Image Compression*, AK Peters, Wellesley, MA, 1993.

[conway 1983] J. Conway and N.J.A. Slone, A fast encoding method for lattice codes and quantizers, IEEE Trans. on Information Theory, Vol. IT-29, 1983, 820–824.

[conway 1991] J. Conway and N.J.A. Slone, *Sphere Packings, Lattices and Groups*, Spring-Verlag, New York, 1991.

[delp 1979] E.J. Delp and D.R. Mitchell, Image compression using block truncation coding, *IEEE Transactions on Communications*, COM-27, 9, 1335–1342, September 1979.

[dunham 1985] M. Dunham and R. Gray, An algorithm for the design of labelled-transition finite-state vector quantizer, *IEEE Transactions on Communications*, COM-33, 83–89, May 1985.

[equitz 1989] W.H. Equits, A new vector quantization clustering algorithm, *IEEE Transactions on ASSP*, 37, 1568–1575, October 1989.

[fischer 1986] T.R. Fischer, A paramid vector quantization, IEEE *Transactions on Information Theory*, IT-32, 568–583, 1986.

[fisher 1994] Y. Fisher, *Fractal Image Compression—Theory and Application*, Springer-Verlag, New York, 1994.

[foster 1985] J. Foster, R.M. Gray, and M.O. Dunham, Finite-state vector quantization for waveform coding, *IEEE Transactions on Information Theory*, IT-31, 348–359, May 1985.

[gersho/ramamuthi 1982] A. Gersho and B. Ramamurthi, Image coding using vector quantization, International Symposium on Circuits and Systems (ICASP'82), May 1982, pp. 428–431.

[gersho 1982] A. Gersho, On the structure of vector quantizer, *IEEE Transactions on Information Theory*, IT-28, 157–166, March 1982.

[hang 1985] H.M. Hang and J.W. Woods, Predictive vector quantization of images, *IEEE Transactions on Communications*, COM-33, 1208–1219, November 1985.

[jacquin 1993] A.E. Jacquin, Fractal image coding: A review, *Proceedings of the IEEE*, 81, 10, 1451–1465, October 1993.

[jang 1990] J. Jang, and S.A. Rajala, Segmentation-based image coding using fractals and the human visual system, IEEE International Conference of Acoustics Speech Signal Processing, 1990, pp. 1957–1960.

[laroia 1993] R. Laroia and N. Favardin, A structured fixed rate vector quantizer derived from a variable length scalar quantier: I & II, *IEEE Transaction on Information Theory*, IT-39, 851–876, 1993.

[li 1994] H. Li, A. Lundmark, and R. Forchheimer, Image sequence coding at very low bitrates: A review, *IEEE Transactions on Image Processing*, 3, 5, 589–604, September 1994.

[li 1995] W. Li, and Ya-qin Zhang, Vector-based signal processing and quantization for image and video compression, *Proceedings of IEEE*, 83, 2, 317–335, February 1995.

[li 1997] W. Li et al., A video coding algorithm using vector-based technique, *IEEE Transactions on Circuits and Systems for Video Technology*, 7, 1, 146–157, February 1997.

[linde 1980] Y. Linde, A. Buzo, and R.M. Gray, An algorithm for vector quantizer design, *IEEE Transactions on Communications*, 28, 84–95, 1980.

[lu 1993] G. Lu, Fractal image compression, *Signal Processing: Image Communications* 5, 327–343, 1993.

[mpeg4 visual] ISO/IEC 14496-2, Coding of audio-visual objects, Part 2, December 18, 1998.

[murakami 1982] T. Murakami, K. Asai, and E. Yamazaki, Vector quantization of video signals, *Electronic Letters*, 7, 1005–1006, November 1982.

[nasrabadi 1988] N.M. Nasrabadi and R.A. King, Image coding using vector quantization: A review, *IEEE Transactions on Communications*, COM-36, 8, 957–971, August 1988.

[stewart 1982] L.C. Stewart, R.M. Gray, and Y. Linde, The design of trellis waveform coders, *IEEE Transactions on Communications*, COM-30, 702–710, April 1982.

[sun 1984] H. Sun and M. Goldberg, Image coding using LPC with vector quantization, Proceedings of the IEEE International Conference on Digital Signal Processing, Florence, Italy, September 1984, pp. 508–512.

[udpikar 1987] V.R. Udpikar and J.P. Raina, BTC image coding using vector quantization, *IEEE Transactions on Communications*, COM-35, 352–356, March 1987.

[walach 1986] E. Walach and E. Karnin, A fractal-based approach to image compression, IEEE International Conference on Acoustics Speech Signal Processing, 1986, pp. 529–532.

[wang 1998] C. Wang, H.Q. Cao, W. Li, and K.K. Tzeng, Lattice labeling algorithm for vector quantization, *IEEE Transactions on Circuits and Systems for Video Technology*, 8, 2, 206–220, April 1998.

Part III

Motion Estimation and Compensation

10
Motion Analysis and Motion Compensation

The basic techniques in image coding, specifically, techniques utilized in still image coding were discussed in the previous chapters. From this chapter, we start to address the issue of video sequence compression. To fulfill the task, in this chapter, we first define the concepts of image and video sequences. Then we address the issue of interframe correlation between successive frames. Later, two techniques in exploitation of interframe correlation, frame replenishment, and motion compensated (MC) coding, are discussed. The rest of the chapter covers the concepts of motion analysis and motion compensation in general.

10.1 Image Sequences

In this section, the concept of various image sequences is defined in a theoretical and systematic manner. The relationship between image sequences and video sequences is also discussed.

It is well known that in the 1960s, the advent of the semiconductor computer and the space program swiftly brought the field of digital image processing into public focus. Since then, the field has experienced rapid growth and has entered every aspect of modern technology. Since the early 1980s, digital image sequence processing has been an attractive research area [huang 1981a, 1983]. This is not surprising, because an image sequence, as a collection of images, may provide more information than a single image frame. The increased computational complexity and memory space associated with image sequence processing are becoming more affordable due to more advanced, achievable computational capability. With the tremendous advancements continuously made in VLSI computer and information processing, image and video sequences are evermore indispensable elements of modern life. Although the pace and the future of this development cannot be predicted, one thing is certain: this process is going to drastically change all aspects of our world in the next several decades.

As far as image sequence processing is concerned, it is noted that in addition to temporal image sequences, stereo image pair and stereo image sequences also obtained attention in the mid-1980s [waxman 1986]. The concepts of temporal and spatial image sequences, and the imaging space (which may be considered as a next higher-level unification of temporal and spatial image sequences) may be illustrated as follows.

Consider a sensor located in a specific position in the three-dimensional (3-D) world space. It generates images about the scene, one after another. As time goes by, the images form a sequence. The set of these images can be represented with a brightness function $g(x,y,t)$, where x and y are coordinates on the image planes. This is referred to as a temporal image sequence. This is the basic outline about the brightness function $g(x,y,t)$ dealt with by researchers in both computer vision [e.g., horn 1980] and signal processing fields [e.g., pratt 1979].

Now consider a generalization of the above basic outline. A sensor, as a solid article, can be translated (in three free dimensions) and rotated (in two free dimensions). It is noted that the rotation of a sensor about its optical axis is not counted because the images generated will remain unchanged when this type of rotation takes place. Thus, we can obtain a variety of images when a sensor is translated to different coordinates and rotated to different angles in the 3-D world space. Equivalently, we can imagine that there is an infinite number of sensors in the 3-D world space that occupies all possible spatial coordinates and assumes all possible orientations at each coordinate; i.e., they are located on all possible positions. At one specific moment, all of these images form a set, which can be referred to as a spatial image sequence. When time varies, these sets of images form a much larger set of images, called an imaging space.

Clearly, it is impossible to describe such a set of images by using the above-mentioned $g(x,y,t)$. Instead, it should be described by a more general brightness function,

$$g(x,y,t,\vec{s}), \qquad (10.1)$$

where $\vec{s}$ indicates the sensor's position in the 3-D world space; i.e., the coordinates of the sensor center and the orientation of the optical axis of the sensor. Hence $\vec{s}$ is a five-dimensional (5-D) vector. That is,

$$\vec{s} = (\tilde{x}, \tilde{y}, \tilde{z}, \beta, \gamma), \qquad (10.2)$$

where $\tilde{x}$, $\tilde{y}$, and $\tilde{z}$ represent the coordinates of the optical center of the sensor in the 3-D world space; and β and γ represent the orientation of the optical axis of the sensor in the 3-D world space. More specifically, each sensor in the 3-D world space may be considered associated with a 3-D Cartesian coordinate system such that its optical center is located on the origin and its optical axis is aligned with the OZ axis. In the 3-D world space, we choose a 3-D Cartesian coordinate system as the reference coordinate system. Hence, a sensor with its Cartesian coordinate system coincident with the reference coordinate system has its position in the 3-D world space denoted by $\vec{s} = (0,0,0,0,0)$. An arbitrary sensor position denoted by $\vec{s} = (\tilde{x}, \tilde{y}, \tilde{z}, \beta, \gamma)$ can be described as follows. The sensor's associated Cartesian coordinate system is first shifted from the reference coordinate system in the 3-D world space with its origin settled at $(\tilde{x}, \tilde{y}, \tilde{z})$ in the reference coordinate system. Then it is rotated with the rotation angles β and γ being the same as Euler angles [shu 1991; shi 1994]. Figure 10.1 shows the reference coordinate system and an arbitrary Cartesian coordinate system (indicating an arbitrary sensor position). There, oxy and $o'x'y'$ represent, respectively, the related image planes.

Assume now a world point P in the 3-D space that is projected onto the image plane as a pixel with the coordinates x_P and y_P. Then, x_P and y_P are also dependent on t and $\vec{s}$. That is, the coordinates of the pixel can be denoted by $x_P = x_P(t,\vec{s})$ and $y_P = y_P(t,\vec{s})$. So generally speaking, we have

$$g = g(x_P(t,\vec{s}), y_P(t,\vec{s}), t, \vec{s}). \qquad (10.3)$$

As far as temporal image sequences are concerned, let us take a look at the framework of Pratt [pratt 1979], and Horn and Schunck [horn 1980]. There, $g = g(x_P(t), y_P(t), t)$ is actually a special case of Equation 10.3. That is,

$$g = g(x_P(t, \vec{s} = \text{constant vector}), y_P(t, \vec{s} = \text{constant vector}), t, \vec{s} = \text{constant vector}).$$

In other words, the variation of $\vec{s}$ is restricted to be zero, i.e., $\Delta \vec{s} = 0$. This means the sensor is fixed in a certain position in the 3-D world space.

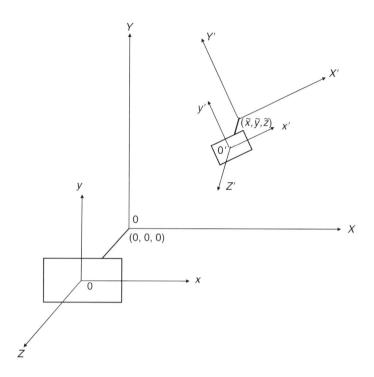

FIGURE 10.1
Two sensors' positions: $\bar{s} = (0,0,0,0,0)$ and $\bar{s} = (\tilde{x}, \tilde{y}, \tilde{z}, \beta, \gamma)$.

Obviously, an alternative is to define the imaging space as a set of all temporal image sequences; i.e., those taken by sensors located at all possible positions in the 3-D world space. Stereo image sequences can thus be viewed as a proper subset of the imaging space, just like a stereo pair of images can be considered as a proper subset of a spatial image sequence.

In summary, the imaging space is a collection of all possible forms assumed by the general brightness function $g(x, y, t, \bar{s})$. Each picture, taken by a sensor located on a particular position at a specific moment, is merely a special cross section of this imaging space. Both temporal and spatial image sequences are special proper subsets of the imaging space. They are in the middle level, between the imaging space and the individual images. This hierarchical structure is depicted in Figure 10.2.

Before concluding this section, we will discuss the relationship between image sequences and video sequences. It is noted that the term video is used very often nowadays in addition to the terms image frames and sequences. It is necessary to pause for a while to discuss the relationship between these terms. Image frames and sequences have been defined clearly above with the introduction of the concept of the imaging space. Video can mean an individual video frame or video sequences. It refers, however, to those frames and sequences that are associated with the visible frequency band in the electromagnetic spectrum. For image frames and sequences, there is no such restriction. For instance, infrared image frames and sequences correspond to a band outside the visible band in the spectrum. From this point of view, the scope of image frames and sequences is wider than that of video frames and sequences. When the visible band is concerned, the terms image frame and sequence are interchangeable with that of video frame and sequence.

Another point we would like to bring to readers' attention is as follows. Although video is referred to as visual information, which includes both a single frame and frame

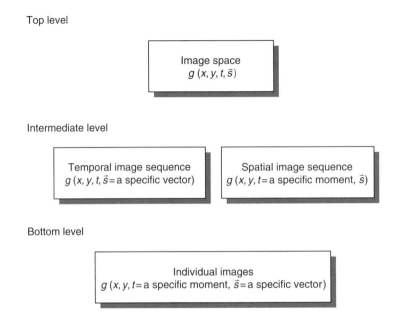

FIGURE 10.2
A hierarchical structure.

sequences, in practice it is often used to mean sequences exclusively. Such an example can be found in a book entitled *Digital Video Processing* by Tekalp [tekalp 1995].

In this book, we use image compression to indicate still image compression, and video compression to indicate video sequence compression. Readers should keep in mind, however, that first, video can mean a single frame or sequences of frames; second, the scope of image is wider than that of video, and video is more pertinent to multimedia engineering.

10.2 Interframe Correlation

As far as video compression is concerned, all the techniques discussed in the previous chapters are applicable. By this we mean two classes of techniques. The first class, which is also the most straightforward way to handle video compression, is to code each frame separately. That is, individual frames are coded independently on each other. For instance, using a JPEG compression algorithm to code each frame in a video sequence results in motion JPEG [westwater 1997]. In the second class, methods utilized for still image coding can be generalized for video compression. For instance, discrete cosine transform (DCT) coding can be generalized and applied to video coding by extending two-dimensional (2-D) DCT to 3-D DCT. That is, instead of 2-D DCT, say, 8×8, applied to a single image frame, we can apply 3-D DCT, say, $8 \times 8 \times 8$, to a video sequence; see Figure 10.3. That is, eight blocks of 8×8 each located, respectively, at the same position in one of the eight successive frames from a video sequence are coded together with the 3-D DCT. It was reported that this 3-D DCT technique is quite efficient [lim 1990; westwater 1997]. In addition, the differential pulse code modulation (DPCM) technique and the hybrid technique can be generalized and applied to video compression in a similar fashion [jain 1989; lim 1990]. It is noted that in the second class of techniques, several successive frames are grouped and coded together, whereas in the first class each frame is coded independently.

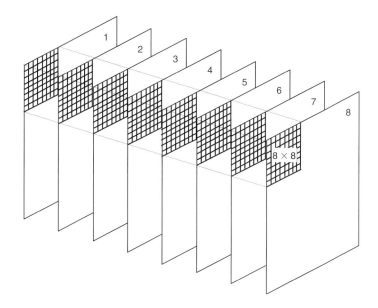

FIGURE 10.3
A 3-D discrete cosine transform (DCT) of $8 \times 8 \times 8$.

Video compression has its own characteristics; however, these make it quite different from still image compression. The major difference lies on the exploitation of interframe correlation that exists between successive frames in video sequences in addition to the intraframe correlation that exists within each frame. As mentioned in Chapter 1, the interframe correlation is also referred to as temporal redundancy, while the intraframe correlation is referred to as spatial redundancy. To achieve coding efficiency, we need to remove these redundancies for video compression. To do so we must first understand these redundancies.

Consider a video sequence taken in a videophone service, where the camera is static most of the time. A typical scene is a head and shoulders view of a person imposed on a background. In this type of video sequence the background is usually static. Only the speaker experiences motion, which is not severe. Therefore, there is a strong similarity between successive frames, that is, a strong adjacent-frame correlation. In other words, there is a strong interframe correlation. It was reported in [mounts, 1969] that when using videophone-like signals with moderate motion in the scene, on average, less than one-tenth of the elements change between frames by an amount which exceeds 1% of the peak signal. Here, a 1% change is regarded as significant. Our experiment on the first 40 frames of the Miss America sequence supports this observation. Two successive frames of the sequence, frames 24 and 25, are shown in Figure 10.4.

Now, consider a video sequence generated in a television broadcast. It is well known that television signals are generated with a scene scanned in a particular manner to maintain a steady picture for a human being to view regardless of whether there is a scenery change or not. That is, although there is no change from one frame to the next, the scene is still scanned constantly. Hence there is a great deal of frame-to-frame correlation [haskell 1972b; netravali 1979]. In TV broadcasts, the camera is most likely not static, and it may be panned, tilted, and zoomed. Furthermore, more movement is involved in the scene. As long as the TV frames are taken densely enough, most of the time we think the changes between successive frames are due mainly to the apparent motion of the objects in the scene that takes place during the frame intervals. This implies that there is also a high

(a) Frame 24   (b) Frame 25

FIGURE 10.4
Two frames of the Miss America sequence.

correlation between sequential frames. In other words, there is an interframe redundancy (interpixel redundancy between pixels in successive frames). There is more correlation between television picture elements along the frame-to-frame temporal dimension than there is between adjacent elements in a single frame along the spatial dimension. That is, there is generally more interframe correlation than intraframe correlation. Taking advantage of the interframe correlation, i.e., eliminating or decreasing the uncertainty of successive frames, leads to video data compression. This is analog to the case of still image coding with the DPCM technique, where we can predict part of an image by knowing the other part. Now the knowledge of the previous frames can remove the uncertainty of the next frame. In both cases, knowledge of the past removes the uncertainty of the future, leaving less actual information to be transmitted [kretzmer 1952]. In Chapter 16, the words "past" and "future" used here are changed respectively, to "some frames" and "some other frames" in advanced video coding techniques, such as MPEG. There, a frame might be predicted from both its earlier frames and its future frames.

At this point, it becomes clear that the second class of techniques (Section 10.2), which generalizes techniques originally developed for still image coding and applies them to video coding, exploits interframe correlation. For instance, in the case of the 3-D DCT technique, a strong temporal correlation causes an energy compaction within the low temporal frequency region. The 3-D DCT technique drops transform coefficients associated with high temporal frequency, thus achieving data compression.

The two techniques specifically developed to exploit interframe redundancy, i.e., frame replenishment and MC coding, are introduced below. The former is the early work, whereas the latter is the more popular recent work.

10.3 Frame Replenishment

As mentioned in Chapter 3, frame-to-frame redundancy has long been recognized in TV signal compression. The first few experiments of a frame sequence coder exploiting interframe redundancy may be traced back to the 1960s [seyler 1962, 1965; mounts, 1969]. In [mounts, 1969] the first real demonstration was presented and was termed conditional replenishment. This frame replenishment technique can be briefly described as follows. Each pixel in a frame is classified into changing or unchanging areas depending on whether or not the intensity difference between its present value and its previous one (the intensity value at the same position on the previous frame) exceeds a threshold. If the difference does exceed the threshold, i.e., a significant change has been identified, the address and intensity of this pixel are coded and stored in a buffer and then transmitted to the receiver to replenish intensity. For those unchanging pixels, nothing is coded and transmitted. Their earlier intensities are repeated in the receiver. It is noted that the buffer is utilized to make the information presented to the transmission channel occur at a smooth bit rate. The threshold is to make the average replenishment rate match the channel capacity.

Since the replenishment technique only encodes those pixels whose intensity value has changed significantly between successive frames, its coding efficiency is much higher than the coding techniques, which encode every pixel of every frame, say, the DPCM technique applied to each single frame. In other words, utilizing interframe correlation, the replenishment technique achieves a lower bit rate, while keeping the equivalent reconstructed image quality.

Much effort had been made to further improve this type of simple replenishment algorithm. As mentioned in the discussion of 3-D DPCM in Chapter 3, for instance, it was soon realized that intensity values of pixels in a changing area need not be transmitted independently on one another. Instead, using both spatial and temporal neighbors' intensity values to predict the intensity value of a changing pixel leads to a frame-difference predictive coding technique. There, the differential signal is coded instead of the original intensity values, thus achieving a lower bit rate (refer Section 3.5.2 for more detail). Another example of the improvements is that measures have been taken to distinguish the intensity difference caused by noise from those associated with changing to avoid the dirty window effect, whose meaning is given in the next paragraph. For more detailed information on these improvements over the simple frame replenishment technique, readers are referred to two excellent reviews [haskell 1972b, 1979].

The main drawback associated with the frame replenishment technique is that it is difficult to handle frame sequences containing more rapid changes. When there are more rapid changes, the number of pixels whose intensity values need to be updated increases. To maintain the transmission bit rate in a steady and proper level, the threshold has to be raised, thus causing many slow changes that cannot show up in the receiver. This poorer reconstruction in the receiver is somewhat analogous to viewing a scene through a dirty window. This is referred to as the dirty window effect. The result of one experiment on the dirty window effect is displayed in Figure 10.5. From frame 22–25 of the Miss America sequence, there are 2166 pixels (less than 10% of the total pixels) which change their gray level values by more than 1% of the peak signal. When we only update the gray level values for 25% (randomly chosen) of these changing pixels, we can clearly see the dirty window effect. When rapid scene changes exceed a certain level, buffer saturation will result, causing picture breakup [mounts, 1969]. MC coding, which is discussed below, has been proved to be able to provide better performance than the replenishment technique in situations with rapid changes.

10.4 Motion Compensated Coding

In addition to the frame-difference predictive coding technique (a variant of the frame replenishment technique discussed above), another technique, displacement-based predictive coding, was developed at almost the same time [rocca 1969; haskell 1972a]. In this

FIGURE 10.5
Dirty window effect.

technique, a motion model is assumed. That is, the changes between successive frames are considered due to the translation of moving objects in the image planes. Displacement vectors of objects are first estimated. Differential signals between the intensity value of the picture elements in the moving areas and that of their counterpart in the previous frame, which are translated by the estimated displacement, are encoded. This approach, which takes motion into account to compress video sequences, is referred to as motion compensated predictive coding. It was found to be much more efficient than the frame-difference prediction technique.

To understand the above statement, let us take a look at the diagram shown in Figure 10.6. Assume a car translating from the right side to the left side in the image planes in a uniform speed during the time interval between the two consecutive image frames. Other than this, there are no movements or changes in the frames. Under this circumstance, if we know the displacement vector of the car on the image planes during the time interval between two consecutive frames, we can then predict the position of the car in the latter frame from its position in the former frame. One may think that if the translation vector is estimated well, then so is the prediction of the car position. This is true. In reality, however, estimation

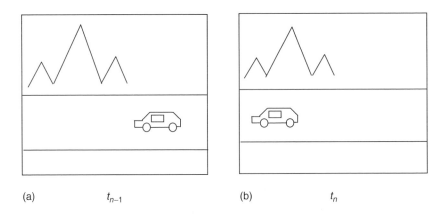

FIGURE 10.6
Two consecutive frames of a video sequence.

errors occurring in determination of the motion vector, which may be caused by various noises existing in the frames, may cause the predicted position of the car in the latter frame to differ from the actual position of the car in the latter frame.

The above translational model is a very simple one; it cannot accommodate motions other than translation, say, rotation, and camera zooming. Occlusion and disocclusion of objects make the situation even more complicated because in the case of occlusion, some portions of the images may disappear, whereas in the case of disocclusion, some newly exposed areas may appear. Therefore, the prediction error is almost inevitable. To have good-quality frames in the receiver, we can find the prediction error by subtracting the predicted version of the latter frame from the actual version of latter frame. If we encode both the displacement vectors and the prediction error, and transmit the data to the receiver, we may be able to obtain high-quality reconstructed images in the receiver. This is because in the receiving end, using the displacement vectors transmitted from the transmitter and the reconstructed former frame, we can predict the latter frame. Adding the transmitted prediction error to the predicted frame, we may reconstruct the latter frame with satisfactory quality. Furthermore, if manipulating the procedure properly, we are able to achieve data compression.

The displacement vectors are referred to as side or overhead information to indicate their auxiliary nature. It is noted that motion estimation drastically increases the computational complexity of the coding algorithm. In other words, the higher coding efficiency is obtained in MC coding, but with a higher computational burden. As pointed out in Section 10.1, this is both technically feasible and economically desired because the cost of digital signal processing decreases much faster than that of transmission [dubois 1981].

MC video compression has been a major development in coding since then. For more information, readers should refer to several excellent survey papers [musmann 1985; zhang 1995; kunt 1995].

The common practice of MC coding in video compression can be split into the following three stages: First, the motion analysis stage; that is, displacement vectors for either every pixel or a set of pixels in image planes from sequential images are estimated. Second, the present frame is predicted by using estimated motion vectors and the previous frame. The prediction error is then calculated. This stage is called prediction and differentiation. The third stage is encoding. The prediction error (difference between the present and the predicted present frames) and the motion vectors are encoded. Through an appropriate manipulation, the total amount of data for both the motion vectors and prediction error is expected to be much less than the raw data existing in the image frames, thus resulting in data compression. A block diagram of MC coding is shown in Figure 10.7.

Before concluding this section, we compare the frame replenishment technique with the MC coding technique. Qualitatively speaking, from the above discussion, we see that the

FIGURE 10.7
Block diagram of motion compensated (MC) coding.

replenishment technique is also a kind of predictive coding in nature. This is particularly true if we consider the frame-difference predictive technique used in frame replenishment. There, it uses a pixel's intensity value in the previous frame as an estimator of its intensity value in the present frame. Now let us take a look at MC coding. Consider a pixel on the present frame. Through motion analysis, the MC technique finds its counterpart in the previous frame. That is, a pixel in the previous frame is identified such that it is supposed to translate to the position on the present frame of the pixel under consideration during the time interval between successive frames. This counterpart's intensity value is used as an estimator of that of the pixel under consideration. Therefore, we see the model used for MC coding is much more advanced than that used for frame replenishment; therefore, it achieves much higher coding efficiency. An MC coding technique that utilized the first pel recursive algorithm for motion estimation [netravali 1979] was reported to achieve a bit rate 22%–50% lower than that obtained by simple frame-difference prediction, a version of frame replenishment.

The more advanced model utilized in MC coding, on the other hand, leads to higher computational complexity. Consequently, both the coding efficiency and the computational complexity in MC coding are higher than that in frame replenishment.

10.5 Motion Analysis

As discussed above, we usually conduct motion analysis in video sequence compression. There, 2-D displacement vectors of a pixel or a group of pixels on image planes are estimated from given image frames. Motion analysis can be viewed from a much broader point of view. It is well known that the vision systems of both human beings and animals observe the outside world to ascertain motion and to navigate themselves in the 3-D world space. Two groups of scientists study vision. Scientists in the first group, including psychophysicists, physicians, and neurophysiologists, study human and animal vision. Their goal is to understand biological vision systems—their operation, features, and limitations. Computer scientists and electrical engineers form the second group. As pointed out in [aggarwal 1988], their ultimate goal is to develop computer vision systems with the ability to navigate, recognize, and track objects, and estimate their speed and direction. Each group benefits from the research results of the other group. The knowledge and results of research in psychophysics, physiology, and neurophysiology have influenced the design of computer vision systems. Simultaneously, the research results achieved in computer vision have provided a framework in modeling biological vision systems and have helped in remedying faults in biological vision systems. This process will continue to advance research in both groups, hence benefiting human beings.

10.5.1 Biological Vision Perspective

In the field of biological vision, most scientists consider motion perception as a two-step process, even though there is no ample biological evidence to support this view [singh 1991]. The two steps are measurement and interpretation. The first step measures the 2-D motion projected on the imaging surfaces. The second step interprets the 2-D motion to induce the 3-D motion and structure on the scene.

10.5.2 Computer Vision Perspective

In the field of computer vision, motion analysis from image sequences is traditionally split into two steps. In the first step, intermediate variables are derived. By intermediate

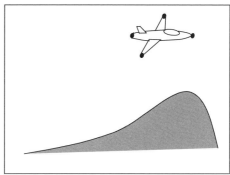

FIGURE 10.8
Feature exaction and correspondence from two consecutive frames in a temporal image sequence.

variables, we mean 2-D motion parameters in image planes. In the second step, 3-D motion variables, say, speed, displacement, position, and direction are determined.

Depending on the different intermediate results, all approaches to motion analysis can be basically classified into two categories: feature correspondence and optical flow. In the former category, a few distinct features are first extracted from image frames. For instance, consider an image sequence containing an aircraft. Two consecutive frames are shown in Figure 10.8. The head and tail of an aircraft and the tips of its wings may be chosen as features. The correspondence of these features on successive image frames needs to be established. In the second step, 3-D motion can then be analyzed from the extracted features and their correspondence in successive frames. In the latter category, the intermediate variables are optical flow. An optical flow vector is defined as a velocity vector of a pixel on an image frame. An optical flow field is referred to as the collection of the velocity vectors of all the pixels on the frame. In the first step, optical flow vectors are determined from image sequences as the intermediate variables. In the second step, 3-D motion is estimated from optical flow. It is noted that optical flow vectors are closely related to displacement vectors in that a velocity vector multiplying by the time interval between two consecutive frames results in the corresponding displacement vector. Optical flow and its determination are discussed in detail in Chapter 13.

It is noted that there is a so-called direct method in motion analysis. Contrary to the above optical flow approach, instead of determining 2-D motion variables, (i.e., the intermediate variables), prior to 3-D motion estimation, the direct method attempts to estimate 3-D motion without explicitly solving for the intermediate variables. In [huang 1981b], the equation characterizing displacement vectors in the 2-D image plane and the equation characterizing motion parameters in 3-D world space are combined so that the motion parameters in 3-D world space can be directly derived. This method has been utilized to recover structure (object surfaces) in 3-D world space as well [negahdaripour 1987; horn 1988; shu 1993]. The direct method has certain limitations. That is, if the geometry of object surfaces is unknown in advance then the method fails.

The feature correspondence approach is sometimes referred to as the discrete approach, while the optical flow approach is sometimes referred to as the continuous approach. This is because the correspondence approach concerns only a set of relatively sparse but highly discriminatory 2-D features on image planes. The optical flow approach is concerned with a dense field of motion vectors.

It has been found that both feature extraction and correspondence establishment are not trivial tasks. Occlusion and disocclusion, which cause some features to disappear and some features to reappear, respectively, make feature correspondence even more difficult.

The development of robust techniques to solve the correspondence problem is an active research area and is still in its infancy. So far, only partial solutions suitable for simplistic situations have been developed [aggarwal 1988]. Hence the feature correspondence approach is rarely used in video compression. Therefore, we do not discuss this approach any further.

Motion analysis (sometimes referred to as motion estimation or motion interpretation) from image sequences is necessary in automated navigation. It has played a central role in the field of computer vision since the late 1970s and early 1980s. A great deal of papers presented at the International Conference on Computer Vision cover motion analysis and related topics. Many workshops, symposiums, and special sessions are organized around this subject [thompson 1989].

10.5.3 Signal Processing Perspective

In the field of signal processing, motion analysis is mainly considered in the context of bandwidth reduction and data compression in the transmission of visual signals. Therefore, instead of the motion in 3-D world space, only the 2-D motion in the image plane is concerned.

Because of the real-time nature in visual transmission, the motion model cannot be very complicated. So far, the 2-D translational model is most frequently assumed in the field. In the 2-D translational model, it is assumed that the change between a frame and its previous one is due to the motion of objects in the frame plane during the time interval between two consecutive frames. In many cases, as long as frames are taken densely enough, this assumption is valid. By motion analysis, we mean the estimation of translational motion—either the displacement vectors or velocity vectors. Using this kind of motion analysis, one can apply the MC coding discussed above, making coding more efficient.

Basically, there are three techniques in 2-D motion analysis: correlation, recursive, and differential techniques. Philosophically speaking, the first two techniques belong to the same group: region matching.

Refer to Figure 10.6, where the moving car is the object under investigation. By motion analysis, we mean finding the displacement vector, i.e., a vector representing the relative positions of the car in the two consecutive frames. With region matching, one may consider the car (or a portion of the car) as a region of interest, and seek the best matching between the two regions in the two frames: specifically, the region in the present frame and the region in the previous frame. For identifying the best matching, two techniques, the correlation and the recursive methods, work differently in methodology. The correlation technique finds the best matching by searching the maximum correlation between the two regions in a predefined search range, whereas the recursive technique estimates the best matching by recursively minimizing a nonlinear measure of the dissimilarity between the two regions.

A couple of comments are in order. First, it is noted that the most frequently used technique in motion analysis is called block matching, which is a type of the correlation technique. There, a video frame is divided into nonoverlapped rectangular blocks with each block having the same size, usually 16×16. Each block thus generated is assumed to move as one, i.e., all pixels in a block share the same displacement vector. For each block, we find its best matching in the previous frame with correlation. That is, the block in the previous frame, which gives the maximum correlation, is identified. The relative position of these two best matched blocks produces a displacement vector. This block matching technique is simple and very efficient, and will be discussed in detail in Chapter 11. Second, as multimedia finds more and more applications, the regions occupied by arbitrary-shaped objects (no longer always rectangular blocks) become increasingly important in content-based video retrieval and manipulation. Motion analysis in this case is discussed in Chapter 18. Third, although the recursive technique is categorized as a region matching technique, it may be used for finding displacement vectors for individual pixels. In fact, the recursive technique was originally developed for determining displacement vectors of pixels, and hence, it is called pel recursive. This technique is discussed in Chapter 12. Fourth, both correlation and recursive techniques can be utilized for determining optical flow vectors. Optical flow is discussed in Chapter 13.

The third technique in 2-D motion analysis is a differential technique. This is one of the main techniques utilized in determining optical flow vectors. It is named after the term of differential because it uses the partial differentiation of an intensity function with respect to the spatial coordinates x and y, as well as the temporal coordinate t. This technique is also discussed in Chapter 13.

10.6 Motion Compensation for Image Sequence Processing

Motion analysis has long been considered a key issue in image sequence processing [huang 1981a; shi 1997]. Obviously, in an area like automated navigation, motion analysis plays a central role. From the discussion in this chapter, we see that motion analysis also plays a key role in video data compression. Specifically, we have discussed the concept of motion compensated video coding in Section 10.4. In this section, we would like to consider motion compensation for image sequence processing, in general. Let us first consider motion compensated interpolation. Then, we will discuss motion compensated enhancement, restoration, and down-conversion.

10.6.1 Motion Compensated Interpolation

Interpolation is a simple yet efficient and important method in image and video compression. In image compression, we may only transmit, say, every other row. We then try to interpolate these missing rows from the other half transmitted rows in the receiver. In this way, we compress the data to half. As the interpolation is carried out within a frame, it is referred to as spatial interpolation. In video compression, for instance, in videophone service, instead of transmitting 30 frames/s, we may choose a lower frame rate, say, 10 frames/s. In the receiver, we may try to interpolate the dropped frames from the transmitted frames. This strategy immediately drops the transmitted data to one-third. Another example is the conversion of a motion picture into an NTSC (national television system commission) TV signal. There, every first frame in the motion picture is repeated three times and the next frame twice, thus converting a 24 frame/s motion picture to a 60 field/s NTSC signal. This is commonly referred to as 3:2 pulldown. In these two examples concerning video, interpolation is along the temporal dimension, which is referred to as temporal interpolation.

For basic concepts of zero-order interpolation, bilinear interpolation, and polynomial interpolation, readers are referred to signal processing texts, for instance, [lim 1990]. In temporal interpolation, the zero-order interpolation means creation of a frame by copying its nearest frame along the time dimension. The conversion of a 24 frame/s motion picture to a 60 field/s NTSC signal can be classified into this type of interpolation. Weighted linear interpolation can be illustrated with Figure 10.9.

There, the weights are determined according to the lengths of time intervals, which is similar to the bilinear interpolation widely used in spatial interpolation, except that here only one index (along the time axes) is used, while two indexes (along two spatial axes) are used in spatial bilinear interpolation. That is,

$$f(x,y,t) = \frac{l_2}{l_1+l_2} f(x,y,t_1) + \frac{l_1}{l_1+l_2} f(x,y,t_2). \qquad (10.4)$$

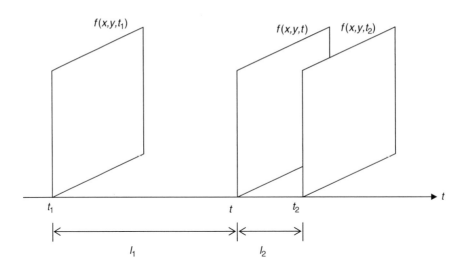

FIGURE 10.9
Weighted linear interpolation.

If there are one or multiple moving objects existing in successive frames, however, the weighted linear interpolation will blur the interpolated frames. Taking motion into account in the interpolation results in MC interpolation. In Figure 10.10, we still use three frames shown in Figure 10.9 to illustrate the concept of MC interpolation. First, motion between two given frames is estimated. That is, the displacement vectors for each pixel are determined. Second, we choose a frame that is nearer to the frame we want to interpolate. Third, the displacement vectors determined in the first step are proportionally converted to the frame to be created. Each pixel in this frame is projected via the determined motion trajectory to the frame chosen in step 2. In the process of MC interpolation, spatial interpolation in the frame chosen in step 2 is usually needed.

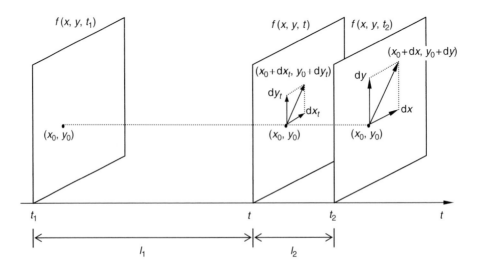

FIGURE 10.10
Motion compensated (MC) interpolation.

10.6.2 Motion Compensated Enhancement

It is well known that when an image is corrupted by additive white Gaussian noise (AWGN) or burst noise, linear low-pass filtering such as simple averaging or nonlinear low-pass filtering such as a median filter performs well in removing the noise. When an image sequence is concerned, we may apply such types of filtering along the temporal dimension to remove noise. This is called temporal filtering. These types of low-pass filtering may blur images, an effect that may become quite serious when motion exists in image planes. The enhancement, which takes motion into account, is referred to as MC enhancement, and it was found very efficient in temporal filtering [huang 1981c].

To facilitate the discussion, we consider simple averaging as a means for noise filtering in what follows. It is understood that other filtering techniques are possible, and that everything discussed here is applicable there. Instead of simply averaging n successive image frames in a video sequence, MC temporal filtering will first analyze the motion existing in these frames. That is, we estimate the motion of pixels in successive frames first. Then averaging will be conducted only on those pixels along the same motion trajectory. In Figure 10.11, three successive frames are shown and denoted by $f(x, y, t_1)$, $f(x, y, t_2)$, and $f(x, y, t_3)$, respectively. Assume that three pixels, denoted by (x_1, y_1), (x_2, y_2), and (x_3, y_3), respectively, are identified to be perspective projections of the same object point in the 3-D world space on the three frames. The averaging is then applied to these three pixels. It is noted that the number of successive frames, n, may not necessarily have to be three. Motion analysis can be any one of several techniques discussed in Section 10.5. MC temporal filtering is not necessarily implemented pixelwise; it can also be objectwise or regionwise.

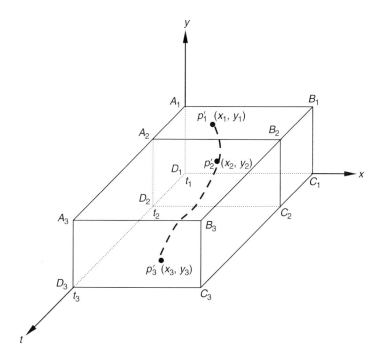

FIGURE 10.11
Motion compensated (MC) temporal filtering.

10.6.3 Motion Compensated Restoration

Extensive attention has been paid to the restoration of full-length feature films. There, typical artifacts are due to dirt and sparkle. Earlier study in the detection of these artifacts ignored motion information completely. Late motion estimation has been utilized to detect these artifacts based on the assumption that the artifacts occur occasionally along the temporal dimension. Once the artifacts have been found, MC temporal filtering and/or interpolation will be used to remove the artifacts. One successful algorithm for the detection and removal of anomalies in digitized animation film can be found in [tom 1998].

10.6.4 Motion Compensated Down-Conversion

Here we present one more example in which motion compensation finds application in digital video processing.

It is believed that there will be a need to down-convert a high definition television (HDTV) image sequence for display onto an NTSC monitor during the upcoming transition to digital television broadcast. The most straightforward approach is to fully decode the image sequence first, then apply a prefiltering and subsampling process to each field of the interlaced sequence. This is referred to as a full-resolution decoder (FRD). The merit of this approach is the high quality achieved, whereas the drawback is a high cost in terms of the large amount of memory required to store the reference frames. To reduce the required memory space, another approach is considered. In this approach, the down-conversion is conducted within the decoding loop and is referred to as a low-resolution decoder (LRD). It can significantly reduce the required memory and still achieve a reasonably good picture quality.

The prediction drift is a major type of artifact existing in the down-conversion. It is defined as the successive blurring of forward-predicted frames with a group of pictures. It is caused mainly by nonideal interpolation of sub-pixel intensities and the loss of high-frequency data within the block. An optimal set of filters to perform low-resolution motion compensation has been derived to effectively minimize the drift. For details on an algorithm in the down-conversion, utilizing an optimal motion compensation scheme, readers are referred to [vetro 1998].

10.7 Summary

After Part II, still image compression, we shift our attention to video compression. Before Part IV, where we discuss various video compression algorithms and standards, however, we first address the issue of motion analysis and motion compensation in this chapter that starts Part III, motion estimation and compensation. This is because video compression has its own characteristics, which are different from that of still image compression. The main difference lies on interframe correlation.

In this chapter, the concept of various image sequences is discussed in a broad scope. In doing so, a single image sequence, temporal image sequences, and spatial image sequences are all unified under the concept of imaging space. The redundancy between pixels in successive frames is analyzed for both videoconferencing and TV broadcast cases. In these applications, there is more interframe correlation than intraframe correlation in general. Therefore, the utilization of interframe correlation becomes a key issue in video compression.

There are two major techniques in exploitation of interframe correlation: frame replenishment and motion compensation. In the conditional replenishment technique, only those pixels' gray level values, whose variation from their counterparts in the previous frame exceeds a threshold, are encoded and transmitted to the receiver. These pixels are called changing pixels. For the pixels other than the changing pixels, their gray values are just repeated in the receiver. This simplest frame replenishment technique achieves higher coding efficiency than coding each pixel in each frame due to utilization of interframe redundancy. In the more advanced frame replenishment techniques, say, frame-difference predictive coding technique, both temporal and spatial neighboring pixels' gray values are used to predict that of a changing pixel. Instead of the intensity values of the changing pixels, the prediction error is encoded and transmitted. Because the variance of the prediction error is smaller than that of the intensity values, this more advanced frame replenishment technique is more efficient than the conditional replenishment technique.

The main drawback of the frame replenishment techniques is associated with rapid motion and/or intensity variation occurring on the image planes. Under these circumstances, frame replenishment will suffer from dirty window effect, and even buffer saturation.

In MC coding, the motion of pixels is first analyzed. On the basis of previous frames and the estimated motion, the current frame is predicted. The prediction error together with motion vectors are encoded and transmitted to the receiver. Due to more accurate prediction based on motion model, MC coding achieves higher coding efficiency compared with frame replenishment. This is conceivable because frame replenishment basically uses the intensity value of a pixel in the previous frame to predict that of the pixel in the same location in the present frame, whereas the prediction in MC coding uses motion trajectory. This implies that higher coding efficiency is obtained in motion compensation at the cost of higher computational complexity. This is technically feasible and economically desired since the cost of digital signal processing decreases much faster than that of transmission.

Because of the real-time requirement in video coding, only simple 2-D translational model is used. There are mainly three types of motion analysis techniques used in MC coding. They are block matching, pel recursion, and optical flow. By far, block matching is used most frequently. These three techniques are discussed in detail in Chapters 11 through 13.

Motion compensation is also widely utilized in other tasks of digital video sequence processing. Examples include MC interpolation, MC enhancement, MC restoration, and MC down-conversion.

Exercises

1. Explain the analogy between a stereo image sequence versus the imaging space, and a stereo image pair versus the spatial image sequence, to which the stereo image pair belongs.
2. Explain why the imaging space can be considered as a unification of image frames, spatial image sequences, and temporal image sequences.
3. Give the definitions of the following several concepts: image, image sequence, and video. Discuss the relation between them.
4. What feature causes video compression quite different from still image compression?
5. Describe the conditional replenishment technique. Why can it achieve higher coding efficiency in video coding than those techniques encoding each pixel in each frame?
6. Describe the frame-difference predictive coding technique. Refer to Section 3.5.2.
7. What is the main drawback of frame replenishment?

8. Both the frame-difference predictive coding and MC coding are predictive coding in nature.
   (a) What is the main difference between the two?
   (b) Explain why MC coding is usually more efficient.
   (c) What is the price paid for higher coding efficiency with MC coding?
9. Motion analysis is an important task encountered in both computer vision and video coding. What is the major different requirement for motion analysis in these two fields?
10. Work on the first 40 frames of a video sequence other than the Miss America. Determine, on an average basis, how many percentages of the total pixels change their gray level values by more than 1% of the peak signal between two consecutive frames.
11. Similar to the experiment associated with Figure 10.5, do your own experiment to observe the dirty window effect. That is, work on two successive frames of a video sequence chosen by yourself, and only update a part of those changing pixels.
12. Take two frames from the Miss America sequence or from other sequences with your own choice, between which a relatively large motion is involved.
    (a) Using the weighted linear interpolation defined in Equation 10.4, create an interpolated frame, which is located in the one-third of the time interval from the second frame (i.e., $l_2 = \frac{1}{3}(l_1 + l_2)$) according to Figure 10.9).
    (b) Using MC interpolation, create an interpolated frame at the same position along the temporal dimension.
    (c) Compare the two interpolated frames and make your comments.

References

[aggarwal 1988] J.K. Aggarwal and N. Nandhakumar, On the computation of motion from sequences of images—a review, *Proceedings of the IEEE*, 76, 8, 917–935, 1988.

[dubois 1981] E. Dubois, B. Prasada, and M.S. Sabri, Image sequence coding, chapter 3, in *Image Sequence Analysis*, T.S. Huang (Ed.), Springer-Verlag, Berlin, 1981.

[haskell 1972a] B.G. Haskell and J.O. Limb, Predictive video encoding using measured subject velocity, U.S. Patent 3,632,865, January 1972.

[haskell 1972b] B.G. Haskell, F.W. Mounts, and J.C. Candy, Interframe coding of videotelephone pictures, *Proceedings of IEEE*, 60, 7, 792–800, July 1972.

[haskell 1979] B.G. Haskell, Frame replenishment coding of television, chapter 6 in *Image Transmission Techniques*, W.K. Pratt (Ed.), Academic Press, New York, 1979.

[horn 1980] B.K.P. Horn and B.G. Schunck, Determining optical flow, *Artificial Intelligence*, 17, 185–203, 1981.

[horn 1988] B.K.P. Horn and E.J. Weldon Jr., Direct methods for recovering motion, *International Journal of Computer Vision*, 2, 51–76, 1988.

[huang 1981a] T.S. Huang (Ed.), *Image Sequence Analysis*, Springer-Verlag, 1981.

[huang 1981b] T.S. Huang and R.Y. Tsai, Image sequence analysis: Motion estimation, chapter 1 in *Image Sequence Analysis*, T.S. Huang (Ed.), Springer-Verlag, Berlin, 1981.

[huang 1981c] T.S. Huang and Y.P. Hsu, Image Sequence Enhancement, chapter 4 in *Image Sequence Analysis*, T.S. Huang (Ed.), Springer-Verlag, Berlin, 1981.

[huang 1983] T.S. Huang (Ed.), *Image Sequence Processing and Dynamic Scene Analysis*, Springer-Verlag, Berlin, 1983.

[jain 1989] A.K. Jain, *Fundamentals of Digital Image Processing*, Prentice-Hall, Englewood Cliffs, NJ, 1989.

[kretzmer 1952] E.R. Kretzmer, Statistics of television signal, *The Bell System Technical Journal*, 31, 4, 751–763, July 1952.

[kunt 1995] M. Kunt (Ed.), Special issue on digital television part 1: Technologies, *Proceedings of the IEEE*, 83, 6, June 1995.

[mounts 1969] F.W. Mounts, A video encoding system with conditional picture-element replenishment, *The Bell System Technical Journal*, 48, 7, 2545–2554, September 1969.

[musmann 1985] H.G. Musmann, P. Pirsch, and H.J. Grallert, Advances in picture coding, *Proceedings of the IEEE*, 73, 4, 523–548, 1985.

[negahdaripour 1987] S. Negahdaripour and B.K.P. Horn, Direct passive navigation, *IEEE Transactions on Pattern Analysis and Machine Intelligence*, PAMI-9, 1, 168–176, January 1987.

[netravali 1979] A.N. Netravali and J.D. Robbins, Motion compensated television coding: Part I, *The Bell System Technical Journal*, 58, 3, 631–670, March 1979.

[lim 1990] J.S. Lim, *Two-Dimensional Signal and Image Processing*, Prentice-Hall, Englewood, NJ, 1990.

[pratt 1979] W.K. Pratt, (Ed.), *Image Transmission Techniques*, Academic Press, New York, 1979.

[rocca 1969] F. Rocca, Television bandwidth compression utilizing frame-to-frame correlation and movement compensation, *Symposium on Picture Bandwidth Compression*, MIT Cambridge, MA, 1969, Gordon and Breach, 1972.

[seyler 1962] A.J. Seyler, The coding of visual signals to reduce channel-capacity requirements, *The Institution of Electrical Engineers Monograph*, no. 533E, July 1962.

[seyler 1965] A.J. Seyler, Probability distributions of television frame difference, Proceedings of IREE, (Australia), 26, 335, November 1965.

[singh 1991] A. Singh, *Optical flow computation: A unified perspective*, IEEE Computer Society Press, CA, 1991.

[shi 1994] Y.Q. Shi, C.Q. Shu, and J.N. Pan, Unified optical flow field approach to motion analysis from a sequence of stereo images, *Pattern Recognition*, 27, 12, 1577–1590, 1994.

[shi 1997] Y.Q. Shi, Editorial introduction to special issue on image sequence processing, *International Journal on Imaging Systems and Technology*, 9, 4, 189–191, August 1998.

[shu 1991] C.Q. Shu and Y.Q. Shi, On unified optical flow field, *Pattern Recognition*, 24, 6, 579–586, 1991.

[shu 1993] C.Q. Shu and Y.Q. Shi, Direct recovering of Nth order surface structure using unified optical flow field, *Pattern Recognition*, 26, 8, 1137–1148, 1993.

[tekalp 1995] A.M. Tekalp, *Digital Video Processing*, Prentice-Hall PTR, Upper Saddle River, NJ, 1995.

[thompson 1989] W.B. Thompson, Introduction to special issue on visual motion, *IEEE Transactions on Pattern Analysis and Machine Intelligence*, 11, 5, 449–450, 1989.

[tom 1998] B.C. Tom, M.G. Kang, M.C. Hong, and A.K. Katsaggelos, Detection and removal of anomalies in digitized animation film, in Y.Q. Shi (Ed.), special issue on image sequence processing, *International Journal of Imaging Systems and Technology*, 9, 4, 283–293, 1998.

[vetro 1998] A. Vetro and H. Sun, Frequency domain down-conversion of HDTV using an optimal motion compensation scheme, in Y.Q. Shi (Ed.), special issue on image sequence processing, *International Journal of Imaging Systems and Technology*, 9, 4, 274–282, 1998.

[waxman 1986] A.M. Waxman and J.H. Duncan, Binocular image flow: Steps towards stereo-motion fusion, *IEEE Transactions on Pattern Analysis and Machine Intelligence*, PAMI-8, 6, 715–729, 1986.

[westwater 1997] R. Westwater and B. Furht, *Real-time Video Compression*, Kluwer Academic Publishers, Dordrecht, 1997.

[zhang 1995] Y.-Q. Zhang, W. Li, and M.L. Liou (Eds.), Special Issue on Advances in Image and Video Compression, *Proceedings of the IEEE*, 83, 2, 133–340, February 1995.

11
Block Matching

As mentioned in Chapter 10, displacement vector measurement and its usage in motion compensation in interframe coding for a TV signal can be traced back to the 1970s. Netravali and Robbins [netravali 1979] developed a pel recursive technique, which estimates the displacement vector for each pixel recursively from its neighboring pixels using an optimization method. Limb and Murphy [limb 1975], Rocca and Zanoletti [rocca 1972], Cafforio and Rocca [cafforio 1976], and Brofferio and Rocca [brofferio 1977] developed techniques for estimation of displacement vectors of a block of pixels. In the latter approach, an image is first segmented into areas with each having an approximately uniform translation. Then the motion vector is estimated for each area. The segmentation and motion estimation associated with these arbitrarily shaped blocks are very difficult. When there are multiple moving areas in images, the situation becomes more challenging. In addition to motion vectors, the shape information of these areas needs to be coded. Hence, when moving areas have various complicated shapes, both computational complexity and coding load will increase remarkably.

In contrast, the block matching technique, which is the focus of this chapter, is simple, straightforward, and yet very efficient. It has been by far the most popularly utilized motion estimation technique in video coding. In fact, it has been adopted by all the international video coding standards: ISO MPEG-1 and MPEG-2, and ITU H.261, H.263 and H.264. These standards will be introduced in detail in Chapters 16 through 20, respectively.

It is interesting to note that nowadays, with the tremendous advancements in multimedia engineering, object-based and/or content-based manipulation of audiovisual information is very demanding, particularly in audiovisual data storage, retrieval, and distribution. The applications include digital library, video-on-demand, audiovisual database, and so on. Therefore, the coding of arbitrarily shaped objects has regained great research attention these days. It is included in the MPEG-4 activities [iscas 1997], and hence will be discussed in Chapter 18.

In this chapter various aspects of block matching are addressed. They include the concept and algorithm, matching criteria, searching strategies, limitations, and new improvements.

11.1 Nonoverlapped, Equally Spaced, Fixed Size, Small Rectangular Block Matching

To avoid the kind of difficulties encountered in motion estimation and motion compensation with arbitrarily shaped blocks, the block matching technique was proposed by Jain and Jain in 1981 [jain 1981] based on the following simple motion model.

An image is partitioned into a set of nonoverlapped, equally spaced, fixed size, small rectangular blocks; and the translation motion within each block is assumed to be uniform. Although this simple model considers only translation motion, other types of motions, such as rotation and zooming of large objects, may be closely approximated by the piecewise translation of these small blocks provided that these blocks are small enough. This observation, originally made by Jain and Jain, has been confirmed again and again since then.

Displacement vectors for these blocks are estimated by finding their best-matched counterparts in the previous frame. In this manner, motion estimation is significantly easier than that for arbitrarily shaped blocks. Since the motion of each block is described by only one displacement vector, the side information on motion vectors decreases. Furthermore, the rectangular shape information is known to both the encoder and the decoder, and hence does not need to be encoded, which saves both computation load and side information.

The block size needs to be chosen properly. In general, the smaller the block size, the more accurate the approximation is. It is apparent, however, that the smaller block size leads to more motion vectors to be estimated and encoded, which means an increase in both computation and side information. As a compromise, a size of 16×16 is considered to be a good choice. (This has been specified in international video coding standards such as H.261, H.263; and MPEG-1, MPEG-2.) Note that for finer estimation sometimes a block size of 8×8 is used.

Figure 11.1 is utilized to illustrate the block matching technique. In Figure 11.1a an image frame at moment t_n is segmented into nonoverlapped $p \times q$ rectangular blocks. As mentioned above, in common practice, square blocks of $p=q=16$ are used most often. Consider one of the blocks centered at (x, y). It is assumed that the block is translated as a whole. Consequently, only one displacement vector needs to be estimated for this block. Figure 11.1b shows the previous frame: the frame at moment t_{n-1}. In order to estimate the displacement vector, a rectangular search window is opened in the frame t_{n-1} and centered at the pixel (x, y). Consider a pixel in the search window, a rectangular correlation window

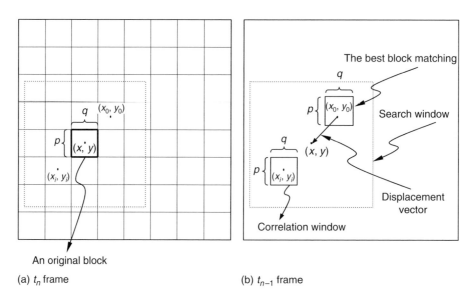

FIGURE 11.1
Block matching.

Block Matching

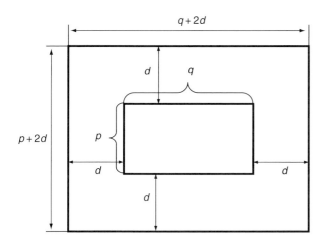

FIGURE 11.2
Search window and correlation window.

of the same size $p \times q$ is opened with the pixel located in its center. A certain type of similarity measure (correlation) is calculated. After this matching process has been completed for all candidate pixels in the search window, the correlation window corresponding to the largest similarity becomes the best match of the block under consideration in frame t_n. The relative position between these two blocks (the block and its best match) gives the displacement vector (Figure 11.1b).

The size of the search window is determined by the size of the correlation window and the maximum possible displacement along four directions: upwards, downwards, rightwards, and leftwards. In Figure 11.2 these four quantities are assumed to be the same and are denoted by d. Note that d is estimated from a priori knowledge about the translation motion, which includes the largest possible motion speed and the temporal interval between two consecutive frames, i.e., $t_n - t_{n-1}$.

11.2 Matching Criteria

Block matching belongs to image matching and can be viewed from a wider perspective. In many image processing tasks, we need to examine two images or two portions of images on a pixel-by-pixel basis. These two images or two image regions can be selected from a spatial image sequence, i.e., from two frames taken at the same time with two different sensors aiming at the same object, or from a temporal image sequence, i.e., from two frames taken at two different moments by the same sensor. The purpose of the examination is to determine the similarity between the two images or two portions of images. Examples of this type of application include image registration [pratt 1974] and template matching [jain 1989]. Image registration deals with spatial registration of images, while template matching extracts and recognizes an object in an image by matching the object template and a certain area of the image.

The similarity measure, or correlation measure, is a key element in the matching process. The basic correlation measure between two images t_n and t_{n-1}, $C(s, t)$, is defined as follows [anuta 1969].

$$C(s,t) = \frac{\sum_{j=1}^{p} \sum_{k=1}^{q} f_n(j,k) f_{n-1}(j+s, k+t)}{\sqrt{\sum_{j=1}^{p} \sum_{k=1}^{q} f_n(j,k)^2} \sqrt{\sum_{j=1}^{p} \sum_{k=1}^{q} f_{n-1}(j+s, k+t)^2}}. \qquad (11.1)$$

This is also referred to as a normalized two-dimensional (2-D) cross-correlation function [musmann 1985].

Instead of finding the maximum similarity, or correlation, an equivalent yet more computationally efficient way of block matching is to find the minimum dissimilarity, or matching error. The dissimilarity (sometimes referred to as the error, distortion, or distance) between two images t_n and t_{n-1}, $D(s, t)$ is defined as follows:

$$D(s,t) = \frac{1}{lm} \sum_{j=1}^{p} \sum_{k=1}^{q} M(f_n(j,k), f_{n-1}(j+s, k+t)), \quad (11.2)$$

where $M(u,v)$ is a metric that measures the dissimilarity between the two arguments u and v. The $D(s, t)$ is also referred to as the matching criterion or the D values.

In the literature there are several types of matching criteria, among which the mean square error (MSE) [jain 1981] and mean absolute difference (MAD) [koga 1981] are used most often. It is noted that the sum of squared difference (SSD) [anandan 1987] or the sum of squared error (SSE) [chan 1990] is essentially the same as MSE. The MAD is sometimes referred to as the mean absolute error (MAE) in the literature [nogaki 1992].

In the MSE matching criterion, the dissimilarity metric $M(u, v)$ is defined as

$$M(u, v) = (u - v)^2. \quad (11.3)$$

In the MAD,

$$M(u, v) = |u - v|. \quad (11.4)$$

Obviously, both criteria are simpler than the normalized 2-D cross-correlation measure defined in Equation 11.1.

Before proceeding to Section 11.3, a comment on the selection of the dissimilarity measure is due. A study based on experimental works reported that the matching criterion does not significantly affect the search [srinivasan 1984]. The MAD is hence preferred due to its simplicity in implementation [musmann 1985].

11.3 Searching Procedures

Searching strategy is another important issue to deal with in block matching. Several searching strategies are discussed below.

11.3.1 Full Search

Figure 11.2 shows a search window, a correlation window, and their sizes. In searching for the best matching, the correlation window is moved to each candidate position within the search window. That is, there are a total $(2d+1) \times (2d+1)$ positions that need to be examined. The minimum dissimilarity gives the best matching. Apparently, this full search procedure is brute force in nature. While the full search delivers good accuracy in searching for the best matching (thus, good accuracy in motion estimation), a large amount of computation is involved.

In order to lower computational complexity, several fast searching procedures have been developed. They are introduced below.

11.3.2 2-D Logarithm Search

Jain and Jain developed a 2-D logarithmic searching procedure in [jain 1981]. Based on a 1-D logarithm search procedure [knuth 1973], the 2-D procedure successively reduces the search area, thus reducing the computational burden. The first step computes the matching criterion for five points in the search window. These five points are as follows: the central point of the search window and the four points surrounding it, with each being a midpoint between the central point and one of the four boundaries of the window. Among these five points, the one corresponding to the minimum dissimilarity is picked as the winner. In the next step, surrounding this winner, another set of five points are selected in a similar fashion to that in the first step, with the distances between the five points remaining unchanged. The exception takes place when either a central point of a set of five points or a boundary point of the search window gives a minimum D value. In these circumstances, the distances between the five points need to be reduced. The procedure continues until the final step, in which a set of candidate points are located in a 3×3 2-D grid. Figure 11.3 demonstrates two cases of the procedure. Figure 11.3a shows that the minimum D value takes place on a boundary, while Figure 11.3b the minimum D value in the central position.

A convergence proof of the procedure is presented in [jain 1981], under the assumption that the dissimilarity monotonically increases as the search point moves away from the point corresponding to the minimum dissimilarity.

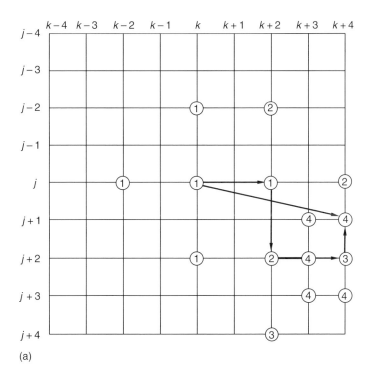

FIGURE 11.3
(a) 2-D logarithm search procedure: points at $(j, k+2)$, $(j+2, k+2)$, $(j+2, k+4)$, and $(j+1, k+4)$ are found to give the minimum dissimilarity in steps 1, 2, 3, and 4, respectively.

(continued)

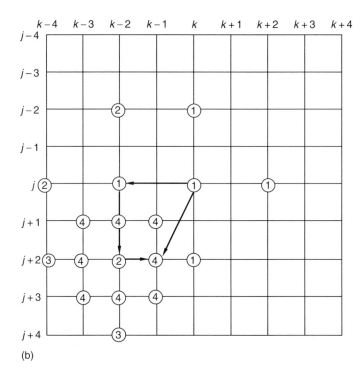

(b)

FIGURE 11.3 (continued)
(b) 2-D logarithm search procedure: points at $(j, k - 2)$, $(j + 2, k - 2)$, and $(j + 2, k - 1)$ are found to give the minimum dissimilarity in steps 1, 2, 3, and 4, respectively.

11.3.3 Coarse–Fine Three-Step Search

Another important work on the block matching technique was completed at almost the same time by Koga, Linuma, Hirano, Iijima, and Ishiguro. A coarse–fine three-step procedure was developed for fast searching [koga 1981].

The three-step search is very similar to the 2-D logarithm search. There are, however, three main differences between the two procedures. First, each step in the three-step search compares a set of nine points that form a 3×3 2-D grid structure. Second, the distances between the points in the 3×3 2-D grid structure in the three-step search decrease monotonically in steps 2 and 3. Third, a total of only three steps are carried out. Obviously, these three items are different from the 2-D logarithm search described in Section 11.3.2.

An illustrative example of the three-step search is shown in Figure 11.4.

11.3.4 Conjugate Direction Search

The conjugate direction search is another fast search algorithm, developed by Srinivasan and Rao. In principle, the procedure consists of two parts. In the first part, it finds the minimum dissimilarity along the horizontal direction with the vertical coordinate fixed at an initial position. In the second part, it finds the minimum D value along the vertical direction with the horizontal coordinate fixed in the position determined in the first part. Starting with the vertical direction followed by the horizontal direction is, of course, functionally equivalent. It was reported that this search procedure works quite efficiently [srinivasan 1984].

Figure 11.5 illustrates the principle of the conjugate direction search. In this example, each step involves a comparison between three testing points. If a point assumes the

Block Matching

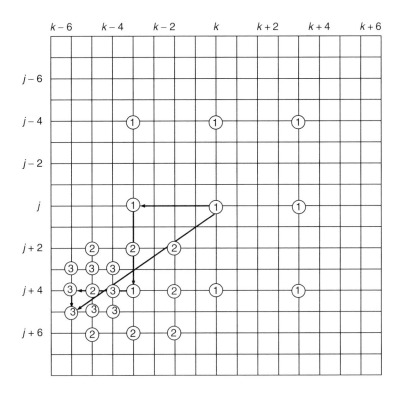

FIGURE 11.4
Three-step search procedure: points $(j+4, k-4)$, $(j+4, k-6)$, and $(j+5, k-7)$ give the minimum dissimilarity in steps 1, 2, and 3, respectively.

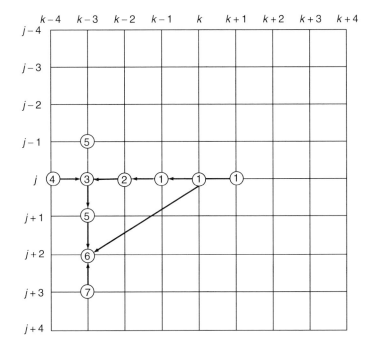

FIGURE 11.5
Conjugate direction search.

minimum D value compared with both of its two immediate neighbors (in one direction), then it is considered to be the best matching along this direction, and the search along another direction is started. Specifically, the procedure starts to compare the D values for three points $(j, k-1)$, (j, k), and $(j, k+1)$. If the D value of point $(j, k-1)$ appears to be the minimum among the three, then points $(j, k-2)$, $(j, k-1)$, and (j, k) are examined. The procedure continues, finding point $(j, k-3)$ as the best matching along the horizontal direction since its D value is smaller than that of points $(j, k-4)$ and $(j, k-2)$. The procedure is then conducted along the vertical direction. In this example the best matching is finally found at point $(j+2, k-3)$.

11.3.5 Subsampling in the Correlation Window

In the evaluation of the matching criterion, either MAD or MSE, all pixels within a correlation window at the t_{n-1} frame and an original block at the t_n frame are involved in the computation. Note that the correlation window and the original block are the same size (refer to Figure 11.1). In order to further reduce the computational effort, a subsampling inside the window and the block is performed [bierling 1988]. Aliasing effects can be avoided by using low-pass filtering. For instance, only every second pixel, both horizontally and vertically, inside the window and the block, is taken into account for the evaluation of the matching criterion. Obviously, by using this subsampling technique, the computational burden is reduced by a factor of 4. Since 3/4 of the pixels within the window and the block are not involved in the matching computation, however, the use of such a subsampling procedure may affect the accuracy of the estimated motion vectors, especially in the case of small size blocks. Therefore, the subsampling technique is recommended only for the cases with a large enough block size so that the matching accuracy will not be seriously affected. Figure 11.6 shows an example of 2×2 subsampling applied to both an original block of 16×16 at the t_n frame and a correlation window of the same size at the t_{n-1} frame.

11.3.6 Multiresolution Block Matching

It is well known that multiresolution structure, also known as pyramid structure, is a very powerful computational configuration for various image processing tasks. To save computation in block matching, it is natural to resort to the pyramid structure. In fact,

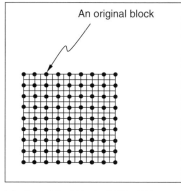
(a) An original block of 16 × 16 in frame at t_n

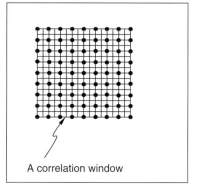
(b) A correlation window of 16 × 16 in frame at t_{n-1}

FIGURE 11.6
An example of 2×2 subsampling in the original block and correlation window for fast search.

Block Matching

multiresolution technique has been regarded as one of the most efficient methods in block matching [tzovaras 1994]. In a named top-down multiresolution technique, a typical Gaussian pyramid is formed first.

Before diving into further description, let us give a short introduction to the concept of Gaussian pyramid. Gaussian pyramid can be understood as a set of images with different resolutions related to an original image in a certain way. The original image has the highest resolution and is considered as the lowest level, sometimes called the bottom level, in the set. From the bottom level to the top level, the resolution decreases monotonically. Specifically, between two consecutive levels, the upper level is half as large as the lower level in both horizontal and vertical directions. The upper level is generated by applying a low-pass filter (which has a group of weights) to the lower level, followed by a 2×2 subsampling. That is, each pixel in the upper level is a weighted average of some pixels in the lower level. In general, this iterative procedure of generating a level in the set is equivalent to convolving a specific weight function with the original image at the bottom level followed by an appropriate subsampling. Under certain conditions, these weight functions can closely approximate the Gaussian probability density function (pdf), which is why the pyramid is named after Gaussian. (For a detailed discussion, readers are referred to [burt 1983, 1984].) A diagram of a Gaussian pyramid structure is depicted in Figure 11.7. Note that the Gaussian pyramid depicted in Figure 11.7 resembles the so-called quad-tree structure in which each node has four children nodes. In the simplest quad-tree pyramid, each pixel in an upper level is assigned an average value of its corresponding four pixels in the next lower level.

Now let us return to our discussion on the top-down multiresolution technique. After a Gaussian pyramid has been constructed, motion search ranges are allocated among the different pyramid levels. Block matching is initiated at the lowest resolution level to obtain an initial estimation of motion vectors. These computed motion vectors are then propagated to the next higher resolution level, where they are corrected, and then propagated to the next level. This procedure continues until the highest resolution level is reached. As a result, a large amount of computation can be saved. In [tzovaras 1994] it was shown that a two-level Gaussian pyramid outperforms a three-level pyramid. Compared with full search block matching, the top-down multiresolution block search saves up to 67% computation without seriously affecting the quality of the reconstructed images.

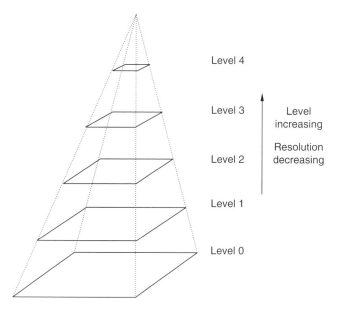

FIGURE 11.7
A Gaussian pyramid structure.

In conclusion, it has been demonstrated that multiresolution is indeed an efficient computational structure in block matching. This once again confirms the high computational efficiency of the multiresolution structure.

11.3.7 Thresholding Multiresolution Block Matching

With the multiresolution technique discussed earlier, the computed motion vectors at any intermediate pyramid level are projected to the next higher resolution level. In reality, some computed motion vectors at the lower resolution level may be inaccurate and have to be refined further, while others may be relatively accurate and able to provide satisfactory motion compensation for the corresponding block. From a computation saving point of view, for the latter class it may not be worth propagating the motion vectors to the next higher resolution level for further processing.

Motivated by the above observation, a new multiresolution block matching method with a thresholding technique was developed [shi 1997]. With the thresholding technique, it prevents those blocks, whose estimated motion vectors provide satisfactory motion compensation, from further processing, thus saving a lot of computation. In what follows, this technique is presented in detail so as to provide readers with an insight to both multiresolution block matching and thresholding multiresolution block matching techniques.

11.3.7.1 Algorithm

Let $f_n(x, y)$ be the frame of an image sequence at current moment n. First, two Gaussian pyramids are formed, pyramids n and $n-1$, from image frames $f_n(x, y)$ and $f_{n-1}(x, y)$, respectively. Let the levels of the pyramids be denoted by l, $l = 0, 1, \ldots, L$, where 0 is the lowest resolution level (top level), L is the full resolution level (bottom level), and $L + 1$ is the total number of layers in the pyramids. (Note that this way of numbering the levels of the pyramid structure [shi 1997] is different from the way depicted in Figure 11.7.) If (i, j) are the coordinates of the upper left corner of a block at level l of pyramid n, the block is referred to as block $(i, j)_n^l$. The horizontal and vertical dimensions of a block at level l are denoted by b_x^l and b_y^l, respectively. Like the variable block size method (refer to method 1 in [tzovaras 1994]), the size of the block in this work varies with the pyramid levels. That is, if the size of a block at level l is $b_x^l \times b_y^l$, then the size of the block at level $l-1$ becomes $2b_x^l \times 2b_y^l$. The variable block size method is used because it gives more efficient motion estimation than the fixed block size method. Here, the matching criterion used for motion estimation is the MAD because it does not require multiplication and gives similar performance as the MSE does. The MAD between block $(i, j)_n^l$ of the current frame and block $(i + v_x, j + v_y)_{n-1}^l$ of the previous frame at level l can be calculated as

$$MAD_{(i,j)_n^l}(v_x^l, v_y^l) = \frac{1}{b_x^l \times b_y^l} \sum_{k=0}^{b_x^l-1} \sum_{m=0}^{b_y^l-1} \left| f_n^l(i+k, j+m) - f_{n-1}^l\left(i+k+v_x^l, j+m+v_y^l\right) \right| \quad (11.5)$$

where
$V^l = (v_x^l, v_y^l)$ is one of the candidates of the motion vector of block $(i, j)_n^l$
v_x^l, v_y^l are the two components of the motion vector along the x and y directions, respectively

A block diagram of the algorithm is shown in Figure 11.8. The threshold in terms of MAD needs to be determined in advance according to the accuracy requirement of the motion estimation. How to determine the threshold is discussed below in Section 11.3.7.2. Gaussian pyramids are formed for two consecutive frames of an image sequence from

Block Matching

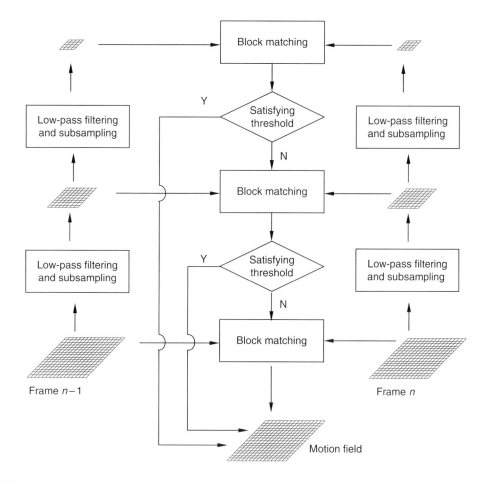

FIGURE 11.8
Block diagram for a three-level threshold multiresolution block matching.

which motion estimation is desired. Block matching is then performed at the top level with the full search scheme. The estimated motion vectors are checked to see if they provide satisfactory motion compensation. If the accuracy requirement is met, then the motion vectors will be directly transformed to the bottom level of the pyramid. Otherwise, the motion vectors will be propagated to the next higher resolution level for further refinement. This thresholding process is discussed below in Section 11.3.7.3. The algorithm continues in this fashion until either the threshold has been satisfied or the bottom level has been reached. The skipping of some intermediate level computation provides for computational saving. Experimental work with quite different motion complexities demonstrates that the proposed algorithm reduces the processing time from 14% to 20%, while maintaining almost the same quality in the reconstructed image compared with the fastest existing multiresolution block matching algorithm [tzovaras 1994].

11.3.7.2 Threshold Determination

The MAD accuracy criterion is used in this work for the sake of saving computation. The threshold value has a direct impact on the performance of the proposed algorithm. A small threshold value can improve the reconstructed image quality at the expense of increased computational effort. On the other hand, a large threshold value can reduce the

TABLE 11.1

Parameters Used in Experiments

Parameters at Level	Low Resolution Level	Full Resolution Level
"Miss America" sequence		
Search range	3 × 3	1 × 1
Block size	4 × 4	8 × 8
Thresholding value	2	None (not applicable)
"Train" sequence		
Search range	4 × 4	1 × 1
Block size	4 × 4	8 × 8
Thresholding value	3	None (not applicable)
"Football" sequence		
Search range	4 × 4	1 × 1
Block size	4 × 4	8 × 8
Thresholding value	4	None (not applicable)

computational complexity, but the quality of the reconstructed image may be degraded. One possible way to determine a threshold value, which is used in many experiments in [shi 1997], is as follows.

The peak signal-to-noise ratio (PSNR) is commonly used as a measure of the quality of the reconstructed image. As introduced in Chapter 1, it is defined as

$$\text{PSNR} = 10 \log_{10} \frac{255^2}{\text{MSE}}. \tag{11.6}$$

From the given required PSNR, one can find out the necessary MSE value. A square root of this MSE value can be chosen as a threshold value, which is applied to the first two images from the sequence. If the resulting PSNR and required processing time are satisfactory, it is then used for the rest of the sequence. Otherwise, the threshold can be slightly adjusted accordingly and applied to the second and third images to check the PSNR and processing time. It was reported that this adjusted threshold value has been good enough, and that there is no need for further adjustment in numerous experiments. As shown in Table 11.1, the threshold values used for the "Miss America," "Train," and "Football" sequences (note that three sequences have quite different motion complexities) are 2, 3, and 4, respectively. They are all determined in this fashion and give satisfactory performance, as shown in the three rows marked "New Method (TH = 2)," "New Method (TH = 3)," and "New Method (TH = 4)," respectively (Table 11.2). That is, the PSNR experiences only about 0.1 dB loss and the processing time decreases drastically. In the experiments, the threshold value of 3, i.e., the average value of 2, 3, and 4, was also tried. Refer to the three rows marked "New Method (TH = 3)" in Table 11.2. It is noted that this average threshold value 3 has already given satisfactory performance for all three sequences. Specifically, for the Miss America sequence, since the criterion increases from 2 to 3, the PSNR loss increases from 0.12 to 0.48 dB, and the reduction in processing time increases from 20% to 38%. For the Football sequence, since the criterion decreases from 4 to 3, the PSNR loss decreases from 0.08 to 0.05 dB, and the reduction in processing time decreases from 14% to 9%. Obviously, for the Train sequence, the criterion, as well as the performance, remains the same. One can therefore conclude that the threshold determination may not require much computation at all.

11.3.7.3 Thresholding

Motion vectors estimated at each pyramid level will be checked to see if they provide satisfactory motion compensation. Assume $V^i(i,j) = (v_x^l, v_y^l)$ is the estimated motion vector

Block Matching

TABLE 11.2

Experimental Results (I)

	PSNR (dB)	Error Image Entropy (bits/pixel)	Vector Entropy (bits/vector)	Block Stopped at Top Level/ Total Block	Processing Times (Number of Additions, 10^6)
"Miss America" sequence					
Method 1 [tzoraras 1994]	38.91	3.311	6.02	0/1280	10.02
New method (TH = 2)	38.79	3.319	5.65	679/1280	8.02
New method (TH = 3)	38.43	3.340	5.45	487/1280	6.17
"Train" sequence					
Method 1 [tzoraras 1994]	27.37	4.692	6.04	0/2560	22.58
New method (TH = 3)	27.27	4.788	5.65	1333/2560	18.68
"Football" sequence					
Method 1 [tzoraras 1994]	24.26	5.379	7.68	0/3840	30.06
New method (TH = 4)	24.18	5.483	7.58	1464/3840	25.90
New method (TH = 3)	24.21	5.483	7.57	1128/3840	27.10

for block $(i,j)_n^l$ at level l of pyramid n. For thresholding, $V^l(i,j)$ should be directly projected to the bottom level L. The corresponding motion vector for the same block at the bottom level of pyramid n will be $V^L(2^{(L-l)}i, 2^{(L-l)}j)$, and is given as

$$V^L(2^{(L-l)}i, 2^{(L-l)}j) = 2^{(L-l)}V^l(i,j). \quad (11.7)$$

The MAD between the block at the bottom pyramid level of the current frame and its counterpart in the previous frame can be determined according to Equation 11.5, where the motion vector is $V^L = V^L(2^{(L-l)}i, 2^{(L-l)}j)$.

This computed MAD value can be compared with the predefined threshold. If this MAD value is less than the threshold, the computed motion vector $V^L(2^{(L-l)}i, 2^{(L-l)}j)$ will be assigned to block $(2^{(L-l)}i, 2^{(L-l)}j)_n^L$ at level L in the current frame and motion estimation for this block will be stopped. If not, the estimated motion vector $V^l(i,j)$ at level l will be propagated to level $l+1$ for further refinement. Figure 11.9 gives an illustration of the above thresholding process.

11.3.7.4 Experiments

To verify the effectiveness of the proposed algorithm, extensive experiments have been conducted. The performance of the new algorithm is evaluated and compared with that of

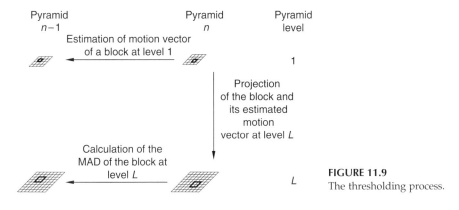

FIGURE 11.9
The thresholding process.

method 1, one of the most efficient multiresolution block matching methods [tzovaras 1994], in terms of PSNR, error image entropy, motion vector entropy, the number of blocks stopped at the top level versus the total number of blocks, and processing time. The number of blocks stopped at the top level is the number of blocks withheld from further processing, while the total number of blocks is the number of blocks existing at the top level. It is noted that the total number of blocks is the same for each level in the pyramid. The processing time is the sum of the total number of additions involved in the evaluation of the MAD and the thresholding operation.

In the experiments, two-level pyramids are used since they give better performance for motion estimation purposes [tzovaras 1994]. The algorithms are tested on three video sequences with different motion complexities, i.e., the Miss America, Train, and Football. The Miss America sequence has a speaker imposed on a static background and contains less motion. The Train sequence has more detail and contains a fast moving object (train). The 20th frame of the sequence is shown in Figure 11.10. The Football sequence contains the most complicated motion compared with the other two sequences. The 20th frame is shown in Figure 11.11. Table 11.1 is the list of implementing parameters used in the experiments. Tables 11.2 and 11.3 give the performance of the proposed algorithm compared with method 1. In all three cases, the motion estimation has a half-pixel accuracy, the meaning of which will be explained in the next section. All performance measures listed there are averaged for the first 25 frames of the testing sequences.

Each frame of the Miss America sequence is of 360×288 pixels. For convenience, only the central portion, 320×256 pixels, is processed. With the operational parameters listed in Table 11.1 (with a criterion value of 2), 38% of the total blocks at the top level satisfy the predefined criterion and are not propagated to the bottom level. The processing time needed by the proposed algorithm is 20% less than method 1, while the PSNR, the error image entropy, and the vector entropy are almost the same. Compared with method 1, an extra amount of computation (around 0.16×10^6 additions) is conducted on the thresholding operation, but a large computational savings (around 2.16×10^6 additions) are achieved through withholding from further processing those blocks whose MAD values at the full resolution level are less than the predefined accuracy criterion.

The frames of the Train sequence are 720×288 pixels, and only the central portion, 640×256 pixels, is processed. With the operational parameters listed in Table 11.1 (with a

FIGURE 11.10
The 20th frame of the "Train" sequence.

Block Matching 255

FIGURE 11.11
The 20th frame in the "Football" sequence.

criterion value of 3), about 52% of the total blocks are stopped at the top level. The processing time is reduced by about 17% by the new algorithm, compared with method 1. The PSNR, the error image entropy, and the vector entropy are almost the same.

The frames of the Football sequence are 720 × 480 pixels, and only the central portion, 640 × 384 pixels, is processed. With the operational parameters listed in Table 11.1 (with a criterion value of 4), about 38% of the total blocks are stopped at the top level. The processing time is about 14% less than that required by method 1, while the PSNR, the error image entropy, and the vector entropy are almost the same.

As discussed, the experiments with a single accuracy criterion of 3 also produce similarly good performance for the three different image sequences.

In summary, it is clear that with the three different testing sequences, the thresholding multiresolution block matching algorithm works faster than the fastest existing top-down multiresolution block matching algorithm while achieving almost the same quality of the reconstructed image.

TABLE 11.3

Experimental Results (II)

	Number of % of Total Blocks Stopped at Top Level (%)	Number of % Saved Processing Time Compared with Method 1 in [tzovaras 1994] (%)
"Miss America" sequence (TH = 2)	38	20
"Train" sequence (TH = 3)	52	17
"Football" sequence (TH = 4)	38	14

11.4 Matching Accuracy

Apparently, the two components of the displacement vectors obtained using the technique described above are an integer multiple of pixels. This is referred to as one-pixel accuracy. If a higher accuracy is desired, i.e., the components of the displacement vectors may be a non-integer multiple of pixels, then spatial interpolation is required. Not only will more computation be involved, but also will more bits be required to represent motion vectors. The gain is more accurate motion estimation, hence less prediction error. In practice, half-pixel and quarter-pixel accuracy are two widely utilized accuracies other than one-pixel accuracy.

11.5 Limitations with Block Matching Techniques

Although very simple, straightforward, and efficient, hence, utilized most widely in video coding, the block matching motion compensation technique has its drawbacks. First, it has an unreliable motion vector field with respect to the true motion in 3-D world space, in particular, it has unsatisfactory motion estimation and compensation along moving boundaries. Second, it causes block artifacts. Third, it needs to handle side information. That is, it needs to encode and transmit motion vectors as an overhead to the receiving end, thus making it difficult to use smaller block size to achieve higher accuracy in motion estimation.

All these drawbacks are due to its simple model: each block is assumed to experience a uniform translation, the motion vectors of partitioned blocks are estimated independently of each other.

Unreliable motion estimation, particularly along moving boundaries, causes more prediction error, hence reduced coding efficiency.

The block artifacts do not cause severe perceptual degradation to the human visual system (HVS) when the available coding bit rate is adequately high. This is because, with a high bit rate, a sufficient amount of the motion compensated (MC) prediction error can be transmitted to the receiving end, hence improving the subjective visual effect to such an extent that the block artifacts do not appear to be annoying. However, when the available bit rate is low, particularly less than 64 kbits/s (kilobits per second), the artifacts become visually unpleasant. In Figure 11.12, a reconstructed frame of the Miss America sequence at a low bit rate is shown. Obviously, block artifacts are very annoying, especially where mouth and hair are. The sequence was coded and decoded by using a codec following ITU-T Recommendations H.263, an international standard in which block matching is utilized for motion estimation.

The assumption that motion within each block is uniform requires a small block size such as 16×16 and 8×8. A small block size leads to a large number of motion vectors, however, resulting in a large overhead of side information. A study in [chan 1990] indicates that 8×8 block matching performs much better than 16×16 in terms of decoded image quality due to better motion estimation and compensation. The bits used for encoding motion vectors, however, increase significantly (about four times), which may be prohibitive for very low bit rate coding since the total bit rate needed for both prediction error and motion vectors may exceed the available bit rate. It is noted that when coding bit rate is quite low, say, on the order of 20 kbits/s, the side information becomes compatible with the main information (prediction error) [lin 1997].

Tremendous research efforts have been made to overcome the limitations of block matching techniques. Some improvements have been achieved and are discussed next. It should be kept in mind that, however, so far block matching is still by far the most popular and

Block Matching 257

FIGURE 11.12
The 21st reconstructed frame of the "Miss America" sequence by using a codec following H.263.

efficient motion estimation and compensation technique utilized for video coding, and it has been adopted by various international coding standards. In other words, block matching is the most appropriate in the framework of first-generation video coding [dufaux 1995].

11.6 New Improvements

11.6.1 Hierarchical Block Matching

Bierling developed the hierarchical search in 1988 [bierling 1988] based on the following two observations. On the one hand, for a relatively large displacement, accurate block matching requires a relatively large block size. This is conceivable if one considers its opposite case: a large displacement with a small correlation window. Under this circumstance, the search range is large. Therefore, the probability of finding multiple matching is high, resulting in unreliable motion estimation. On the other hand, however, a large block size may violate the assumption that all pixels in the block share the same displacement vector. Hence a relatively small block size is required to meet the assumption. These observations shed light on the problem of using a fixed block size, which may lead to unreliable motion estimation.

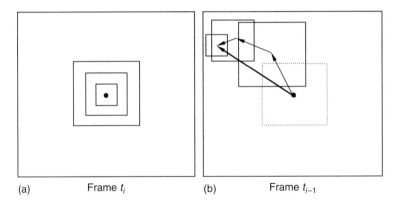

FIGURE 11.13
Hierarchical block matching.

To satisfy these two contradicting requirements simultaneously, in a hierarchical search procedure, a set of different sizes of blocks and correlation windows are utilized. To facilitate the discussion, consider a three-level hierarchical block matching algorithm, in which three block matching procedures are conducted, each with its own parameters. Block matching is first conducted with respect to the largest size of blocks and correlation windows. Using the estimated displacement vector as an initial vector at the second level, a new search is carried out with respect to the second largest size of blocks and correlation windows. The third search procedure is carried out similarly based on the results of the second search. An example with three correlation windows is illustrated in Figure 11.13. It is noted that the resultant displacement vector is the sum of the three displacement vectors determined by three searches.

The parameters in these three levels are listed in Table 11.4. The algorithm is described below with an explanation of the various parameters in Table 11.3. Before each block matching, a separate low-pass filter is applied to the whole image to achieve reliable block matching. The low-pass filtering used is simply a local averaging. That is, the gray value of every pixel is replaced by the mean value of the gray values of all pixels within a square area centered at the pixel to which the mean value is assigned. In calculating the matching criterion D value, a subsampling is applied to the original block and the correlation window in order to save computation, which was discussed in Section 11.3.5.

In the first level, for every eighth pixel horizontally and vertically (a step size of 8×8) block matching is conducted with the maximum displacement being ± 7 pixels, a correlation window size of 64×64, and a subsampling factor of 4×4. A 5×5 averaging low-pass filtering is applied before first-level block matching. Second-level block matching is

TABLE 11.4

Parameters Used in a Three-Level Hierarchical Block Matching

Hierarchical Level	Maximum Displacement (pel)	Correlation Window Size	Step Size	LPF Window Size	Subsampling
1	±7	64 × 64	8	5 × 5	4 × 4
2	±3	28 × 28	4	5 × 5	4 × 4
3	±1	12 × 12	2	3 × 3	2 × 2

Source: M. Bierling, Displacement estimation by hierarchical blockmatching, *Proceedings of Visual Communications and Image Processing*, SPIE 1001, pp. 942–951, 1988. With permission.

Block Matching

conducted with respect to every fourth pixel horizontally and vertically (a step size of 4 × 4). Note that for a pixel whose displacement vector estimate has not been determined in first-level block matching, an average of the four nearest neighboring estimates will be taken as its estimate. All the parameters for the second level are listed in Table 11.3. One thing that needs to be emphasized is that in block matching at this level the search window should be displaced by the estimated displacement vector obtained in the first level. Third-level block matching is dealt with accordingly for every second pixel horizontally and vertically (a step size of 2 × 2). The different parameters are listed in Table 11.4.

In each of the three levels, the three-step search discussed in Section 11.3.3 is utilized. Experimental work has demonstrated a more reliable motion estimation due to the usage of a set of different sizes for both the original block and the correlation window. The first level with a large window size and a large displacement range determines a major portion of the displacement vector reliably. The successive levels with smaller window sizes and smaller displacement ranges are capable of adaptively estimating motion vectors more locally.

Figure 11.14 shows a portion of an image with pixels processed in the three levels, respectively. It is noted that it is possible to apply one more interpolation after these three levels so that a motion vector field of full resolution is available. Such a full resolution motion vector field is useful in such applications as MC interpolation in the context of videophony. There, to maintain a low bit rate some frames are skipped for transmission. At the receiving end these skipped frames need to be interpolated. As discussed in

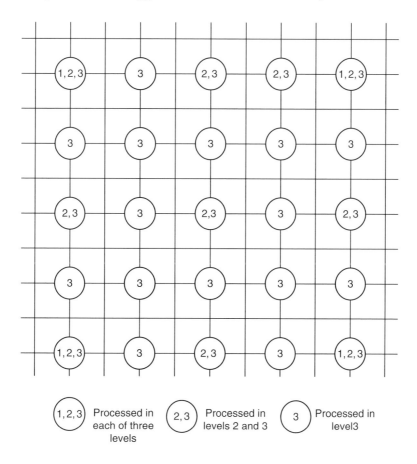

FIGURE 11.14
A portion of an image with pixels processed in the three levels, respectively.

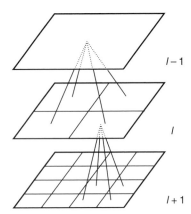

FIGURE 11.15
An illustration of a three-level hierarchical structure.

Chapter 10, MC interpolation is able to produce better frame quality than that achievable by using weighted linear interpolation.

11.6.2 Multigrid Block Matching

Multigrid theory was developed originally in mathematics [hackbusch 1982]. It is a useful computational structure in image processing besides the multiresolution one described in Section 11.3.6. A diagram with three different levels is used to illustrate a multigrid structure (Figure 11.15). Although it is also a hierarchical structure, each level within the hierarchy is of the same resolution. A few algorithms based on multigrid structure have been developed to improve the block matching technique. Two advanced methods are introduced below.

11.6.2.1 Thresholding Multigrid Block Matching

Realizing that the simple block-based motion model (assuming a uniform motion within a fixed-size block) in the block matching technique causes several drawbacks, Chan, Yu, and Constantinides proposed a variable size block matching technique. The main idea is using a split-and-merge strategy with a multigrid structure in order to segment an image into a set of variable size blocks, each of which has an approximately uniform motion. A binary tree (also known as bin-tree) structure is used to record the relationship between these blocks with different size.

Specifically, an image frame is initially split into a set of square blocks by cutting the image alternately horizontally and vertically. With respect to each block thus generated, a block matching is performed in conjunction with its previous frame. Then the matching accuracy in terms of the SSE is compared with a preset threshold. If it is smaller than or equal to the threshold, the block remains unchanged in the whole process and the estimated motion vector is final. Otherwise, the block will be split into two blocks, and a new run of block matching is conducted for each of these two children blocks. The process continues until either the estimated vector satisfies a preset accuracy requirement or the block size has reached a predefined minimum. At this point, a merge process is proposed by Chan et al.: Neighboring blocks under the same intermediate nodes in the bin-tree are checked to see if they can be merged, i.e., if the merged block can be approximated by a block in the reconstructed previous frame with adequate accuracy. It is noted that the merge operation may be optional depending on the specific application.

A block diagram of the multigrid block matching is shown in Figure 11.16. Note that it is similar to that shown in Figure 11.8 for the thresholding multiresolution block matching

Block Matching

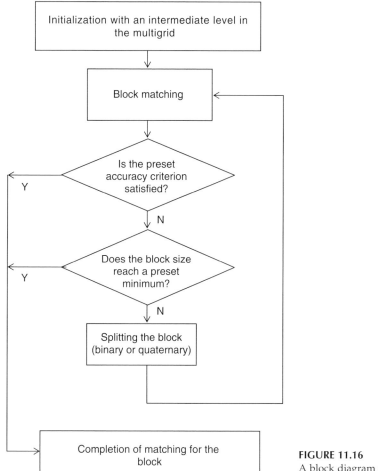

FIGURE 11.16
A block diagram of multigrid block matching.

discussed in Section 11.3.6. This observation reflects the similarities between multigrid and multiresolution structures: both are hierarchical in nature and the split and merge can be easily performed.

An example of an image decomposition and its corresponding bin-tree are shown in Figure 11.17.

It was reported in [chan 1990] that, with respect to a picture of a computer mouse and a coin, the proposed variable size block matching achieves up to 6 dB improvement in SNR and about 30% reduction in required bits compared with fixed-size (16 × 16) block matching. For several typical videoconferencing sequences, the proposed algorithm constantly performs better than the fixed-size block matching technique in terms of improved SNR of reconstructed frames with the same bit rate.

A similar algorithm was reported in [xia 1996], where a quad-tree based segmentation is used. The thresholding technique is similar to that used in [shi 1997], and the emphasis is placed on the reduction of computational complexity. It was found that for head–shoulder type of videophony sequences the thresholding multigrid block matching algorithm [xia 1996] performs better than the thresholding multiresolution block matching algorithm [shi 1997]. For video sequences that contain more complicated details and motion, however, the performance comparison turns out to be reversed.

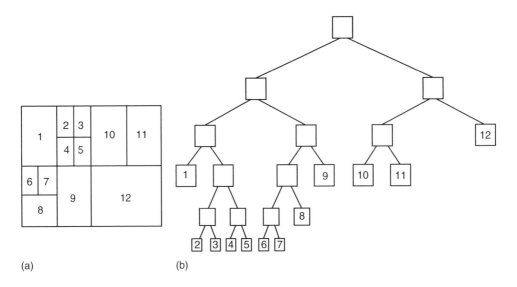

FIGURE 11.17
Thresholding multigrid block matching. (a) An example of a decomposition. (b) The corresponding bin-tree.

A few remarks can be made as a conclusion for the thresholding technique. Although it needs to encode and transmit the bin-tree or quad-tree as a portion of side information, and it has to resolve the preset threshold issue, overall speaking, the proposed algorithms achieve better performance compared with fixed-size block matching. With the flexibility provided through the variable-size methodology, the proposed approach is capable of making the motion model of the uniform motion within each block more accurate than fixed-size block matching can do.

11.6.2.2 Optimal Multigrid Block Matching

As pointed out in Chapter 10, the ultimate goal of motion estimation and motion compensation in the context of video coding is to provide high code efficiency in real time. In other words, accurate true motion estimation is not the final goal, although accurate motion estimation is certainly desired. This point was presented in [bierling 1988] as well. There, the different requirements with respect to MC coding and MC interpolation were discussed. While the former requires motion vector estimation leading to minimum prediction error and at the same time a low amount of motion vector information, the latter requires accurate estimation of true vectors and a high resolution the motion vector field.

This point was very much emphasized in [dufaux 1995]. There, Dufaux and Moscheni clearly stated that in the context of video coding, estimation of true motion in 3-D world space is not the ultimate goal. Instead, motion estimation should be able to provide good temporal prediction and at the same time require low overhead information. In a word, the total amount of information that needs to be encoded should be minimized. Based on this observation, a multigrid block matching technique with an advanced entropy criterion was proposed.

Since it belongs to the category of thresholding multigrid block matching, it shares many similarities with that in [chan 1990; xia 1996]. It also bears some resemblance to thresholding multiresolution block matching [shi 1997]. What really distinguishes this approach from other algorithms is its segmentation decision rule. Instead of a preset threshold, the algorithm works with an adaptive entropy criterion, which aims at controlling the segmentation in order to achieve an optimal solution in such a way that the total bits

Block Matching

needed for representing both the prediction error and motion overhead are minimized. The decision of splitting a block is made only when the extra motion overhead involved in the splitting is lower than the gain obtained from less prediction error due to more accurate motion estimation. Not only is it optimal in the sense of bit saving, but it also eliminates the need for setting a threshold.

The amount of bits needed for encoding motion information can be estimated in a straightforward manner. As far as the prediction error is concerned, the amount of bits required can be represented by a total entropy of the prediction error, which can be estimated by using an analytical expression presented in [moscheni 1993, dufaux 1992, 1994]. Note that the coding cost for quad-tree segmentation information is negligible compared with that used for encoding prediction error and motion vectors and, hence, is omitted in determining the criterion.

In addition to this entropy criterion, a more advanced procedure is adopted in the algorithm for down-projecting the motion vectors between two consecutive grids in the coarse-to-fine iterative refinement process.

Both qualitative and quantitative assessments in experiments demonstrate its good performance. It was reported that, when the PSNR is fixed, the bit rate saving for the sequence "Flower Garden" is from 10% to 20%, for "Mobile Calendar" from 6% to 12%, and for "Table Tennis" up to 8%. This can be translated into a gain in the PSNR ranging from 0.5 to 1.5 dB. Subjectively, the visual quality is improved greatly. In particular, moving edges become much sharper. Figures 11.18 through 11.20 show a frame from Flower Garden, Mobile Calendar, and Table Tennis sequences, respectively.

11.6.3 Predictive Motion Field Segmentation

As pointed at the beginning of Section 11.5, the block-based model, which assumes constant motion within each block, leads to unreliable motion estimation and compensation. This

FIGURE 11.18
The 20th frame of the "Flower garden" sequence.

FIGURE 11.19
The 20th frame of the "Mobile and Calendar" sequence.

FIGURE 11.20
The 20th frame of the "Table tennis" sequence.

block effect becomes more obvious and severe for motion discontinuous areas in image frames. This is because there are two or more regions in a block in the areas, each having a different motion. Using one motion vector to represent and compensate for the whole block results in significant prediction error increase.

Orchard proposed a predictive motion field segmentation technique to improve motion estimation and compensation along boundaries of moving objects in [orchard 1993]. The significant improvement in the accuracy of the MC frame was achieved through relaxing the restrictive block-based model along moving boundaries. That is, for those blocks involving moving boundaries, the motion field assumes pixel resolution instead of block resolution.

Two key issues have to be resolved in order to realize the idea. One is the segmentation issue. It is known that the segmentation information is needed at the receiving end for motion compensation. This gives rise to a large increase in side information. To maintain almost the same amount of coding cost as the conventional block matching technique, the motion field segmentation was proposed to be conducted based on previously decoded frames. This scheme is based on the following observation: the shape of an moving object does not change from frame to frame.

This segmentation is similar to the pel recursive technique (discussed in detail in Chapter 12) in the sense that both techniques operate backwards: based on previously decoded frames. The segmentation is different from the pel recursive method in that it only uses previously decoded frames to predict the shape of discontinuity in the motion field; not the whole motion field itself. Motion vectors are still estimated using the current frame at the encoder. Consequently, this scheme is capable of achieving high accuracy in motion estimation, and at the same time it does not cause a large increase in side information due to the motion field segmentation.

Another key issue is how to achieve a reconstructed motion field with pixel resolution along moving boundaries. In order to avoid extra motion vectors that need to be encoded and transmitted, the motion vectors applied to these segmented regions in the areas of motion discontinuity are selected from a set of neighboring motion vectors. As a result, the proposed technique is capable of reconstructing discontinuities in the motion field at pixel resolution whereas maintaining the same amount of motion vectors as the conventional block matching technique.

A number of algorithms using this type of motion field segmentation technique have been developed and their performance has been tested and evaluated on some real video sequences [orchard 1993]. Two of the 40 frame test sequences used were the Table Tennis and the Football. The former contains fast ball motion and camera zooming, while the latter contains small objects with relatively moderate amounts of motion and camera panning. Several proposed algorithms were compared with conventional block matching in terms of average pixel prediction error energy and bits per frame required for coding prediction error. For the average pixel prediction error energy, the proposed algorithms achieve a significant reduction, ranging from -0.7 to -2.8 dB with respect to the Table Tennis sequence, and from -1.3 to -4.8 dB with the Football sequence. For bits per frame required for coding prediction error, a reduction of 20%–30% was reported.

11.6.4 Overlapped Block Matching

All the techniques discussed so far in this section aim at more reliable motion estimation. As a result, they also alleviate annoying block artifacts to a certain extent. In this section we discuss a group of techniques, termed overlapped block matching, developed to alleviate or eliminate block artifacts [watanabe 1991; auyeung 1992; nogaki 1992].

The idea is to relax the restriction of a nonoverlapped block partition imposed in the block-based model in block matching. After the nonoverlapped, fixed size, small rectangular

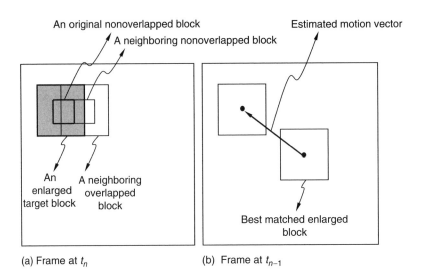

FIGURE 11.21
Overlapped block matching. (a) Frame at t_n. (b) Frame at t_{n-1}.

block partition has been made, each block is enlarged along all four directions from the center of the block (refer to Figure 11.21). Both motion estimation (block matching) and MC prediction are conducted in the same manner as that in block matching except for the inclusion of a window function. That is, a 2-D window function is utilized in order to maintain an appropriate quantitative level along the overlapped portion. The window function decays toward the boundaries. In [nogaki 1992] a sine shaped window function was used.

Next, we use the algorithm proposed by Nogaki and Ohta as an example to specifically illustrate this type of technique. Consider one of the enlarged, overlapped original (also known as target) blocks, $T(x, y)$, with a dimension of $l \times l$. Assume that a vector v_i is one of the candidate displacement vectors under consideration. The predicted version of the target block with v_i is denoted by $P_{v_i}(x, y)$. Thus, the prediction error with v_i, $E_{v_i}(x, y)$ can be calculated according to Equation 11.8

$$E_{v_i}(x, y) = P_{v_i}(x, y) - T(x, y). \tag{11.8}$$

The window function $W(x, y)$ is applied at this stage as follows, resulting in a window-operated prediction error with v_i, WE_{v_i}.

$$WE_{v_i}(x, y) = E_{v_i}(x, y) \times W(x, y). \tag{11.9}$$

Assume that the MAD is used as the matching criterion. It can then be determined as usual by using the window-operated prediction error $WE_{v_i}(x, y)$. That is,

$$\text{MAD} = \frac{1}{l^2} \sum_{x=1}^{l} \sum_{y=1}^{l} |WE_{v_i}(x, y)|. \tag{11.10}$$

The best matching, which corresponds to the minimum MAD, produces the displacement vector v.

In MC prediction, the predicted version of the enlarged target block $P_v(x, y)$ is derived from the frame at t_{i-1} by using estimated vector v. The same window function $W(x, y)$ is used to generate the final window-operated predicted version of the target block. That is,

$$WP_v(x,y) = P_v(x,y) \times W(x,y). \tag{11.11}$$

It was reported in [nogaki 1992] that the luminance signal of an HDTV sequence was used in computer simulation. A block size of 16×16 was used for conventional block matching, while a block size of 32×32 was employed for the proposed overlapped block matching. The maximum displacement range d was taken as $d = 15$, i.e., from -15 to $+15$ in the both horizontal and vertical directions. The simulation indicated a reduction in the power of prediction error by about 19%. Subjectively, it was observed that the blocking edges originally existing in the prediction error signal with conventional block matching were largely removed with the proposed overlapped block matching technique.

11.7 Summary

By far, block matching is used more frequently than any other motion estimation techniques in MC coding. By partitioning a frame into nonoverlapped, equally spaced, fixed size, small rectangular blocks, and assuming that all the pixels in a block experience the same translational motion, block matching avoids the difficulty encountered in motion estimation of arbitrarily shaped blocks. Consequently, block matching is much simpler and involves less side information compared with motion estimation with arbitrarily shaped blocks.

Although this simple model considers translation motion only, other types of motions, such as rotation and zooming of large objects, may be closely approximated by the piecewise translation of these small blocks provided that these blocks are small enough. This important observation, originally made by Jain and Jain, has been confirmed again and again since then.

Various issues related to block matching such as selection of block sizes, matching criteria, search strategies, matching accuracy, its limitations and improvements are discussed in this chapter. Specifically, a block size of 16×16 is used most often. For more accurate motion estimation, the size of 8×8 is used sometimes. In the latter case, more accurate motion estimation is obtained at the cost of more side information and higher computational complexity.

There are several different types of matching criteria that can be used in block matching. Since it was shown that the different criteria do not cause significant difference in block matching, the MAD is hence preferred due to its simplicity in implementation.

On the one hand, full search procedure delivers good accuracy in searching for the best matching. On the other hand, it requires a large amount of computation. In order to lower computational complexity, several fast searching procedures were developed: 2-D logarithm search, coarse–fine three-step search, conjugate direction search, etc.

Besides these suboptimum search procedures, there are some other measures developed to lower computation. One of them is subsampling in the original blocks and the correlation windows. By the subsampling, the computational burden in block matching can be reduced drastically, while the accuracy of the estimated motion vectors may be affected. Therefore, the subsampling procedure is only recommended for the case with a large block size.

Naturally, multiresolution structure, a powerful computational configuration in image processing, lends itself well to fast search in block matching. It significantly reduces computation involved in block matching. Thresholding multiresolution block matching further saves computation.

In terms of matching accuracy, several common choices are one-pixel, half-pixel, and quarter-pixel accuracies. Spatial interpolation is usually required for half-pixel and quarter-pixel accuracies. That is, a higher accuracy is achieved with more computation.

Limitations with block matching techniques are mainly: unreliable motion vector field, and block artifacts. Both are caused by the simple model: each block is assumed to experience a uniform translation. Much efforts have been made to improve these drawbacks. Several techniques that have made improvement over the conventional block matching technique are discussed in this chapter.

In the hierarchical block matching technique, a set of different sizes for both the original block and the correlation window are used. The first level in the hierarchy with a large window size and a large displacement range determines a major portion of the displacement vector reliably. The successive levels with smaller window sizes and smaller displacement ranges are capable of adaptively estimating motion vectors more locally.

The multigrid block matching technique uses multigrid structure, another powerful computational structure in image processing, to provide a variable size block matching. With a split-and-merge strategy, the thresholding multigrid block matching technique segment an image into a set of variable size block, each of which experiences an approximately uniform motion. A tree structure (bin-tree or quad-tree) is used to record the relationship between these variable size blocks. With the flexibility provided through the variable-size methodology, the thresholding block matching technique is capable of making the motion model of the uniform motion within each block more accurate than the fixed-size block matching can do.

As pointed out in Chapter 10, the ultimate goal of motion compensation in video coding is to achieve a high coding efficiency. In other words, accurate true motion estimation is not the final goal. From this point of view, in the above-mentioned multigrid block matching, the decision of splitting a block is made only when the bits used to encode extra motion vectors involved in the splitting are less than the bits saved from encoding reduced prediction error due to more accurate estimation. To this end, an adaptive entropy criterion is proposed and used in the optimal multigrid block matching technique. Not only is it optimal in the sense of bit saving, but it also eliminates the need for setting a threshold.

Apparently the block-based model encounters more severe problem along moving boundaries. To solve the problem, the predictive motion field segmentation technique make the blocks involving moving boundaries to have the motion field with pixel resolution instead of block resolution. In order to save shape overhead, segmentation is carried out backwards, i.e., based on previously decoded frames. In order to avoid a large increase of side information associated with extra motion vectors, the motion vectors applied to these segmented regions along moving boundaries are selected from a set of neighboring motion vectors. As a result, the technique is capable of reconstructing discontinuities in the motion field at pixel resolution whereas maintaining the same amount of motion vectors as the conventional block matching technique.

The last improvement over the conventional block matching discussed in this chapter is overlapped block matching. In contrast to dealing with blocks independently of each other, the overlapped block matching technique enlarges blocks so as to make them overlapped. A window function is then constructed and used in both motion estimation and motion compensation. Because it relaxes the restriction of a nonoverlapped block partition imposed by the conventional block matching, it achieves better performance than the conventional block matching.

Exercises

1. Refer to Figure 11.2, it is said that there are a total of $(2d+1) \times (2d+1)$ positions that need to be examined in block matching with full search if one-pixel accuracy is required. How many positions are there that need to be examined in block matching with full search if half-pixel and quarter-pixel accuracies are required?
2. What are the two effects that the subsampling in the original block and the correlation block may bring out?
3. Read [burt 1983, 1984] and explain why the pyramid is named after Gaussian.
4. Read [burt 1983, 1984] and explain why a pyramid structure is considered as a powerful computational configuration. Specifically, in multiresolutional block matching, how and to what extent does it save computation dramatically compared with the conventional block matching technique? (Refer to Section 11.3.7.)
5. How is the threshold determined in the thresholding multidimensional block matching technique? (Refer to Section 11.3.7.) It is said that the square root of the MSE value, derived from the given PSNR (Equation 11.6), is used as an initial threshold value. Justify the necessity of the square root operation.
6. Refer to Section 11.6.1 or [bierling 1988]. State the different requirements in the applications of MC interpolation and MC coding. Discuss where a full resolution of translational motion vector field may be used?
7. Read [dufaux 1995] and explain the main feature of the optimal multigrid block matching. State how the adaptive entropy criterion is established. Implement algorithm and compare its performance with that presented in [chan 1990].
8. Learn the predictive motion field segmentation technique [orchard 1993]. Explain how the algorithms avoid a large increase in overhead due to motion field segmentation.
9. Implement the overlapped block matching algorithm introduced in [nogaki 1992]. Compare its performance with that of the conventional block matching technique.

References

[anandan 1987] P. Anandan, Measurement visual motion from image sequences, Ph.D. thesis, COINS Department, University of Massachusetts, Amherst, 1987.

[anuta 1969] P.F. Anuta, Digital registration of multispectral video imagery, *Society of Photo-Optical Instrumentation Engineers Journal*, 7, 168–175, September 1969.

[auyeung 1992] C. Auyeung, J. Kosmach, M. Orchard, and T. Kalafatis, Overlapped block motion compensation, *SPIE Proceedings of Visual Communication and Image Processing'92*, Vol. 1818, Boston, MA, pp. 561–571, November 1992.

[bierling 1988] M. Bierling, Displacement estimation by hierarchical blockmatching, *Proceedings of Visual Communications and Image Processing*, SPIE 1001, pp. 942–951, 1988.

[brofferio 1977] S. Brofferio and F. Rocca, Interframe redundancy reduction of video signals generated by translating objects, *IEEE Transactions on Communications*, COM-25, 448–455, April 1977.

[burt 1983] P.J. Burt and E.H. Adelson, The Laplacian pyramid as a compact image code, *IEEE Transactions on Communications*, COM-31, 4, 532–540, April 1983.

[burt 1984] P.J. Burt, The pyramid as a structure for efficient computation, in *Multiresolution Image Processing and Analysis*, A. Rosenfeld (Ed.), Springer-Verlag, Germany, pp. 6–37, 1984.

[cafforio 1976] C. Cafforio and F. Rocca, Method for measuring small displacement of television images, *IEEE Transactions on Information Theory*, IT-22, 573–579, September 1976.

[chan 1990] M.H. Chan, Y.B. Yu, and A.G. Constantinides, Variable size block matching motion compensation with applications to video coding, *IEE Proceedings*, 137, Part I, 4, 205–212, August 1990.

[dufaux 1992] F. Dufaux and M. Kunt, Multigrid block matching motion estimation with an adaptive local mesh refinement, *SPIE Proceedings of Visual Communications and Image Processing'92*, Vol. 1818, Boston, MA, pp. 97–109, November 1992.

[dufaux 1994] F. Dufaux, Multigrid block matching motion estimation for generic video coding, Ph.D. dissertation, Swiss Federal Institute of Technology, Lausanne, Switzerland, 1994.

[dufaux 1995] F. Dufaux and F. Moscheni, Motion estimation techniques for digital TV: A review and a new contribution, *Proceedings of the IEEE*, 83, 6, 858–876, 1995.

[hackbusch 1982] W. Hackbusch and U. Trottenberg, Eds., *Multigrid Methods*, Springer-Verlag, New York, 1982.

[haskell 1972] B.G. Haskell and J.O. Limb, Predictive video encoding using measured subject velocity, U.S. Patent 3, 632, 865, January 1972.

[iscas 1997] J. Brailean, Universal accessibility and object-based functionality, *ISCAS Tutorial on MPEG 4*, Chapter 3.3, June 1997.

[jain 1981] J.R. Jain and A.K. Jain, displacement measurement and its application in interframe image coding, *IEEE Transactions on Communications*, COM-29, 12, 1799–1808, December 1981.

[jain 1989] A.K. Jain, *Fundamentals of Digital Image Processing*, Prentice-Hall, Englewood Cliffs, NJ, 1989.

[koga 1981] T. Koga, K. Linuma, A. Hirano, Y. Iijima, and T. Ishiguro, Motion compensated interframe coding for video conferencing, *Proceedings of NTC'81*, New Orleans, LA, pp. G5.3.1–G5.3.5, December 1981.

[knuth 1973] D.E. Knuth, *Searching and Sorting, Vol. 3: The Art of Computer Programming*. Addison-Wesley, Reading, MA, 1973.

[limb 1975] J.O. Limb and J.A. Murphy, Measuring the speed of moving objects from television signals, *IEEE Transactions on Communications*, COM-23, 474–478, April 1975.

[lin 1997] S. Lin, Y.Q. Shi, and Y.-Q. Zhang, An optical flow based motion compensation algorithm for very low bit-rate video coding, *Proceedings of 1997 IEEE International Conference on Acoustics, Speech and Signal Processing*, Munich, Germany, pp. 2869–2872, April 1997; Y.Q. Shi, S. Lin, and Y.-Q. Zhang, Optical flow-based motion compensation algorithm for very low-bit-rate video coding, *International Journal of Imaging Systems and Technology*, 9, 4, 230–237, 1998.

[moscheni 1993] F. Moscheni, F. Dufaux, and H. Nicolas, Entropy criterion for optimal bit allocation between motion and prediction error information, *SPIE 1993 Proceedings of Visual Communications and Image Processing*, Cambridge, MA, pp. 235–242, November 1993.

[musmann 1985] H.G. Musmann, P. Pirsch, and H.J. Grallert, Advances in picture coding, *Proceedings of the IEEE*, 73, 4, 523–548, 1985.

[netravali 1979] A.N. Netravali and J.D. Robbins, Motion compensated television coding, Part I, *The Bell System Technical Journal*, 58, 3, 631–670, March 1979.

[nogaki 1992] S. Nogaki and M. Ohta, An overlapped block motion compensation for high quality motion picture coding, *Proceedings of IEEE International Symposium on Circuits and Systems*, 1, 184–187, May 1992.

[orchard 1993] M.T. Orchard, Predictive motion-field segmentation for image sequence coding, *IEEE Transactions on Circuits and Systems for Video Technology*, 3, 1, 54–69, February 1993.

[pratt 1974] W.K. Pratt, Correlation techniques of image registration, *IEEE Transactions on Aerospace and Electronic Systems*, AES-10, 3, 353–358, May 1974.

[rocca 1972] F. Rocca and S. Zanoletti, Bandwidth reduction via movement compensation on a model of the random video process, *IEEE Transactions on Communications*, COM-20, 960–965, October 1972.

[shi 1997] Y.Q. Shi and X. Xia, A thresholding multidimensional block matching algorithm, *IEEE Transactions on Circuits and Systems for Video Technology*, 7, 2, 437–440, April 1997.

[srinivasan 1984] R. Srinivasan and K.R. Rao, Predictive coding based on efficient motion estimation, *Proceedings of ICC*, pp. 521–526, May 1984.

[tzovaras 1994] D. Tzovaras, M.G. Strintzis, and H. Sahinolou, Evaluation of multiresolution block matching techniques for motion and disparity estimation, *Signal Processing: Image Communication*, 6, 56–67, 1994.

[xia 1996] X. Xia and Y.Q. Shi, A thresholding hierarchical block matching algorithm, *Proceedings of IEEE 1996 International Symposium on Circuits and Systems*, Vol. II, Atlanta, GA, pp. 624–627, May 1996; X. Xia, Y.Q. Shi, and Y. Shi, A thresholding hierarchical block matching algorithm, *Journal of Computer Science and Information Management*, 1, 2, 83–90, 1998.

12

Pel Recursive Technique

As discussed in Chapter 10, the pel recursive technique is one of the three major approaches to two-dimensional (2-D) displacement estimation in image planes for the signal processing community. Conceptually speaking, it is one type of region matching technique. In contrast to block matching (which was discussed in Chapter 11), it *recursively* estimates displacement vector for each pixel in an image frame. The displacement vector of a pixel is estimated by recursively minimizing a nonlinear function of the dissimilarity between two certain regions located in two consecutive frames. Note that region means a group of pixels, but it could be as small as a single pixel. Also note that the terms "pel" and "pixel" have the same meaning. Both terms are used frequently in the field of signal and image processing.

This chapter is organized as follows. A general description of the recursive technique is provided in Section 12.1. Some fundamental techniques in optimization are covered in Section 12.2. Section 12.3 describes the Netravali and Robbins algorithm, the pioneering work in this category. Several other typical pel recursive algorithms are introduced in Section 12.4. In Section 12.5, a performance comparison between these algorithms is made.

12.1 Problem Formulation

In 1979 Netravali and Robbins published the first pel recursive algorithm to estimate displacement vectors for motion compensated interframe image coding. In [netravali 1979], a quantity, called the displaced frame difference (DFD), was defined as follows:

$$\text{DFD}(x, y; d_x, d_y) = f_n(x, y) - f_{n-1}(x - d_x, y - d_y), \tag{12.1}$$

where
  n and $n-1$ indicate two moments associated with two successive frames based on which motion vectors are to be estimated
  x, y are coordinates in image planes
  d_x, d_y are the two components of the displacement vector, $\vec{d}$, along the horizontal and vertical directions in the image planes, respectively

$\text{DFD}(x, y; d_x, d_y)$ can also be expressed as $\text{DFD}(x, y; \vec{d})$. Whenever it does not cause confusion, it can be written as DFD for the sake of brevity. Obviously, if there is no error in the estimation, i.e., the estimated displacement vector is exactly equal to the true motion vector then DFD will be zero.

A nonlinear function of the DFD was then proposed as a dissimilarity measure in [netravali 1979], which is a square function of DFD, i.e., DFD^2.

Netravali and Robbins thus converted displacement estimation into a minimization problem. That is, each pixel corresponds to a pair of integers (x, y), denoting its spatial position in the image plane. Therefore, the DFD is a function of $\vec{d}$. The estimated displacement vector $\vec{d} = (d_x, d_y)^T$, where $(.)^T$ denotes the transposition of the argument vector or matrix, can be determined by minimizing the DFD^2. This is a typical nonlinear programming problem, on which a large body of research has been reported in the literature. In the next section, several techniques that rely on a method, called descent method, in optimization are introduced. The Netravali and Robbins algorithm can be applied to a pixel once or iteratively applied several times for displacement estimation. Then the algorithm moves to the next pixel. The estimated displacement vector of a pixel can be used as an initial estimate for the next pixel. This recursion can be carried out horizontally, vertically, or temporally. By temporally, we mean that the estimated displacement vector can be passed to the pixel of the same spatial position within image planes in a temporally neighboring frame. Figure 12.1 illustrates these three different types of recursion.

(a) Horizontal

(b) Vertical

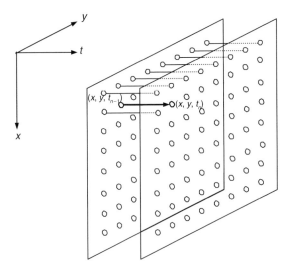

(c) Temporal

FIGURE 12.1
Three types of recursions.

12.2 Descent Methods

Consider a nonlinear real-valued function z of a vector variable $\bar{x}$,

$$z = f(\bar{x}) \tag{12.2}$$

with $\bar{x} \in R^n$, where R^n represents the set of all n-tuples of real numbers. The question we face now is how to find such a vector denoted by $\bar{x}^*$ that the function z is minimized. This is classified as an unconstrained nonlinear programming problem.

12.2.1 First-Order Necessary Conditions

According to the optimization theory, if $f(\bar{x})$ has continuous first-order partial derivatives, then the first-order necessary conditions that $\bar{x}^*$ has to satisfy are

$$\nabla f(\bar{x}^*) = 0, \tag{12.3}$$

where ∇ denotes the gradient operation with respect to $\bar{x}$ evaluated at $\bar{x}^*$. Note that whenever there is only one vector variable in the function z to which the gradient operator is applied, the sign ∇ would remain without a subscript, as in Equation 12.3. Otherwise, i.e., if there is more than one vector variable in the function, we will explicitly write out the variable, to which the gradient operator is applied, as a subscript of the sign ∇. In the component form, Equation 12.3 can be expressed as

$$\begin{cases} \dfrac{\partial f(\bar{x})}{\partial x_1} = 0 \\ \dfrac{\partial f(\bar{x})}{\partial x_2} = 0 \\ \vdots \\ \dfrac{\partial f(\bar{x})}{\partial x_n} = 0 \end{cases} \tag{12.4}$$

12.2.2 Second-Order Sufficient Conditions

If $f(\bar{x})$ has second-order continuous derivatives, then the second-order sufficient conditions for $f(\bar{x}^*)$ to reach the minimum are known as

$$\nabla f(\bar{x}^*) = 0 \tag{12.5}$$

and

$$H(\bar{x}^*) > 0, \tag{12.6}$$

where H denotes the Hessian matrix and is defined as follows:

$$H(\bar{x}) = \begin{bmatrix} \frac{\partial^2 f(\bar{x})}{\partial^2 x_1} & \frac{\partial^2 f(\bar{x})}{\partial x_1 \partial x_2} & \cdots & \frac{\partial^2 f(\bar{x})}{\partial x_1 \partial x_n} \\ \frac{\partial^2 f(\bar{x})}{\partial x_2 \partial x_1} & \frac{\partial^2 f(\bar{x})}{\partial^2 x_2} & \cdots & \frac{\partial^2 f(\bar{x})}{\partial x_2 \partial x_n} \\ \vdots & \vdots & \vdots & \vdots \\ \frac{\partial^2 f(\bar{x})}{\partial x_n \partial x_1} & \frac{\partial^2 f(\bar{x})}{\partial x_n \partial x_2} & \cdots & \frac{\partial^2 f(\bar{x})}{\partial^2 x_n} \end{bmatrix} \quad (12.7)$$

We can thus see that the Hessian matrix consists of all the second-order partial derivatives of f with respect to the components of $\bar{x}$. Equation 12.6 means that H is positive definite.

12.2.3 Underlying Strategy

Our aim is to derive an iterative procedure for the minimization. That is, we want to find a sequence

$$\bar{x}_0, \bar{x}_1, \bar{x}_2, \ldots, \bar{x}_n, \ldots \quad (12.8)$$

such that

$$f(\bar{x}_0) > f(\bar{x}_1) > f(\bar{x}_2) > \cdots > f(\bar{x}_n) > \cdots \quad (12.9)$$

and the sequence converges to the minimum of $f(\bar{x}), f(\bar{x}^*)$.

A fundamental underlying strategy for almost all the descent algorithms [luenberger 1984] is described as follows. We start with an initial point in the space; we determine a direction to move according to a certain rule; then we move along the direction to a relative minimum of the function z. This minimum point becomes the initial point for the next iteration.

This strategy can be better visualized using a 2-D example, shown in Figure 12.2. There, $\bar{x} = (x_1, x_2)^T$. Several closed curves are referred to as contour curves or level curves. That is, each of the curves represents

$$f(x_1, x_2) = c \quad (12.10)$$

with c being a constant.

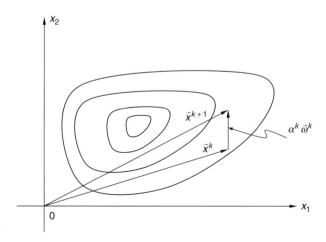

FIGURE 12.2
Descent method.

Assume that at the kth iteration, we have a guess: $\vec{x}^k$. For the $(k+1)$th iteration, we need to

1. Find a search direction, pointed by a vector $\vec{\omega}^k$
2. Determine an optimal step size α^k with $\alpha^k > 0$

such that the next guess $\vec{x}^{k+1}$ is

$$\vec{x}^{k+1} = \vec{x}^k + \alpha^k \vec{\omega}^k \tag{12.11}$$

and $\vec{x}^{k+1}$ satisfies $f(\vec{x}^k) > f(\vec{x}^{k+1})$.

In Equation 12.11, $\vec{x}^k$ can be viewed as a prediction vector for $\vec{x}^{k+1}$, while $\alpha^k \vec{\omega}^k$ an update vector, $\vec{v}^k$. Hence, using the Taylor series expansion, we can have

$$f(\vec{x}^{k+1}) = f(\vec{x}^k) + \langle \nabla f(\vec{x}^k), \alpha^k \vec{\omega}^k \rangle + \varepsilon, \tag{12.12}$$

where $\langle \vec{s}, \vec{t} \rangle$ denotes the inner product between vectors $\vec{s}$ and $\vec{t}$, and ε represents the higher-order terms in the expansion. Consider that the increment of $\alpha^k \vec{\omega}^k$ is small enough and, thus, ε can be ignored. From Equation 12.10, it is obvious that to have $f(\vec{x}^{k+1}) < f(\vec{x}^k)$ we must have $\langle \nabla f(\vec{x}^k), \alpha^k \vec{\omega}^k \rangle < 0$. That is,

$$f(\vec{x}^{k+1}) < f(\vec{x}^k) \Rightarrow \langle \nabla f(\vec{x}^k), \alpha^k \vec{\omega}^k \rangle < 0 \tag{12.13}$$

Choosing a different update vector, i.e., the product of the $\vec{\omega}^k$ vector and the step size α^k results in a different algorithm in implementing descent method.

In the same category of the descent method, a variety of techniques have been developed. The reader may refer to [luenberger 1984] or the many other existing books on optimization. Two commonly used techniques of the descent method are discussed. One is called the steepest descent method, in which the search direction represented by the $\vec{\omega}$ vector is chosen to be opposite to that of the gradient vector, and a real parameter of the step size α^k is used; the other is the Newton–Raphson method, in which the update vector in estimation, determined jointly by the search direction and the step size, is related to the Hessian matrix (defined in Equation 12.7). These two techniques are further discussed in Sections 12.2.5 and 12.2.6, respectively.

12.2.4 Convergence Speed

Speed of convergence is an important issue in discussing the descent method. It is utilized to evaluate the performance of different algorithms.

12.2.4.1 Order of Convergence

Assume a sequence of vectors $\{\vec{x}^k\}$, with $k = 0, 1, \ldots, \infty$, converges to a minimum denoted by $\vec{x}^*$. We say that the convergence is of order p if the following formula holds [luenberger 1984]:

$$0 \leq \overline{\lim}_{k \to \infty} \frac{|\vec{x}^{k+1} - \vec{x}^*|}{|\vec{x}^k - \vec{x}^*|^p} < \infty \tag{12.14}$$

where
  p is positive
  $\overline{\lim}$ is the limit superior
  $|.|$ is the magnitude or norm of a vector argument

For the two latter notions, more descriptions follow.

The concept of the limit superior is based on the concept of supremum. Hence, let us first discuss the supremum. Consider a set of real numbers, denoted by Q that is bounded above. Then there must exist a smallest real number o such that for all the real numbers in the set Q, i.e., $q \in Q$, we have $q \leq o$. This real number o is referred to as the least upper bound or the supremum of the set Q, and it is denoted by

$$\sup\{q : q \in Q\} \text{ or } \sup_{q \in Q}(q). \tag{12.15}$$

Now turn to a real bounded above sequence r^k, $k = 0, 1, \ldots, \infty$. If $s^k = \sup\{r^j : j \geq k\}$, then the sequence $\{s^k\}$ converges to a real number s^*. This real number s^* is referred to as the limit superior of the sequence $\{r^k\}$, and is denoted by

$$\overline{\lim}_{k \to \infty}(r^k). \tag{12.16}$$

The magnitude or norm of a vector $\vec{x}$, denoted by $|\vec{x}|$, is defined as

$$|\vec{x}| = \langle \vec{x}, \vec{x} \rangle, \tag{12.17}$$

where $\langle \vec{s}, \vec{t} \rangle$ is the inner product between the vectors $\vec{s}$ and $\vec{t}$. Throughout this discussion, when we say vector we mean column vector. (Row vectors can be handled accordingly.) The inner product is therefore defined as

$$\langle \vec{s}, \vec{t} \rangle = \vec{s}\vec{t}^T, \tag{12.18}$$

with the superscript T indicating the transposition operator.

With the definitions of the limit superior and the magnitude of a vector introduced, we are now in a position to easily understand the concept of the order of convergence defined in Equation 12.14. Since the sequences generated by the descent algorithms behave quite well in general [luenberger 1984], the limit superior is rarely necessary. Hence, roughly speaking, instead of the limit superior, the limit may be used in considering the speed of convergence.

12.2.4.2 Linear Convergence

Among the various orders of convergence, the order of unity is important, and is referred to as linear convergence. Its definition is as follows. If a sequence $\{\vec{x}^k\}$, $k = 0, 1, \ldots, \infty$, converges to $\vec{x}^*$ with

$$\lim_{k \to \infty} \frac{|\vec{x}^{k+1} - \vec{x}^*|}{|\vec{x}^k - \vec{x}^*|} = \gamma < 1, \tag{12.19}$$

then we say that this sequence converges linearly with a convergence ratio γ. The linear convergence is also referred to as geometric convergence because a linear convergent sequence with convergence ration γ converges to its limit at least as fast as the geometric sequences $c\gamma^k$, with c being a constant.

12.2.5 Steepest Descent Method

The steepest descent method, often referred to as the gradient method, is the oldest and simplest one among various techniques in the descent method. As Luenberger pointed out in his book, it remains to be the fundamental method in the category for the following two reasons. First, owing to its simplicity, it is usually the first method attempted for solving a new problem. This observation is very true. As we shall see soon, when handling the displacement estimation as a nonlinear programming problem in the pel recursive technique, the first algorithm developed by Netravali and Robbins is essentially the steepest descent method. Second, owing to the existence of a satisfactory analysis for the steepest descent method, it continues to serve as a reference for comparing and evaluating various newly developed and more advanced methods.

12.2.5.1 Formulae

In the steepest descent method, $\vec{\omega}^k$ is chosen as

$$\vec{\omega}^k = -\nabla f(\vec{x}^k), \tag{12.20}$$

resulting in

$$f(\vec{x}^{k+1}) = f(\vec{x}^k) - \alpha^k \nabla f(\vec{x}^k), \tag{12.21}$$

where the step size α^k is a real parameter, and, with our rule mentioned before, the sign ∇ here denotes a gradient operator with respect to $\vec{x}^k$. Since the gradient vector points to the direction along which the function $f(\vec{x})$ has greatest increases, it is naturally expected that the selection of the negative direction of the gradient as the search direction will lead to the steepest descent of $f(\vec{x})$. This is where the term steepest descent originated.

12.2.5.2 Convergence Speed

It can be shown that if the sequence $\{\vec{x}\}$ is bounded above, then the steepest descent method will converge to the minimum. Furthermore, it can be shown that the steepest descent method is linear convergent.

12.2.5.3 Selection of Step Size

It is worth noting that the selection of the step size α^k has significant influence on the algorithm's performance. In general, if it is small, it produces an accurate estimate of $\vec{x}^*$. But a smaller step size means it will take longer for the algorithm to reach the minimum. Although a larger step size will make algorithm converge faster, it may lead to an estimate with large error. This situation can be demonstrated in Figure 12.3. There, for the sake of an easy graphical illustration, $\vec{x}$ is assumed to be one-dimensional (1-D). Two cases of too small (with subscript 1) and too large (with subscript 2) step sizes are shown for comparison.

12.2.6 Newton–Raphson's Method

The Newton–Raphson method is the next most popular method among various descent methods.

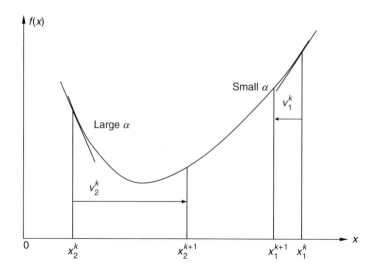

FIGURE 12.3
An illustration of effect of selection of step size on minimization performance. Too small α requires more steps to reach x^*. Too large α may cause overshooting.

12.2.6.1 Formulae

Consider $\vec{x}^k$ at the kth iteration. The $k+1$th guess, $\vec{x}^{k+1}$, is the sum of $\vec{x}^k$ and $\vec{v}^k$,

$$\vec{x}^{k+1} = \vec{x}^k + \vec{v}^k, \tag{12.22}$$

where $\vec{v}^k$ is an update vector as shown in Figure 12.4. Now expand the $\vec{x}^{k+1}$ into the Taylor series explicitly containing the second-order term.

$$f(\vec{x}^{k+1}) = f(\vec{x}^k) + \langle \nabla f, \vec{v} \rangle + \frac{1}{2} \langle H(\vec{x}^k)\vec{v}, \vec{v} \rangle + \varphi, \tag{12.23}$$

where
  φ denotes the higher-order terms
  ∇ denotes the gradient
  H denotes the Hessian matrix

If $\vec{v}$ is small enough, we can ignore the φ. According to the first-order necessary conditions for $\vec{x}^{k+1}$ to be the minimum, discussed in Section 12.2.1, we have

$$\nabla_{\vec{v}} f(\vec{x}^k + \vec{v}) = \nabla f(\vec{x}^k) + H(\vec{x}^k)\vec{v} = 0, \tag{12.24}$$

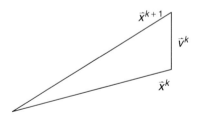

FIGURE 12.4
Derivation of Newton–Raphson's method.

where $\nabla_{\vec{v}}$ denotes the gradient operator with respect to $\vec{v}$. This leads to

$$\vec{v} = -H^{-1}(\vec{x}^k)\nabla f(\vec{x}^k). \tag{12.25}$$

The Newton–Raphson method is thus derived as follows.

$$f(\vec{x}^{k+1}) = f(\vec{x}^k) - H^{-1}(\vec{x}^k)\nabla f(\vec{x}^k). \tag{12.26}$$

Another loose and intuitive way to view the Newton–Raphson method is that its format is similar to the steepest descent method, except that the step size α^k is now chosen as $H^{-1}(\vec{x}^k)$, the inverse of the Hessian matrix evaluated at $\vec{x}^k$.

The idea behind the Newton–Raphson method is that the function being minimized is approximated locally by a quadratic function and this quadratic function is then minimized. It is noted that any function will behave like a quadratic function when it is close to the minimum. Hence, the closer to the minimum, the more efficient is the Newton–Raphson method. This is the exact opposite of the steepest descent method, which works more efficiently at the beginning, and less efficiently when close to the minimum. The price paid with the Newton–Raphson method is the extra calculation involved in evaluating the inverse of the Hessian matrix at $\vec{x}^k$.

12.2.6.2 Convergence Speed

Assume that the second-order sufficient conditions discussed in Section 12.2.2 are satisfied. Furthermore, assume that the initial point $\vec{x}^0$ is sufficiently close to the minimum $\vec{x}^*$. Then it can be shown that the Newton–Raphson method converges with an order of at least two. This indicates that the Newton–Raphson method converges faster than the steepest descent method.

12.2.6.3 Generalization and Improvements

In [luenberger 1984], a general class of algorithms is defined as

$$\vec{x}^{k+1} = \vec{x}^k - \alpha^k G \nabla f(\vec{x}^k), \tag{12.27}$$

where G denotes an $n \times n$ matrix, and α^k a positive parameter. Both the steepest descent and the Newton–Raphson methods fall into this framework. It is clear that if G is an $n \times n$ identical matrix $\mathbf{I}$, then this general form reduces to the steepest descent method. If $G = H$ and $\alpha = 1$ then this is the Newton–Raphson method.

Although it descends rapidly near the solution, the Newton–Raphson method may not descend for points far away from the minimum because the quadratic approximation may not be valid there. The introduction of the α^k, which minimizes f, can guarantee the descent of f at the general points. Another improvement is to set $G = [\zeta^k \mathbf{I} + H(\vec{x}^k)]^{-1}$ with $\zeta \geq 0$. Obviously this is a combination of the steepest descent method and the Newton–Raphson method. Two extreme ends are the steepest method (very large ζ^k) and the Newton–Raphson method ($\zeta^k = 0$). For most cases, the selection of the parameter ζ^k aims at making the G matrix positive definite.

12.2.7 Other Methods

There are other gradient methods such as the Fletcher–Reeves method (also known as the conjugate gradient method) and the Fletcher–Powell–Davidon method (also known as the variable metric method). Readers may refer to [luenberger 1984] or other optimization texts.

12.3 Netravali–Robbins' Pel Recursive Algorithm

Having had an introduction to some basic nonlinear programming theory, we now turn to the pel recursive technique in displacement estimation from the perspective of the descent methods. Let us take a look at the first pel recursive algorithm, the Netravali–Robbins pel recursive algorithm. It actually estimates displacement vectors using the steepest descent method to minimize the squared DFD. That is,

$$\vec{d}^{k+1} = \vec{d}^k - \frac{1}{2}\alpha \nabla_{\vec{d}} \text{DFD}^2(x, y, \vec{d}^k), \tag{12.28}$$

where $\nabla_{\vec{d}} \text{DFD}^2(x, y, \vec{d}^k)$ denotes the gradient of DFD^2 with respect to $\vec{d}$ evaluated at $\vec{d}^k$, the displacement vector at the kth iteration, and α is positive. Equation 12.28 can be further written as

$$\vec{d}^{k+1} = \vec{d}^k - \alpha \text{DFD}(x, y, \vec{d}^k) \nabla_{\vec{d}} \text{DFD}(x, y, \vec{d}^k). \tag{12.29}$$

Owing to Equation 12.1, the above equation leads to

$$\vec{d}^{k+1} = \vec{d}^k - \alpha \text{DFD}(x, y, \vec{d}^k) \nabla_{x,y} f_{n-1}(x - d_x, y - d_y), \tag{12.30}$$

where $\nabla_{x,y}$ means a gradient operator with respect to x and y. In [netravali 1979], a constant of $1/1024$ is assigned to α.

12.3.1 Inclusion of a Neighborhood Area

To make displacement estimation more robust, Netravali and Robbins considered an area for evaluating the DFD^2 in calculating the update term. More precisely, they assume the displacement vector is constant within a small neighborhood Ω of the pixel for which the displacement is being estimated. That is,

$$\vec{d}^{k+1} = \vec{d}^k - \frac{1}{2}\alpha \nabla_{\vec{d}} \sum_{i,x,y \in \Omega} w_i \text{DFD}^2(x, y, \vec{d}^k), \tag{12.31}$$

where
 i represents an index for the ith pixel (x, y) within Ω
 w_i is the weight for the ith pixel in Ω

All the weights satisfy the following two constraints:

$$\begin{cases} w_i \geq 0 & (12.32) \\ \sum_{i \in \Omega} w_l = 1 & (12.33) \end{cases}$$

This inclusion of a neighborhood area also explains why pel recursive technique is classified into the category of region matching techniques as we discussed at the beginning of this chapter.

12.3.2 Interpolation

It is noted that interpolation will be necessary when displacement vectors' components d_x and d_y are not integer number of pixels. A bilinear interpolation technique is used in [netravali 1979]. For the bilinear interpolation, readers may refer to Chapter 10.

12.3.3 Simplification

To make the proposed algorithm more efficient in computation, Netravali and Robbins also proposed simplified versions of the displacement estimation and interpolation algorithms in their paper.

One simplified version of the Netravali and Robbins algorithm is as follows:

$$\vec{d}^{k+1} = \vec{d}^k - \alpha \, \text{sign}\{\text{DFD}(x,y;\vec{d}^k)\} \, \text{sign}\{\nabla_{x,y} f_{n-1}(x - d_x, y - d_y)\} \quad (12.34)$$

where $\text{sign}\{s\} = 0, 1, -1$ depending on $s = 0$, $s > 0$, $s < 0$, respectively, while the sign of a vector quantity is the vector of signs of its components. In this version the update vectors can only assume an angle which is an integer multiple of 45°. As shown in [netravali 1979], this version is effective.

12.3.4 Performance

The performance of the Netravali and Robbins algorithm has been evaluated using computer simulation [netravali 1979]. Two video sequences with different amounts and different types of motion are tested. In either case, the proposed pel recursive algorithm displays superior performance over the replenishment algorithm [mounts 1969; haskell 1979], discussed briefly in Chapter 10. The Netravali and Robbins algorithm achieves a bit rate which is 22% to 50% lower than that required by the replenishment technique with the simple frame difference prediction.

12.4 Other Pel Recursive Algorithms

The progress and success of the Netravali and Robbins algorithm stimulated great research interests in pel recursive techniques. Many new algorithms have been developed. Some of them are discussed in this section.

12.4.1 Bergmann's Algorithm (1982)

Bergmann modified Netravali and Robbins algorithm by using the Newton–Raphson method [bergmann 1982]. In doing so, the following difference between the fundamental framework of the descent methods discussed in Section 12.2 and the minimization problem in displacement estimation discussed in Section 12.3 needs to be noticed. That is, the object function $f(\vec{x})$ discussed in Section 12.2 now becomes $\text{DFD}^2(x,y;\vec{d})$. The Hessian matrix H, consisting of the second-order partial derivatives of the $f(\vec{x})$ with respect to the components of $\vec{x}$ now become the second-order derivatives of DFD^2 with respect to d_x and d_y. Since the vector $\vec{d}$ is a 2-D column vector now, the H matrix is hence a 2×2 matrix. That is,

$$H = \begin{bmatrix} \dfrac{\partial^2 \text{DFD}^2(x,y,\vec{d})}{\partial^2 d_x} & \dfrac{\partial^2 \text{DFD}^2(x,y,\vec{d})}{\partial d_x \partial d_y} \\ \dfrac{\partial^2 \text{DFD}^2(x,y,\vec{d})}{\partial d_y \partial d_x} & \dfrac{\partial^2 \text{DFD}^2(x,y,\vec{d})}{\partial^2 d_y} \end{bmatrix}. \quad (12.35)$$

As expected, the Bergmann algorithm (1982) converges to the minimum faster than the steepest descent method since the Newton–Raphson method converges with an order of at least two.

12.4.2 Bergmann's Algorithm (1984)

Based on Burkhard and Moll's algorithm [burkhard 1979], Bergmann developed an algorithm, which is similar to the Newton–Raphson algorithm. The primary difference is that an average of two second-order derivatives is used to replace those in the Hessian matrix. In this sense, it can be considered as a variation of the Newton–Raphson algorithm.

12.4.3 Cafforio and Rocca's Algorithm

Based on their earlier study, Cafforio and Rocca proposed an algorithm in 1982, which is essentially the steepest descent method. That is, the step size α is defined as follows [cafforio 1983]:

$$\alpha = \frac{1}{|\nabla f_{n-1}(x - d_x, y - d_y)|^2 + \eta^2} \qquad (12.36)$$

with $\eta^2 = 100$. The addition of η^2 is intended to avoid the problem that would have occurred in a uniform region where the gradients are very small.

12.4.4 Walker and Rao's Algorithm

Walker and Rao developed an algorithm based on the steepest descent method [walker 1984; tekalp 1995], and also with a variable step size. That is,

$$\alpha = \frac{1}{2|\nabla f_{n-1}(x - d_x, y - d_y)|^2}, \qquad (12.37)$$

where

$$|\nabla f_{n-1}(x - d_x, y - d_y)|^2 = \left[\frac{\partial f_{n-1}(x - d_x, y - d_y)}{\partial d_x}\right]^2 + \left[\frac{\partial f_{n-1}(x - d_x, y - d_y)}{\partial d_y}\right]^2 \qquad (12.38)$$

It is observed that this step size is variable instead of being a constant. Furthermore, this variable step size is reverse proportional to the norm square of the gradient of $f_{n-1}(x - d_x, y - d_y)$ with respect to x, y. This means that this type of step size will be small in the edge or rough area, and will be large in the relatively smooth area. These features are desirable.

Although it is quite similar to the Cafforio and Rocca algorithm, the Walker and Rao algorithm differs in the following two aspects: (i) α is selected differently, (ii) implementation of the algorithm is different. For instance, instead of putting an η^2 in the denominator of α, the Walker and Rao algorithm uses a logic.

As a result of using the variable step size α, the convergence rate is improved substantially. This implies fast implementation and accurate displacement estimation. It was reported that usually one to three iterations are able to achieve quite satisfactory results in most cases.

Another contribution is that the Walker and Rao algorithm eliminates the need to transmit explicit address information so as to bring out higher coding efficiency.

12.5 Performance Comparison

A comprehensive survey of various algorithms using the pel recursive technique can be found in a study by Musmann, Pirsch, and Grallert [musmann 1985]. There, two performance features are compared among the algorithms. One is the convergence rate and hence the accuracy of displacement estimation. The other is the stability range. By *stability range*, we mean a range starting from which an algorithm can converge to the minimum of DFD^2, or the true displacement vector.

Compared with the Netravali and Robbins algorithm, those improved algorithms discussed in Section 12.4 do not use a constant step size, thus providing better adaptation to local image statistics. Consequently, they achieve a better convergence rate and more accurate displacement estimation. According to [bergmann 1984] and [musmann 1985], Bergmann's algorithm (1984) performs best among these various algorithms in terms of convergence rate and accuracy.

According to [musmann 1985], the Newton–Raphson algorithm has a relatively smaller stability range than the other algorithms. This agrees with our discussion in Section 12.2.2. That is, the performance of the Newton–Raphson method improves when it works in the area close to the minimum. The choice of the initial guess, however, is relatively more restricted.

12.6 Summary

The pel recursive technique is one of three major approaches to displacement estimation for motion compensation. It recursively estimates displacement vectors in a pixel-by-pixel fashion. There are three types of recursion: horizontal, vertical, and temporal. Displacement estimation is carried out by minimizing the square of the displaced frame difference (DFD). Therefore, the steepest descent method and the Newton–Raphson method, the two most fundamental methods in optimization, naturally find their application in pel recursive techniques. The pioneering Netravali and Robbins algorithm and several other algorithms, such as Bergmann's (1982), Cafforio and Rocca, Walker and Rao, and Bergmann's (1984) are discussed in this chapter. They can be classified into one of two categories: The steepest descent-based algorithms or the Newton–Raphson-based algorithms. Table 12.1 contains a classification of these algorithms.

Note that the DFD can be evaluated within a neighborhood of the pixel for which a displacement vector is being estimated. The displacement vector is assumed constant within this neighborhood. This makes the displacement estimation more robust against various noises.

TABLE 12.1

Classification of Several Pel Recursive Algorithms

Algorithms	Category I Steepest Descent-Based Algorithm	Category II Newton–Raphson Based Algorithm
Netravali and Robbins	Steepest descent	
Bergmann (1982)		Newton–Raphson
Walker and Rao	Variation of steepest descent	
Cafforio and Rocca	Variation of steepest descent	
Bergmann (1984)		Variation of Newton–Raphson

Compared with the replenishment technique with simple frame difference prediction (the first real interframe coding algorithm), the Netravali and Robbins algorithm (the first pel recursive technique) achieves much higher coding efficiency. Specifically, a 22% to 50% saving in bit rate has been reported for some computer simulations. Several new pel recursive algorithms have made further improvements in terms of the convergence rate and the estimation accuracy owing to the replacement of the fixed step size utilized in the Netravali and Robbins algorithms, which make these algorithms more adaptive to the local statistics in image frames.

Exercises

1. What is the definition of displaced frame difference? Justify Equation 12.1.
2. Why does the inclusion of a neighborhood area make the pel recursive algorithm more robust against noise?
3. Compare the performance of the steepest descent method with that of the Newton–Raphson method.
4. Explain the function of η^2 in the Cafforio and Rocca algorithm.
5. What is the advantage you expect to have from the Walker and Rao algorithm?
6. What is the difference between the Bergmann algorithm (1982) and the Bergmann algorithm (1984)?
7. Why does the Newton–Raphson method have a smaller stability range?

References

[bergmann 1982] H.C. Bergmann, Displacement estimation based on the correlation of image segments, IEEE Proceedings of International Conference on Electronic Image Processing, York, England, pp. 215–219, July 1982.

[bergmann 1984] H.C. Bergmann, Ein schnell konvergierendes Displacement-Schätzverfahrenfür die interpolation von Fernsehbildsequenzen, Ph.D. dissertation, Technical University of Hannover, Hannover, Germany, February 1984.

[biemond 1987] J. Biemond, L. Looijenga, D.E. Boekee, and R.H.J.M. Plompen, A pel recursive Wiener-based displacement estimation algorithm, *Signal Processing*, 13, 399–412, December 1987.

[burkhard 1979] H. Burkhard and H. Moll, A modified Newton-Raphson search for the model-adaptive identification of delays, in *Identification and System Parameter Identification*, R. Isermann (Ed.), Pergamon Press, New York/Oxford, England, 1979, pp. 1279–1286.

[cafforio 1983] C. Cafforio and F. Rocca, The differential method for image motion estimation, in *Image Sequence Processing and Dynamic Scene Analysis*, T.S. Huang (Ed.), Springer-Verlag, Berlin, Germany, 1983, pp. 104–124.

[haskell 1979] B.G. Haskell, Frame replenishment coding of television, in *Image Transmission Techniques*, W.K. Pratt (Ed.), Academic Press, New York, 1979.

[luenberger 1984] D.G. Luenberger, *Linear and Nonlinear Programming*, Addison Wesley, Reading, MA, 1984.

[mounts 1969] F.W. Mounts, A video encoding system with conditional picture-element replenishment, *The Bell System Technical Journal*, 48, 7, 2545–2554, September 1969.

[musmann 1985] H.G. Musmann, P. Pirsch, and H.J. Grallert, Advances in picture coding, *Proceedings of the IEEE*, 73, 4, 523–548, 1985.

[netravali 1979] A.N. Netravali and J.D. Robbins, Motion compensated television coding, Part I, *The Bell System Technical Journal*, 58, 3, 631–670, March 1979.

[tekalp 1995] A.M. Tekalp, *Digital Video Processing*, Prentice-Hall, Englewood Cliffs, NJ, 1995.

[walker 1984] D.R. Walker and K.R. Rao, Improved pel recursive motion compensation, *IEEE Transactions on Communications*, COM-32, 1128–1134, October 1984.

13
Optical Flow

As mentioned in Chapter 10, optical flow is one of the three major techniques that can be used to estimate displacement vectors from successive image frames. As opposed to the other two displacement estimation techniques, block matching and pel recursive method, discussed in Chapters 11 and 12, however, the optical flow technique was developed primarily for 3-D motion estimation in the computer vision community. Although it provides a relatively more accurate displacement estimation than the other two techniques, as we shall see in this chapter and the next chapter, optical flow has not yet found wide applications for motion compensated (MC) video coding. This is mainly due to the fact that there is a large number of motion vectors (one vector per pixel) involved, hence, the more side information that needs to be encoded and transmitted. As emphasized in Chapter 11, we should not forget the ultimate goal in MC video coding: to encode video data with a *total* bit rate as low as possible, while maintaining a satisfactory quality of reconstructed video frames at the receiving end. If the extra bits required for encoding a large amount of optical flow vectors counterbalance the bits saved in encoding the prediction error (owing to more accurate motion estimation), then the usage of optical flow in MC coding is not worthwhile. Besides, more computation is required in optical flow determination. These factors have prevented optical flow from being practically utilized in MC video coding. With the continued advance in technologies, however, we believe this problem may be resolved in the near future. In fact, an initial, successful attempt has been made [shi 1998].

On the other hand, in theory, the optical flow technique is of great importance in understanding the fundamental issues in 2-D motion determination, such as the aperture problem, the conservation and neighborhood constraints, and the distinction and relationship between 2-D motion and 2-D apparent motion.

In this chapter, we will focus on the optical flow technique. In Section 13.1, as stated above, some fundamental issues associated with optical flow are addressed. Section 13.2 discusses the differential method, while the correlation method is covered in Section 13.3. In Section 13.4, a multiple attributes approach is presented. Some performance comparisons between various techniques are included in Sections 13.3 and 13.4. A summary is given in Section 13.5.

13.1 Fundamentals

Optical flow is referred to as the 2-D distribution of apparent velocities of movement of intensity patterns in an image plane [horn 1981]. In other words, an optical flow field consists of a dense velocity field with one velocity vector for each pixel in the image plane.

If we know the time interval between two consecutive images, which is usually the case, then velocity vectors and displacement vectors can be converted from one to another. In this sense, optical flow is one of techniques used for displacement estimation.

13.1.1 2-D Motion and Optical Flow

In the above definition, it is noted that the word *apparent* is used and nothing about 3-D motion in the scene is stated. The implication behind this observation is discussed in this section, beginning with the definition of 2-D motion. 2-D motion is referred to as motion in a 2-D image plane caused by 3-D motion in the scene. That is, 2-D motion is the projection (commonly perspective projection) of 3-D motion in the scene onto the 2-D image plane. This can be illustrated by using a very simple example shown in Figure 13.1. There the world coordinate system O-XYZ and the camera coordinate systems o-xyz are aligned. The point C is the optical center of the camera. A point A_1 moves to A_2, while its perspective projection moves correspondingly from a_1 to a_2. We then see that a 2-D motion (from a_1 to a_2) in image plane is invoked by a 3-D motion (from A_1 to A_2) in 3-D space. By a 2-D motion field, or sometimes image flow, we mean a dense 2-D motion field: one velocity vector for each pixel in the image plane.

Optical flow, according to its definition, is caused by movement of intensity patterns in an image plane. Therefore, 2-D motion (field) and optical flow (field) are generally different. To support this conclusion, let us consider the following two examples. One is given by Horn and Schunck [horn 1981]. Imagine a uniform sphere rotating with a constant speed in the scene. Assume that the luminance and all other conditions do not change at all when pictures are taken. Then, there is no change in brightness patterns in the images. According to the definition of optical flow, the optical flow is zero, whereas the 2-D motion field is obviously not zero. At the other extreme, consider a stationary scene; all objects in 3-D world space are still. If illuminance changes when pictures are taken in such a way that there is movement of intensity patterns in image planes, as a consequence, optical flow may be nonzero. This confirms a statement made by Singh: the scene does not have to be in motion relative to the image for the optical flow field to be nonzero [singh 1991]. It can be shown that the 2-D motion field and the optical flow field are equal under certain conditions. Understanding the difference between the two quantities and the conditions under which they are equal is important.

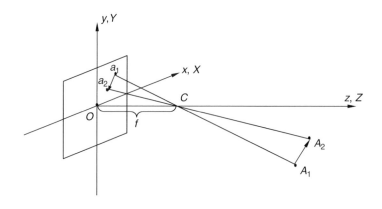

FIGURE 13.1
2-D motion versus 3-D motion.

Optical Flow

This understanding can provide us with some sort of guide to evaluate the reliability of estimating 3-D motion from optical flow. This is because, in practice, time-varying image sequences are the only ones what we have at hand. The task in computer vision is to interpret 3-D motion from time-varying sequences. Therefore, we can only work with optical flow in estimating 3-D motion. Since the main focus of this book is on image and video coding, we do not cover these equality conditions here. (Interested readers may refer to [singh 1991].) In motion compensated video coding, it is likewise true that the image frames and video data are only what we have at hand. We also, therefore, have to work with optical flow. Our attention is thus turned to optical flow determination and its usage in video data compression.

13.1.2 Aperture Problem

Aperture problem is an important issue, originating in optics. Since it is inherent in the local estimation of optical flow, we address this issue in this section. In optics, apertures are openings in flat screens [bracewell 1995]. Therefore, apertures can have various shapes, such as circular, semicircular, and rectangular. Examples of apertures include a thin slit or array of slits in a screen. A circular aperture, a round hole made on the shutter of a window, was used by Newton to study the composition of sunlight. It is also well known that the circular aperture is of special interest in studying the diffraction pattern [sears 1986].

Roughly speaking, the aperture problem in motion analysis refers to the problem that occurs when viewing motion via an aperture, i.e., a small opening in a flat screen. In [marr 1982], it is stated that when a straight moving edge is observed through an aperture only the component of motion orthogonal to the edge can be measured. Let us examine some simple examples depicted in Figure 13.2. In Figure 13.2a, a large rectangular ABCD is located in the *XOZ* plane. A rectangular screen EFGH with a circular aperture is perpendicular to the *OY* axis. Figure 13.2b and c shows, respectively, what is observed through the aperture when the rectangular ABCD is moving along the positive *X* and *Z* directions with a uniform speed. Since the circular opening is small and the line AB is very long, no motion will be observed in Figure 13.2b. Obviously, in Figure 13.2c the upward movement can be observed clearly. In Figure 13.2d, the upright corner of the rectangle ABCD, angle B, appears. At this time the translation along any direction in the *XOZ* plane can be observed clearly. The phenomena observed in this example demonstrate that it is sometimes impossible to estimate motion of a pixel by only observing a small neighborhood surrounding it. The only motion that can be estimated from observing a small neighborhood is the motion orthogonal to the underlying moving contour. In Figure 13.2b, there is no motion orthogonal to the moving contour AB, the motion is aligned with the moving contour AB, which cannot be observed through the aperture. Therefore, no motion can be observed through the aperture. In Figure 13.2c, the observed motion is upward, which is perpendicular to the horizontal moving contour AB. In Figure 13.2d, any translation in the *XOZ* plane can be decomposed into horizontal and vertical components. Either of these two components is orthogonal to one of the two moving contours: AB or BC.

A more accurate statement on the aperture problem needs a definition of the so-called normal optical flow. The normal optical flow refers to the component of optical flow along the direction pointed by the local intensity gradient. Now we can make a more accurate statement: the only motion in an image plane that can be determined is the normal optical flow.

In general, the aperture problem becomes severe in image regions where strong intensity gradients exist, such as at the edges. In image regions with strong higher-order intensity variations, such as corners or textured areas, the true motion can be estimated.

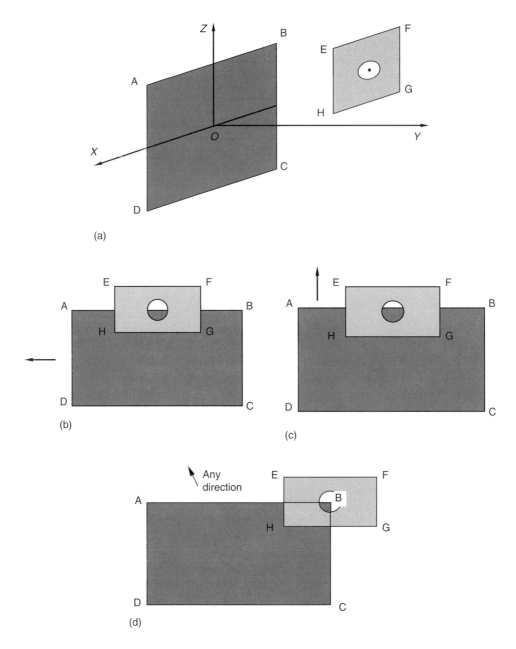

FIGURE 13.2
(a) Aperture problem: A large rectangle ABCD is located in the XOZ plane. A rectangular screen EFGH with a circular aperture is perpendicular to the OY axis. (b) Aperture problem: No motion can be observed through the circular aperture when the rectangular ABCD is moving along the positive X direction. (c) Aperture problem: The motion can be observed through the circular aperture when the ABCD is moving along the positive Z direction. (d) Aperture problem: The translation of ABCD along any direction in the XOZ plane can be observed through the circular aperture when the upright corner of the rectangle ABCD, angle B, appears in the aperture.

Singh provides a more elegant discussion on the aperture problem, in which he argues that the aperture problem should be considered as a continuous problem (it always exists, but in varying degrees of acuteness) instead of a binary problem (either it exists or it does not) [singh 1991].

(a)

FIGURE 1.2
(a) A picture of boy and girl.

(a)

FIGURE 1.8
The bridge in Vancouver: (a) Original [Courtesy of Minhuai Shi].

(a)

FIGURE 1.9
Christmas in Winorlia: (a) Original.

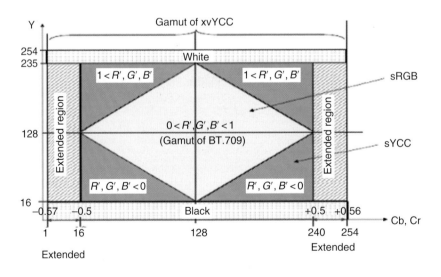

FIGURE 15.2
Two-dimensional (2-D) view of xvYCC.

FIGURE 15.3
Gamut coverage of sRGB, sYCC, and xvYCC color spaces.

FIGURE 15.4
(a) An example of progressive video and (b) an example of interlaced video.

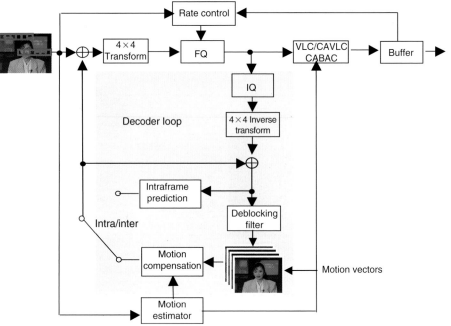

FIGURE 20.3
Block diagram of H.264 encoder.

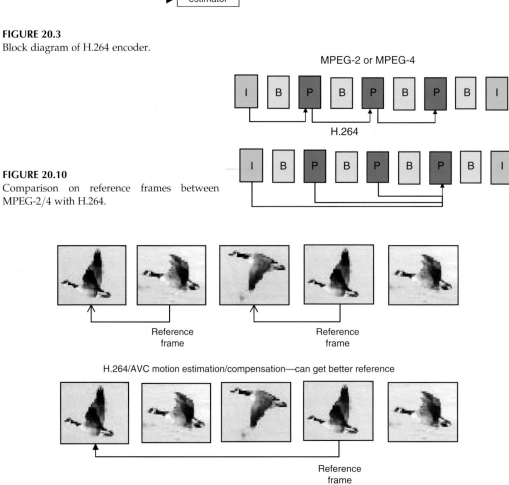

FIGURE 20.10
Comparison on reference frames between MPEG-2/4 with H.264.

FIGURE 20.11
An example to explain the benefit by using multiple reference frames, it is noted that the better reference can be obtained by using multiple reference pictures for the video sequences with periodic changes.

Optical Flow

13.1.3 Ill-Posed Problem

Motion estimation from image sequences, including optical flow estimation, belongs to the category of inverse problems. This is because we want to infer motion from given 2-D images, which is the perspective projection of 3-D motion. According to Hadamard [bertero 1988], a mathematical problem is well-posed if it possesses the following three characteristics:

1. Existence (the solution exists)
2. Uniqueness (the solution is unique)
3. Continuity (when the error in the data tends toward zero, then the induced error in the solution tends toward zero as well)

Inverse problems usually are not well-posed in that the solution may not exist. In the example discussed in Section 13.1.1, i.e., a uniform sphere rotated with illuminance fixed, the solution to motion estimation does not exist since no motion can be inferred from given images. The aperture problem discussed in Section 13.1.2 is the case, where the solution to the motion may not be unique. Let us take a look at Figure 13.2b. From the given picture, one cannot tell whether the straight line AB is static, or is moving horizontally. If it is moving horizontally, one cannot tell the moving speed. In other words, infinitely many solutions exist for the case. In optical flow determination, we will see that computations are noise sensitive. That is, even a small error in the data can produce an extremely large error in the solution. Hence, we see that the motion estimation from image sequences suffers from all the three aspects just mentioned: nonexistence, nonuniqueness, and discontinuity. The last term is also referred to as the instability of the solution.

It is pointed out in [bertero 1988] that all the low-level processing (also known as early vision) in computational vision are inverse problems and are often ill-posed. Examples in the low-level processing include motion recovery, computation of optical flow, edge detection, structure from stereo, structure from motion, structure from texture, shape from shading, and so on. Fortunately, the problem with early vision is mildly ill-posed in general. By *mildly*, we mean that a reduction of errors in the data can significantly improve the solution.

Since the early 1960s, the demand for accurate approximates and stable solutions in areas such as optics, radioastronomy, microscopy, and medical imaging has stimulated great research efforts in inverse problems, resulting in a unified theory: the regularization theory of ill-posed problems [tikhonov 1977]. In the discussion of optical flow methods, we shall see that some regularization techniques have been posed and have improved accuracy in flow determination. More advanced algorithms continue to come.

13.1.4 Classification of Optical Flow Techniques

Optical flow in image sequences provides important information regarding both motion and structure, and it is useful in such diverse fields as robot vision, autonomous navigation, and video coding. Although this subject has been studied for more than a decade, reducing the error in the flow estimation remains a difficult problem. A comprehensive review and a comparison of the accuracy of various optical flow techniques have recently been made [barron 1994]. So far, most of the techniques in the optical flow computations use one of the following basic approaches:

- Gradient-based [horn 1981; lucas 1981; nagel 1986; uras 1988; szeliski 1995; black 1996]
- Correlation-based [anandan 1989; singh 1992; pan 1998]
- Spatiotemporal energy-based [adelson 1985; heeger 1988; bigun 1991]
- Phase-based [waxman 1988; fleet 1990]

Besides these deterministic approaches, there is the stochastic approach to optical flow computation [konrad 1992]. In this chapter, we focus our discussion of optical flow on the gradient-based and correlation-based techniques because of their frequent applications in practice and because of their fundamental importance in theory. We also discuss multiple attribute techniques in optical flow determination. The other two approaches will be explained briefly when we discuss new techniques in motion estimation in Chapter 14.

13.2 Gradient-Based Approach

It is noted that before the methods of optical flow determination were actually developed, optical flow had been discussed and exploited for motion and structure recovery from image sequences in computer vision for years. That is, the optical flow field was assumed to be available in the study of motion recovery. The first type of methods in optical flow determination is referred to as gradient-based techniques. This is because the spatial and temporal partial derivatives of intensity function are utilized in these techniques. In this section, we shall present the Horn and Schunck algorithm. It is regarded as the most prominent representative of this category. Other methods in this category are briefly discussed after presenting the basic concepts.

13.2.1 Horn and Schunck's Method

We shall begin with a very general framework [shi 1994] to derive a brightness time-invariance equation. We will then introduce Horn and Schunck's method.

13.2.1.1 Brightness Invariance Equation

As stated in Chapter 10, the imaging space can be represented by

$$f(x, y, t, \bar{s}) \tag{13.1}$$

where $\bar{s}$ indicates the sensor's position in 3-D world space, i.e., the coordinates of the sensor center and the orientation of the optical axis of the sensor. The $\bar{s}$ is a 5-D vector. That is, $\bar{s} = (\tilde{x}, \tilde{y}, \tilde{z}, \beta, \gamma)$ where $\tilde{x}, \tilde{y},$ and $\tilde{z}$ represent the coordinate of the optical center of the sensor in 3-D world space; and β and γ represent the orientation of the optical axis of the sensor in 3-D world space, the Euler angles: pan and tilt, respectively.

With this very general notion, each picture, taken by a sensor located on a particular position at a specific moment, is merely a special cross-section of this imaging space. Both temporal and spatial image sequences become a proper subset of the imaging space.

Assume now a world point P in 3-D space that is perspectively projected onto the image plane as a pixel with the coordinates x_P and y_P. Then, x_P and y_P are also dependent on t and $\bar{s}$. That is,

$$f = f(x_P(t, \bar{s}), y_P(t, \bar{s}), t, \bar{s}) \tag{13.2}$$

Optical Flow

If the optical radiation of the world point P is invariant with respect to the time interval from t_1 to t_2, we then have

$$f(x_P(t_1, \bar{s}_1), y_P(t_1, \bar{s}_1), t_1, \bar{s}_1) = f(x_P(t_2, \bar{s}_1), y_P(t_2, \bar{s}_1), t_2, \bar{s}_1) \tag{13.3}$$

This is the brightness time-invariance equation.

At a specific moment t_1, if the optical radiation of P is isotropical we then get

$$f(x_P(t_1, \bar{s}_1), y_P(t_1, \bar{s}_1), t_1, \bar{s}_1) = f(x_P(t_1, \bar{s}_2), y_P(t_1, \bar{s}_2), t_1, \bar{s}_2). \tag{13.4}$$

This is the brightness space-invariance equation.

If both conditions are satisfied, we get the brightness time-and-space-invariance equation, i.e.,

$$f(x_P(t_1, \bar{s}_1), y_P(t_1, \bar{s}_1), t_1, \bar{s}_1) = f(x_P(t_2, \bar{s}_2), y_P(t_2, \bar{s}_2), t_2, \bar{s}_2). \tag{13.5}$$

Consider two brightness functions $f(x(t, \bar{s}), y(t, \bar{s}), t, \bar{s})$ and $f(x(t + \Delta t, \bar{s} + \Delta \bar{s}), y(t + \Delta t, \bar{s} + \Delta \bar{s}), t + \Delta t, \bar{s} + \Delta \bar{s})$ in which the variation in time, Δt, and the variation in the spatial position of the sensor, $\Delta \bar{s}$, are very small. Due to the time-and-space-invariance of brightness, we can get

$$f(x(t, \bar{s}), y(t, \bar{s}), t, \bar{s}) = f(x(t + \Delta t, \bar{s} + \Delta \bar{s}), y(t + \Delta t, \bar{s} + \Delta \bar{s}), t + \Delta t, \bar{s} + \Delta \bar{s}) \tag{13.6}$$

The expansion of the right-hand side of Equation 13.6 in the Taylor series at $(t, \bar{s})$, and the use of Equation 13.5 lead to

$$\left(\frac{\partial f}{\partial x} u + \frac{\partial f}{\partial y} v + \frac{\partial f}{\partial t}\right) \Delta t + \left(\frac{\partial f}{\partial x} u^{\bar{s}} + \frac{\partial f}{\partial y} v^{\bar{s}} + \frac{\partial f}{\partial \bar{s}}\right) \Delta \bar{s} + \varepsilon = 0 \tag{13.7}$$

where

$$u \triangleq \frac{\partial x}{\partial t}, \quad v \triangleq \frac{\partial y}{\partial t}, \quad u^{\bar{s}} \triangleq \frac{\partial x}{\partial \bar{s}}, \quad v^{\bar{s}} \triangleq \frac{\partial y}{\partial \bar{s}}$$

If $\Delta \bar{s} = 0$, i.e., the sensor is static in a fixed spatial position (in other words, both the coordinate of the optical center of the sensor and its optical axis direction remain unchanged), dividing both sides of the equation by Δt and evaluating the limit as $\Delta t \to 0$ degenerate Equation 13.7 into

$$\frac{\partial f}{\partial x} u + \frac{\partial f}{\partial y} v + \frac{\partial f}{\partial t} = 0 \tag{13.8}$$

If $\Delta t = 0$, both its sides are divided by $\Delta \bar{s}$ and $\Delta \bar{s} \to 0$ is examined, Equation 13.7 then reduces to

$$\frac{\partial f}{\partial x} u^{\bar{s}} + \frac{\partial f}{\partial y} v^{\bar{s}} + \frac{\partial f}{\partial \bar{s}} = 0 \tag{13.9}$$

when $\Delta t = 0$, i.e., at a specific time moment, the images generated with sensors at different spatial positions can be viewed as a spatial sequence of images. Equation 13.9 is, then, the equation for the spatial sequence of images.

For the sake of brevity, we shall focus on the gradient-based approach to optical flow determination with respect to temporal image sequences. That is, in the rest of this section we shall address only Equation 13.8. It is noted that the derivation can be extended to spatial image sequences. The optical flow technique for spatial image sequences is useful in stereo image data compression. It plays an important role in motion and structure recovery. Interested readers are referred to [shi 1994; shu 1993].

13.2.1.2 Smoothness Constraint

Careful examination of Equation 13.8 reveals that we have two unknowns: u and v, i.e., the horizontal and vertical components of an optical flow vector at a three-tuple (x, y, t), but only one equation to relate them. This once again demonstrates the ill-posed nature of optical flow determination. This also indicates that there is no way to compute optical flow by considering a single point of the brightness pattern moving independently. As stated in Section 13.1.3, some regularization measure—here an extra constraint—must be taken to overcome the difficulty.

A most popularly used constraint was proposed by Horn and Schunck and is referred to as the smoothness constraint. As the name implies, it constrains flow vectors to vary from one to another smoothly. Clearly, this is true for points in the brightness pattern most of the time, particularly for points belonging to the same object. It may be violated, however, along moving boundaries. Mathematically, the smoothness constraint is imposed in optical flow determination by minimizing the square of the magnitude of the gradient of the optical flow vectors:

$$\left(\frac{\partial u}{\partial x}\right)^2 + \left(\frac{\partial u}{\partial y}\right)^2 + \left(\frac{\partial v}{\partial x}\right)^2 + \left(\frac{\partial v}{\partial y}\right)^2 \tag{13.10}$$

It can be easily verified that the smoother the flow vector field, the smaller these quantities. Actually, the square of the magnitude of the gradient of intensity function with respect to the spatial coordinates, summed over a whole image or an image region, has been used as a smoothness measure of the image or the image region in the digital image processing literature [gonzalez 1992].

13.2.1.3 Minimization

Optical flow determination can then be converted into a minimization problem.

The square of the left-hand side of Equation 13.8, which can be derived from the brightness time-invariance equation, represents one type of error. It may be caused by quantization noise or other noises and can be written as

$$\varepsilon_b^2 = \left(\frac{\partial f}{\partial x}u + \frac{\partial f}{\partial y}v + \frac{\partial f}{\partial t}\right)^2 \tag{13.11}$$

The smoothness measure expressed in Equation 13.10 denotes another type of error, which is

$$\varepsilon_s^2 = \left(\frac{\partial u}{\partial x}\right)^2 + \left(\frac{\partial u}{\partial y}\right)^2 + \left(\frac{\partial v}{\partial x}\right)^2 + \left(\frac{\partial v}{\partial y}\right)^2 \tag{13.12}$$

Optical Flow

The total error to be minimized is

$$\varepsilon^2 = \sum_x \sum_y \varepsilon_b^2 + \alpha^2 \varepsilon_s^2$$

$$= \sum_x \sum_y \left(\frac{\partial f}{\partial x}u + \frac{\partial f}{\partial y}v + \frac{\partial f}{\partial t}\right)^2 + \alpha^2 \left[\left(\frac{\partial u}{\partial x}\right)^2 + \left(\frac{\partial u}{\partial y}\right)^2 + \left(\frac{\partial v}{\partial x}\right)^2 + \left(\frac{\partial v}{\partial y}\right)^2\right] \quad (13.13)$$

where α is a weight between these two types of errors. The optical flow quantities u and v can be found by minimizing the total error. Using the calculus of variation, Horn and Schunck derived the following pair of equations for two unknown u and v at each pixel in the image.

$$\begin{cases} f_x^2 u + f_x f_y v = \alpha^2 \nabla^2 u - f_x f_t \\ f_x f_y u + f_y^2 v = \alpha^2 \nabla^2 v - f_y f_t \end{cases} \quad (13.14)$$

where $f_x = \frac{\partial f}{\partial x}, f_y = \frac{\partial f}{\partial y}, f_t = \frac{\partial f}{\partial t}; \nabla^2$ denotes the Laplacian operator. The Laplacian operator of u and v is defined below.

$$\nabla^2 u = \frac{\partial^2 u}{\partial x^2} + \frac{\partial^2 u}{\partial y^2}$$

$$\nabla^2 v = \frac{\partial^2 v}{\partial x^2} + \frac{\partial^2 v}{\partial y^2} \quad (13.15)$$

13.2.1.4 Iterative Algorithm

Instead of using a classical algebraic method to solve the pair of equations for u and v, Horn and Schunck adopted the Gaussian Seidel [ralston 1978] method to have the following iterative procedure:

$$u^{k+1} = \bar{u}^k - \frac{f_x[f_x \bar{u}^k + f_y \bar{v}^k + f_t]}{\alpha^2 + f_x^2 + f_y^2}$$

$$v^{k+1} = \bar{v}^k - \frac{f_y[f_x \bar{u}^k + f_y \bar{v}^k + f_t]}{\alpha^2 + f_x^2 + f_y^2} \quad (13.16)$$

where the superscripts k and $k+1$ are indexes of iteration and $\bar{u}, \bar{v}$ are the local averages of u and v, respectively.

Horn and Schunck define $\bar{u}, \bar{v}$ as follows:

$$\bar{u} = \frac{1}{6}\{u(x, y+1) + u(x, y-1) + u(x+1, y) + u(x-1, y)\}$$

$$+ \frac{1}{12}\{u(x-1, y-1) + u(x-1, y+1) + u(x+1, y-1) + u(x+1, y+1)\}$$

$$\bar{v} = \frac{1}{6}\{v(x, y+1) + v(x, y-1) + v(x+1, y) + v(x-1, y)\}$$

$$+ \frac{1}{12}\{v(x-1, y-1) + v(x-1, y+1) + v(x+1, y-1) + v(x+1, y+1)\} \quad (13.17)$$

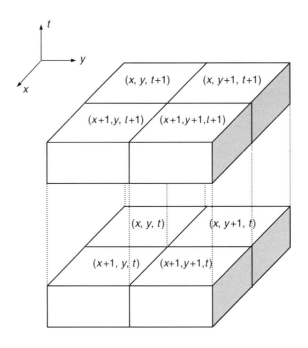

$$f_x = \frac{1}{4}\{[f(x+1,y,t) - f(x,y,t)] + [f(x+1,y,t+1) - f(x,y,t+1)]$$
$$+ [f(x+1,y+1,t) - f(x,y,t)] + [f(x+1,y+1,t+1) - f(x,y+1,t+1)]\}$$

$$f_y = \frac{1}{4}\{[f(x,y+1,t) - f(x,y,t)] + [f(x+1,y+1,t) - f(x+1,y,t)]$$
$$+ [f(x,y+1,t+1) - f(x,y,t+1)] + [f(x+1,y+1,t+1) - f(x+1,y,t+1)]\}$$

$$f_t = \frac{1}{4}\{[f(x,y,t+1) - f(x,y,t)] + [f(x+1,y,t+1) - f(x+1,y,t)]$$
$$+ [f(x,y+1,t+1) - f(x,y+1,t)] + [f(x+1,y+1,t+1) - f(x+1,y+1,t)]\}$$

FIGURE 13.3
Estimation of $f_x, f_y,$ and f_t.

The estimation of the partial derivatives of intensity function and the Laplacian of flow vectors need to be addressed. Horn and Schunck considered a $2 \times 2 \times 2$ spatiotemporal neighborhood, shown in Figure 13.3, for estimation of partial derivatives $f_x, f_y,$ and f_t. Note that replacing the first-order differentiation by the first-order difference is a common practice in managing digital images. The arithmetic average can remove the noise effect, thus making the obtained first-order differences less sensitive to various noises.

The Laplacian operator of u and v is approximated by

$$\nabla^2 u = \bar{u}(x,y) - u(x,y)$$
$$\nabla^2 v = \bar{v}(x,y) - v(x,y) \quad (13.18)$$

Equivalently, the Laplacian of u and v, $\nabla^2(u)$ and $\nabla^2(v)$, can be obtained by applying a 3×3 window operator, shown in Figure 13.4, to each point in the u and v planes, respectively.

Similar to the pel recursive technique discussed in Chapter 12, there are two different ways to iterate. One way is to iterate at a pixel until a solution is steady. Another way is to

Optical Flow

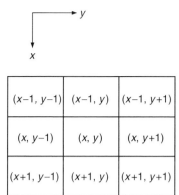

$$\nabla^2 u \approx \frac{1}{6}[u(x-1, y) + u(x, y-1) + u(x, y+1) + u(x+1, y)]$$
$$+ \frac{1}{12}[u(x-1, y-1) + u(x-1, y+1) + u(x+1, y-1) + u(x+1, y+1)]$$
$$- u(x,y)$$
$$\nabla^2 v \approx \frac{1}{6}[v(x-1, y) + v(x, y-1) + v(x, y+1) + v(x+1, y)]$$
$$+ \frac{1}{12}[v(x-1, y-1) + v(x-1, y+1) + v(x+1, y-1) + v(x+1, y+1)]$$
$$- v(x,y)$$

FIGURE 13.4
A 3 × 3 window operation for estimation of the Laplacian of flow vector.

iterate only once for each pixel. In the latter case, a good initial flow vector is required and is usually derived from the previous pixel.

13.2.2 Modified Horn and Schunck Method

Observing that the first-order difference is used to approximate the first-order differentiation in Horn and Schunck's original algorithm, and regarding this as a relatively crude form and a source of error, Barron et al. [barron 1994] developed a modified version of the Horn and Schunck method.

It features a spatiotemporal presmoothing and a more advanced approximation of differentiation. Specifically, it uses a Gaussian filter as a spatiotemporal prefilter. By the term Gaussian filter, we mean a low-pass filter with a mask shaped similar to that of the Gaussian probability density function (pdf). This is similar to what was utilized in the formulation of the Gaussian pyramid discussed in Chapter 11. The term spatiotemporal means that the Gaussian filter is used for low-pass filtering in both spatial and temporal domains.

With respect to the more advanced approximation of differentiation, a four-point central difference operator is used, which has a mask, shown in Figure 13.5.

As we shall see later in this chapter, this modified Horn and Schunck algorithm has achieved better performance than the original one owing to the two above-mentioned measures. This success indicates that a reduction of noise in image (data) leads to a

FIGURE 13.5
Four-point central difference operator mask.

$-\frac{1}{12}$	$\frac{8}{12}$	0	$-\frac{8}{12}$	$\frac{1}{12}$

significant reduction of noise in optical flow (solution). This example supports the statement we mentioned earlier that the ill-posed problem in low-level computational vision is mildly ill-posed.

13.2.3 Lucas and Kanade's Method

Lucas and Kanade assume that a flow vector is constant within a small neighborhood of a pixel, denoted by Ω. Then they form a weighted object function as follows:

$$\sum_{(x,y)\in\Omega} w^2(x,y) \left[\frac{\partial f(x,y,t)}{\partial x}u + \frac{\partial f(x,y,t)}{\partial v}v + \frac{\partial f(x,y,t)}{\partial t}\right]^2 \quad (13.19)$$

where $w(x, y)$ is a window function that gives more weight to the central portion than the surrounding portion of the neighborhood Ω.

The flow determination thus becomes a problem of a least square fit of the brightness invariance constraint. We observe that the smoothness constraint has been implied in Equation 13.19, where the flow vector is assumed to be constant within Ω.

13.2.4 Nagel's Method

Nagel first used the second-order derivatives in optical flow determination in the very early days [nagel 1983]. Since the brightness function $f(x,y,t,\bar{s})$ is a real-valued function of multiple variables (or a vector of variables), the Hessian matrix, discussed in Chapter 12, is used for the second-order derivatives.

An oriented-smoothness constraint was developed by Nagel that prohibits imposition of the smoothness constraint across edges (Figure 13.6). In Figure 13.6, an edge AB separates two different moving regions: region 1 and region 2. The smoothness constraint is imposed in these regions separately. That is, no smoothness constraint is imposed across the edge. Obviously, it would be a disaster if we smoothen the flow vectors across the edge. As a result, this reasonable treatment effectively improves the accuracy of optical flow estimation [nagel 1989].

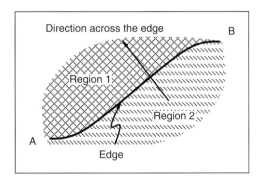

FIGURE 13.6
Oriented-smoothness constraint.

Optical Flow

13.2.5 Uras, Girosi, Verri, and Torre's Method

The Uras, Girosi, Verri, and Torre method is another method that uses second-order derivatives. Based on a local procedure, it performs quite well [uras 1988].

13.3 Correlation-Based Approach

The correlation-based approach to optical flow determination is similar to block matching, covered in Chapter 11. As may be recalled, the conventional block matching technique partitions an image into nonoverlapped, fixed-size, rectangle blocks. Then, for each block, the best matching in the previous image frame is found. In doing so, a search window is opened in the previous frame according to some a priori knowledge: the time interval between the two frames and the maximum possible moving velocity of objects in frames. Centered on each of the candidate pixels in the search window, a rectangle correlation window of the same size as the original block is opened. The best matched block in the search window is chosen so that either the similarity measure is maximized or the dissimilarity measure is minimized. The relative spatial position between these two blocks (the original block in the current frame and the best matched one in the previous frame) gives a translational motion vector to the original block. In the correlation-based approach to optical flow computation, the mechanism is very similar to that in the conventional block matching. The only difference is that for each pixel in an image, we open a rectangle correlation window centered on this pixel for which an optical flow vector needs to be determined. It is for this correlation window that we find the best match in the search window in its temporal neighboring image frame (Figure 13.7). A comparison between Figures 13.7 and 11.1 can convince us about the above observation. In this section, we first briefly discuss Anandan's method, which is a pioneer work in this category, and then Singh's method is described. His unified view of optical flow computation is introduced. We then present a correlation-feedback method by Pan, Shi, and Shu, which uses the feedback technique in flow calculation.

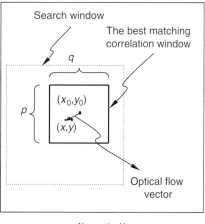

FIGURE 13.7
Correlation-based approach to optical flow determination.

13.3.1 Anandan's Method

As mentioned in Chapter 11, the sum of squared difference (SSD) is used as a dissimilarity measure in [anandan 1987]. It is essentially a simplified version of the well-known mean square error (MSE). Due to its simplicity, it is used in the methods developed by Singh [singh 1992] as well as Pan, Shi, and Shu [pan 1998].

In Anandan's method [anandan 1989], a pyramid structure is formed, and it can be used for an efficient coarse–fine search. This is very similar to the multiresolution block matching techniques discussed in Chapter 11. In the higher levels (with lower resolution) of the pyramid, a full search can be performed without a substantial increase in computation. The estimated velocity (or displacement) vector can be propagated to the lower levels (with higher resolution) for further refinement. As a result, a relatively large motion vector can be estimated with a certain degree of accuracy.

Instead of the Gaussian pyramid discussed in Chapter 11, a Laplacian pyramid is used here. To understand the Laplacian pyramid let us take a look at Figure 13.8a. There two consecutive levels are shown in a Gaussian pyramid structure: level k, denoted by $f^k(x, y)$ and level $k + 1$, $f^{k+1}(x, y)$. Figure 13.8b shows how level $k + 1$ can be derived from level k in the Gaussian pyramid. That is, as stated in Chapter 11, level $k + 1$ in the Gaussian pyramid can be obtained through low-pass filtering applied to level k, followed by subsampling. In Figure 13.8c, level $k + 1$ is first interpolated, thus producing an estimate of level k, $\hat{f}^k(x, y)$. The difference between the original level k and the interpolated estimate of level k generates an error at level k, denoted by $e^k(x, y)$. If there are no quantization errors involved, then level k, $f^k(x, y)$ can be recovered completely from the interpolated estimate of level k, $\hat{f}^k(x, y)$, and the error at level k, $e^k(x, y)$. That is,

$$f^k(x,y) = \hat{f}^k(x,y) + e^k(x,y) \qquad (13.20)$$

With quantization errors, however, the recovery of level k, $f^k(x, y)$ is not error free. It can be shown that coding $\hat{f}^k(x, y)$ and $e^k(x, y)$ is more efficient than directly coding $f^k(x, y)$.

A set of images $e^k(x, y)$, $k = 0, 1, \ldots, K - 1$ and $f^k(x, y)$ forms a Laplacian pyramid. Figure 13.8d displays a Laplacian pyramid with $K = 5$. It can be shown that Laplacian pyramids provide an efficient way for image coding [burt 1983]. A more detailed description of Gaussian and Laplacian pyramids can be found in [burt 1984; lim 1990].

13.3.2 Singh's Method

Singh presented a unified point of view on optical flow computation in [singh 1991, 1992]. He classified the information available in image sequences for optical flow determination into two categories: conservation information and neighborhood information. Conservation information is the information assumed to be conserved from one image frame to the next in flow estimation. Intensity is an example of conservation information, which is used most frequently in flow computation. Clearly, the brightness invariance constraint in the Horn and Schunck method is another way to state this type of conservation. Some functions of intensity may be used as conservation information as well. In fact, Singh uses the Laplacian of intensity as conservation information for computational simplicity. More examples can be found later in Section 13.4. Other information, different from intensity, such as color, can be used as conservation information. Neighborhood information is the information available in the neighborhood of the pixel from which optical flow is estimated.

These two different types of information correspond to two steps in flow estimation. In the first step, conservation information is extracted, resulting in an initial estimate of flow

Optical Flow

(a) Two consecutive levels in a pyramid structure

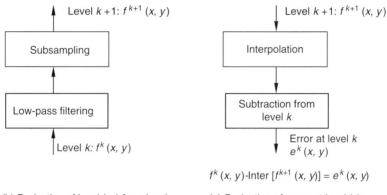

(b) Derivation of level $k+1$ from level k in a Gaussian pyramid

(c) Derivation of error at level k in a Laplacian pyramid

$$f^k(x, y) - \text{Inter}[f^{k+1}(x, y)] = e^k(x, y)$$

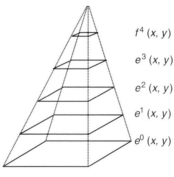

(d) Structure of Laplacian pyramid

FIGURE 13.8
Laplacian pyramid (level k in a Gaussian Pyramid).

vector. In the second step, this initial estimate is propagated into a neighborhood area and is iteratively updated. Obviously, in the Horn and Schunck method, the smoothness constraint is essentially one type of neighborhood information. Iteratively, estimates of flow vectors are refined with neighborhood information so that flow estimators from areas having sufficient intensity variation, such as the intensity corners as shown Figure 13.2d and areas with strong texture, can be propagated into areas with relatively small intensity variation or uniform intensity distribution.

With this unified point of view on optical flow estimation, Singh treated flow computation as parameter estimation. By applying estimation theory to flow computation, he developed an estimation-theoretical method to determine optical flow. It is a correlation-based method and consists of the above-mentioned two steps.

13.3.2.1 Conservation Information

In the first step, for each pixel (x, y) in the current frame $f_n(x, y)$, a correlation window of $(2l + 1) \times (2l + 1)$ is opened, centered on the pixel. A search window of $(2N + 1) \times (2N + 1)$ is opened in the previous frame $f_{n-1}(x, y)$ centered on (x, y). An error distribution of those $(2N + 1) \times (2N + 1)$ samples are calculated by using SSD as follows:

$$E_c(u,v) = \sum_{s=-l}^{l} \sum_{t=-l}^{l} [f_n(x+s, y+t) - f_{n-1}(x-u+s, y-v+t)]^2 \quad -N \leq u, v \leq N \quad (13.21)$$

A response distribution for these $(2N + 1) \times (2N + 1)$ samples is then calculated.

$$R_c(u, v) = e^{-\beta E_c(u,v)} \quad (13.22)$$

where β is a parameter, whose function and selection will be described in Section 13.3.3.1.

According to the weighted-least-square estimation, the optical flow can be estimated in this step as follows:

$$u_c = \frac{\sum_u \sum_v R_c(u,v) u}{\sum_u \sum_v R_c(u,v)}$$

$$v_c = \frac{\sum_u \sum_v R_c(u,v) v}{\sum_u \sum_v R_c(u,v)} \quad (13.23)$$

Assuming errors are additive and zero-mean random noise, we can also find the covariance matrix associated with the above estimate:

$$S_c = \begin{pmatrix} \dfrac{\sum_u \sum_v R_c(u,v)(u-u_c)^2}{\sum_u \sum_v R_c(u,v)} & \dfrac{\sum_u \sum_v R_c(u,v)(u-u_c)(v-v_c)}{\sum_u \sum_v R_c(u,v)} \\ \dfrac{\sum_u \sum_v R_c(u,v)(u-u_c)(v-v_c)}{\sum_u \sum_v R_c(u,v)} & \dfrac{\sum_u \sum_v R_c(u,v)(v-v_c)^2}{\sum_u \sum_v R_c(u,v)} \end{pmatrix} \quad (13.24)$$

13.3.2.2 Neighborhood Information

After step 1, all initial estimates are available. In step 2, they need to be refined according to neighborhood information. For each pixel, the method considers a $(2w + 1) \times (2w + 1)$ neighborhood centered on it. The optical flow of the center pixel is updated from the estimates in the neighborhood. A set of Gaussian coefficients is used in this method such that the closer the neighbor pixel is to the center pixel, the more influence the neighbor pixel has on the flow vector of the center pixel. The weighted-least-square-based estimate in this step is

$$\bar{u} = \frac{\sum_u \sum_v R_n(u,v) u}{\sum_u \sum_v R_n(u,v)}$$

$$\bar{v} = \frac{\sum_u \sum_v R_n(u,v) v}{\sum_u \sum_v R_n(u,v)} \quad (13.25)$$

Optical Flow

(0.25×0.25) $\frac{1}{16}$	(0.5×0.25) $\frac{1}{8}$	(0.25×0.25) $\frac{1}{16}$
(0.5×0.25) $\frac{1}{8}$	(0.5×0.5) $\frac{1}{4}$	(0.5×0.25) $\frac{1}{8}$
(0.25×0.25) $\frac{1}{16}$	(0.5×0.25) $\frac{1}{8}$	(0.25×0.25) $\frac{1}{16}$

FIGURE 13.9
3×3 Gaussian mask.

and the associated covariance matrix is

$$S_c = \begin{pmatrix} \dfrac{\sum_i R_n(u_i,v_i)(u_i-\bar{u})^2}{\sum_i R_n(u_i,v_i)} & \dfrac{\sum_i R_n(u_i,v_i)(u_i-\bar{u})(v_i-\bar{v})}{\sum_i R_n(u_i,v_i)} \\ \dfrac{\sum_i R_n(u_i,v_i)(u_i-\bar{u})(v_i-\bar{v})}{\sum_i R_n(u_i,v_i)} & \dfrac{\sum_i R_n(u_i,v_i)(v_i-\bar{v})^2}{\sum_i R_n(u_i,v_i)} \end{pmatrix} \quad (13.26)$$

where

$$1 \leq i \leq (2w+1)^2$$

In implementation, Singh uses a 3×3 neighborhood (i.e., $w=1$) centered on the pixel under consideration. The weights are depicted in Figure 13.9.

13.3.2.3 Minimization and Iterative Algorithm

According to estimation theory [beck 1977], two covariance matrices, expressed in Equations 13.24 and 13.26, respectively, are related to the confidence measure. That is, the reciprocals of the eigenvalues of the covariance matrix reveal confidence of the estimate along the direction represented by the corresponding eigenvectors. Moreover, conservation error and neighborhood error can be represented as the following two quadratic terms, respectively:

$$(U - U_c)^T S_c^{-1}(U - U_c) \quad (13.27)$$

$$(U - \bar{U})^T S_n^{-1}(U - \bar{U}) \quad (13.28)$$

where

$$\bar{U} = (\bar{u}, \bar{v}),\ U_c = (u_c, v_c),\ U = (u, v)$$

The minimization of the sum of these two errors over the image area leads to an optimal estimate of optical flow. That is, find (u, v) such that the following error is minimized:

$$\sum_x \sum_y [(U - U_c)^T S_c^{-1}(U - U_c) + (U - \bar{U})^T S_n^{-1}(U - \bar{U})] \quad (13.29)$$

An iterative procedure according to the Gauss–Siedel algorithm [ralston 1978] is used by Singh:

$$\begin{aligned} U^{k+1} &= [S_c^{-1} + S_n^{-1}]^{-1}[S_c^{-1}U_c + S_n^{-1}\bar{U}^k] \\ U^0 &= U_c \end{aligned} \quad (13.30)$$

Note that U_c, S_c are calculated once and remain unchanged in all the iterations. On the contrary, $\bar{U}$ and S_n vary with each iteration. This agrees with the description of the method in Section 13.3.2.2.

13.3.3 Pan, Shi, and Shu's Method

Applying feedback (a powerful technique widely used in automatic control and many other fields) to a correlation-based algorithm, Pan, Shi, and Shu developed a correlation-feedback method to compute optical flow. The method is iterative in nature. In each iteration, the estimated optical flow and its several variations are fed back. For each of the varied optical flow vectors, the corresponding sum of squared displaced frame difference (DFD) (Chapter 12), which often involves bilinear interpolation, is calculated. This useful information is then utilized in a revised version of a correlation-based algorithm [singh 1992]. They choose to work with this algorithm because it has several merits, and its estimation-theoretical computation framework lends itself to the application of the feedback technique.

As expected, the repeated usage of two given images via the feedback iterative procedure improves the accuracy of optical flow considerably. Several experiments on real image sequences in the laboratory and some synthetic image sequences demonstrate that the correlation-feedback algorithm performs better than some standard gradient- and correlation-based algorithms in terms of accuracy.

13.3.3.1 Proposed Framework

Block diagram of the proposed framework shown in Figure 13.10 is described next.

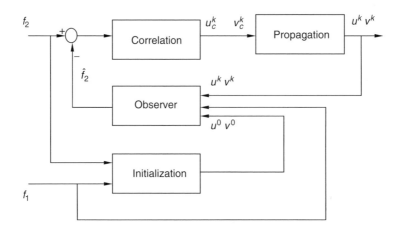

FIGURE 13.10
Block diagram of correlation-feedback technique.

13.3.3.1.1 Initialization

Although any flow algorithms can be used to generate an initial optical flow field $\bar{u}^o = (u^o, v^o)$ (even a nonzero initial flow field without applying any flow algorithm may work, but slowly), the Horn and Schunck algorithm [horn 1981], discussed in Section 13.2.1 (usually 5–10 iterations) is used to provide an appropriate starting point after preprocessing (involving low-pass filtering), since the algorithm is fast and the problem caused by the smoothness constraint is not serious in the first 10–20 iterations. The modified Horn and Schunck method, discussed in Section 13.2.2, may also be used for the initialization.

13.3.3.1.2 Observer

The DFD at the kth iteration is observed as $f_n(\bar{x}) - f_{n-1}(\bar{x} - \bar{u}^k)$, where f_n and f_{n-1} denote two consecutive digital images, $\bar{x} = (x,y)$ denotes the spatial coordinates of the pixel under consideration, and $\bar{u}^k = (u^k, v^k)$ denotes the optical flow of this pixel estimated at the kth iteration. (Note that the vector representation of the spatial coordinates in image planes is used quite often in the literature, owing to its brevity in notation.) Demanding fractional pixel accuracy usually requires interpolation. In Pan et al.'s work, the bilinear interpolation is adopted. The bilinearly interpolated image is denoted by $\hat{f}_{n-1}$.

13.3.3.1.3 Correlation

Once the bilinearly interpolated image is available, a correlation measure needs to be selected to search for the best match of a given pixel in $f_n(\bar{x})$ in a search area in the interpolated image. In their work, the sum-of-square-differences (SSD) is used. For each pixel in f_n, a correlation window W_c of size $(2l + 1) \times (2l + 1)$ is formed, centered on the pixel.

The search window in the proposed approach is quite different from that used in the correlation-based approach, say, in [singh 1992]. Let u be a quantity chosen from the following five quantities:

$$u \in \left\{ u^k - \frac{1}{2}u^k, u^k - \frac{1}{4}u^k, u^k, u^k + \frac{1}{4}u^k, u^k + \frac{1}{2}u^k \right\} \quad (13.31)$$

Let v be a quantity chosen from the following five quantities:

$$v \in \left\{ v^k - \frac{1}{2}v^k, v^k - \frac{1}{4}v^k, v^k, v^k + \frac{1}{4}v^k, v^k + \frac{1}{2}v^k \right\} \quad (13.32)$$

Hence, there are 25 (i.e., 5×5) possible combinations for (u, v). (It is noted that the restriction of the nonzero initial flow field mentioned above in Section 13.3.3.1.1 comes from here.) Note that other choices of variations around (u^k, v^k) are possible. Each of them corresponds to a pixel, $(x-u, y-v)$, in the bilinearly interpolated image plane. A correlation window is formed and centered in this pixel. The 25 samples of error distribution around (u^k, v^k) can be computed by using the SSD. That is,

$$E(u,v) = \sum_{s=-l}^{l} \sum_{t=-l}^{l} (f_n(x+s, y+t) - \hat{f}_{n-1}(x-u+s, y-v+t))^2 \quad (13.33)$$

The 25 samples of response distribution can be computed as follows:

$$R_c(u,v) = e^{-\beta E(u,v)} \quad (13.34)$$

where β is chosen so as to make the maximum R_c among the 25 samples of response distribution be a number close to unity. The choice of an exponential function for converting the error distribution into the response distribution is based primarily on the following consideration: the exponential function is well behaved when the error approaches zero and all the response distribution values are positive. The choice of β mentioned above is motivated by the following observation: in this way, the R_c values, which are the weights used in Equation 13.35, will be more effective. That is, the computation in Equation 13.35 will be more sensitive to the variation of the error distribution defined in Equation 13.33.

The optical flow vector derived at this correlation stage is then calculated as follows, according to the weighted-least-square estimation [singh 1992].

$$u^k(x,y) = \frac{\sum_u \sum_v R_c(u,v)u}{\sum_u \sum_v R_c(u,v)}, \quad v_c^k(x,y) = \frac{\sum_u \sum_v R_c(u,v)v}{\sum_u \sum_v R_c(u,v)} \qquad (13.35)$$

13.3.3.1.4 Propagation

Except in the vicinity of motion boundaries, the motion vectors associated with neighboring pixels are expected to be similar. Therefore, this constraint can be used to regularize the motion field. That is,

$$u^{k+1}(x,y) = \sum_{i=-w}^{w} \sum_{j=-w}^{w} w_1(i,j)\, u_c^k(x+i, y+j),$$

$$v^{k+1}(x,y) = \sum_{i=-w}^{w} \sum_{j=-w}^{w} w_1(i,j)\, u_c^k(x+i, y+j) \qquad (13.36)$$

where $w_1(i,j)$ is a weighting function. The Gaussian mask shown in Figure 13.9 is chosen as the weighting function $w_1(i,j)$ used in our experiments. By using this mask, the velocity of various pixels in a pixel's neighborhood will be weighted according to their distance from the pixel: the larger the distance the smaller the weight. The mask smooths the optical flow field as well.

13.3.3.1.5 Convergence

Under the assumption of the symmetric response distribution with a single maximum value assumed by the ground-truth optical flow, the convergence of the correlation-feedback technique is justified in [pan 1995].

13.3.3.2 Implementation and Experiments

13.3.3.2.1 Implementation

To make the algorithm more robust against noise, three consecutive images in an image sequence, denoted by f_1, f_2, and f_3, respectively, are used to implement their algorithm instead of the two images in the above principle discussion. This implementation was proposed in [singh 1992]. Assume the time interval between f_1 and f_2 is the same as that between f_2 and f_3. Also assume the apparent 2-D motion is uniform during these two intervals along the motion trajectories. From images f_1 and f_2, (u^0, v^0) can be computed. From (u^k, v^k), the optical flow estimated during the kth iteration, and f_1 and f_2, the response distribution, $R_c^+(u^k, v^k)$, can be calculated as

$$R_c^+(u^k, v^k) = \exp\left\{-\beta \sum_{s=-l}^{l} \sum_{t=-l}^{l} [f_2(x+s, y+t) - \hat{f}_1(x - u^k + s, y - v^k + t)]^2\right\} \quad (13.37)$$

Similarly, from images f_3 and f_2, $(-u^k, -v^k)$ can be calculated. Then $R_c^-(-u^k, -v^k)$ can be calculated as

$$R_c^-(-u^k, -v^k) = \exp\left\{-\beta \sum_{s=-l}^{l} \sum_{t=-l}^{l} [f_2(x+s, y+t) - \hat{f}_3(x + u^k + s, y + v^k + t)]^2\right\} \quad (13.38)$$

The response distribution $R_c(u^k, v^k)$ can then be determined as the sum of $R_c^+(u^k, v^k)$ and $R_c^-(-u^k, -v^k)$. The size of the correlation window and the weighting function is chosen to be 3×3, i.e., $l = 1, w = 1$. In each search window, β is chosen so as to make the larger one among R_c^+ and R_c^- a number close to unity. In the observer stage, the bilinear interpolation is used, which is shown to be faster and better than the B-spline in Pan et al.'s many experiments.

13.3.3.2.2 Experiment I

Figure 13.11 shows the three successive image frames f_1, f_2, and f_3 about a square post. They were taken by a CCD video camera and a DATACUBE real time image processing system supported by a Sun workstation. The square post is moving horizontally, perpendicular to the optical axis of the camera, in a uniform speed of 2.747 pixels per frame. To remove various noises to a certain extent and to speed up processing, these three 256×256 images are low-pass filtered and then subsampled prior to optical flow estimation. That is, the intensities of every 16 pixels in a block of 4×4 are averaged and the average value is assigned to represent this block. Note that the choice of other low-pass filters is also possible. In this way, these three images are compressed into three 64×64 images. The "ground-truth" 2-D motion velocity vector is hence known as $u^a = -0.6868; v^a = 0$.

To compare the performance of the correlation-feedback approach with that of the gradient-based and correlation-based approaches, Horn and Schunck's algorithm is chosen to represent the gradient-based approach and Singh's framework to represent the correlation-based approach. Table 13.1 shows the results of the comparison. There, l, w, and N indicate the sizes of the correlation window, weighting function, and search window, respectively. The program that implements Singh's algorithm is provided by the authors of [barron 1994]. In the correlation-feedback algorithm, 10 iterations of Horn and Schunck's algorithm with $\alpha = 5$ are used in the initialization. (Recall that α is a regularization parameter used in [horn 1981].) Only the central 40×40 flow vector array is used to compute u_{error}, which is the root mean square (RMS) error in the vector magnitudes between the ground-truth and estimated optical flow vectors. It is noted that the relative error in Experiment I is greater than 10%. This is because the denominator in the formula calculating the RMS error is too small due to the static background and, hence, many zero ground-truth 2-D motion velocity vectors in this experiment. Relatively speaking, the correlation-feedback algorithm performs best in determining optical flow for a texture post in translation. The correct optical flow field and those calculated by using three different algorithms are shown in Figure 13.12.

13.3.3.2.3 Experiment II

The images in Figure 13.13 were obtained by rotating a CCD camera with respect to the center of a ball. The rotating velocity is 2.5° per frame. Similarly, three 256×256 images are compressed into three 64×64 images by using the averaging and subsampling discussed above. Only the central 40×40 optical vector arrays are used to compute u_{error}. Table 13.2 reports the results for this experiment. There, u_{error}, l, w, and N have the same meaning as

FIGURE 13.11
(a) Texture square A. (b) Texture square B. (c) Texture square C.

that discussed in Experiment I. It is obvious that our correlation-feedback algorithm performs best in determining optical flow for this rotating ball case.

13.3.3.2.4 Experiment III

To compare the correlation-feedback algorithm with other existing techniques in a more objective, quantitative manner, Pan et al. cite some results reported in [barron 1994], which were obtained by applying some typical optical flow techniques to some image sequences

TABLE 13.1

Comparison in Experiment I

Techniques	Gradient-Based Approach	Correlation-Based Approach	Correlation-Feedback Approach
Conditions	*Iteration number* = 128 $\alpha = 5$	*Iteration number* = 25 $l=2, w=2, N=4$	*Iteration number* = 10 $l=1, w=1, N=5$
u_{error}	56.37%	80.97%	44.56%

Optical Flow

chosen with deliberation. In the meantime they report the results obtained by applying their feedback technique to the identical image sequences with the same accuracy measurement as used in [barron 1994].

Three image sequences used in [barron 1994] were utilized here: namely "Translating Tree," "Diverging Tree," and "Yosemite." The first two simulate translational camera motion with respect to a textured planar surface (see Figure 13.14), and are sometimes referred to as "Tree 2-D" sequence. Therefore, there are no occlusions and no motion

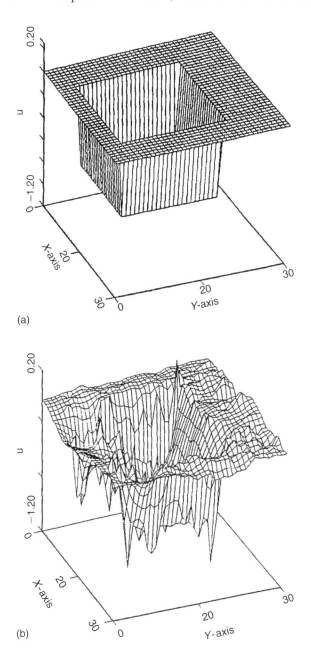

FIGURE 13.12
(a) Correct optical flow field. (b) Optical flow field calculated by the gradient-based approach.

(continued)

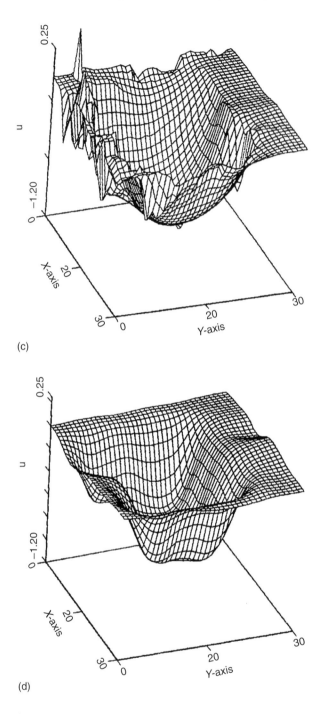

FIGURE 13.12 (continued)
(c) Optical flow field calculated by the correlation-based approach. (d) Optical flow field calculated by the correlation-feedback approach.

discontinuities in these two sequences. In the Translating Tree sequence, the camera moves normally to its line of sight, with velocities between 1.73 and 2.26 pixels/frame parallel to the X-axis in the image plane. In the Diverging Tree sequence, the camera moves along its line of sight. The focus of expansion is at the center of the image. The speeds vary from 1.29

Optical Flow

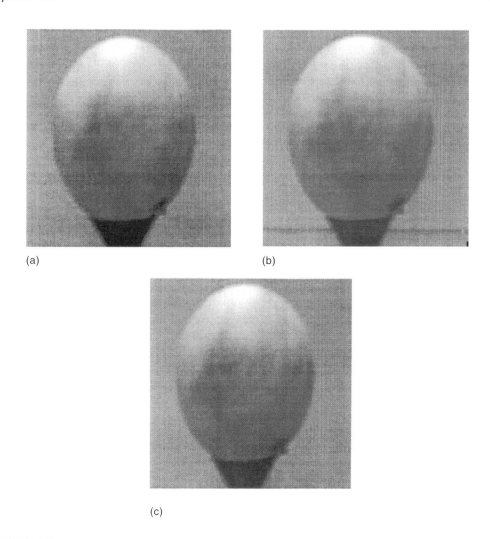

FIGURE 13.13
(a) Ball A. (b) Ball B. (c) Ball C.

pixels/frame on the left side to 1.86 pixels/frame on the right. The "Yosemite" sequence is a more complex test case (see Figure 13.15). The motion in the upper right is mainly divergent. The clouds translate to the right with a speed of 1 pixel/frame, while velocities in the lower left are about 4 pixels/frame. This sequence is challenging because of the range of velocities and the occluding edges between the mountains and at the horizon. There is severe aliasing in the lower portion of the images, causing most methods to produce poorer velocity measurements. Note that this synthetic sequence is for quantitative study purposes

TABLE 13.2

Comparison in Experiment II

Techniques	Gradient-Based Approach	Correlation-Based Approach	Correlation-Feedback Approach
Conditions	Iteration number = 128 $\alpha = 5$	Iteration number = 25 $l=2, w=2, N=4$	Iteration number = 10 $l=1, w=1, N=5$
u_{error}	65.67%	55.29%	49.80%

FIGURE 13.14
A frame of the "Tree 2-D" sequence.

FIGURE 13.15
A frame of the "Yosemite" sequence.

TABLE 13.3

Summary of the "Translating Tree" 2-D Velocity Results

Techniques	Average Error (degree)	Standard Deviation (degree)	Density (%)
Horn and Schunck (original)	38.72	27.67	100
Horn and Schunck (modified)	2.02	2.27	100
Uras et al. (unthresholded)	0.62	0.52	100
Nagel	2.44	3.06	100
Anandan	4.54	3.10	100
Singh (step 1, $l=2$, $w=2$)	1.64	2.44	100
Singh (step 2, $l=2$, $w=2$)	1.25	3.29	100
Pan, Shi and Shu ($l=1$, $w=1$)	1.07	0.48	100

since its ground-truth flow field is known and is, otherwise, far less complex than many real world outdoor sequences processed in the literature.

The angular measure of the error used in [barron 1994] is utilized here, as well. Let image velocity $\vec{u} = (u,v)$ be represented as 3-D direction vectors,

$$\vec{V} \equiv \frac{1}{\sqrt{u^2 + v^2 + 1}} (u, v, 1) \quad (13.39)$$

The angular error between the correct image velocity $\vec{V}$ and an estimate $\vec{V}_e$ is $\psi_E = \text{across}(\vec{V}_c \cdot \vec{V}_e)$. It is obvious that the smaller the angular error ψ_E, the more accurate the estimation of the optical flow field will be. Despite the fact that the confidence measurement can be used in the correlation-feedback algorithm as well, Pan et al. did not consider the usage of the confidence measurement in their work. Therefore, only the results with 100% density in Tables 4.6, Table 4.7, and Table 4.10 in [barron 1994] were used in Tables 13.3 through 13.5, respectively.

Before computation of the optical flow field, the Yosemite and Tree 2-D test sequences were compressed by a factor of 16 and 4, respectively, using the averaging and subsampling method discussed earlier.

As mentioned in [barron 1994] the optical flow field for the "Yosemite" sequence is complex, and Table 13.5 indicates that the correlation-feedback algorithm evidently performs best. In [black 1996], a robust method was developed and applied to a cloudless Yosemite sequence. It is noted that the performance of flow determination algorithms will be improved if the sky is removed from consideration [barron 1994; black 1996]. Still, it is clear that the algorithm in [black 1996] achieved very good performance in terms of accuracy. In order to make a comparison with their algorithm, the correlation-feedback algorithm was applied to the same cloudless "Yosemite" sequence. The results were reported in Table 13.6, from which it can be observed that the results obtained by Pan et al. are slightly

TABLE 13.4

Summary of the "Diverging Tree" 2-D Velocity Results

Techniques	Average Error (degree)	Standard Deviation (degree)	Density (%)
Horn and Schunck (original)	12.02	11.72	100
Horn and Schunck (modified)	2.55	3.67	100
Uras et al. (unthresholded)	4.64	3.48	100
Nagel	2.94	3.23	100
Anandan (frames 19 and 21)	7.64	4.96	100
Singh (step 1, $l=2$, $w=2$)	17.66	14.25	100
Singh (step 2, $l=2$, $w=2$)	8.60	5.60	100
Pan, Shi, and Shu ($l=1$, $w=1$)	5.12	2.16	100

TABLE 13.5

Summary of the "Yosemite" 2-D Velocity Results

Techniques	Average Error (degree)	Standard Deviation (degree)	Density (%)
Horn and Schunck (original)	32.43	30.28	100
Horn and Schunck (modified)	11.26	16.41	100
Uras et al. (unthresholded)	10.44	15.00	100
Nagel	11.71	10.59	100
Anandan (frames 19 and 21)	15.84	13.46	100
Singh (step 1, $l=2, w=2$)	18.24	17.02	100
Singh (step 2, $l=2, w=2$)	13.16	12.07	100
Pan, Shi, and Shu ($l=1, w=1$)	7.93	6.72	100

better. Tables 13.3 and 13.4 indicate that the feedback technique also performs very well in translating and diverging texture post cases.

13.3.3.2.5 Experiment IV

Here, the correlation-feedback algorithm is applied to a real sequence named "Hamburg Taxi," which is used as a testing sequence in [barron 1994]. There are four moving objects in the scene: a moving pedestrian in the upper left portion, a turning car in the middle, a car moving toward right at the left side and a car moving toward left at the right side. A frame of the sequence and the needle diagram of flow vectors estimated by using 10 iterations of the correlation-feedback algorithm (with 10 iterations of Horn and Schunck's algorithm for initialization) are shown in Figures 13.16 and 13.17, respectively. The needle diagram is printed in the same fashion as those shown in [barron 1994]. It is noted that the moving pedestrian in the upper left portion cannot be shown because of the scale used in the needle diagram. The other three moving vehicles in the sequence are shown very clearly. The noise level is low. Compared with those diagrams reported in [barron 1994], the correlation-feedback algorithm achieves very good results.

For a comparison on a local basis, the portion of the needle diagram associated with the area surrounding the turning car (a sample of the velocity fields), obtained by 50 iterations of the correlation-feedback algorithm with 5 iterations of Horn and Schunck's algorithm as initialization, is provided in Figure 13.18c. Its counterparts obtained by applying Horn and Schunck's (50 iterations), and Singh's (50 iterations) algorithms are displayed in Figure 13.18a and b, respectively. It is observed that the correlation-feedback algorithm achieves the best results among the three algorithms.

13.3.3.3 Discussion and Conclusion

Although it uses a revised version of a correlation-based algorithm [singh 1992], the correlation-feedback technique is quite different from the correlation-based algorithm [singh 1992] in the following four aspects. First, different optimization criteria: the algorithm does not use the iterative minimization procedure used in [singh 1992]. Instead,

TABLE 13.6

Summary of the Cloudless "Yosemite" 2-D Velocity Results

Techniques	Average Error (degree)	Standard Deviation (degree)	Density (%)
Robust formulation	4.46	4.21	100
Pan, Shi, and Shu ($l=1, w=1$)	3.79	3.44	100

Optical Flow

FIGURE 13.16
Hamburg taxi.

some variations of the estimated optical flow vectors are generated and fed back. The associated bilinearly interpolated displaced frame difference (DFD) for each variation is calculated and utilized. In essence, the feedback approach utilizes two given images repeatedly, while the Singh method uses two given images only once (u_c and v_c derived from the two given images are only calculated once). The best local matching between the displaced image, generated via feedback of the estimated optical flow, and the given image is actually used as the ultimate criterion for improving optical flow accuracy in the iterative process. Second, the search window in the algorithm is an adaptive "rubber" window, having a variable size depending on (u^k, v^k). In the correlation-based approaches [singh 1992], the search window has a fixed size. Third, the algorithm uses a bilinear interpolation technique in the observation stage and provides the correlation stage with a virtually

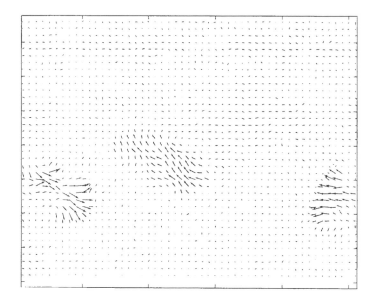

FIGURE 13.17
Needle diagram of flow field of Hamburg taxi sequence obtained by using the correlation-feedback algorithm.

continuous image field for more accurate motion vector computation, while that in [singh 1992] does not. Fourth, different performances are achieved when image intensity is a linear function of image coordinates. In fact, in the vicinity of a pixel, the intensity can usually be considered as such a linear function. Except if the optical flow vectors happen to have only an integer multiple of pixels as their components, an analysis in [pan 1994] shows that the correlation-based approach [singh 1992] will not converge to the apparent

(a)

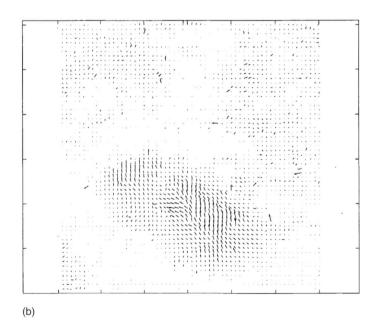

(b)

FIGURE 13.18
A portion of the needle diagram obtained by using (a) Horn and Schunk's algorithm, and (b) Singh's algorithm.
(continued)

Optical Flow

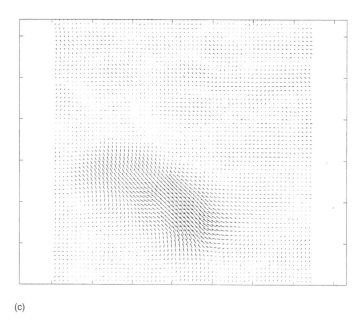

(c)

FIGURE 13.18 (continued)
(c) The correlation-feedback algorithm.

2-D motion vectors and will easily have error much greater than 10%. In [pan 1994] it is also shown that the linear intensity function guarantees the assumption of the symmetric response distribution with a single maximum value assumed by the ground-truth optical flow. As discussed in Section 13.3.3.1, under this assumption the convergence of the correlation-feedback technique is justified.

Numerous experiments have demonstrated the correlation-feedback algorithm's convergence and accuracy, and usually it is more accurate than some standard gradient- and correlation-based approaches. In the complicated optical flow cases, specifically in the case of the "Yosemite" image sequence (regarded as the most challenging quantitative test image sequence in [barron 1994]), it performs better than all other techniques.

13.4 Multiple Attributes for Conservation Information

As stated at the beginning of this chapter, there are many algorithms in optical flow computation reported in the literature, and many new algorithms are to be developed. In Sections 13.2 and 13.3, we introduced some typical algorithms using gradient- and correlation-based approaches, and will not explore various algorithms any further here. It is hoped that the fundamental concepts and algorithms introduced above have provided a solid base for readers to study more advanced techniques.

We would like to discuss optical flow from another point of view, however: multiple image attributes versus a single image attribute. All of methods discussed so far use only one kind of image attributes as conservation information in flow determination. Most methods use intensity. Singh's method uses the Laplacian of intensity, which is calculated by using the difference of the Gaussian operation [burt 1984]. It was reported by Weng, Ahuja, and Huang that using a single attribute as conservation information may result in ambiguity in matching two perspective views, while multiple attributes, which are

FIGURE 13.19
Multiple attributes versus single attribute. (a) With intensity information only, points D, E, and F tend to match to points A, B, and C, respectively. (b) With intensity, edge, and corner information points D and E tend to match points B and C, respectively.

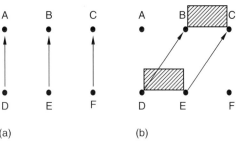

motion insensitive, may reduce ambiguity remarkably, resulting in better matching [weng 1992]. An example is shown in Figure 13.19 to illustrate this argument. In this section, Weng et al.'s method is discussed first. Then we introduce Xia and Shi's method, which uses multiple attributes in a framework based on weighted-least-square estimation and feedback techniques.

13.4.1 Weng, Ahuja, and Huang's Method

Weng, Ahuja, and Huang proposed a quite different approach to image point matching [weng 1992]. Note that the image matching amounts to flow field computation since it calculates a displacement field for each point in image planes, which is essentially a flow field if the time interval between two image frames is known.

Based on an analysis indicating that using image intensity as a single attribute is not enough in accurate image matching, Weng, Ahuja, and Huang utilize multiple attributes associated with images in estimation of the dense displacement field. These image attributes are motion insensitive, i.e., they generally sustain only small change under motion assumed to be locally rigid. The image attributes used are image intensity, edgeness, and cornerness. For each image attribute, the algorithm forms a residual function, reflecting the inaccuracy of the estimated matching. The matching is then determined via an iterative procedure to minimize the weighted sum of these residual functions. In handling neighborhood information, a more advanced smoothness constraint is used to take care of moving discontinuities. The method considers uniform regions and the occlusion issue as well.

In addition to using multiple image attributes, the method is point-wise processing. There is no need for calculation of correlation within two correlation windows, which saves computation dramatically. However, the method also has some drawbacks. First, the edgeness and cornerness involve calculation of the spatial gradient, which is noise sensitive. Second, in solving for minimization, the method resorts to numerical differentiation again: the estimated displacement vectors are updated based on the partial derivatives of the noisy attribute images. In a word, the computational framework heavily relies on numerical differentiation, which is considered to be impractical for accurate computation [barron 1994].

On the other hand, the Pan, Shi, and Shu method, discussed in Section 13.3.3 in the category of correlation-based approaches, seems to have some complementary features. It is correlation-based. It uses intensity as a single attribute. In these two aspects the method of Pan, Shi, and Shu is inferior to the method of Weng, Ahuja, and Huang. The feedback technique and the weighted-least-square computation framework used in the Pan et al. method are, however, superior compared with the Weng et al. method. Motivated by the above observations, an efficient, multiattribute feedback method was developed by Xia and Shi [xia 1995, 1996], discussed in the next section. It is expected that more insight of the Weng, Ahuja, and Huang method will become clear in the discussion as well.

Optical Flow 317

13.4.2 Xia and Shi's Method

This method uses multiple attributes that are motion insensitive. The following five attributes are used: image intensity, horizontal edgeness, vertical edgeness, contrast, and entropy. The first three are used in [weng 1992] as well, and can be considered as structural attributes, while the last two, which are not used in [weng 1992], can be considered as textural attributes according to [haralick 1979].

Instead of the computational framework presented in [weng 1992], which, as discussed above, may be not practical for accurate computation, the method uses the computational framework of [pan 1994, 1998]. That is, the weighted-least-squared estimation technique used in [singh 1992] and the feedback technique used in [pan 1994, 1998] are utilized here. Unlike in [weng 1992], sub pixel accuracy is considered and a confidence measure is generated in the method.

The Xia and Shi method is also different from those algorithms presented in [singh 1992; pan 1995, 1998]. First, there is no correlation in the method, while both [singh 1992; pan 1995, 1998] are correlation-based. Specifically, the method is a point-wise processing. Second, the method uses multiple attributes, while both [singh 1992; pan 1995, 1998] use image intensity as a single attribute.

In summary, the Xia and Shi method to compute optical flow is motivated by several existing algorithms mentioned above. It does, however, differ from each of them significantly.

13.4.2.1 Multiple Image Attributes

As mentioned earlier, there are five image attributes in the Xia and Shi method. They are defined below.

13.4.2.1.1 Image Intensity

The intensity at a pixel (x, y) in an image $f_n(x, y)$, denoted by $A_i(x, y)$, i.e., $A_i(x, y) = f_n(x, y)$.

13.4.2.1.2 Horizontal Edgeness

The horizontal edgeness at a pixel (x, y), denoted by $A_h(x, y)$, is defined as

$$A_h(x, y) = \frac{\partial f(x, y)}{\partial y} \tag{13.40}$$

i.e., the partial derivative of $f(x, y)$ with respect to y, the second component of the gradient of intensity function at the pixel.

13.4.2.1.3 Vertical Edgeness

The vertical edgeness at a pixel (x, y), denoted by $A_v(x, y)$, is defined as

$$A_v(x, y) = \frac{\partial f(x, y)}{\partial x} \tag{13.41}$$

i.e., the first component of the gradient of intensity function at the pixel. Note that the partial derivatives in Equations 13.40 and 13.41 are computed by applying a Sobel operator [gonzalez 1992] in a 3×3 neighborhood of the pixel.

13.4.2.1.4 Contrast

The local contrast at a pixel (x, y), denoted by $A_c(x, y)$, is defined as

$$A_c(x, y) = \sum_{i,j \in S} (i - j)^2 C_{i,j} \tag{13.42}$$

where S is a set of all the distinct gray levels within a 3×3 window centered at pixel (x, y). $C_{i,j}$ specifies a relative frequency with which two neighboring pixels separated horizontally by a distance 1 occur in the 3×3 window, one with gray level i and the other with gray level j.

13.4.2.1.5 Entropy

The local entropy at a point (x, y), denoted by $A_e(x, y)$, is given by

$$A_e(x, y) = -\sum_{i \in S} p_i \log p_i \qquad (13.43)$$

where S was defined above, and p_i is the probability of occurrence of the gray level i in the 3×3 window.

Since the intensity is assumed to be invariant to motion, so are the horizontal edgeness, vertical edgeness, contrast, and entropy.

As mentioned above, the intensity and edgeness are used as attributes in Weng et al.'s algorithm as well. Compared with the negative and positive cornerness used in Weng et al.'s algorithm, the local contrast and entropy need no differentiation and therefore are less sensitive to various noises in original images. In addition, these two attributes are inexpensive in terms of computation. They reflect the textural information about the local neighborhood of the pixel for which flow vector is to be estimated.

13.4.2.2 Conservation Stage

In the Xia et al. algorithm, this stage is similar to that in the Pan et al. algorithm. That is, for a flow vector estimated at the kth iteration, denoted by (u^k, v^k), we find its 25 variations, (u, v), according to

$$u \in \left\{ u^k - \frac{u^k}{2}, u^k - \frac{u^k}{4}, u^k, u^k + \frac{u^k}{4}, u^k + \frac{u^k}{2} \right\}$$
$$v \in \left\{ v^k - \frac{v^k}{2}, v^k - \frac{v^k}{4}, u^k, v^k + \frac{v^k}{4}, v^k + \frac{v^k}{2} \right\} \qquad (13.44)$$

For each of these 25 variations, the matching error is computed as

$$E(u, v) = r_{A_i}^2(x, y, u, v) + r_{A_h}^2(x, y, u, v) + r_{A_v}^2(x, y, u, v) + r_{A_c}^2(x, y, u, v) + r_{A_e}^2(x, y, u, v) \qquad (13.45)$$

where $r_{A_i}, r_{A_h}, r_{A_v}, r_{A_c}, r_{A_e}$ denote the residual function with respect to the five attributes, respectively.

The residual function of intensity is defined as

$$r_{A_i}(x, y, u, v) = A_{i_n}(x, y) - A_{i_{n-1}}(x - u, y - v) = f_n(x, y) - f_{n-1}(x - u, y - v) \qquad (13.46)$$

where $f_n(x, y)$ and $f_{n-1}(x, y)$ are defined as before, i.e., the intensity function at t_n and t_{n-1}, respectively; $A_{i_n}, A_{i_{n-1}}$ denote the intensity attributes on f_n and f_{n-1}, respectively.

It is observed that the residual error of intensity is essentially the DFD (Chapter 12). The rest of the residual functions are defined similarly. When sub-pixel accuracy is required, spatial interpolation in the attribute images generally is necessary. Thus, the flow vector estimation is now converted to a minimization problem. That is, find u and v at pixel (x, y) such that the matching error defined in Equation 13.45 is minimized. The weighted-least-square method [singh 1992; pan 1998] is then used. That is,

Optical Flow

$$R(u,v) = e^{-\beta E(u,v)} \tag{13.47}$$

$$u_c^{k+1} = \frac{\sum_u \sum_v R(u,v)u}{\sum_u \sum_v R(u,v)}, \quad v_c^{k+1} = \frac{\sum_u \sum_v R(u,v)v}{\sum_u \sum_v R(u,v)} \tag{13.48}$$

Since the weighted-least-square method has been discussed in detail in Sections 13.3.2 and 13.3.3, we will not go into more detail here.

13.4.2.3 Propagation Stage

Similar to what was proposed in the Pan et al. algorithm, in this stage Xia et al. form a window W of size $(2w + 1) \times (2w + 1)$ centered at the pixel (x, y) in the image $f_n(x, y)$. The flow estimate at the pixel (x, y) in this stage, denoted by (u^{k+1}, v^{k+1}), is calculated as a weighted sum of the flow vectors of the pixel within the window W.

$$u^{k+1} = \sum_{s=-w}^{w} \sum_{t=-w}^{w} w_1[f_n(x,y), f_n(x+s, y+t)] \cdot u_c^{k+1}(x+s, y+t)$$

$$v^{k+1} = \sum_{s=-w}^{w} \sum_{t=-w}^{w} w_1[f_n(x,y), f_n(x+s, y+t)] \cdot v_c^{k+1}(x+s, y+t) \tag{13.49}$$

where $w_1[.,.]$ is a weight function. For each point in the window W, a weight is assigned according to the weight function. Let $(x + s, y + t)$ denote a pixel within the window W, then the weight of the pixel $(x + s, y + t)$ is given by

$$w_1[f_n(x,y), f_n(x+s; y+t)] = \frac{c}{\varepsilon + |f_n(x,y) - f_n(x+s, y+t)|} \tag{13.50}$$

where ε is a small positive number to prevent the denominator from vanishing, c is a normalization constant that makes the summation of all the weights in the W equal 1.

From Equation 13.50, we see that the weight is determined based on the intensity difference between the pixel under consideration and its neighboring pixel. The larger the difference in the intensity, the more likely the two points belong to different regions. Therefore, the weight will be small in this case. On the other hand, the flow vector in the same region will be similar since the corresponding weight is large. Thus, the weighting function implicitly takes flow discontinuity into account and is more advanced than that in [singh 1992; pan 1994, 1998].

13.4.2.4 Outline of Algorithm

The following points summarize the procedures of the algorithm:

1. Perform a low-pass prefiltering on two input images to remove various noises.
2. Generate attribute images: intensity, horizontal edgeness, vertical edgeness, local contrast, and local entropy. Those attributes are computed at each grid point of both images.
3. Set the initial flow vectors to zero. Set the maximum iteration number and/or estimation accuracy.
4. For each pixel under consideration, generate flow variations according to Equation 13.44. Compute matching error for each flow variation according to Equation 13.45

TABLE 13.7

Summary of the "Translating Tree" 2-D Velocity Results

Techniques	Average Error (degree)	Standard Deviation (degree)	Density (%)
Horn and Schunck (original)	38.72	27.67	100
Horn and Schunck (modified)	2.02	2.27	100
Uras et al. (unthresholded)	0.62	0.52	100
Nagel	2.44	3.06	100
Anandan	4.54	3.10	100
Singh (step 1, $n=2$, $w=2$)	1.64	2.44	100
Singh (step 2, $n=2$, $w=2$)	1.25	3.29	100
Pan, Shi, and Shu ($n=1$, $w=1$)	1.07	0.48	100
Weng, Ahuja, and Huang	1.81	2.03	100
Xia and Shi	0.55	0.52	100

and transform them to the corresponding response distribution R using Equation 13.47. Compute the flow estimation u^c, v^c using Equation 13.48.

5. Form a $(2w+1) \times (2w+1)$ neighborhood window W centered at the pixel. Compute the weight for each pixel within the window W using Equation 13.50. Update the flow vector using Equation 13.49.

6. Decrease the preset iteration number. If the iteration number is zero, the algorithm returns with the resultant optical flow field. Otherwise, go to the next step.

7. If the change in flow vector over two successive iterations is less than the predefined threshold, the algorithm returns with the estimated optical flow field. Otherwise, go to step 4.

13.4.2.5 Experimental Results

To compare the method with other methods existing in the literature, similar to what has been done in [pan 1998] (Section 13.3.3), the method was applied to three test sequences used in [barron 1994]: the "Translating Tree" sequence, the "Diverging Tree" sequence, and the "Yosemite" sequence. The same accuracy criterion is used as that in [barron 1994]. Only those results reported in [barron 1994] with 100% density are listed in Tables 13.7 through 13.9 for a fair and easy comparison. The algorithm of Weng et al. was implemented by Xia et al. and the results were reported in [xia 1995].

TABLE 13.8

Summary of the "Diverging Tree" 2-D Velocity Results

Techniques	Average Error (degree)	Standard Deviation (degree)	Density (%)
Horn and Schunck (original)	32.43	30.28	100
Horn and Schunck (modified)	11.26	16.41	100
Uras et al. (unthresholded)	10.44	15.00	100
Nagel	11.71	10.59	100
Anandan	15.84	13.46	100
Singh (step 1, $n=2$, $w=2$, $N=4$)	18.24	17.02	100
Singh (step 2, $n=2$, $w=2$, $N=4$)	13.16	12.07	100
Pan, Shi, and Shu ($n=1$, $w=1$)	7.93	6.72	100
Weng, Ahuja, and Huang	8.41	8.22	100
Xia and Shi	7.54	6.61	100

TABLE 13.9

Summary of the "Yosemite" 2-D Velocity Results

Techniques	Average Error (degree)	Standard Deviation (degree)	Density (%)
Horn and Schunck (original)	12.02	11.72	100
Horn and Schunck (modified)	2.55	3.67	100
Uras et al. (unthresholded)	4.64	3.48	100
Nagel	2.94	3.23	100
Anandan (frames 19 and 21)	7.64	4.96	100
Singh (step 1, $n=2, w=2, N=4$)	17.66	14.25	100
Singh (step 2, $n=2, w=2, N=4$)	8.60	5.60	100
Pan, Shi, and Shu ($n=1, w=1$)	5.12	2.16	100
Weng, Ahuja, and Huang	8.01	9.71	100
Xia and Shi	4.04	3.82	100

13.4.2.6 Discussion and Conclusion

The above experimental results demonstrate that the Xia and Shi method outperforms both Pan, Shi, and Shu method and Weng, Ahuja, and Huang's method in terms of accuracy of optical flow determined. Computationally speaking, Xia and Shi's method is less expensive than Pan et al.'s, since there is no correlation involved and the correlation is known to be computationally expensive.

13.5 Summary

The optical flow field is a dense 2-D distribution of apparent velocities of movement of intensity patterns in image planes, while the 2-D motion field can be understood as the perspective projection of 3-D motion in the scene onto image planes. They are different. Only under certain circumstances, are they equal to each other. In practice, however, they are closely related in that image sequences are usually the only data we have in motion analysis. Hence, we can only deal with the optical flow in motion analysis, instead of the 2-D motion field. The aperture problem in motion analysis refers to the problem that occurs when viewing motion via an aperture. Specifically, the only motion we can observe from local measurement is the motion component orthogonal to the underlying moving contour. That is another way to manifest the ill-posed nature of optical flow computation. In general, motion analysis from image sequences is an inverse problem, which is ill-posed. Fortunately, low-level computational vision problems are only mildly ill-posed. Hence, lowering the noise in image data leads to a possible significant reduction of errors in flow determination.

Numerous flow determination algorithms have appeared over the course of more than one decade. Most of the techniques take one of the following approaches: the gradient-based approach, the correlation-based approach, the energy-based approach, and the phase-based approach. In addition to these deterministic approaches, there is also a stochastic approach. A unification point of view of optical flow computation is presented in Section 13.3. That is, for any algorithm in optical flow computation, there are two types of information that need to be extracted—conservation information and neighborhood information.

Several techniques are introduced for the gradient-based approach, particularly the Horn and Schunck algorithm, which is a pioneer work in flow determination. There, the brightness invariant equation is used to extract conservation information; the smoothness constraint is used to extract neighborhood information. The modified Horn and Schunck algorithm shows significant error reduction in flow determination, owing to a reduction of noise in image data, which confirms the mildly ill-posed nature of optical flow computation.

Several techniques are discussed for the correlation-based approach. The Singh algorithm is emphasised due to its estimation–theoretical framework. The Pan, Shi, and Shu algorithm that applies the feedback technique to the correlation method, demonstrates an accuracy enhancement in flow estimation.

Section 13.4 addresses the usage of multiple image attributes versus that of a single image attribute in the flow determination technique. It is found that the usage of multiple motion insensitive attributes can help reduce the ambiguity in motion analysis. The application of multiple image attributes to conservation information turns out to be promising for flow computation.

Some experimental works were presented in Sections 13.3 and 13.4. With Barron et al.'s recent comprehensive survey of various existing optical flow algorithms, we can have a quantitative assessment on various optical flow techniques.

FIGURE 13.20
(a) The 21st original frame of the "Miss America" sequence, (b) the reconstructed 21st frame with H.263, and (c) the reconstructed 21st frame with the proposed technique.

Optical Flow

Optical flow finds application in areas such as computer vision, image interpolation, temporal filtering, and video coding. In computational vision, raising the accuracy of optical flow estimation is important. In video coding, however, lowering the bit rate for both prediction error and motion overhead, while keeping certain quality of reconstructed frames is the ultimate goal. Proper handling of the large amount of velocity vectors is a key issue in this regard. It is noted that the optical flow-based motion estimation for video compression has been applied for many years. However, the high bit overhead and computational complexity prevent it from practical usage in video coding. With the continued advance in technologies, however, we believe this problem may be resolved in the near future. In fact, an initial, successful attempt has been made and reported in [shi 1998]. There, based on a study which demonstrates that flow vectors are highly correlated and can be modeled by a first-order autoregressive (AR) model, the DCT is applied to flow vectors. An adaptive threshold technique is developed to match optical flow motion prediction and minimize the residual errors. Consequently, this optical flow-based motion compensated video coding algorithm achieves good performance for very low bit rate video coding. It obtains a bit rate compatible with that obtained by an H.263 standard algorithm, which uses block matching for motion estimation. (Note that the video coding standard H.263 is covered in Chapter 19.) Furthermore, the reconstructed video frames by using this flow-based algorithm are free of annoying blocking artifacts. This effect is demonstrated in Figure 13.20. Note that Figure 13.20b has appeared in Figure 11.12, where the same picture is displayed in a larger size and the blocking artifacts are hence clearer.

Exercises

1. What is an optical flow field? What is a 2-D motion field? What is the difference between the two? How are they related to each other?
2. What is an aperture problem? Give two of your own examples.
3. What is the ill-posed problem? Why do we consider motion analysis from image sequences an ill-posed problem?
4. Is the relationship between the optical flow in an image plane and the velocities of objects in 3-D world space necessarily obvious? Justify your answer.
5. What does the smoothness constraint imply? Why is it required?
6. How are the derivatives of intensity function and the Laplacian of flow components estimated in the Horn and Schunck method?
7. What are the differences between the Horn and Schunck original method and the modified Horn and Schunck method? What do you observe from these differences?
8. What is the difference between the smoothness constraint proposed by Horn and Schunck and the oriented smoothness constraint proposed by Nagel? Provide comments.
9. In your own words, describe the Singh method. What is the weighted-least-square estimation technique?
10. In your own words, describe conservation information and neighborhood information. Using this perspective, take a new look at the Horn and Schunck algorithm.
11. How is the feedback technique applied in the Pan et al. algorithm?
12. In your own words, tell the difference between the Singh method and the Pan et al. method.
13. Give two of your own examples to show that multiple image attributes are able to reduce ambiguity in image matching.

14. How does the Xia et al. method differ from the Weng et al. method?
15. How does the Xia et al. method differ from the Pan et al. method?

References

[adelson 1985] E.H. Adelson and J.R. Bergen, Spatiotemporal energy model for the perception of motion, *Journal of the Optical Society of America A*, 2, 2, 284–299, 1985.

[anandan 1987] P. Anandan, Measurement visual motion from image sequences, Ph.D. thesis, COINS Department, University of Massachusetts, Amherst, MA, 1987.

[anandan 1989] P. Anandan, A computational framework and an algorithm for the measurement of visual motion, *International Journal of Computer Vision*, 2, 283–310, 1989.

[barron 1994] J.L. Barron, D.J. Fleet, and S.S. Beauchemin, Systems and experiment performance of optical flow techniques, *International Journal of Computer Vision*, 12, 1, 43–77, 1994.

[beck 1977] J.V. Beck and K.J. Arnold, *Parameter. Estimation in Engineering and Science*, John Wiley & Sons, New York, 1977.

[bertero 1988] M. Bertero, T.A. Poggio, and V. Torre, Ill-posed problems in early vision, *Proceedings of the IEEE*, 76, 8, 869–889, August 1988.

[bigun 1991] J. Bigun, G. Granlund, and J. Wiklund, Multidimensional orientation estimation with applications to texture analysis and optical flow, *IEEE Transactions on Pattern Analysis and Machine Intelligence*, 13, 775–790, 1991.

[black 1996] M.J. Black and P. Anandan, The robust estimation of multiple motions: Parametric and piecewise-smooth flow fields, *Computer Vision and Image Understanding*, 63, 1, 75–104, 1996.

[bracewell 1995] R.N. Bracewell, *Two-Dimensional Imaging*, Prentice Hall, Englewood, NJ, 1995.

[burt 1983] P.J. Burt and E.H. Adelson, The Laplacian pyramid as a compact image code, *IEEE Transactions on Communications*, 31, 4, 532–540, April 1983.

[burt 1984] P.J. Burt, The pyramid as a structure for efficient computation, in *Multiresolution Image Processing and Analysis*, A. Rosenfeld (Ed.), Springer-Verlag, Germany, pp. 6–37, 1984.

[gonzalez 1992] R.C. Gonzalez and R.E. Woods, *Digital Image Processing*, Addison-Wesley, Reading, MA, 1992.

[fleet 1990] D.J. Fleet and A.D. Jepson, Computation of component image velocity from local phase information, *International Journal of Computer Vision*, 5, 77–104, 1990.

[haralick 1979] R.M. Haralick, Statistical and structural approaches to texture, *Proceedings of the IEEE*, 67, 5, 786–804, May 1979.

[heeger 1988] D.J. Heeger, Optical flow using spatiotemporal filters, *International Journal of Computer Vision*, 1, 279–302, 1988.

[horn 1981] B.K.P. Horn and B.G. Schunck, Determining optical flow, *Artificial Intelligence*, 17, 185–203, 1981.

[konrad 1992] Stochastic approach J. Konrad and E. Dubois, Bayesian estimation of motion vector fields, *IEEE Transactions on Pattern Analysis and Machine Intelligence*, 14, 9, 910–927, 1992.

[lim 1990] J.S. Lim, *Two-Dimensional Signal and Image Processing*, Prentice Hall, Englewood Cliffs, NJ, 1990.

[lucas 1981] B. Lucas and T. Kanade, An iterative image registration technique with an application to stereo vision, *Proceedings of DARPA Image Understanding Workshop*, pp. 121–130, 1981.

[marr 1982] D. Marr, *Vision*, Freeman, Boston, MA, 1982.

[nagel 1983] H.H. Nagel, Displacement vectors derived from second-order intensity variations in image sequences, *Computer Graphics and Image Processing*, 21, 85–117, 1983.

[nagel 1986] H.H. Nagel and W. Enkelmann, An investigation of smoothness constraints for the estimation of displacement vector fields from image sequences, *IEEE Transactions on Pattern Analysis and Machine Intelligence*, 8, 565–593, 1986.

[nagel 1989] H.H. Nagel, On a constraint equation for the estimation of displacement rates in image sequences, *IEEE Transactions on Pattern Analysis and Machine Intelligence*, 11, 13–30, 1989.

[pan 1994] J.N. Pan, Motion estimation using optical flow field, Ph.D. dissertation, Electrical and Computer Engineering, New Jersey Institute of Technology, Newark, NJ, April 1994.

[pan 1995] J.N. Pan, Y.Q. Shi, and C.Q. Shu, A convergence justification of the correlation-feedback algorithm in optical flow determination, *Technical Report, Electronic Imaging Laboratory*, Electrical and Computer Engineering Department, New Jersey Institute of Technology, Newark, NJ, May 1995.

[pan 1998] J.N. Pan, Y.Q. Shi, and C.Q. Shu, Correlation-feedback technique in optical flow determination, *IEEE Transactions on Image Processing*, 7, 7, 1061–1067, July 1998.

[ralston 1978] A. Ralston and P. Rabinowitz, *A First Course in Numerical Analysis*, McGraw-Hill, New York, 1978.

[sears 1986] F.W. Sears, M.W. Zemansky, and H.D. Young, *University Physics*, Addison-Wesley, Readings, MA, 1986.

[shi 1994] Y.Q. Shi, C.Q. Shu, and J.N. Pan, Unified optical flow field approach to motion analysis from a sequence of stereo images, *Pattern Recognition*, 27, 12, 1577–1590, 1994.

[shi 1998] Y.Q. Shi, S. Lin, and Y.Q. Zhang, Optical flow-based motion compensation algorithm for very low-bit-rate video coding, *International Journal of Imaging Systems and Technology*, 9, 4, 230–237, 1998.

[shu 1993] C.Q. Shu and Y.Q. Shi, Direct recovering of Nth order surface structure using UOFF approach, *Pattern Recognition*, 26, 8, 1137–1148, 1993.

[singh 1991] A. Singh, *Optical Flow Computation: A Unified Perspective*, IEEE Computer Society Press, Los Alamitos, CA, 1991.

[singh 1992] A. Singh, An estimation-theoretic framework for image-flow computation, *CVGIP: Image Understanding*, 56, 2, 152–177, 1992.

[szeliski 1995] R. Szeliski, S.B. Kang, and H.-Y. Shum, A parallel feature tracker for extended image sequences, *Proceedings of the International Symposium on Computer Vision*, pp. 241–246, Florida, November 1995.

[tikhonov 1977] A.N. Tikhonov and V.Y. Arsenin, *Solutions of Ill-posed Problems*, Winston & Sons, Washington, DC, 1977.

[uras 1988] S. Uras, F. Girosi, A. Verri, and V. Torre, A computational approach to motion perception, *Biological Cybernetics*, 60, 79–97, 1988.

[waxman 1988] A.M. Waxman, J. Wu, and F. Bergholm, Convected activation profiles and receptive fields for real time measurement of short range visual motion, *Proceedings of IEEE Computer Vision and Pattern Recognition*, pp. 717–723, Ann Arbor, 1988.

[weng 1992] J. Weng, N. Ahuja, and T.S. Huang, Matching two perspective views, *IEEE Transactions on PAMI*, 14, 8, 806–825, August 1992.

[xia 1995] X. Xia and Y.Q. Shi, A multiple attributes algorithm to compute optical flow, *Proceedings of the Twenty-ninth Annual Conference on Information Sciences and Systems*, p. 480, The John Hopkins University, Baltimore, MD, March 1995.

[xia 1996] X. Xia, Motion estimation and video coding, Ph.D. dissertation, Electrical and Computer Engineering, New Jersey Institute of Technology, Newark, NJ, October, 1996.

14

Further Discussion and Summary on 2-D Motion Estimation

Since Chapter 10, we have been devoting our discussion to motion analysis and motion compensated (MC) coding. Following a general description in Chapter 10, three major techniques, block matching, pel recursion, and optical flow, are covered in Chapters 11, 12, and 13, respectively.

Before concluding this subject, in this chapter, we provide further discussion and a summary. A general characterization of 2-D motion estimation is given in Section 14.1. In Section 14.2, different classifications of various methods for 2-D motion analysis are given in a wider scope. Section 14.3 is concerned with a performance comparison among the three major techniques. More advanced techniques and new trends in motion analysis and motion compensation are introduced in Section 14.4.

14.1 General Characterization

A few common features characterizing all three major techniques are discussed in this section.

14.1.1 Aperture Problem

The aperture problem, discussed in Chapter 13, describes phenomena that occur when observing motion through a small opening in a flat screen. That is, one can only observe normal velocity. It is essentially a form of ill-posed problem since it is concerned with existence and uniqueness issues, as illustrated in Figure 13.2a and b. This problem is inherent with the optical flow technique.

We note, however, that the aperture problem also exists in block matching and pel recursive techniques. Consider an area in an image plane having strong intensity gradients. According to our discussion in Chapter 13, the aperture problem does exist in this area no matter what type of technique is applied to determine local motion. That is, motion perpendicular to the gradient cannot be determined as long as only a local measure is utilized. It is noted that, in fact, the steepest descent method of the pel recursive technique only updates the estimate along the gradient direction [tekalp 1995].

14.1.2 Ill-Posed Inverse Problem

In Chapter 13, when we discussed the optical flow technique, a few fundamental issues were raised. It is stated that optical flow computation from image sequences is an inverse problem, which is usually ill-posed. Specifically, there are three problems: nonexistence, nonuniqueness, and instability. That is, the solution may not exist; if it exists, it may not be unique; the solution may not be stable in the sense that a small perturbation in the image data may cause a huge error in the solution.

Now we can extend our discussion to both block matching and pel recursion techniques. This is because both the techniques are intended for determining 2-D motion from image sequences, and are therefore inverse problems.

14.1.3 Conservation Information and Neighborhood Information

Because of the ill-posed nature of 2-D motion estimation, a unified point of view regarding various optical flow algorithms is also applicable for block matching and pel recursive techniques. That is, all three major techniques involve extracting conservation information and extracting neighborhood information.

Take a look at the block matching technique. There, conservation information is a distribution of some sort of features (usually intensity or functions of intensity) within blocks. Neighborhood information manifests itself in that all pixels within a block share the same displacement. If the latter constraint is not imposed, block matching cannot work. One example is the following extreme case. Consider a block size of 1×1, i.e., a block containing only a single pixel. It is well known that there is no way to estimate the motion of a pixel whose movement is independent of all its neighbors [horn 1981].

With the pel recursive technique, say, the steepest descent method, conservation information is the intensity of the pixel for which the displacement vector is to be estimated. Neighborhood information manifests itself as recursively propagating displacement estimates to neighboring pixels (spatially or temporally) as initial estimates.

In Section 12.3, it is pointed out that Netravali and Robbins suggested an alternative, called "inclusion of a neighborhood area." That is, to make displacement estimation more robust, they consider a small neighborhood Ω of the pixel for evaluating the square of displaced frame difference (DFD) in calculating the update term. They assume a constant displacement vector within the area. The algorithm thus becomes

$$\vec{d}^{k+1} = \vec{d}^k - \frac{1}{2}\alpha \nabla_{\vec{d}} \sum_{i,x,y \in \Omega} w_i \text{DFD}^2(x,y;\vec{d}^k) \tag{14.1}$$

where
  i represents an index for the ith pixel (x, y) within Ω
  w_i is the weight for the ith pixel in Ω

All the weights satisfy certain conditions; i.e., they are nonnegative, and their sum equals 1. Obviously, in this more advanced algorithm, the conservation information is the intensity distribution within the neighborhood of the pixel, the neighborhood information is imposed more explicitly, and it is stronger than that in the steepest descent method.

14.1.4 Occlusion and Disocclusion

The problems of occlusion and disocclusion make motion estimation more difficult and hence more challenging. Here we give a brief description about these and other related concepts.

Let us consider Figure 14.1. There, the rectangle ABCD represents an object in an image taken at the moment of t_{n-1}, $f(x, y, t_{n-1})$. The rectangle EFGH denotes the same object, which has been translated, in the image taken at t_n moment, $f(x, y, t_n)$. In the image $f(x, y, t_n)$,

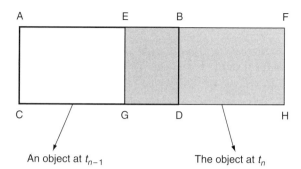

FIGURE 14.1
Occlusion and disocclusion.

the area BFDH is occluded by the object that newly moves in. On the other hand, in $f(x, y, t_n)$, the area of AECG resurfaces and is referred to as a newly visible area, or a newly exposed area.

Clearly, when occlusion and disocclusion occur, all three major techniques discussed in this part will encounter a fatal problem, since conservation information may be lost making motion estimation fail in the newly exposed areas. If image frames are taken densely enough along the temporal dimension, however, occlusion and disocclusion may not cause serious problems, since the failure in motion estimation may be restricted to some limited areas. An extra bit rate paid for the corresponding increase in encoding prediction error is another way to resolve the problem. If high quality and low bit rate are both desired, then some special measures have to be taken.

One of the techniques suitable for handling the situation is Kalman filtering, which is known as the best, by almost any reasonable criterion, working in the Gaussian white noise case [brown 1992]. If we consider the system that estimates the 2-D motion to be contaminated by the Gaussian white noise, we can use Kalman filtering to increase the accuracy of motion estimation, particularly along motion discontinuities. It is powerful in doing incremental, dynamic, and real-time estimation.

In estimating 3-D motion, the Kalman filtering was applied in [matthies 1989; pan 1994a]. Kalman filters were also utilized in optical flow computation [singh 1991; pan 1994b]. In using Kalman filter technique, the question of how to handle the newly exposed areas was raised in [matthies 1989]. In [pan 1994a], one way to handle this issue was proposed, and some experimental work demonstrated its effectiveness.

14.1.5 Rigid and Nonrigid Motion

There are two types of motion: rigid motion and nonrigid motion. Rigid motion refers to motion of rigid objects. It is known that our human vision system is capable of perceiving 2-D projections of 3-D moving rigid bodies as 2-D moving rigid bodies. Most cases in computer vision are concerned with rigid motion. Perhaps this is due to the fact that most applications in computer vision fall into this category. On the other hand, rigid motion is easier to handle than nonrigid motion. This can be seen in the following discussion.

Consider a point P in 3-D world space with the coordinates (X, Y, Z) that can be represented by a column vector $\vec{v}$:

$$\vec{v} = (X, Y, Z)^T \tag{14.2}$$

Rigid motion involves rotation and translation and has six free motion parameters. Let R denote the rotation matrix and T the translational vector. The coordinates of point P in the 3-D world after the rigid motion are denoted by $\vec{v}'$. Then we have

$$\vec{v}' = R\vec{v} + T \tag{14.3}$$

Nonrigid motion is more complicated. It involves deformation in addition to rotation and translation, and thus cannot be characterized by Equation 14.3. According to the Helmholtz theory [sommerfeld 1950], the counterpart of Equation 14.3 becomes

$$\vec{v}' = R\vec{v} + T + D\vec{v} \tag{14.4}$$

where D is a deformation matrix. Note that R, T, and D are pixel dependent. Handling nonrigid motion, hence, is very complicated.

In videophony and videoconferencing applications, a typical scene might be a head and shoulder view of a person imposed on a background. The facial expression is nonrigid in nature. Model-based facial coding has been studied extensively [aizawa 1989, 1995; li 1993]. There, a 3-D wire frame model is used for handling rigid head motion. In [li 1993], the facial nonrigid motion is analyzed as a weighted linear combination of a set of *action units*, instead of determining $D\vec{v}$ directly. Since the number of action units is limited, the computation becomes less expensive. In [aizawa 1989], the portions in the human face with rich expression, such as lips, are *cut* and then transmitted out. At the receiving end, the portions are *pasted* back in the face.

Among the three types of techniques, block matching may be used to manage rigid motion, while pel recursive and optical flow may be used to handle either rigid or nonrigid motion.

14.2 Different Classifications

There are various methods in motion estimation. They can be classified in many different ways. We will discuss some of the classifications here.

14.2.1 Deterministic Methods versus Stochastic Methods

Most algorithms are deterministic in nature. To see this, let us take a look at the most prominent algorithm for each of the three major 2-D motion estimation techniques. That is, the Jain and Jain algorithm for the block matching technique [jain 1981]; the Netravali and Robbins algorithm for the pel recursive technique [netravali 1979]; and the Horn and Schuck algorithm for the optical flow technique [horn 1981]. All are deterministic methods. There are also stochastic methods in 2-D motion estimation, such as the Konrad and Dubois algorithm [konrad 1992] that estimates 2-D motion using the maximum *a posteriori* probability (MAP).

14.2.2 Spatial Domain Methods versus Frequency Domain Methods

While most techniques in 2-D motion analysis are spatial domain methods, there are also frequency domain methods [kughlin 1975; heeger 1988; porat 1990; girod 1993; kojima 1993; koc 1998]. In [heeger 1988], a method to determine optical flow in the frequency domain, which is based on spatiotemporal filters, was developed. The basic idea and principle of the method is introduced in this section. A very new and effective frequency method for 2-D motion analysis [koc 1998] is presented in Section 14.4, where we discuss new trends in 2-D motion estimation.

14.2.2.1 Optical Flow Determination Using Gabor Energy Filters

The frequency domain method of optical flow computation developed by Heeger is suitable for highly textured image sequences. First let us take a look at how motion can be detected in the frequency domain.

14.2.2.1.1 Motion in the Spatiotemporal Frequency Domain

We begin our discussion with a one-dimensional (1-D) case. The spatial frequency of a (translationally) moving sinusoidal signal, ω_x, is defined as cycles per distance (usually cycles per pixel), while temporal frequency, ω_t, is defined as cycles per time unit (usually cycles per frame). Hence, the velocity of (translational) motion, defined as distance per time unit (usually pixels per frame), can be related to the spatial and temporal frequencies as follows:

$$v = \omega_t/\omega_x \quad (14.5)$$

A 1-D moving signal with a velocity v may have multiple spatial frequency components. Each spatial frequency component ω_{xi}, $i = 1, 2, \ldots$ has a corresponding temporal frequency component ω_{ti} such that

$$\omega_{ti} = v\omega_{xi}. \quad (14.6)$$

This relation is shown in Figure 14.2. Thus, we see that in the spatiotemporal frequency domain, velocity is the slope of a straight line relating temporal and spatial frequencies. For 2-D moving signals, we denote spatial frequencies by ω_x and ω_y, and velocity vector by $\bar{v} = (v_x, v_y)$ The above 1-D result can be extended in a straightforward manner as follows:

$$\omega_t = v_x\omega_x + v_y\omega_y \quad (14.7)$$

The interpretation of Equation 14.7 is that a 2-D translating texture pattern occupies a plane in the spatiotemporal frequency domain.

14.2.2.1.2 Gabor Energy Filters

As Adelson and Berger pointed out, the translational motion of image patterns is characterized by orientation in the spatiotemporal domain [adelson 1985] as shown in Figure 14.3. Therefore, motion can be detected by using spatiotemporally oriented filters. One of this type of filter, suggested by Heeger, is the Gabor filter.

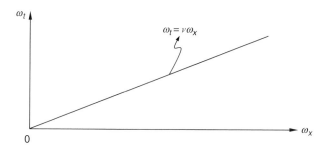

FIGURE 14.2
Velocity in 1-D spatiotemporal frequency domain.

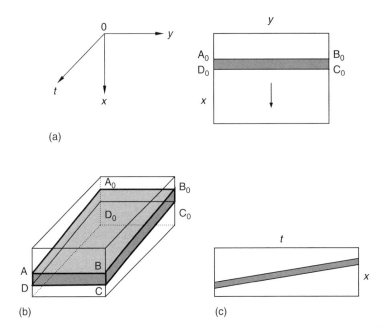

FIGURE 14.3
Orientation in spatiotemporal domain. (a) A horizontal bar translating downwards. (b) A spatiotemporal cube. (c) A slice of the cube perpendicular to y axis. The orientation of the slant edges represents the motion.

A 1-D sine phase Gabor filter is defined as follows:

$$g(t) = \frac{1}{\sqrt{2\pi}\sigma} \sin(2\pi\omega t) \exp\left\{-\frac{t^2}{2\sigma^2}\right\} \quad (14.8)$$

Obviously, this is a product of a sine function and a Gaussian probability density function (pdf). In the frequency domain, this is the convolution between a pair of impulses located in ω and $-\omega$, and the Fourier transform of the Gaussian, which is itself again a Gaussian function. Hence, the Gabor function is localized in a pair of Gaussian windows in the frequency domain. This means that the Gabor filter is able to selectively pick up some frequency components.

A 3-D sine Gabor function is

$$g(x,y,t) = \frac{1}{\sqrt{2\pi^{\frac{3}{2}}}\sigma_x \sigma_y \sigma_t} \cdot \exp\left\{-\frac{1}{2}\left(\frac{x^2}{\sigma_x^2} + \frac{y^2}{\sigma_y^2} + \frac{t^2}{\sigma_t^2}\right)\right\}$$
$$\cdot \sin[2\pi(\omega_{x_0} x + \omega_{y_0} y + \omega_{t_0} t)] \quad (14.9)$$

where σ_x, σ_y, and σ_t are, respectively, the spreads of the Gaussian window along the spatiotemporal dimensions and ω_{x_0}, ω_{y_0}, and ω_{t_0} are, respectively, the central spatiotemporal frequencies. The actual Gabor energy filter used by Heeger is the sum of a sine-phase filter (which is defined above), and a cosine-phase filter (which shares the same spreads and central frequencies as that in the sine-phase filter, and replaces sine by cosine in Equation 14.9). Its frequency response, therefore, is as follows:

$$G(\omega_x, \omega_y, \omega_t) = \frac{1}{4} \exp\left\{-4\pi^2 \left[\sigma_x^2(\omega_x - \omega_{x_0})^2 + \sigma_y^2(\omega_y - \omega_{y_0})^2 \right.\right.$$
$$\left.\left. + \sigma_t^2(\omega_t - \omega_{t_0})^2\right]\right\} + \frac{1}{4} \exp\left\{-4\pi^2 \left[\sigma_x^2(\omega_x + \omega_{x_0})^2 \right.\right.$$
$$\left.\left. + \sigma_y^2(\omega_y + \omega_{y_0})^2 + \sigma_t^2(\omega_t + \omega_{t_0})^2\right]\right\} \tag{14.10}$$

This indicates that the Gabor filter is motion-sensitive in that it responds largely to motion that has more power distributed near the central frequencies in the spatiotemporal frequency domain, while it responds poorly to motion that has little power near the central frequencies.

14.2.2.1.3 Flow Extraction with Motion Energy

Using a vivid example, Heeger explains in his paper why one such filter is not sufficient in detection of motion. Multiple Gabor filters must be used. In fact, a set of 12 Gabor filters are utilized in Heeger's algorithm. The 12 Gabor filters in the set have one thing in common:

$$\omega_0 = \sqrt{\omega_{x_0}^2 + \omega_{y_0}^2} \tag{14.11}$$

In other words, the 12 filters are tuned to the same spatial frequency band but to different spatial orientation and temporal frequencies.

Briefly speaking, optical flow is determined as follows. Denote the measured motion energy by n_i, $i = 1, 2, \ldots, 12$. Here i indicates one of the 12 Gabor filters. The summation of all n_i is denoted by

$$\bar{n} = \sum_{i=1}^{12} n_i \tag{14.12}$$

Denote the predicted motion energy by $P_i(v_x, v_y)$, and the sum of predicted motion energy by

$$\bar{P} = \sum_{i=1}^{12} P_i(v_x, v_y) \tag{14.13}$$

Similar to what many algorithms do, optical flow determination is then converted to a minimization problem. That is, optical flow should be able to minimize error between the measured and predicted motion energies.

$$J(v_x, v_y) = \sum_{i=1}^{12} \left[n_i - \bar{n}_i \frac{P_i(v_x, v_y)}{\bar{P}_i(v_x, v_y)}\right]^2 \tag{14.14}$$

Similarly, many readily available numerical methods can be used for solving this minimization problem.

14.2.3 Region-Based Approaches versus Gradient-Based Approaches

As stated in Chapter 10, methodologically speaking, there are generally two approaches to 2-D motion analysis for video coding: region-based and gradient-based approaches. As we have gone through the three major techniques, we can see this classification more clearly.

The region-based approach can be characterized as follows. For a region in an image frame, we find its best match in another image frame. The relative spatial position between these two regions produces a displacement vector. The best matching is found by minimizing a dissimilarity measure between the two regions, which is defined as

$$\sum_{(x,y)\in R} \sum M[f(x,y,t), f(x-dx, y-dy, t-\Delta t)], \quad (14.15)$$

where
R denotes a spatial region, on which the displacement vector $(d_x, d_y)^T$ estimate is based
$M[\alpha, \beta]$ denotes a dissimilarity measure between two arguments α and β
Δt is the time interval between two consecutive frames

Block matching certainly belongs to the region-based approach (here region means a rectangular block). For an original block in a (current) frame, block matching searches for its best match in another (previous) frame among candidates. Several dissimilarity measures are utilized, among which the mean absolute difference (MAD) is used most often.

Although it uses the spatial gradient of intensity function, the pel recursive method with inclusion of a neighborhood area assumes the same displacement vector within a neighborhood region. A weighted sum of the squared DFD within the region is used as a dissimilarity measure. By using numerical methods such as various descent methods, the pel recursive method iteratively minimizes the dissimilarity measure, thus delivering displacement vectors. The pel recursive technique is therefore in the category of region-based approaches.

In optical flow computation, the two most frequently used techniques discussed in Chapter 13 are the gradient method and the correlation method. Clearly, the correlation method is region-based. In fact, as we pointed out in Chapter 13, it is very similar to block matching.

As far as the gradient-based approach is concerned, we start its characterization with the brightness invariant equation, covered in Chapter 13. That is, we assume that brightness is conserved during the time interval between two consecutive image frames.

$$f(x, y, t) = f(x - d_x, y - d_y, t - \Delta t) \quad (14.16)$$

By expanding the right-hand side of Equation 14.16 into the Taylor series, applying Equation 14.16, and some mathematical manipulation, we can derive Equation 14.17.

$$f_x u + f_y v + f_t = 0 \quad (14.17)$$

where f_x, f_y, f_t are partial derivatives of intensity function with respect to x, y, and t, respectively; and u and v are two components of pixel velocity. Equation 14.17 contains gradients of intensity function with respect to spatial and temporal variables and links two components of the displacement vector. The square of the left-hand side in Equation 14.17 is an error that needs to be minimized. Through the minimization, we can estimate displacement vectors.

Clearly, the gradient method in optical flow determination, discussed in Chapter 13, falls into the above framework. There, an extra constraint is imposed and included into the error represented in Equation 14.17.

Table 14.1 summarizes what we discussed in this section.

TABLE 14.1
Region-Based versus Gradient-Based Approaches

	Block Matching	Pel Recursive	Optical Flow	
			Gradient-Based Method	Correlation-Based Method
Regional-based approaches	√	√		√
Gradient-based approaches			√	

14.2.4 Forward versus Backward Motion Estimation

Motion compensated predictive video coding may be done in two different ways: forward and backward [boroczky 1991]. These ways are depicted in Figures 14.4 and 14.5, respectively. With the forward manner, motion estimation is carried out by using the original input video frame and the reconstructed previous input video frame. With the backward manner, motion estimation is implemented with two successive reconstructed input video frames.

Forward manner provides relatively higher accuracy in motion estimation and hence more efficient motion compensation than the backward manner, owing to the fact that the original input video frames are utilized. However, the backward manner does not need to transmit motion vectors to the receiving end as an overhead, while the forward manner does.

Block matching is used in almost all the international video coding standards, such as H.261, H.263, and MPEG 1 and MPEG 2 (covered in the next part of this book), as forward motion estimation. The pel recursive technique is used as backward motion estimation. In this way, the pel recursive technique avoids encoding a large amount of motion vectors. On the other hand, however, it provides relatively less accurate motion estimation than

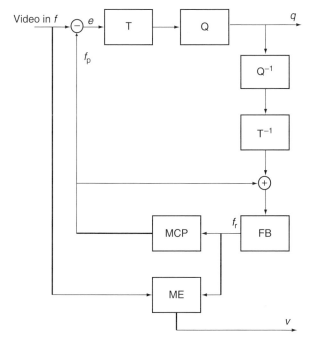

FIGURE 14.4
Forward motion estimation and compensation. T: transformer, Q: quantizer, FB: frame buffer, MCP: motion compensated predictor, ME: motion estimator, e: prediction error, f: input video frame, f_p: predicted video frame, f_r: reconstructed video frame, q: quantized transform coefficients, v: motion vector.

FIGURE 14.5
Backward motion estimation and compensation. T: transformer, Q: quantizer, FB: frame buffer, MCP: motion compensated predictor, ME: motion estimator, e: prediction error, f: input video frame, f_p: predicted video frame, f_{r1}: reconstructed video frame, f_{r2}: reconstructed previous video frame, q: quantized transform coefficients.

block matching. Optical flow is usually used as forward motion estimation in motion compensated video coding. Therefore, as expected, it achieves higher motion estimation accuracy on the one hand and needs to handle a large amount of motion vectors as overhead on the other hand. This will be discussed in Section 14.3.

It is noted that one of the new improvements in the block matching technique is described in Section 11.6.3. It is called the predictive motion field segmentation technique [orchard 1993], and it is motivated by backward motion estimation. There, segmentation is conducted *backwards*, i.e., based on previously decoded frames. The purpose of this is to save overhead for shape information of motion discontinuities.

14.3 Performance Comparison between Three Major Approaches

14.3.1 Three Representatives

A performance comparison between the three major approaches, block matching, pel recursion, and optical flow, was provided in a review paper by Dufaux and Moscheni [dufaux 1995]. Experimental work was carried out as follows. The conventional full search block matching is chosen as a representative for the block matching approach, while the Netravali and Robbins algorithm and the modified Horn and Schunck algorithm are chosen to represent the pel recursion and optical flow approaches, respectively.

14.3.2 Algorithm Parameters

In full search block matching, the block size is chosen as 16×16 pixels, the maximum displacement is ± 15 pixels, and the accuracy is half pixels. In the Netravali and Robbins pel recursion, $\varepsilon = 1/1024$, the update term is averaged in an area of 5×5 pixels and clipped to a maximum of 1/16 pixels per frame, and the algorithm iterates one iteration per pixel. In the modified Horn and Schunck algorithm, the weight α^2 is set to 100, and 100 iterations of the Gauss and Seidel procedure are carried out.

14.3.3 Experimental Results and Observations

The three test video sequences are the "Mobile and Calendar," "Flower Garden," and "Table Tennis." Both subjective criterion (in terms of needle diagrams showing displacement vectors) and objective criteria (in terms of DFD error energy) are applied to access the quality of motion estimation.

It turns out that the pel recursive algorithm gives the worst accuracy in motion estimation. In particular, it cannot follow the fast and large motions. Both block matching and optical flow algorithms give better motion estimation.

It is noted that we must be cautious in drawing conclusions from these tests. This is because different algorithms in the same category and the same algorithm under different implementation conditions will provide quite different performances. In the above experiments, the full search block matching with half-pixel accuracy is one of the better block matching techniques. On the contrary, there are many improved pel recursive and optical flow algorithms, which outperform the chosen representatives in the reported experiments.

The experiments do, however, provide an insight about the three major approaches. Pel recursive algorithms are seldom used in video coding now, mainly due to their inaccurate motion estimation, although they do not require transmitting motion vectors to the receiving end. Although they can provide relatively accurate motion estimation, optical flow algorithms require a large amount of overhead for handling dense motion vectors. This prevents the optical flow techniques from wide and practical usage in video coding. Block matching is simple, yet very efficient for motion estimation. It provides quite accurate and reliable motion estimation for most practical video sequences in spite of its simple piecewise translational model. At the same time it does not require much overhead. Therefore, for first-generation video coding, block matching is considered to be the most suitable among the three approaches.

14.4 New Trends

In Chapters 11, 12, and 13, many new, effective improvements within the three major approaches were discussed. These techniques include multiresolution block matching, (locally adaptive) multigrid block matching, overlapped block matching, thresholding techniques, (predictive) motion field segmentation, feedback and multiple attributes in optical flow computation, sub-pixel accuracy, and so on. Some improvements will be discussed in Part IV, where various international video coding standards such as H.263 and MPEG 2, and 4 are introduced.

As pointed out in [orchard 1998], today our understanding of motion analysis and video compression is still based on an ad hoc framework, in general. What today's standards have achieved is not near the ideally possible performance. Therefore, more efforts are continuously made in this field, seeking much more simple, practical, and efficient algorithms.

As an example of such developments, we conclude this chapter by presenting a novel method for 2-D motion estimation: the discrete cosine transform (DCT)-based motion estimation [koc 1998].

14.4.1 DCT-Based Motion Estimation

As pointed out in Section 14.2.2, as opposed to the conventional 2-D motion estimation techniques, this method is carried out in the frequency domain. It is also different from the Gabor energy filter method by Heeger (Section 14.2.2.1). Without introducing Gabor filters, this method is directly DCT-based. The fundamental concepts and techniques of this method are discussed below.

14.4.1.1 DCT and DST Pseudophases

The underlying idea behind this method is to estimate 2-D translational motion by determining the DCT and DST (discrete sine transform) pseudophases. Let us use the simpler 1-D case to illustrate this concept. Once it is established, it can be easily extended to the 2-D case.

Consider a 1-D signal sequence $\{f(n), n \in (0, 1, \ldots, N-1)\}$ of length N. Its translated version is denoted by $\{g(n), n \in (0, 1, \ldots, N-1)\}$. The translation is defined as follows:

$$g(n) = \begin{cases} f(n-d), & \text{if } (n-d) \in (0, 1, \ldots, N-1) \\ 0, & \text{otherwise} \end{cases} \quad (14.18)$$

In Equation 14.18, d is the amount of the translation and it needs to be estimated. Let us define the following several functions before introducing the pseudophases. The DCT and the DST of the second kind of $g(n)$, $G^C(k)$ and $G^S(k)$, are defined as follows:

$$G^C(k) = \frac{2}{N} C(k) \sum_{n=0}^{N-1} g(n) \cos\left[\frac{k\pi}{N}(n+0.5)\right] \quad k \in \{0, 1, \ldots, N-1\} \quad (14.19)$$

$$G^S(k) = \frac{2}{N} C(k) \sum_{n=0}^{N-1} g(n) \sin\left[\frac{k\pi}{N}(n+0.5)\right] \quad k \in \{1, \ldots, N\} \quad (14.20)$$

The DCT and DST of the first kind of $f(n)$, $F^C(k)$ and $F^S(k)$, are defined as

$$F^C(k) = \frac{2}{N} C(k) \sum_{n=0}^{N-1} f(n) \cos\left[\frac{k\pi}{N}n\right] \quad k \in \{0, 1, \ldots, N-1\} \quad (14.21)$$

$$F^S(k) = \frac{2}{N} C(k) \sum_{n=0}^{N-1} f(n) \sin\left[\frac{k\pi}{N}n\right] \quad k \in \{1, \ldots, N\} \quad (14.22)$$

In the above equations, $C(k)$ is defined as

$$C(k) = \begin{cases} \frac{1}{\sqrt{2}} & \text{for } n = 0 \text{ or } N \\ 1 & \text{otherwise} \end{cases} \quad (14.23)$$

Now we are in a position to introduce Equation 14.24, which relates the translational amount d to the DCT and DST of the original sequence and its translated version, defined above. That is,

$$\begin{bmatrix} G^C(k) \\ G^S(k) \end{bmatrix} = \begin{bmatrix} F^C(k) & -F^S(k) \\ F^S(k) & F^C(k) \end{bmatrix} \begin{bmatrix} D^C(k) \\ D^S(k) \end{bmatrix} \quad (14.24)$$

where $D^C(k)$ and $D^S(k)$ are referred to as the pseudophases and defined as follows:

$$D^C(k) \triangleq \cos\left[\frac{k\pi}{N}\left(d+\frac{1}{2}\right)\right]$$
$$D^S(k) \triangleq \sin\left[\frac{k\pi}{N}\left(d+\frac{1}{2}\right)\right] \quad (14.25)$$

Equation 14.24 can be solved for the amount of translation d, thus motion estimation. This becomes clearer when we rewrite the equation in a matrix–vector format. Denote the 2×2 matrix in Equation 14.24 by $\mathbf{F}(k)$, the 2×1 column vector at the left-hand side of the equation by $\vec{G}(k)$, and the 2×1 column vector at the right-hand side by $\vec{D}(k)$. It is easy to verify that the matrix $\mathbf{F}(k)$ is orthogonal by observing the following:

$$\lambda \mathbf{F}^T(k)\mathbf{F}(k) = \mathbf{I} \quad (14.26)$$

where $\mathbf{I}$ is a 2×2 identity matrix, the constant λ is

$$\lambda = \frac{1}{[\mathbf{F}^C(k)]^2 + [\mathbf{F}^S(k)]^2} \quad (14.27)$$

We then derive the matrix–vector format of Equation 14.24 as follows:

$$\vec{D}(k) = \lambda \mathbf{F}^T(k)\vec{G}(k) \quad k \in \{1,\ldots,N-1\} \quad (14.28)$$

14.4.1.2 Sinusoidal Orthogonal Principle

It was shown that the pseudophases, which contain the translation information, can be determined in the DCT and DST frequency domain. But how the amount of the translation can be found has not been mentioned. Here, the algorithm uses the sinusoidal principle to pick up this information. That is, the inverse DST of the second kind of scaled pseudophase, $C(k)D^S(k)$, is found to equal an algebraic sum of the following two discrete impulses according to the sinusoidal orthogonal principle:

$$ISDT\{C(k)D^S(k)\} \triangleq \frac{2}{N}\sum_{k=1}^{N} C^2(k)D^S(k)\sin\left[\frac{k\pi}{N}\left(n+\frac{1}{2}\right)\right]$$
$$= \delta(d-n) - \delta(d+n+1) \quad (14.29)$$

Since the inverse DST is limited to $n \in \{0, 1, \ldots, N-1\}$, the only peak value among this set of N values indicates the amount of the translation d. Furthermore, the direction of the translation (positive or negative) can be determined from the polarity (positive or negative) of the peak value.

The block diagram of the algorithm is shown in Figure 14.6. This technique can be extended to the 2-D case in a straightforward manner. Interested readers can refer to [koc 1998].

14.4.1.3 Performance Comparison

The algorithm was applied to several typical testing video sequences, such as the "Miss America" and "Flower Garden" sequences, and an "Infrared Car" sequence. The results were compared with the conventional full search block matching technique

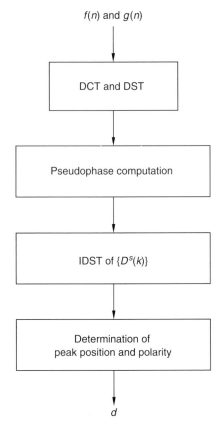

FIGURE 14.6
Block diagram of DCT-based motion estimation (1-D case).

and several fast search block matching techniques such as the 2-D logarithm search, three-step search, search with subsampling in the original block, and the correlation windows.

Before applying the algorithm, one of the following preprocessing procedures is implemented: frame differentiation or edge extraction. It was reported that for the "Flower Garden" and "Infrared Car" sequences, the DCT-based algorithm achieves higher coding efficiency than all three fast search block matching methods, while for the Miss America sequence it obtains a lower efficiency. It was also reported that it performs well even in a noisy situation.

A lower computational complexity, $O(M^2)$ for an $M \times M$ search range, is one of the major advantages possessed by the DCT-based motion estimation algorithm compared with conventional full search block matching, $O(M^2N^2)$ for an $M \times M$ search range and an $N \times N$ block size.

With DCT-based motion estimation, a fully DCT-based motion compensated coder structure becomes possible, which is expected to achieve a higher throughput and a lower system complexity.

14.5 Summary

In this chapter, which concludes the motion analysis and compensation portion of the book, we first generalize the discussion of the aperture problem, the ill-posed nature, and the conservation and neighborhood information unified point of view, previously made with respect to the optical flow technique in Chapter 13, to cover block matching and pel recursive techniques. Then, the occlusion and disocclusion, and rigidity and nonrigidity are discussed with respect to the three techniques. The difficulty of nonrigid motion estimation is analyzed. Its relevance in visual communications is addressed.

Different classifications of various methods in the three major 2-D motion estimation techniques, block matching, pel recursion, and optical flow, are presented. Besides the frequently utilized deterministic methods, spatial domain methods, region-based methods, and forward motion estimation, their counterparts: stochastic methods, frequency domain methods, gradient methods, and backward motion estimation are introduced. In particular, two frequency domain methods are presented with some detail. They are the method using the Gabor energy filter and the DCT-based method.

A performance comparison between the three techniques is also introduced here, based on which observations are drawn. A main point is that block matching is at present the most suitable technique for 2-D motion estimation among the three techniques.

Exercises

1. What is the difference between the rigid motion and nonrigid motion? In facial encoding, what is the nonrigid motion? How is the nonrigid motion handled?
2. How is 2-D motion estimation carried out in the frequency domain? What are the underlying ideas behind the Heeger method, and the Koc and Liu method?
3. Why is one Gabor energy filter not sufficient in motion estimation? Draw the power spectrum of a 2-D sine-phase Gabor function.
4. Show the correspondence of a positive (negative) peak value in the inverse DST of the second kind of DST pseudophase to a positive (negative) translation in the 1-D spatial domain.
5. How does neighborhood information manifest itself in the pel recursive technique?
6. Using your own words and some diagrams, state that the translational motion of an image pattern is characterized by orientation in the spatiotemporal domain.

References

[adelson 1985] E.H. Adelson and J.R. Bergen, Spatiotemporal energy models for the perception of motion, *Journal of Optical Society of America*, A2, 2, 284–299, 1985.

[aizawa 1989] K. Aizawa and H. Harashima, Model-based analysis synthesis image coding (MBASIC) system for a person's face, *Signal Processing: Image Communications*, 139–152, 1989.

[aizawa 1995] K. Aizawa and T.S. Huang, Model-based image coding: Advanced video coding techniques for very low bit rate applications, *Proceedings of the IEEE*, 83, 2, 259–271, February 1995.

[boroczky 1991] L. Boroczky, Pel recursive motion estimation for image coding, Ph.D. dissertation, Delft University of Technology, The Netherlands, 1991.

[brown 1992] R.G. Brown and P.Y.C. Hwang, *Introduction to Random Signals*, 2nd edn., John Wiley & Sons, New York, 1992.

[dufaux 1995] F. Dufaux and F. Moscheni, Motion estimation techniques for digital TV: A review and a new contribution, *Proceedings of the IEEE*, 83, 6, 858–876, 1995.

[girod 1993] B. Girod, Motion-compensating prediction with fractional-pel accuracy, *IEEE Transactions on Communications*, 41, 604, 1993.

[**heeger 1988**] D.J. Heeger, Optical flow using spatiotemporal filters, *International Journal of Computer Vision*, 1, 279–302, 1988.

[**horn 1981**] B.K.P. Horn and B.G. Schunck, Determining optical flow, *Artificial Intelligence*, 17, 185–203, 1981.

[**jain 1981**] J.R. Jain and A.K. Jain, Displacement measurement and its application in interframe image coding, *IEEE Transactions on Communications*, COM-29, 12, 1799–1808, December 1981.

[**koc 1998**] U.-V. Koc and K.J.R. Liu, DCT-based motion estimation, *IEEE Transactions on Image Processing*, 7, 7, 948–865, July 1998.

[**kojima 1993**] A. Kojima, N. Sakurai, and J. Kishigami, Motion detection Using 3D FFT Spectrum, *Proceedings of International Conference on Acoustics, Speech, and Signal Processing*, V, 213–216, April 1993.

[**konrad 1992**] J. Konrad and E. Dubois, Bayesian estimation of motion vector fields, *IEEE Transactions on Pattern Analysis and Machine Intelligence*, 14, 9, 910–927, 1992.

[**kughlin 1975**] C.D. Kughlin and D.C. Hines, The phase correlation image alignment method, Proceedings of 1975 IEEE International Conference on Systems, Man, and Cybernetics, 163–165, 1975.

[**li 1993**] H. Li, P. Roivainen, and R. Forchheimer, 3-D motion estimation in model-based facial image coding, *IEEE Transactions on Pattern Analysis and Machine Intelligence*, 6, 545–555, 1993.

[**matthies 1989**] L. Matthies, T. Kanade, and R. Szeliski, Kalman filter-based algorithms for estimating depth from image sequences, *International Journal of Computer Vision*, 3, 209–236 (1989).

[**netravali 1979**] A.N. Netravali and J.D. Robbins, Motion compensated television coding: Part I, *The Bell System Technical Journal*, 58, 3, 631–670, March 1979.

[**orchard 1993**] M.T. Orchard, Predictive motion-field segmentation for image sequence coding, *IEEE Transactions on Circuits and Systems for Video Technology*, 3, 1, 54–69, February 1993.

[**orchard 1998**] M.T. Orchard, Visual coding standards: A research community's midlife crisis? *IEEE Signal Processing Magazine*, 43, March 1998.

[**pan 1994a**] J.N. Pan, Y.Q. Shi, and C.Q. Shu, A Kalman filter in motion analysis from stereo image sequences, Proceedings of IEEE 1994 International Conference on Image Processing, Vol. 3, Austin, TX, pp. 63–67, November 1994.

[**pan 1994b**] J.N. Pan and Y.Q. Shi, A Kalman filter for improving optical flow accuracy along moving boundaries, Proceedings of SPIE 1994 Visual Communication and Image Processing, Vol. 1, Chicago, IL, pp. 638–649, September 1994.

[**porat 1990**] B. Porat and B. Friedlander, A frequency domain algorithm for multiframe detection and estimation of dim targets, *IEEE Transactions on Pattern Recognition and Machine Intelligence*, 12, 398–401, April 1990.

[**singh 1991**] A. Singh, Incremental estimation of image-flow using a Kalman filter, Proceedings of 1991 IEEE Workshop on Visual Motion, Princeton, NJ, 36–43, 1991.

[**sommerfeld 1950**] A. Sommerfeld, *Mechanics of Deformable Bodies*, Academic Press, 1950.

[**tekalp 1995**] A.M. Tekalp, *Digital Video Processing*, Prentice-Hall PTR, Upper Saddle River, NJ, 1995.

Part IV

Video Compression

15
Fundamentals of Digital Video Coding

In this chapter, we introduce the fundamentals of digital video coding which include digital video representation, rate distortion theory, and digital video formats. We also give a brief overview of image and video coding standards, which will be discussed in the subsequent chapters.

15.1 Digital Video Representation

As we discussed in previous chapters, a digital image is obtained by quantizing a continuous image both spatially and in amplitude. Digitization of the spatial coordinates is called image sampling, whereas digitization of the amplitude is called gray-level quantization. Suppose that a continuous image is denoted by $g(x, y)$, where the amplitude or value of g at the point (x, y) is the intensity or brightness of an image at that point. The transformation of a continuous image to a digital image can then be expressed as

$$f(m, n) = Q[g(x_0 + m\Delta x, y_0 + n\Delta y)] \tag{15.1}$$

where
  Q is a quantization operator
  x_0 and y_0 are the origins of image plane
  m and n are the discrete values $0, 1, 2, \ldots$, Δx and Δy are the sampling intervals in the horizontal and vertical directions, respectively

If the sampling process is extended to a third temporal direction (or the original signal in the temporal direction is discrete format), a sequence, $f(m, n, t)$, is obtained as introduced in Chapter 10,

$$f(m, n, t) = Q[g(x_0 + m\Delta x, y_0 + n\Delta y, t_0 + t\Delta t)] \tag{15.2}$$

where
  t takes the values $0, 1, 2, \ldots$
  Δt is the time interval

Each point of the image or each basic element of the image is called as a pixel or pel. Each individual image is called a frame. According to the sampling theory, the original continuous signal can be recovered exactly from its samples if the sampling frequency is two times higher than the bandwidth of original signal [oppenheim 1989]. The frames are normally presented at a regular time interval so that the eye can perceive fluid motion. For example, the NTSC (National Television Systems Committee) specified a temporal sampling rate of 30 frames/s and interlace 2 to 1. Therefore, as a result of this spatiotemporal sampling, the digital signals exhibit high spatial and temporal correlation, just as the analog signals did before video data compression. In Section 15.2, we will discuss the theoretical basis of video digitization. An important notion is the strong dependence between values of neighboring pixels within the same frame, and between the frames themselves; this can be regarded as statistical redundancy of the image sequence. In the following section, we explain how this statistical redundancy is exploited to achieve compression of the digitized image sequence.

15.2 Information Theory Results: Rate Distortion Function of Video Signal

The principal goal in the design of a video coding system is to reduce the transmission rate requirements of the video source subject to some picture quality constraint. There are only two ways to accomplish this goal: reduction of the statistical redundancy and psychophysical redundancy of the video source. The video source is normally very highly correlated, both spatially and temporally; that is, strong dependence can be regarded as statistical redundancy of the data source. If the video source to be coded in a transmission system is viewed by a human observer, the perceptual limitations of human vision can be used to reduce transmission requirements. Human observers are subject to perceptual limitations in amplitude, spatial resolution, and temporal acuity. By proper design of the coding system, it is possible to discard information without affecting perception, or at least, with only minimal degradation. In summary, we can use two factors: the statistical structure of the data source and the fidelity requirements of the end user, which make the compression possible. The performance of the video compression algorithm depends on several factors. First, and also a basic one, is the amount of redundancy contained in the video data source. In other words, if the original source contains a large amount of information, or high complexity then more bits are needed to represent the compressed data. Second, if a lossy coding technique is used, by which some amount of loss is permitted in the reconstructed video data then the performance of the coding technique depends on the compression algorithm and distortion measurements. In lossy coding, different distortion measurements will perceive the loss in different ways, giving different subjective results. The development of a distortion measure that can provide consistent numerical and subjective results is a very difficulty task. Moreover, the majority of the video compression applications do not require lossless coding, i.e., it is not required that the reconstructed and original images be identical or reversible.

This intuitive explanation of how redundancy and lossy coding methods can be used to reduce source data is made more precise by Shannon's rate distortion theory [berger 1971], which addresses the problem of how to characterize both the source and the distortion measure. Let us consider the model of a typical visual communication system depicted in Figure 15.1. The source video data is fed to the encoder system, which consists of two parts: source coding and channel coding. The function of the source coding is to remove the redundancy in both the spatial and temporal domain, whereas the function of channel coding is to insert the controlled redundancy, which is used to protect the transmitted data from the interference of channel noise. It should be noted that according to Shannon [shannon 1948] certain conditions allow the source and channel coding operations to be separated without any loss of optimality, such as when the sources are ergodic. However, Shannon did not indicate the complexity constraint on the coder involved. In practical systems, which are limited by the complexity, this separation may not be possible

FIGURE 15.1
A typical visual communication system.

[viterbi 1979]. There is still some work on the joint optimization of the source and channel coding [modestino 1981; sayood 1991]. Coming back to rate distortion theory, the problem addressed here is the minimization of the channel capacity requirement, while maintaining the average distortion at or below an acceptable level.

The rate distortion function $R(D)$ is the minimum average rate (bits per element), and hence minimum channel capacity, required for a given average distortion level D. To make this more quantitative, we suppose that the source is a sequence of pixels, and these values are encoded by successive blocks of length N. Each block of pixels is then described by one of a denumerable set of messages, $\{X_i\}$, with probability function, $P(X_i)$. For a given input source, $\{X_i\}$, and output, $\{Y_j\}$, the decoder system can be described mathematically by the conditional probability, $Q(Y_j/X_i)$. Therefore, the probability of the output message is

$$T(Y_j) = \sum_i P(X_i)Q(Y_j/X_i) \tag{15.3}$$

The information transmitted is called the average mutual information between Y and X and is defined for a block of length N as follows:

$$I_N(X,Y) = \sum_i \sum_j P(X_i)Q(Y_j/X_i)\log_2[Q(Y_j/X_i)/T(Y_j)] \tag{15.4}$$

In the case of error-free encoding, $Y = X$ and then

$$Q(Y_j/X_i) = \begin{cases} 1, & j=i, \\ 0, & j \neq i, \end{cases} \quad \text{and} \quad T(Y_j) = T(Y_i) \tag{15.5}$$

In this case, Equation 15.4 becomes

$$I_N(X,Y) = \sum_i \sum_j P(X_i)\log_2 P(X_i) = H_N(X) \tag{15.6}$$

which is the Nth-order entropy of the data source. This can also be seen as the information contained in the data source under the assumption that no correlation exists between blocks and all the correlation between elements of each N length block is considered. Therefore, it requires at least $H_N(X)$ bits to code the data source without any information loss. In other words, the optimal error-free encoder requires $H_N(X)$ bits for the given data source. In the most general case, noise in the communication channel will result in error at least some of the time, causing $Y \neq X$. As a result,

$$I_N(X,Y) = H_N(X) - H_N(X/Y) \tag{15.7}$$

where $H_N(X/Y)$ is the entropy of the source data at the condition of decoder output Y. Because the entropy is a positive quantity, the source entropy is the upper bound to the mutual information; i.e.,

$$I_N(X,Y) \leq H_N(X) \tag{15.8}$$

Let $d(X,Y)$ be the average distortion between X and Y. Then, the average distortion per pixel is defined as

$$D(Q) = \frac{1}{N}E\{d(X,Y)\} = \frac{1}{N}\sum_i\sum_j d(X_i,Y_j)P(X_i)Q(X_i/Y_j) \tag{15.9}$$

The set of all conditional probability assignments, $Q(Y/X)$ that yield average distortion less than or equal to D^*, can be written as

$$\{Q:D(Q) \leq D^*\} \tag{15.10}$$

The N-block $R(D)$ is then defined as the minimum of the average mutual information, $I_N(X, Y)$, per pixel:

$$R_N(D^*) = \min_{Q:D(Q)\leq D^*} \frac{1}{N}I_N(X,Y) \tag{15.11}$$

The limiting value of the N-block $R(D)$ is simply called the rate distortion function,

$$R(D^*) = \lim_{N\to\infty} R_N(D^*) \tag{15.12}$$

It should be clear from the above discussion that Shannon's rate distortion function is a lower bound on the transmission rate required to achieve an average distortion D when the block size is infinite. In other words, when the block size is approaching infinity, the correlation between all elements within the block is considered as the information contained in the data source. Therefore, the rate obtained is the lowest rate or lower bound. Under these conditions, the rate at which a data source produces information, subject to a requirement of perfect reconstruction, is called the entropy of the data source, i.e., the information contained in the data source. It follows that the $R(D)$ is a generalization of the concept of entropy. Indeed, if the distortion measure is a perfect reproduction, it is assigned zero distortion. Then, $R(0)$ is equal to the source entropy $H(X)$. Shannon's coding theorem states that one can design a coding system with rate only negligibly greater than $R(D)$, which achieves the average distortion D. As D increases, $R(D)$ decreases monotonically and usually becomes zero at some finite value of distortion. The $R(D)$ specifies the minimum achievable transmission rate required to transmit a data with average distortion level D. The main value of this function in a practical application is that it potentially gives a measure for judging the performance of a coding system. However, this potential value has not been completely realized for video transmission. There are two reasons for this. First, there currently do not exist tractable and faithful mathematical models for an image source. The $R(D)$ for Gaussian sources under the squared error distortion criterion can be found, but it is not a good model for images. Second, a suitable distortion measure, D, which matches the subjective evaluation of image quality, has not been totally solved.

Some results have been investigated for this task such as JND (just noticeable distortion) [jnd]. The issue of subjective and objective assessments of image quality has been discussed in Chapter 1. In spite of these drawbacks, the rate distortion theorem is still a mathematical basis for comparing the performance of different coding systems.

15.3 Digital Video Formats

15.3.1 Digital Video Color Systems

In practical applications, most video signals are color signals. Various color systems have been discussed in Chapter 1. A color signal can be seen as a summation of light intensities of three primary wavelength bands. In this section, we introduce several extensively used color systems in the video industry. There are two primary color spaces used to represent digital video signal, which are RGB (red, green, blue) and YCbCr. The difference between RGB and YCbCr is that RGB represents color as red, green, and blue, whereas YCbCr represents color as brightness and two color difference signals. In YCbCr, the Y is the brightness (luma), Cb is blue minus luma $(B - Y)$, and Cr is red minus luma $(R - Y)$. We usually use YCC as a short way of saying YCrCb. The standard RGB color space is referred to as sRGB, which is an RGB color space created cooperatively by Hewlett–Packard and Microsoft Corporation. It has been endorsed by many industry players. It is also well accepted by Open Source software such as the GIMP (GNU Image Manipulation Program), and is used in proprietary and open graphics file formats such as PNG (Portable Network Graphic). The sYCC is simply YCC created from sRGB [IEC 61966-2-1]. The YCbCr color representation is used for most of video coding standards in compliance with the ITU-R BT.601 [ITU-R BT.601], BT.709 [ITU-R BT.709], common intermediate format (CIF), and source input format (SIF). ITU-R BT.601 is an ITU-R standard for component digital video. It was designed to provide a common digital standard for interoperability between the three analog video/TV systems (NTSC, PAL, and SECAM). ITU-R BT.601 enables their signals to be converted to digital and then easily converted back again to any of the three formats for distribution. In 1990, BT.709 was introduced for high definition television (HDTV) with specifications for 1125 and 1250 lines. In 2000, BT.709-4 added 1080 lines to conform to the digital television (DTV) standard. Conversion between the YCbCr and RGB formats can be accomplished with the transformations in Chapter 1.

Different color space covers different range of colors. The term "gamut" is used to represent a set of possible colors within a color system. Currently, most of video displays are sRGB gamut-limited displays, while the still image systems widely use the displays with sYCC color space. Recently, various kinds of extended-gamut displays are emerging and used for displaying still images. Users are always enjoying wide-gamut displays. In video signals, there are many unused regions that could store wide-gamut colors. Therefore, recently a new color space standard, IEC 61966-2-4, has been proposed for video displays [IEC 61966-2-4]. In IEC 61966-2-4, the extended-gamut color space for video applications −xvYCC color space has been proposed. The xvYCC is compatible with currently used video signals. It uses the same definition for inside of the sRGB gamut and there is no change necessary for conventional contents. However, it adds an unambiguous definition to the currently undefined or out-of-sRGB gamut regions. The xvYCC has 100% coverage, whereas sRGB has only 55%, which can be seen in Figures 15.2 and 15.3 [katoh 2005].

350 *Image and Video Compression for Multimedia Engineering*

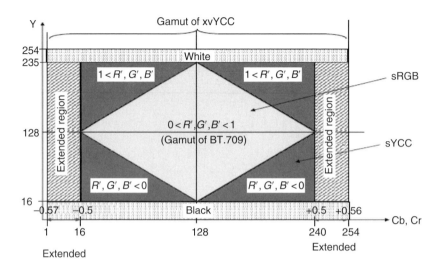

FIGURE 15.2 (See color insert following page 288.)
Two-dimensional (2-D) view of xvYCC.

15.3.2 Progressive and Interlaced Video Signals

Currently, most video signals that are generated by a TV camera are interlaced. These video signals are represented at 30 frames/s for an NTSC system. Each frame consists of two fields, the top field and bottom field, which are 1/60 of a second apart. In the display of an interlaced frame, the top field is scanned first and the bottom field is scanned next. The top and bottom fields are composed of alternating lines of the interlaced frame. Progressive video does not consist of fields, only frames. In an NTSC system, these frames are spaced 1/30 s apart. In contrast to interlaced video, every line within the frame is successively scanned. An example of progressive and interlaced video is shown in Figure 15.4.

FIGURE 15.3 (See color insert following page 288.)
Gamut coverage of sRGB, sYCC, and xvYCC color spaces.

Fundamentals of Digital Video Coding

FIGURE 15.4 (See color insert following page 288.)
(a) An example of progressive video and (b) an example of interlaced video.

15.3.3 Video Formats Used by Video Industry

15.3.3.1 ITU-R

According to ITU-R 601 (earlier ITU-R was CCIR), a color video source has three components: a luminance component (Y) and two color-difference or chrominance components (Cb and Cr or U and V in some documents). The CCIR format has two options: one for the NTSC TV system and another for the PAL TV system, both are interlaced. The NTSC format uses 525 lines per frame at 30 frames/s. The luminance frames of this format have 720×480 active pixels. The chrominance frames have two kinds of formats: one has 360×480 active pixels and is referred as the 4:2:2 format, whereas the other has 360×240 active pixels and is referred as the 4:2:0 format. The PAL format uses 625 lines per frame at 25 frames/s. Its luminance frame has 720×576 active pixels per frame and the chrominance frame has 360×576 active pixels per frame for the 4:2:2 format and 360×288 pixels per frame for the 4:2:0 format, both at 25 frames/s.

The a:b:c notation for sampling ratios, as found in the ITU-R BT.601 [ITU-R BT.601] specifications, has the following meaning:

4:2:2 means 2:1 horizontal downsampling, no vertical downsampling. (Think 4 Y samples for every 2 Cb and 2 Cr samples in a scanline.)

4:2:0 means 2:1 horizontal and 2:1 vertical downsampling. (Think 4 Y samples for every Cb and Cr samples in a scanline.)

15.3.3.2 Source Input Format

Source input format (SIF) has luminance resolution of 360×240 pixels per frame at 30 frames/s or 360×288 pixels per frame at 25 frames/s. For both cases, the resolution of the chrominance components is half the luminance resolution in both horizontal and vertical dimensions. SIF can easily be obtained from a CCIR format using an appropriate anti-aliasing filter followed by subsampling.

15.3.3.3 Common Intermediate Format

Common intermediate format (CIF) is a noninterlaced format. Its luminance resolution has 352×288 pixels per frame at 30 frames/s and the chrominance has half the luminance resolution in both vertical and horizontal dimensions. As its line value, 288, represents half the active lines in the PAL television signal, and its picture rate, 30 frames/s, is the same as the NTSC television signal, it is a common intermediate format for both PAL or PAL-like systems and NTSC systems. In the NTSC systems, only a line number conversion is needed, whereas in the PAL or PAL-like systems only a picture rate conversion is needed. For low bit rate applications, the quarter-SIF (QSIF) or quarter-CIF (QCIF) may be used because these format have only a quarter number of pixels of SIF and CIF formats, respectively.

15.3.3.4 ATSC Digital Television Format

The concept of digital television consists of SDTV (standard-definition television) and HDTV. Recently, in the United States, the FCC (Federal Communication Commission) has approved the ATSC recommended DTV standard [atsc 1995]. The DTV format is not included in the standard due to the divergent opinions of TV and computer manufacturers. Rather, it has been agreed that the picture format will be decided by the future market. The ATSC recommended DTV formats including two kinds of formats: SDTV and HDTV. The ATSC DTV standard includes the following 18 formats:

For HDTV: 1920×1080 pixels at 23.976/24 Hz progressive scan, 29.97/30 Hz interlaced scan, and 59.94/60 Hz progressive scan; 1280×720 pixels at 24, 30, and 60 Hz progressive scan.

For SDTV: 704×480 pixels with 4:3 aspect ratio at 23.976/24, 29.97/30, 59.94/60 Hz progressive scan, 30 Hz interlaced scan; 704×480 pixels with 16:9 aspect ratio at 23.976/24, 29.97/30, 59.94/60 Hz progressive scan, 30 Hz interlaced scan; and 640×480 with 4:3 aspect ratio at 23.976/24, 29.97/30, 59.94/60 Hz progressive scan, 30 Hz interlaced scan.

It is noted that all HDTV formats use square pixels and only part of SDTV formats use square pixels. The number of pixels per line versus the number of lines per frame is known as the aspect ratio.

15.4 Current Status of Digital Video/Image Coding Standards

The fast growth of digital transmission services has generated a great deal of interest in the digital transmission of video signals. Some digitized video source signals require very high bit rates, ranging from more than 100 Mbits/s for broadcast-quality video to more than 1 Gbits/s for HDTV signals. Owing to this, video compression algorithms, which reduce the bit rates to an affordable level on practical communication channels, are required. Digital video coding techniques have been investigated over several decades. There are two factors that make video compression possible: the statistical structure of the data in the video source and the psychophysical redundancy of human vision. Video compression algorithms can remove the spatial and temporal correlation that is normally present in the video source. In addition, human observers are subject to perceptual limitations in amplitude, spatial resolution, and temporal acuity. By proper design of the coding system, it is

Fundamentals of Digital Video Coding

TABLE 15.1

List of Some Organizations for Standardization

Organization	Full Name of Organization
ITU	International Telecommunication Union
JPEG	Joint Photographic Experts Group
MPEG	Moving Picture Experts Group
ISO	International Standards Organization
IEC	International Electrotechnical Commission

possible to discard information without affecting perceived image quality, or at least, with only minimal degradation.

Several traditional techniques have been developed for image and video data compression. Recently, with advances in data compression and VLSI techniques, the data compression techniques have been extensively applied to video signal compression. Video compression techniques have been under development for over 20 years and have recently emerged as the core enabling technology for a new generation of DTV (both SDTV and HDTV) and multimedia applications. Digital video systems currently being implemented (or under active consideration) include terrestrial broadcasting of digital HDTV in the United States [atsc 1993], satellite DBS (Direct Broadcasting System) [isnardi 1993], computer multimedia [ada 1993], and video via packet networks [verbiest 1989]. In response to the needs of these emerging markets for digital video, several national and worldwide standards activities have been started over the last few years. These organizations include ISO (International Standards Organization), ITU (International Telecommunication Union, formally known as CCITT (International Telegraph and Telephone Consultative Committee), JPEG (Joint Photographic Experts Group), and MPEG (Moving Picture Experts Group) as shown in Table 15.1. The related standards include JPEG standards, MPEG-1,2,4 standards, and H.261 and H.263 video teleconferencing coding standards as shown in Table 15.2. It should be noted that the JPEG standards are usually used for still image coding, but they can also be used in video coding. Although the coding efficiency would be lowered, they have shown to be useful in some applications,

TABLE 15.2

Video/Image Coding Standards

Name	Year of Completion	Major Features
JPEG	1992	For still image coding, DCT based
JPEG2000	2000	For still image coding, DWT based
H.261	1990	For videoconferencing, 64 kbits/s–1.92 Mbits/s
MPEG-1	1991	For CD-ROM, 1.5 Mbits/s
MPEG-2 (H.262)	1994	For DTV/DVD, 2–15 Mbits/s; for ATSC HDTV, 19.2 Mbits/s; most extensively used
H.263	1995	For very low bit rate coding, below 64 kbits/s
MPEG-4 Part 2	1999	For multimedia, content-based coding, its simple profile and advanced simple profile is applied to mobile video and streaming
H.264/AVC (MPEG-4 Part 10)	2005	For many applications with significant improved coding performance over MPEG-2 and MPEG-4 Part 2
VC-1	2005	For many applications, coding performance close to H.264
RealVideo	1997	For many applications, coding performance similar to MPEG-4 Part 2
MPEG-7	2000	Content description and indexing
MPEG-21	2002	Multimedia framework

e.g., studio editing systems. Though JPEG standards (discussed in Chapters 7 and 8) are not video coding standards, we include them here to give a full picture of all international image and video coding standards.

15.4.1 JPEG Standard

Since the mid-1980s, the ITU and ISO have been working together to develop a joint international standard for the compression of still images. Officially, JPEG [jpeg] is the ISO/IEC international standard 10918-1, "Digital compression and coding of continuous-tone still images," or the ITU-T recommendation T.81. JPEG became an international standard in 1992. JPEG is a DCT-based coding algorithm. It continues to work on future enhancements, which may adopt wavelet-based algorithms.

15.4.2 JPEG2000

JPEG2000 [jpeg2000] is a new type of image coding system under development by JPEG for still image coding. JPEG2000 is considered using the wavelet transform as its core technique. This is because the wavelet transform can provide not only excellent coding efficiency but also wonderful spatial and quality scalable functionality. This standard is intended to meet a need for image compression with great flexibility and efficient interchangeability. It is also intended to offer unprecedented access into the image while still in compressed domain. Thus, an image can be accessed, manipulated, edited, transmitted, and stored in a compressed form.

15.4.3 MPEG-1

In 1988, ISO established the MPEG to develop standards for the coded representation of moving pictures and associated audio information for digital storage applications. MPEG completed the first phase of its work in 1991. It is known as MPEG-1 [mpeg1] or ISO standard 11172, "Coding of moving picture and associated audio." The target application for this specification is digital storage media at bit rates up to about 1.5 Mbits/s.

15.4.4 MPEG-2

MPEG started its second phase of work, MPEG-2 [mpeg2], in 1990. MPEG-2 is an extension of MPEG-1 that allows for greater input-format flexibility, higher data rate for SDTV or HDTV applications, and better error resilience. This work resulted in the ISO standard 13818 or ITU-T Recommendation H.262, "Generic coding of moving pictures and associated audio."

15.4.5 MPEG-4

Part 2 [mpeg4]. MPEG-4 Part 2 Visual standard has been approved in 1999. The MPEG-4 Part 2 Visual supports object-based coding technology and it aims to provide enabling technology for a variety of functionalities and multimedia applications:

Universal accessibility and robustness in error-prone environments
High interactive functionality
Coding of natural and synthetic data or both
Compression efficiency

15.4.6 H.261

H.261 [h261] was adopted in 1990 and the final revision was approved in 1993 by the ITU-T. It is designed for video teleconferencing and utilizes a DCT-based motion compensation scheme. The target bit rates are from 64 to 1920 kbits/s.

15.4.7 H.263, H.263 Version 2 (H.263+), H.263++, and H.26L

The H.263 [h263] video coding standard is specifically designed for very low bit rate applications such as video conferencing. Its technical content was completed in late 1995 and the standard was approved in early 1996. It is based on the H.261 standard with several added features: unrestricted motion vectors, syntax-based arithmetic coding, advanced prediction, and PB-frames. The H.263 version 2 video coding standard, also known as H.263+, was approved in January 1998 by the ITU-T. H.263+ includes a number of new optional features based on the H.263. These new optional features are added to provide improved coding efficiency, a flexible video format, scalability, and backward-compatible supplemental enhancement information. H.263++ is the extension of H.263+ and was completed in the year 2000. H.26L is a long-term project, which is looking for more efficient video coding algorithms. Finally, the activity of H.26L ended because the joint video team (JVT) of MPEG and ITU-T VCEG developed a new video coding standard H.264, which has greatly improved the coding efficiency over MPEG-2 and H.263.

15.4.8 MPEG-4 Part 10 Advanced Video Coding or H.264/AVC

Recently, the JVT of MPEG and ITU-T VCEG has developed new video coding standard [h264]. Because many new tools have been used, H.264/AVC has achieved higher coding efficiency, which is almost twice better than MPEG-2. The detailed information is introduced in Chapter 20.

15.4.9 VC-1

VC-1 is a video codec developed by Microsoft and late has been standardized by SMPTE (Society of Motion Picture and Television Engineers). It is implemented by Microsoft as Windows Media Video (WMV) 9. Its coding performance is close to the H.264/AVC.

15.4.10 RealVideo

RealVideo is a video codec developed by RealNetWorks. It was first released in 1997 and its version 10 is released in 2006. RealVideo is supported on many platforms, including Windows, Mac, Linux, Solaris, and several mobile phones. Its coding performance is close to MPEG-4 Part 2.

The above organizations and standards are summarized in Tables 15.1 and 15.2, respectively.

It should be noted that MPEG-7 [mpeg-7] and MPEG-21 [mpeg-21] in Table 15.2 are not a coding standard; MPEG-7 is a multimedia content description standard, which can be used to fast indexing and searching for multimedia content; and the MPEG-21 is a multimedia framework, which aims at defining an open framework for multimedia applications. The VC-1 is a SMPTE standard and RealVideo is not an international standard, but it is extensively supported by many platforms.

It is also interesting to note that in terms of video compression methods, there is a growing convergence towards motion compensated (MC), interframe DCT algorithms represented by the video coding standards. However, wavelet-based coding techniques have found recent success in the compression of still image coding in both the JPEG2000 and MPEG-4 standards. This is because it posses unique features in terms of high coding efficiency and excellent spatial and quality scalability. The wavelet transform has not successfully been applied to video coding due to several difficulties. First one, it is not clear how the temporal redundancy can be removed in this domain. Motion compensation is an effective technique for DCT-based video coding scheme, however, it is not so effective for wavelet-based video coding. This is because the wavelet transform uses large block size or full frame, but motion compensation is usually performed on a limited block size. This mismatch would reduce the interframe coding efficiency. Many engineers and researchers are working on these problems.

Among these standards, MPEG-2 has had a great impact on the consumer electronics industry because the DVD (digital versatile disk) and DTV have adopted it as core technology. But recently developed new coding standard is attracted many applications, including HD-DVD, mobile TV, and others.

15.5 Summary

In this chapter, the several fundamental issues of digital video coding have been presented. These include the representation and $R(D)$ of digital video signals and the various video formats, which are widely used by video industry. Finally, existing and emerging video coding standards have been briefly introduced in this chapter.

Exercises

1. Suppose that we have one-dimensional (1-D) digital array (it can be extended to 2-D array that may be an image), $f(i) = X_i$, $(i = 0, 1, 2, \ldots)$. If we use the first-order linear predictor to predict the current component value with the previous component such as $X'_i = \alpha X_{i-1} + \beta$, where α and β are two parameters for this linear predictor. If we want to minimize the mean squared error (MSE) of the prediction $E\{(X_i - X'_i)^2\}$, what values of α and β should we choose? Assuming that $E\{X_i\} = m$, $E\{X_i^2\} = \sigma^2$, and $E\{X_i X_{i-1}\} = \rho$, (for $i = 0, 1, 2, \ldots$), where m, σ, and ρ are constant.

2. To get a 128×128 or 256×256 digital image, write a program to use two 3×3 operators (Sobel operator) such as

$$\begin{bmatrix} -1 & -2 & 0 \\ 1 & 1 & 0 \\ 0 & 0 & 0 \end{bmatrix} \begin{bmatrix} -1 & 0 \\ 1 & 0 \\ -2 & 0 \end{bmatrix}$$

to filter the image, separately. Discuss the resulting image. What will be the result if both the operators are used?

3. The conversion of 2-D array is defined as

$$y(m, n) = \sum_{k=-\infty}^{+\infty} \sum_{l=-\infty}^{+\infty} x(k, l) \, h(m - k, n - l)$$

and

$$\bar{x} = \begin{bmatrix} 1 & 4 & 1 \\ 2 & 5 & 3 \end{bmatrix} \quad \bar{h} = \begin{bmatrix} 1 & 1 \\ 1 & -1 \end{bmatrix}$$

Calculate the convolution $y(m, n)$. If $h(m, n)$ is changed to

$$\begin{bmatrix} 0 & -1 & 0 \\ -1 & 4 & -1 \\ 0 & -1 & 0 \end{bmatrix},$$

recalculate $y(m, n)$.
4. The entropy of an image source is defined as

$$H = -\sum_{k=1}^{M} p_k \log_2 p_k,$$

under assumption that each pixel is an independent random variable. If the image is a binary image, i.e., $M=2$, and the probability $p_1 + p_2 = 1$. If we define $p_1 = p$ then $p_2 = 1 - p$, $(0 \leq p \leq 1)$. The entropy can be rewritten as

$$H = -p \log_2 p - (1-p) \log_2 (1-p).$$

Find several digital binary images and compute their entropies. If one image has almost equal number of "0" and "1" and other has different number of "0" and "1," which image has larger entropy? Prove that the entropy of a binary source is maximum if the number of "0" and "1" is equal.
5. A transformation defined as $y = f(x)$, is applied to a 256 × 256 digital images, where x is the original pixel value and y is transformed pixel value. Obtain new images for (1) f is a linear function, (2) f is logarithm, and (3) f is a square function; compare the results and indicate subjective differences of the resulting images. Repeat the experiments for different images and draw conclusions about possible use of this procedure in image processing applications.

References

[IEC 61966-2-1] Multimedia systems and equipment-Colour measurement and management-Part 2-1: Colour management-Default RGB colour space—sRGB, December 7, 1999.
[IEC 61966-2-4] International Standard: Multimedia Systems and management-Part 2-4: Colour management-Extended-gamut YCC colour space for video applications —xvYCC, January 2006.
[ITU-R BT.601] International Telecommunications Union, ITU-R BT.60, 1987.
[ITU-R BT.709] International Telecommunications Union, ITU-R BT.709, 1990.
[ada 1993] J.A. Ada, Interactive Multimedia, *IEEE Spectrum*, March 1993, 22–31.
[atsc 1995] ATSC Digital Television Standard, Doc. A/53, September 16, 1995.
[berger 1971] T. Berger, *Rate Distortion Theory—A Mathematical Bais for Data Compression*, Prentice-Hall, Englewood Cliffs, NJ, 1971.
[h261] ITU-T Rec. H. 261, Video codec for audio visual services at px64 kbit/s, March 1995.
[h263] ITU-T Rec. H. 263, Video coding for low bit rate communication, May 2, 1996.

[h264] ITU-T Rec. H.264/ISO/IEC 11496-10, Advanced video coding for generic audiovisual services, February 28, 2005.

[isnardi 1993] M. Isnardi, Consumers seek easy to use products, *IEEE Spectrum,* January 1993, 64.

[jnd] www.sarnoff.com/tech_realworld/broadcast/jnd/index.html.

[jpeg] ISO/IEC IS 11544, ITU-T Rec. T.81, 1992.

[jpeg 2000] ISO/IEC 15444-1, ITU-T Rec. T. 800, Information technology-JPEG 2000 image coding system: Core coding system, 2000.

[katoh 2005] Attachment of proposal for MPEG Hong Kong Meeting 2005, IEC 61966-2-4. The Extended-gamut color space for video application –xvYCC color space, by Naoya Katoh and Yoshihide Simpuku, Color Rendering Community, Sony Corporation.

[modestino 1981] J.W. Modestino, D.G. Daut, and A.L. Vickers, Combined source-channel coding of image using the block cosine transform, *IEEE Transactions on Communication,* COM-29, 1262–1274, September 1981.

[mpeg1] ISO/IEC JTC1 IS 11172, Coding of moving picture and coding of continuous audio for digital storage media up to 1.5 Mbps, November 1992.

[mpeg2] ISO/IEC JTC1 IS 13818, Generic coding of moving pictures and associated audio, November 1994.

[mpeg4] ISO/IEC JTC1 FDIS 14496-2, Information technology—generic coding of audio-visual objects, November 19, 1998.

[mpeg7] MPEG-7 Overview v.8, ISO/MPEG N4980, Klagenfurt, Austria, July 2002.

[mpeg21] MPEG-21 Overview v.5, ISO/IEC JTC1/SC29/WG11/N5231, October 2002.

[oppenheim 1989] A.V. Oppenheim and R.W. Schafer, *Discrete-Time Signal Processing,* Prentice-Hall, Englewood Cliffs, NJ, 1989.

[sayood 1991] K. Sayood and J.C. Borkenhagen, Use of residual redundancy in the design of joint source/channel coders, *IEEE Transactions on Communications*, 39, 838–846, June 1991.

[shannon 1948] C.E. Shannon, A mathematical theory of communication, *Bell Systems Technical Journal,* 27, 379–423, 623–656.

[verbiest 1989] W. Verbiest and L. Pinnoo, A variable bit rate video codec for asynchronous transfer mode networks, *IEEE Journal on Selected Areas in Communications*, 7, 5, 761–770, 1989.

[viterbi 1979] A.J. Viterbi and J.K. Omura, *Principles of Digital Communication and Coding*, McGraw-Hill, New York 1979.

16
Digital Video Coding Standards: MPEG-1/2 Video

In this chapter, we introduce the ISO/IEC digital video coding standards, MPEG-1 [mpeg1] and MPEG-2 [mpeg2], which are extensively using in the video industry for television broadcast, visual communications, and multimedia applications.

16.1 Introduction

As we know, MPEG has successfully developed two standards, MPEG-1 and MPEG-2. The MPEG-1 video standard was completed in 1991 with the development of the ISO/IEC specification 11172, which is the standard for coding of moving picture and associated audio for digital storage media at up to about 1.5 Mbits/s. To support a wide range of application profiles, the user can specify a set of input parameters including flexible picture size and frame rate. MPEG-1 was developed for multimedia CD-ROM applications. Important features provided by MPEG-1 include frame-based random access of video, fast forward/fast backward searches through compressed bitstreams, reverse playback of video, and editability of the compressed bitstream. MPEG-2 is formally referred to as ISO/IEC specification 13818, which is the second phase of MPEG video coding solution for applications not originally covered by the MPEG-1 standard. Specifically, MPEG-2 was developed to provide video quality not lower than National Television Systems Committee/phase alternating line (NTSC/PAL) and up to high definition television (HDTV) quality. The MPEG-2 standard was completed in 1994. Its target bit rates for NTSC/PAL are about 2–15 Mbits/s, and it is optimized at about 4 Mbits/s. The bit rates used for HDTV signals are about 19 Mbits/s. In general, MPEG-2 can be seen as a superset of the MPEG-1 coding standard and is backward compatible to MPEG-1 standard. In other words, every MPEG-2 compatible decoder is able to decode a compliant MPEG-1 bitstream.

In this chapter, we briefly introduce the standard itself. As many books and publications exist for the explanation of the standards [haskell 1997; mitchel 1997], we pay more attention to the utility of the standard, how the standard is used, and touch on some interesting research topics that have emerged. In other words, the standards provide the knowledge for of how to design the decoders that are able to successfully decode the compliant MPEG bistreams. But the standards do not specify the means of generating these bitstreams. For instance, given some bit rate, how can one generate a bitstream that provides the best picture quality? To answer this, one needs to understand the encoding process, which is an informative part of standard (referred to as the Test Model), but it is very important for the content and service providers. In this chapter, the issues related to the encoding process are described. The main contents include the following topics: preprocessing, motion compensation, rate control, statistically multiplexing (StatMux) multiple programs, and optimal mode decision. Some of the sections contain the authors' own research results.

These research results are useful in providing examples for the readers to understand how the standard is used.

16.2 Features of MPEG-1/2 Video Coding

It should be noted that MPEG-2 video coding has the feature of being backward compatible with MPEG-1. It turns out that most of the decoders in the market are MPEG-2 compliant decoders. For simplicity, we introduce the technical detail of MPEG-1 and then describe the enhanced features of MPEG-2, which MPEG-1 does not have.

16.2.1 MPEG-1 Features

16.2.1.1 Introduction

The algorithms employed by MPEG-1 do not provide a lossless coding scheme. However, the standard can support a variety of input formats and be applied to a wide range of applications. As we know, the main purpose of MPEG-1 video is to code moving image sequences or video signals. To achieve a high compression ratio, both intraframe and interframe redundancies should be exploited. This implies that it would not be efficient to code the video signal with an intraframe coding scheme, such as JPEG. On the other hand, to satisfy the requirement of random access, we have to use intraframe coding from time to time. Therefore, the MPEG-1 video algorithm is mainly based on discrete cosine transform (DCT) coding and interframe motion compensation. The DCT coding is used to remove the intraframe redundancy and the motion compensation is used to remove the interframe redundancy. With regard to input picture format, MPEG-1 allows progressive pictures only, but offers great flexibility in the size, up to 4095 × 4095 pixels. However, the coder itself is optimized to the extensively used video SIF picture format. The SIF is a simple derivative of the ITU-R 601 video format for digital television applications. According to ITU-R 601, a color video source has three components, a luminance component (Y) and two chrominance components (Cb and Cr), which are in the 4:2:0 subsampling format. Note that the 4:2:0 and 4:2:2 color formats were described in Chapter 15.

16.2.1.2 Layered Structure Based on Group of Pictures

The MPEG coding algorithm is a full-motion compensated DCT and DPCM hybrid coding algorithm. In MPEG coding, first, the video sequence is divided into groups of pictures or frames (GOPs) as shown in Figure 16.1. Each GOP may include three types of pictures

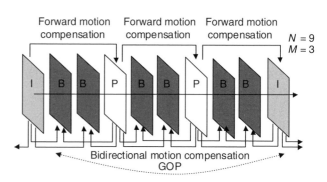

FIGURE 16.1
A group of pictures (GOPs) of video sequence in display order.

or frames: intracoded (I) picture or frame, predictive-coded (P) picture or frame, and bidirectionally predictive-coded (B) picture or frame. I-pictures are coded by intraframe techniques only, without need for previous information. In other words, I-pictures are self-sufficient. They are used as anchors for forward and backward prediction. P-pictures are coded using one-directional motion compensated (MC) prediction from a previous anchor frame, which could be either I- or P-picture. The distance between two nearest I-frames is denoted by N, which is the size of GOP. The distance between two nearest anchor frames is denoted by M. Both Parameters N and M are user selectable parameters, which are selected by user during the encoding. Larger number of N and M will increase the coding performance but cause error propagation or drift. Usually, N is chosen from 12 to 15 and M from 1 to 3. If M is selected to be 1 then no B-picture will be used. Lastly, B-pictures can be coded using predictions from either past or future anchor frames (I or P), or both. Regardless of the type of frame, each frame may be divided into slices; each slice consists of several macroblocks (MBs). There is no rule to decide the slice size. A slice could contain all MBs in a row of a frame or all MBs of a frame. Smaller slice size is favorable for the purpose of error resilience, but will decrease coding performance due to higher overhead. An MB contains a 16×16 Y component and spatially corresponding 8×8 Cb and Cr components. An MB has four luminance blocks and two chrominance blocks (for 4:2:0 sampling format) and the MB is also the basic unit of adaptive quantization and motion compensation. Each block contains 8×8 pixels over which the DCT operation is performed.

To exploit the temporal redundancy in the video sequence, the motion vector for each MB is estimated from two original luminance pictures using a block matching algorithm. The criterion for the best match between the current MB and an MB in the anchor frame is the minimum mean absolute error (MSE). Once the motion vector for each MB is estimated, pixel values for the target MB can be predicted from the previously decoded frame. All MBs in I-frame are coded in intramode with no motion compensation. MBs in P- and B-frames can be coded in several modes. Among the modes are intracoded and intercoded with motion compensation. This decision is made by mode selection. Most encoders depend on the values of predicted differences to make this decision. Within each slice, the values of motion vectors and DC values of each MB are coded using DPCM. The detailed specifications of this coding can be found in the document proposed by the MPEG video committee [mpeg2]. The structure of MPEG implies that if an error occurs within I-frame data, it will be propagated through all frames in the GOP. Similarly, an error in a P-frame will affect the related P- and B-frames, while B-frame errors will be isolated.

16.2.1.3 Encoder Structure

The typical MPEG-1 video encoder structure is shown in Figure 16.2. It should be noted that when B-picture is used as shown in Figure 16.1, two frame memories are needed for bidirectional prediction. However, the encoding order is different from the display order; the input sequence has to be reordered for encoding. For example, if we choose the GOP size (N) to be 12, and the distance between two nearest anchor frames (M) to be 3, the display order and encoding order are as shown in Table 16.1.

It should be noted that in the encoding order or in the bitstream, the first frame in a GOP is always an I-picture. In the display order, the first frame can be either I-picture or the first B-picture of the consecutive series of B-pictures that immediately precedes the first I-picture, and the last picture in a GOP is an anchor picture, either I- or P-picture. The first GOP always starts with an I-picture and as consequence, this GOP will have less than B-pictures than the other GOPs.

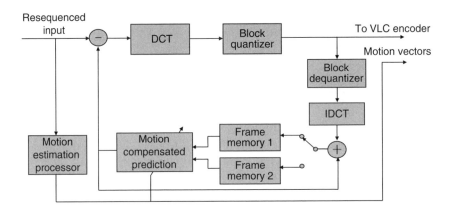

FIGURE 16.2
Typical MPEG-1 encoder structure.

The MPEG-1 video compression technique uses motion compensation to remove the interframe redundancy. The concept of motion compensation is based on the estimation of motion between video frames. The fundamental model, which is used, assumes that a translational motion can approximate the motion of a block. If all elements in a video scene are approximately spatially displaced, the motion between frames can be described by a limited number of motion parameters. In other words, the motion can be described by motion vectors for translatory motion of pixels. Because the spatial correlation between adjacent pixels is usually very high, it is not necessary to transmit motion information for each coded image pixel. This would be too expensive and the coder would never be able to reach a high compression ratio. The MPEG video uses the MB structure for motion compensation, i.e., for each 16 × 16 MB, only one or sometimes two motion vectors are transmitted. The motion vectors for any block are found within a search window that can be up to 512 pixels in each direction. Also, the matching can be done at half-pixel accuracy, where the half-pixel values are computed by averaging the full-pixel values as shown in Figure 16.3.

For interframe coding, the prediction differences or error images are coded and transmitted with motion information. A two-dimensional (2-D) DCT is used for coding both the intraframe pixels and predictive error pixels. The image to be coded is first partitioned into 8 × 8 blocks. Each 8 × 8 pixel block is then subject to an 8 × 8 DCT, resulting in a frequency domain representation of the block as shown in Figure 16.4.

The goal of the transformation is to decorrelate the block data so that the resulting transform coefficients can be coded more efficiently. The transform coefficients are then quantized. During the process of quantization, a weighted quantization matrix is used. The function of quantization matrix is to quantize high frequencies with coarser quantization steps that will suppress high frequencies with no subjective degradation, thus taking advantage of human visual perception characteristics. The bits saved for coding high frequencies are used for lower frequencies to obtain better subjective coded images.

TABLE 16.1

Display Order and Encoding Order

Display order	0	1	2	3	4	5	6	7	8	9	10	11	12
Encoding order	0	3	1	2	6	4	5	9	7	8	12	10	11
Coding type	I	P	B	B	P	B	B	P	B	B	I	B	B

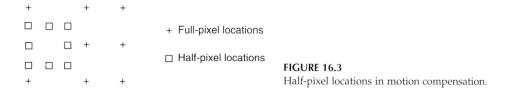

FIGURE 16.3
Half-pixel locations in motion compensation.

There are two quantizer weighting matrices in Test Model 5 (TM5) [tm5], an intra quantizer weighting matrix and a non-intra quantizer weighting matrix; the later is more flat because the energy of coefficients in interframe coding is more uniformly distributed than in intraframe coding.

In intra MBs, the DC value, dc, is an 11 bit value before quantization and it will be quantized to 8, 9, or 10 bits according to the setting of parameter. Thus the quantized DC (QDC) value is calculated as

$$\text{QDC(8 bit)} = dc//8, \quad \text{QDC(9 bit)} = dc//4, \quad \text{or QDC(10 bit)} = dc//2 \quad (16.1)$$

where symbol // means integer division with rounding to the nearest integer and the half integer values are rounded away for zero unless otherwise specified. The AC coefficients, ac(i,j), are first quantized by individual quantization factors to the value of ac~(i,j):

$$ac\sim(i,j) = (16 * ac(i,j))//W_I(i,j) \quad (16.2)$$

where $W_I(i,j)$ is the element at the (i,j) position in the intra quantizer weighting matrix shown in Figure 16.5.

The quantized level QAC(i,j) is given by

$$QAC(i,j) = [ac\sim(i,j) + \text{sign}(ac\sim(i,j)) * ((p * \text{mquant})//q)]/(2 * \text{mquant}) \quad (16.3)$$

where mquant is the quantizer scale or step which is derived for each MB by rate control algorithm, and $p=3$ and $q=4$ in TM5 [tm5]. For non-intra MBs,

$$ac\sim(i,j) = (16 * ac(i,j))//W_N(i,j) \quad (16.4)$$

where $W_N(i,j)$ is non-intra quantizer weighting matrix in Figure 16.5 and

$$QAC(i,j) = ac\sim(i,j)/(2 * \text{mquant}) \quad (16.5)$$

An example of encoding an intrablock is shown in Figure 16.6.

FIGURE 16.4
Example of 8 × 8 discrete cosine transform (DCT).

8	16	19	22	26	27	29	34
16	16	22	24	27	29	34	37
19	22	26	27	29	34	34	38
22	22	26	27	29	34	37	40
22	26	27	29	32	35	40	48
26	27	29	32	35	40	48	58
26	27	29	34	38	46	56	69
27	29	35	38	46	56	69	83

Intra quantizer weighting matrix

16	17	18	19	20	21	22	23
17	18	19	20	21	22	23	24
18	19	20	21	22	23	24	25
19	20	21	22	23	24	26	27
20	21	22	23	25	26	27	28
21	22	23	24	26	27	28	30
22	23	24	26	27	28	30	31
23	24	25	27	28	30	31	33

Non-intra quantizer weighting matrix

FIGURE 16.5
Quantizer matrices for intra- and non-intracoding.

The coefficients are processed in zigzag order because the major part of the energy is usually concentrated in the lower-order coefficients. The zigzag ordering of elements in an 8×8 matrix allows for a more efficient run-length coder. This is illustrated in Figure 16.7.

With the zigzag order, the run-length coder converts the quantized frequency coefficients to pairs of zero runs and nonzero coefficients:

$$34\ 0\ 1\ 0 - 1\ 1\ 0\ 0\ 0\ 0\ 0\ 0 - 1\ 0\ 0\ 0\ 0\ \ldots$$

After parsing, we obtain the pairs of zero runs and values:

$$34 \mid 0\ 1 \mid 0 - 1 \mid 1 \mid 0\ 0\ 0\ 0\ 0\ 0 - 1 \mid 0\ 0\ 0\ 0\ \ldots$$

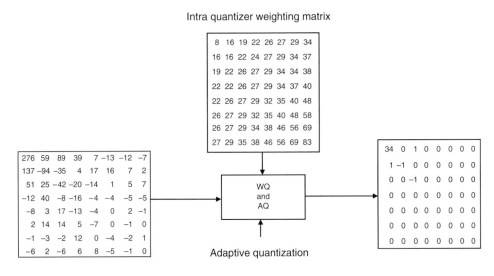

FIGURE 16.6
An example of coding an intrablock.

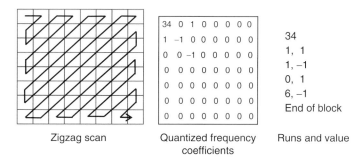

FIGURE 16.7
Zigzag scans to get pairs of zero runs and value.

These pairs of runs and values are then coded by a Huffman-type entropy coder. For example for the above run/value pairs are

Run/Value	VLC (Variable-Length Code)
34	—
1, 1	0110
1, −1	0111
0, 1	110
6, −1	0001011
End of block (EOB)	10

The variable-length code (VLC) tables are obtained by statistically optimizing a large number of training video sequences and are included in the MPEG-2 specification. The same idea is applied to code the DC values, motion vectors, and other information. Therefore, the MPEG video standard contains a number of VLC tables.

16.2.1.4 Structure of the Compressed Bitstream

After coding, all the information is converted to binary bits. The MPEG video bitstream consists of several well-defined layers with headers and data fields. These layers include sequence, GOP (group of pictures), picture, slice, MB, and block. The important syntax elements contained in each layer are summarized in Table 16.2. The typical structure of the MPEG-1 video compressed bitstream is shown in the Figure 16.8. The syntax elements contained in the headers and the amount of bits defined for each element can be found in the standard.

For picture layer, a frame of picture is first partitioned into MBs (16 × 16 for luminance and 8 × 8 for chrominance in the 4:2:0 color representation). The compressed bitstream structure at this layer is shown in Figure 16.9. It is important to note that most elements in the syntax are coded by VLC. The tables of these variable run-length codes (RLCs) are obtained through the simulation of a large number of training video sequences.

16.2.1.5 Decoding Process

The decoding process is an inverse procedure of encoding. The block diagram of a typical decoder is shown in Figure 16.10. The variable-length decoder (VLD) first decodes the coded data or video bitstream. This process yields the quantized DCT coefficients and motion vector data for each MB. The coefficients are inversely scanned and de-quantized.

TABLE 16.2
Summary of Important Syntax of Each Layer

Name of Layer	Important Syntax Elements
Sequence	Picture size and frame rate
	Bit rate and buffering requirement
	Programmable coding parameters
Groups of pictures (GOPs)	Random access unit
	Time code
Picture	Timing information (buffer fullness, temporal reference)
	Coding type (I, P, or B)
Slice	Intraframe addressing information
	Coding re-initialization (error resilience)
MB	Basic coding structure
	Coding mode
	Motion vectors
	Quantization
Block	Discrete cosine transform (DCT) coefficients

The decoded DCT coefficients are then inverse transformed to obtain the spatial-domain pixels. If the MB was intracoded, these pixels represent the reconstructed values, without any further processing. However, if the MB is intercoded then motion compensation is performed to add the prediction from the corresponding reference frame(s).

16.2.2 MPEG-2 Enhancements

The basic coding structure of MPEG-2 video is the same as that of MPEG-1 video, i.e., intraframe and interframe DCT with I-, P-, and B-pictures is used. The most important features of MPEG-2 video coding include

- Field/frame prediction modes for supporting the interlaced video input
- Field/frame DCT coding syntax
- Downloadable quantization matrix and alternative scan order
- Scalability extension

FIGURE 16.8
Description of layered structure of compressed bitstream.

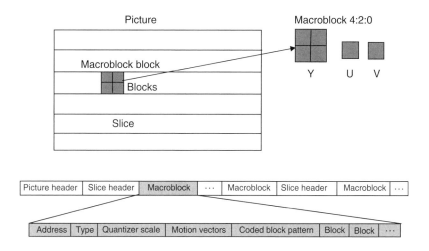

FIGURE 16.9
Picture layer data structure.

The above enhancement items are all coding performance improvements that are related to the support of interlaced material. There are also several non-compression enhancements, which include

- Syntax to facilitate 3:2 pull-down in the decoder
- Pan and scan codes with 1/16 pixel resolution
- Display flags indicating chromaticity, subcarrier amplitude, and phase (for NTSC/PAL/SECAM source material)

In the following, each of these enhancements is introduced.

16.2.2.1 Field/Frame Prediction Mode

In MPEG-1 video, we always code each picture as a frame structure, whether the original material is progressive or interlaced. If the original sequence is interlaced, each frame consists of two fields: top field and bottom field as shown if Figure 16.11. We still can use frame-based prediction if we consider that the two fields as a frame such as shown in Figure 16.11.

In Figure 16.11, three frames are coded as I-, B-, and P-frames and each frame consists of two fields. The P-frame is predicted with the I-frame with one motion vector. The B-frame

FIGURE 16.10
Simplified MPEG video decoder.

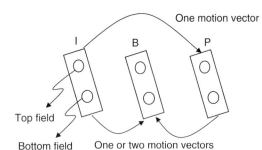

FIGURE 16.11
Frame-based prediction of MPEG-1 video coding.

can be predicted only with I-frame (forward prediction) or only with P-frame (backward prediction) or from both I- and P-frames (bidirectional prediction), the forward and backward prediction needs only one motion vector and the bidirectional prediction needs two motion vectors.

MPEG-2 video provides an enhanced prediction mode to support interlaced material, which uses the adaptive field/frame selection, based on the best match criteria. Each frame consists of two fields: top field and bottom field. Each field can be predicted from either field of the previous anchor frame. The possible prediction modes are shown in Figure 16.12.

In a field-based prediction, the top field of the current frame can be predicted either from the top field or the bottom field of an anchor frame as shown in Figure 16.12. The solid arrow represents the prediction from the top field and the dashed arrow represents the prediction from the bottom field. The same is also true for bottom field of the current frame. If the current frame is a P-frame, there could be up to two motion vectors used to make the prediction (one for top field and one for bottom field); if the current frame is a B-frame, there could be up to four motion vectors (each field could be bidirectional prediction which needs two motion vectors). At the MB level of MPEG-2, several coding modes are added to support these new field-based predictions. Additionally, there is another new prediction mode supported by the MPEG-2 syntax. This is the special prediction mode referred to as dual prime prediction. The basic idea of dual prime prediction is to code a set of field motion vectors with a scaling to a near or far field, plus a transmitted delta vector. Due to the correlation of adjacent pixels, the dual prime coding of field vectors can save the number of bits used for field motion vectors. The dual prime prediction is shown in the Figure 16.13. In Figure 16.13, one field motion vector and the delta motion vector are transmitted, the motion vectors for other field are derived from the above two vectors.

It should be noted that only the P-picture is allowed to use dual prime prediction. In other words, if the dual prime prediction is used in the encoder, there will be no B-pictures. The reason for this restriction is to limit the required memory bandwidth for a real system implementation.

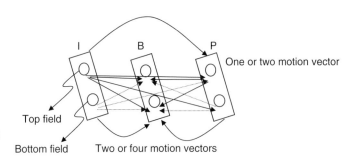

FIGURE 16.12
Field-based prediction of enhanced option of MPEG-2 video coding.

Digital Video Coding Standards: MPEG-1/2 Video

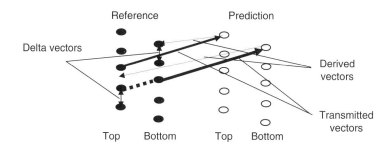

FIGURE 16.13
Dual prime prediction in MPEG-2 video coding.

16.2.2.2 Field/Frame DCT Coding Syntax

Another important feature to support interlaced material is to allow adaptive selection of the field/frame DCT coding as shown in Figure 16.14.

In Figure 16.14, the middle is a luminance MB of 16×16 pixels, the black rectangular block represents the eight pixels in the top field and the white rectangular block represents the eight pixels in the bottom field. The left is the field DCT in which each 8×8 block contains only the pixels from the same field. The right is the frame DCT; each 8×8 block contains the pixels from both top field and bottom field.

At the MB level for interlaced video, the field-type DCT may be selected when the video scene contains less detail and experiences large motion. Moreover, the difference between adjacent fields may be large when there is large motion between fields; it may be more efficient to group the fields together, rather than the frames. In this way, the possibility that there exists more correlation among the fields can be exploited. Ultimately, this can provide much more efficient coding because the block data is represented with fewer coefficients, especially if there is not much detail contained in the scene.

16.2.2.3 Downloadable Quantization Matrix and Alternative Scan Order

The new feature in MPEG-2 regarding the quantization matrix is that it can be downloaded for every frame. This may be helpful if the input video characteristics are very dynamic. In general, the quantizer matrices are different for intracoding and non-intracoding. With 4:2:0 format, only two matrices are used, one for the intrablocks and another for the non-intrablocks. With 4:2:2 or 4:4:4 formats four matrices are used, both an intra- and a non-intra matrix are used for the luminance and chrominance blocks. If the matrix load flags are not

FIGURE 16.14
Frame and field discrete cosine transform (DCT) for interlaced video.

FIGURE 16.15
Two zigzag scan methods for MPEG-2 video coding.

set, the decoder will use default matrices. The formats 4:2:0, 4:2:2 are defined in Chapter 15. In the 4:4:4 format, the luminance and two chrominance pictures have the same picture size.

In the picture layer, there is a flag that can be set for an alternative scan of DCT blocks, instead of using the zigzag scan discussed earlier. Depending on the spectral distribution, the alternative scan can yield run lengths that better exploit the multitude of zero coefficients. The zigzag scan and alternative scan are shown in Figure 16.15.

The normal zigzag scan is used for MPEG-1 and as an option for MPEG-2. The alternative scan is not supported by MPEG-1 and is an option for MPEG-2. For frame-type DCT of interlaced video, more energy may exist at the bottom part of the block; hence the RLC may be better off with the alternative scan.

16.2.2.4 Pan and Scan

In MPEG-2, there are several parameters defined in the sequence display extension and picture display extension for panning and displaying a rectangle around a reconstructed frame. These parameters include display-horizontal-size and display-vertical-size in the sequence display extension, and frame-center-horizontal-offset and frame-center-vertical-offset in the picture display extension. The function of these parameters can be found in the MPEG-2 system specification. A typical example to use pan–scan parameters is the conversion of 16:9 frame to 4:3 frame. The 4:3 region is defined by display-horizontal-size and display-vertical-size and the 16:9 frame is defined by horizontal-size and vertical-size. If we chose the display-horizontal-size be four pixels less than the horizontal-size, and keep the display-vertical-size as the same as the vertical-size, then we can obtain a 4:3 pictures on the display. Figure 16.16 shows the conversion of 16:9 frame to 4:3 frame using pan–scan parameter, but there is no center offset involved in this example.

16.2.2.5 Concealment Motion Vector

The concealment motion vector (CMV) is a new tool supported by MPEG-2. This tool is useful in concealing errors in the noisy channel environment where the transmitted data

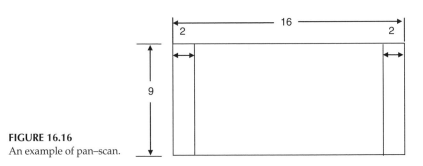

FIGURE 16.16
An example of pan–scan.

may be lost or corrupted. The basic idea of CMV is that the motion vectors are sent for the intracoded MB. These motion vectors are referred to as CMVs, which should be used in MBs immediately below the one in which the CMV occurs. The details are described in Section 17.5.3.2.

16.2.2.6 Scalability

MPEG-2 video has several scalable modes, which includes spatial scalability, temporal scalability, SNR scalability, and data partitioning. These scalability tools allow a subset of any bitstream to be decoded into meaningful imagery. Moreover, scalability is a useful tool for error resilience on prioritized transmission media. The drawback of scalability is that some coding efficiency is lost due to extra overhead. Here, we briefly introduce the basic notions of the above scalability features.

Spatial scalability allows multiresolution coding, which is suitable for video service inter-networking applications. In spatial scalability, a single video source is split into a base layer (lower spatial resolution) and enhancement layers (higher spatial resolution). For example, an ITU-R 601 video can be down-sampled to SIF format with spatial filtering, which can serve as the base layer video. The base layer or low-resolution video can be coded with MPEG-1 or MPEG-2, and the higher resolution layer must be coded by MPEG-2 supported syntax. For the up-sampled lower layer, an additional prediction mode is available in the MPEG-2 encoder. This is a flexible technique in terms of bit rate ratios, and the enhancement layer can be used in high quality service. The problem with spatial scalability is that there exists some bit rate penalty due to overhead and there is also a moderate increase in complexity. A block diagram that illustrates encoding with spatial scalability is shown in Figure 16.17. In Figure 16.17, the output of decoding and spatial up-sampling block provides an additional choice of prediction for the MPEG-2 compatible coder, but not the only choice of prediction. The prediction can be obtained from HDTV input itself also depending on the prediction select criterion such as the minimum prediction difference.

It should be noted that the spatial scalability coding allows the base layer to be coded independently from the enhancement layer. In other words, the base layer or lower layer bitstream is generated without regard for the enhancement layer and can be decoded independently. The enhancement layer bitstream is additional information, which can be seen as the prediction error based on the base layer data. This implies that the enhancement layer is useless without the base. However, this type of structure can find a lot of applications such as error concealment, which is discussed in the Section 17.5.

FIGURE 16.17
Block diagram of spatial scalability encoder.

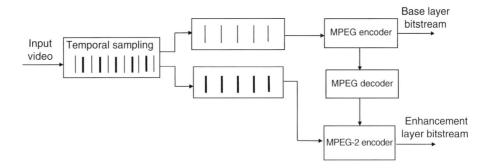

FIGURE 16.18
Block diagram of temporal scalability encoder.

Temporal scalability is a scalable coding technique in the temporal domain. An example of a two-layer temporal scalable coder is shown in Figure 16.18. This example uses temporal scalability to decompose the progressive image sequence to two interlaced image sequences, then one is coded as the base layer and one as the enhancement layer. Of course, the decomposition could be different. For the enhancement layer, there is a choice of making predictions. One choice for prediction is available between one base layer prediction and a temporal prediction from enhancement layer itself. It should be noted that the spatial resolution of two layers is the same and the combined temporal rate of two layers is the full temporal rate of the source. Again, it should be noted that the decoding output of base layer bitstream by the MPEG decoder provides an additional choice of prediction but not the only choice of predictions.

The signal-to-noise ratio (SNR) scalability provides a mechanism for transmitting two-layer service with the same spatial resolution but different quality levels. The low layer is coded at a coarse quantization step at 3–5 Mbits/s to provide NTSC/PAL/SECAM quality video for low capacity channel. In the enhancement layer, the difference between original and coarse-quantized signals is then coded with a finer quantizer to generate an enhancement bitstream for high quality video applications.

The above three scalability schemes generate at least two bitstreams, one for base layer and other for enhancement layer and the lower layer bitstream can be independently decoded to provide low spatial resolution, low quality, or low frame rate video, respectively. There is another scalability scheme, data partitioning, in which the base layer bitstream cannot be independently decoded. In data partitioning, a single video source is split into a high priority portion, which can be better protected, and a low priority portion, which is less important with regard to the reconstructed video quality. The priority breakpoint in the syntax specifies which syntax elements are coded as low priority (for example, the higher-order DCT coefficients in the intercoded blocks).

16.3 MPEG-2 Video Encoding

16.3.1 Introduction

MPEG video compression is a generic standard that is essential for the growth of digital video industry, as mentioned earlier. Although the MPEG video coding standard recommended a general coding methodology and syntax for the creation of a legitimate MPEG bitstream, there are many areas of research left open regarding how to generate high quality MPEG bitstreams. This allows the designers of an MPEG encoder great flexibility in developing and implementing their own MPEG-specific algorithms, leading to product differentiation on the marketplace. To design a performance optimized MPEG-2 encoder system, several major areas of research have to be considered. These include image preprocessing, motion estimation, coding mode decisions, and rate control. Algorithms for all of these areas in an encoder should aim to minimize subjective distortion for a prescribed bit rate and operating delay constraint. The preprocessing includes the noise reduction and the removal of redundant fields, which are contained in the detelecine material. The telecine material is used for the movie industry, which contains 24 progressive frames/s. The TV signal is 30 frames/s. The detelecine process converts the 24 frames/s film signal to the 30 frames/s TV signal. This is also referred to as 3:2 pull-down process. Because the 30 frames/s detelecine material only contains 24 frames/s of unique pictures, the encoder has to detect and remove the redundant fields for obtaining better coding performance. The procession of noise reduction can reduce the bits wasted for coding random noise. Motion compensation is used to remove the temporal redundancy in the video signals. The motion vectors between the anchor picture and the current picture are obtained with motion estimation algorithms. Except for I-pictures each MB can be inter- or intracoded which is determined by the mode decision. The investigation of motion estimation algorithms is an important research topic because different motion estimation schemes may result in different coding efficiency. Rate control is always applied for non-variable bit rate (non-VBR) coding. The purpose of rate control is to properly assign the bits for each MB under the constraints of total bit rate budget and buffer size. This is also an important topic because the optimized bit assignment scheme will result in better coding performance and better subjective reconstruct quality at a given bit rate. In this section, areas of preprocessing and motion estimation are covered. The topics of rate control and optimum mode decision are discussed in later sections.

16.3.2 Preprocessing

For low bit rate video coding, preprocessing is sometimes applied to the video signals before coding to increase the coding efficiency. Usually the preprocessing implies a filtering of the video signals that are corrupted by random and burst noise for various reasons, such as imperfections of the scanner, transmission, or recording medium. Noise reduction not only improves the visual quality but also increases the performance of video coding. Noise reduction can be achieved by filtering each frame independently. There are a variety of spatial filters, which have been developed for image noise filtering and restoration which can be used for noise reduction task [cano 1983; katsaggelos 1991]. On the other hand, it is also possible to filter the video sequence temporally along the motion trajectories using motion compensation [sezan 1991]. However, it was shown that among the recursive stationary methods, the MC spatiotemporal filtering performed better than spatial or MC temporal filtering alone [ozkan 1993].

Another important preprocessing is detelecine processing. As movie material is originally shot at 24 progressive frames/s, standard conversion to television at 30 frames/s is made by a 3:2 pull-down process, which periodically inserts repeated field, giving 30 frames/s telecine source material. Because the 30 frames/s detelecine material only contains 24 frames/s of unique pictures, it is necessary to detect and remove the redundant fields before or during encoding. Rather than directly encoding the 30 frames/s detelecine material, one can remove the redundant fields first and then encode 24 frames/s of unique material, thereby realizing higher coding quality at the same bit rate. The decoder can simply reconstruct the redundant fields before presenting them. Examples of telecine and detelecine process are shown in the Figure 16.19.

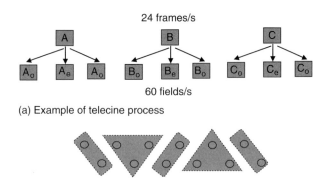

FIGURE 16.19
Examples of telecine and detelecine process. (b) Example of detelecine process

Television broadcast programmers frequently switch between telecine material and natural 30 frames/s material, such as when splicing to and from various sources of movies, ordinary television programs, and commercials. An MPEG-2 encoder should be able to cope with these transitions and consistently produce decent pictures. During movie segments, the encoder should realize the gains from coding at the lower frame rate after detelecine. Ideally, the process of source transition from the lower 24 frames/s rate to the higher 30 frames/s rate should not cause any quality drop of every encoded frame. The quality of encoded frames should maintain the same as the case where the detelecine process is ignored and all material, regardless of source type, is coded at 30 frames/s.

16.3.3 Motion Estimation and Motion Compensation

In principle for coding video signals, if the motion trajectory of each pixel could be measured then only the initial or anchor reference frame and the motion vector information need to be coded. In such a way the interframe redundancy will be removed. To reproduce the pictures, one can simply propagate each pixel along its motion trajectory. Because there is also a cost for transmitting motion vector information, in practice, one can measure only the motion vectors of a group of pixels, which will share the cost for transmission of the motion information. Of course, at the same time, the pixels in the same group are assumed to have the same motion information. This is not always true because the pixels in the block may move in different directions, or some of them belong to the background. Therefore, both motion vectors and the prediction difference have to be transmitted. Usually, the block matching can be considered as the most practical method for motion estimation due to less hardware complexity. In the block matching method, the image frame is divided into fixed size small rectangular blocks such as 16×16 or 16×8 in MPEG video coding. Each block is assumed to undergo a linear translation and the displacement vector of each block and the predictive errors are coded and transmitted. The related issues for motion estimation and compensation include motion vector searching algorithm, searching range, matching criteria, and coding method. Although the matching criteria and searching algorithms have been discussed in Chapter 11, we still briefly introduce them here for the sake of completeness.

16.3.3.1 Matching Criterion

The matching of the blocks can be determined according the various criteria including the maximum cross-correlation, the minimum MSE, the minimum mean absolute difference (MAD), and maximum matching pixel count (MPC). For MSE and MAD, the best matching

Digital Video Coding Standards: MPEG-1/2 Video

block is reached if the MSE or MAD is minimized at that location. In practice, we use MAD instead of MSE as matching criterion due to its computational simplicity. The minimum MSE criterion is not commonly used in hardware implementations because it is difficult to realize the square operation. However, the performance of the MAD criterion deteriorates as the search area becomes larger due to the presence of several local minima. In the maximum MPC criterion, each pixel in the block is classified as either a matching pixel or a mismatching pixel according the difference whether which is smaller than a preset threshold. The best matching is then determined by the maximum number of the matching pixels. However, the MPC criterion requires a threshold comparator and a counter.

16.3.3.2 Searching Algorithm

Finding the best matching block requires optimizing the matching criterion over all possible candidate displacement vectors at each pixel. The so-called full search, logarithmic search, and hierarchical searching algorithms can accomplish this.

16.3.3.2.1 Full Search

The full search algorithm evaluates the matching criterion for all possible values within the predefined searching window. If the search window is restricted to a $[-p, p]$ square, for each motion vector there are $(2p + 1)^2$ search locations. For a block size of $M \times N$ pixels, at each search location we compare $N \times M$ pixels. If we know the matching criterion and the number of operations needed for each comparison then we can calculate the computation complexity of full search algorithm. Full search algorithm is computationally expensive, but guarantees finding the global optimal matching within a defined searching range.

16.3.3.2.2 Logarithmic Search

Actually, the expected accuracy of motion estimation algorithms varies according to the applications. In MC video coding, all one seeks is a matching block in terms of some metric, even if the match does not correlate well with the actual projected motion. Therefore in most cases, search strategies faster than full searches are used, although they lead to suboptimal solutions. These faster search algorithms evaluate the criterion function only at a predetermined subset of the candidate motion vector locations instead of all possible locations. One of these faster search algorithms is the logarithmic search. Its more popular form is referred to as the three-step search. We explain the three-step search algorithm with the help of Figure 16.20 where only the search frame is depicted. Search locations

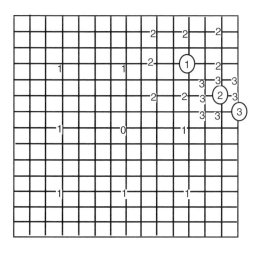

FIGURE 16.20
Three-step search.

corresponding to each of the steps in the three-step search procedure is labeled as 1, 2, and 3. In the first step, starting from pixel 0 we compute MAD for the nine search locations labeled 1. The spacing between these search locations here is 4. Assume that MAD is minimum for the search location (4,4) which is circled 1. In the second step, using the criterion function is evaluated at eight locations around the circled 1 which are labeled 2. The spacing between locations is now two pixels. Assume now the minimum MAD is at the location (6,2) which is also circled. Thus the new search origin is the circled 2 which is located at (6,2). For the third step, the spacing is set to 1 now and the 8 locations labeled 3 are searched. The search procedure is terminated at this point and the output of motion vector is (7,1). Additional steps may be incorporated into the procedure if we wish to obtain subpixel accuracy in the motion estimations. Then the search frame needs to be interpolated to evaluate the criterion function at subpixel locations.

16.3.3.2.3 Hierarchical Motion Estimation

Hierarchical representations of images in the form of a Laplacian pyramid or wavelet transform are also quite often used with block matching method for improved motion estimation. The basic idea of hierarchical block matching is to perform motion estimation at each level successively, starting with the lowest resolution level. The lower resolution levels serve to determine a rough estimate of the motion information using relatively larger blocks. The estimate of the motion vector at a lower resolution level is then passed onto the next higher resolution level as an initial estimate. The higher resolution levels are used to fine-tune the motion vector estimate. At higher resolution levels, relatively smaller window sizes can be used since we start with a good initial estimate. The hierarchical motion estimate can significantly reduce the implementation complexity because its search method is very efficient. However, such a method requires increased storage due to the need to keep pictures at different resolutions. Furthermore, this scheme may yield inaccurate motion vectors for regions containing small objects. When the search starts at the lowest resolution of the hierarchy, regions containing small objects may be eliminated and thus fail to be tracked. On the other hand, the creation of low-resolution pictures provides some immunity of noise. The following table provides some experimental results. The experimental results performed by one of the authors have shown that comparing with full search the two-layer hierarchical motion estimation reduces the search complexity of factor 10 at the price of degrading reconstruction quality from about 0.2 to 0.6 dB for frame-mode coding, from 0.26 to 0.38 dB for field-mode coding and only from 0.16 to 0.37 dB for frame/field adaptive coding, for different video sequences in the case of a fixed bit rate of 4 Mbits/s. In the case of VBR coding, the similar results can be observed from the rate distortion curves.

In the above discussion, we have restricted the motion vector estimation to integer pixel grids, or pixel accuracy. Actually, the motion vectors can be estimated with fractional or sub-pixel accuracy. In MPEG-2 video coding the half-pixel accuracy motion estimation can be used. Half-pixel accuracy can be easily achieved by interpolating the current and reference pictures by a factor 2 and then using any of the motion estimation methods described earlier.

16.3.3.3 Advanced Motion Estimation

Progress has recently been made in several aspects of motion estimation, which are described as follows.

16.3.3.3.1 Motion Estimation Using a Reduced Set of Image Data

The methods to reduce search complexity with subsampling and pyramid processing are well known and around in the literatures [sun 1994]. However, the reduction by lowering the precision of each sample does not appear to have been extensively studied. Some experimental results have shown that the performance degradation of the hierarchical motion estimation algorithm is not serious when each layer up to four-layer pyramid is limited to 6 bits/sample. At 4–5 bits/sample, the performance is degraded 0.2 dB over full precision.

16.3.3.3.2 Overlapped Motion Estimation

A limitation of block matching is that it generates a significant proportion of motion vectors that do not represent the true motion present in the scene. One possible reason is that the motion vectors are estimated without reference to any picture data outside of the nonoverlapping blocks. This problem has been addressed by overlapped motion estimation. In case of the overlapped motion compensation, MC regions translated by the motion vectors are overlapped with each other. Then a window function is used to determine the weighting factors for each vector. This technique has been adopted into the H.263 video coding standard. Some improvements have been clearly identified for low bit rate coding [katto 1994].

16.3.3.3.3 Frequency Domain Motion Estimation

An alternative to spatial-domain block matching methods is to estimate motion vector in frequency domain through calculating the cross-correlation [young 1993]. Most international standards, such as MPEG, H.263, and the proposed HDTV standard, use the DCT and block-based motion estimation as the essential elements to achieve spatial and temporal compression, respectively. The new motion estimation approach is proposed in the DCT-domain [koc 1998]. This method of motion estimation has certain merits over conventional methods. It has very low computational complexity and is robust even in a noise environment. Moreover, the motion compensation loop in the encoder is much simplified due to replacing the IDCT out of the loop [koc 1998].

16.3.3.3.4 Generalized Block Matching

In generalized block matching, the encoded frame is divided into triangular, rectangular, or arbitrary quadrilateral patches. We then search for the best matching triangular or quadrilateral in the search frame under a given spatial transformation. The choice of patch shape and the spatial transform is mutual related. For example, triangular patches offer sufficient degree of freedom with affine transformation, which has only six independent parameters. The bilinear transform has eight free parameters. Hence it is suitable for use with rectangular or quadrilateral patches. The generalized block matching is usually only adaptively used for those blocks where standard block matching is not satisfactory for avoiding imposed computational load.

16.4 Rate Control

16.4.1 Introduction of Rate Control

The purpose of rate control is to optimize the perceived picture quality and to achieve a given constant average bit rate by controlling the allocation of the bits. From the view point of rate control, the encoding can be classified into VBR coding and constant bit rate (CBR) coding. The VBR coding can provide a constant picture quality with variable coding bit rate, whereas the CBR will provide a CBR with a nonuniform picture quality.

Rate control and buffer regulation are important issues for both VBR and CBR applications. In the case of VBR encoding, the rate controller attempts to achieve optimum quality for a given target rate. In the case of CBR encoding and real-time application, the rate control scheme has to satisfy the low-latency and VBV (video buffering verifier) buffer constraints. The VBV is a hypothetical decoder, which is conceptually connected to the output of an encoder (Appendix C of [mpeg2]). The bitstream generated by the encoder is placed to the VBV buffer at the CBR that is being used. The rate control has to assure that the VBV will not be overflow and underflow. In addition, the rate control scheme has to be applicable to a wide variety of sequences and bit rates. At the GOP level, the total number of available bits are allocated among the various picture types, taking into account the constraints of the decoder buffer, so that the perceived quality is balanced. Within each picture, the available bits are allocated among the MBs to maximize the visual quality and to achieve the desired target of encoded bits for the whole picture.

16.4.2 Rate Control of Test Model 5 for MPEG-2

As we described before, the standard only defines the syntax for decoding. The TM5 is an example of encoder, which may not be optimal; however, it can provide a compliant compressed bitstream. Also, the Test Model served as a reference during the development of the standard. The TM5 rate control algorithm consists of three steps to adapting the MB quantization parameter for controlling the bit rate.

Step 1: Target bit allocation
The target bit allocation is the first step of rate control. Before coding a picture, we need to estimate the number of bits available for coding this picture. The estimation is based on several factors. These include the picture type, buffer fullness, and picture complexity. The estimation of picture complexity is based on the number of bits and quantization parameter used for coding the same type previous picture in the GOP. The initial complexity values are given according to the type of picture:

$$X_i = 160 * \text{bit rate}/115$$
$$X_p = 60 * \text{bit rate}/115 \quad (16.6)$$
$$X_b = 42 * \text{bit rate}/115$$

where the subscripts i, p, and b stand for picture types I, P, and B (this will be applied to the formulas in this section). After a picture of a certain type (I, P, or B) is encoded, the respective global complexity measure (X_i, X_p, and X_b) is updated as

$$X_i = S_i Q_i, \ X_p = S_p Q_p, \text{ and } X_b = S_b Q_b \quad (16.7)$$

where S_i, S_p, and S_b are the number of bits generated by encoding this picture, and Q_i, Q_p, and Q_b are the average quantization parameters computed by averaging the actual quantization values used during the encoding of the all the MBs including the skipped MBs. This estimation is very intuitive; however, if the picture is more complicated, more bits are needed to encode it. The quantization parameter (step or interval) is used to normalize this measure because the number of bits generated by encoder is inversely proportional to the quantization step. The quantization step can also be considered as a measure of coded picture quality. The target number of bits for the next picture in the GOP (T_i, T_p, and T_b) is computed as follows:

$$T_i = \max\left\{\frac{R}{1 + \dfrac{N_p X_p}{X_i K_p} + \dfrac{N_b X_b}{X_i K_b}}, \text{bit rate}/(8^* \text{ picture-rate})\right\}$$

$$T_p = \max\left\{\frac{R}{N_p + \dfrac{N_b K_p X_b}{X_b K_p}}, \text{bit rate}/(8^* \text{ picture-rate})\right\} \quad (16.8)$$

$$T_b = \max\left\{\frac{R}{N_b + \dfrac{N_p K_b X_p}{X_p K_b}}, \text{bit rate}/(8^* \text{ picture-rate})\right\}$$

where K_p and K_b are universal constants dependent on the quantization matrices. For the matrices of TM5, $K_p = 1.0$ and $K_b = 1.4$. R is the remaining number of bits assigned to the GOP and after coding the picture this number is updated by subtracting the bit used for the picture. N_p and N_b is the number of P- and B-pictures remaining in the current GOP in the encoding order. The problem of above target bit assignment algorithm is that it does not handle scene changes efficiently.

Step 2: Rate control
Within a picture the bits used for each MB is determined by the rate control algorithm. Then a quantizer step is derived from the number of bits available for the MB to be coded. The following is an example of rate control for P-picture.

In Figure 16.21, d_0^p is initial virtual buffer fullness, the T_p is the target bits for P-picture. B_j is the number of bits generated by encoding all MBs in the picture up to and including jth MB. MB_cnt is the number of MBs in the picture. Before encoding the jth MB, the virtual buffer fullness is adjusted during the encoding according to the following equation for P-picture:

$$d_j^p = d_0^p + B_{j-1} - \frac{T_p(j-1)}{\text{MB\_cnt}} \quad (16.9)$$

Then the quantization step is computed with the equation:

$$Q_j^p = \frac{d_j^p}{r} \quad (16.10)$$

where the reaction parameter r is given by $r = 2 * \text{bit rate}/\text{picture-rate}$ and d_j^p is the fullness of the appropriate virtual buffer. This procedure is shown in Figure 16.21. The fullness of

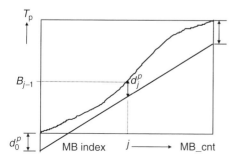

FIGURE 16.21
Rate control for P-picture.

the virtual buffer for the last MB is used for encoding the next picture of the same type as the initial fullness.

The above example can be extended to the general case for all I-, P-, and B-pictures. Before encoding the jth MB, we compute the fullness of the appropriate virtual buffer:

$$d_j^i = d_0^i + B_{j-1} - \frac{T_i(j-1)}{MB\_cnt}$$

or

$$d_j^p = d_0^p + B_{j-1} - \frac{T_p(j-1)}{MB\_cnt} \qquad (16.11)$$

or

$$d_j^b = d_0^b + B_{j-1} - \frac{T_b(j-1)}{MB\_cnt}$$

Depending on the picture type, where d_0^i, d_0^p, d_0^b are initial fullness of the virtual buffers and d_j^i, d_j^p, d_j^b are the fullness of virtual buffer at jth MB—one for each picture type. From the number of bits of the virtual buffer fullness we compute the quantization step Q_j for MB j according to the buffer fullness:

$$Q_j = \frac{d_j * 31}{r} \qquad (16.12)$$

The initial values of the virtual buffer fullness are

$$\begin{aligned} d_0^i &= 10 \cdot r/31 \\ d_0^p &= K_p \cdot d_0^i \\ d_0^b &= K_b \cdot d_0^i \end{aligned} \qquad (16.13)$$

K_p and K_b are constants which are defined in Equation 16.8.

Step 3: Adaptive quantization

Adaptive quantization is the last step of the TM5 rate control. It is noted that for active areas or busy areas, the human eyes are not so sensitive to the quantization noise, whereas the smooth areas are more sensitive to the quantization noise as discussed in Chapter 1. On the basis of this observation, we modulate the quantization step obtained from the previous step in such a way to increase quantization step for active areas and reduce the quantization step for the smooth areas. In other words, we use more bits in the smooth areas and less bits for the active areas. The experiment results have shown that the subjective quality is higher with adaptive quantization step than without this step. The procedure of adaptive quantization in TM5 is as follows. First, the spatial activity measure for the jth MB is calculated from the four luminance frame-organized subblocks and the four luminance field-organized blocks using the intrapixel values:

$$act_j = 1 + \min_{sblk=1,8} (var\_sblk) \qquad (16.14)$$

where var_sblk is the variance of each spatial 8×8 block which value is calculated as

$$var\_sblk = \frac{1}{64} \sum_{k=1}^{64} (P_k - P_{mean})^2 \qquad (16.15)$$

and P_k is the pixel value in the original 8×8 block and P_{mean} is the mean value of the block which is calculated as

$$P_{mean} = \frac{1}{64} \sum_{k=1}^{64} P_k \qquad (16.16)$$

The normalized activity factor N_act_j is

$$N\_act_j = \frac{2 \cdot act_j + avg\_act}{act_j + 2 \cdot avg\_act} \qquad (16.17)$$

where avg_act is the average value of act_j the last picture to be encoded. Therefore, this value will not give good result when a scene change occurs. On the first picture, this parameter takes the value of 400. Finally, we obtain the modulated quantization step for jth MB:

$$mquant_j = Q_j \cdot N\_act_j \qquad (16.18)$$

where Q_j is the reference quantization step value obtained in the last step. The final value of $mquant_j$ is clipped to the range of [1,31] and is used and coded as described in the MPEG standard.

As indicated before, the TM5 rate control provides only a reference model. It is not optimized in many aspects. Therefore, there is still a lot of room for improving the rate control algorithm, such as to provide more precise estimation of average activity by preprocessing. In the following section, we will investigate the optimization problem for mode decision combined with rate control, which can provide a significant quality improvement as shown by experimental results.

16.5 Optimum Mode Decision

16.5.1 Problem Formation

This section addresses the problem of determining the optimal MPEG [mpeg2] coding strategy in terms of the selection of MB coding modes and quantizer scales. In the Test Model [tm5], the rate control operates independently from the coding mode selection for each MB. The coding mode is decided based only upon the energy of predictive residues. Actually, the two processes, coding mode decision and rate control, are intimately related to each other and should be determined jointly to achieve optimal coding performance. A constrained optimization problem can be formulated based on the rate distortion characteristics, or $R(D)$ curves, for all the MBs that compose the picture being coded. Distortion for the entire picture is assumed to be decomposable and expressible as a function of individual MB distortions, with this being the objective function to minimize. The determination of the optimal solution is complicated by the MPEG differential encoding of motion vectors and dc coefficients, which introduce dependencies that carry over from MB to MB for a duration equal to the slice length. As an approximation, a near-optimum greedy algorithm can be developed. Once the upper bound in performance is calculated, it can be used to assess how well practical suboptimum methods perform.

Earlier-related studies dealing with dependent quantization for MPEG include the works done by Ramchandran [ramchandran 1994] and Lee [lee 1994]. These works treated the problem of bit allocation where there is temporal dependency in coding complexity across I, P, and B frames. Although these techniques represent the most proper bit allocation strategies across frames from a theoretical viewpoint, no practical real-time MPEG encoding system will use even those proposed simplified techniques because they require an unwieldy number of pre-analysis encoding passes over the window of dependent frames (one MPEG GOP). To overcome these computational burdens, more pragmatic solutions that can realistically be implemented have been considered by Sun [sun 1997]. In this study, the major emphasis is not on the problem of bit allocation among I-, P-, and B-frames; rather, the authors choose to utilize the frame-level allocation method provided by the Test Model [tm5]. In this way, frame-level coding complexities are estimated from past frames without any forward pre-analysis knowledge of future frames. This type of analysis forms the most reasonable set of assumptions for a practical real-time encoding system. Another method that extends the basic Test Model idea to alter frame budgets heuristically in the case of scene changes, use of dynamic GOP size, and temporal masking effects can be found in [wang 1995]. These techniques also offer very effective and practical solutions for implementation. Given the chosen method for frame-level bit budget allocation, the focus of this section is to jointly optimizing MB coding modes and quantizers within each frame.

There exists many choices for the MB coding mode under the MPEG-2 standard for P- and B-pictures, including intramode, no motion compensation mode, frame/field/dual prime motion compensation intermode, forward/backward/average intermode, and field/frame DCT mode. In the standard Test Model reference [tm5], the coding mode for each MB is selected by comparing the energy of predictive residuals. For example, the intra/inter decision is determined by a comparison of the variance of the MB pixels against the variance of the predictive residuals; the inter prediction mode is selected to be the intermode that has the least predictive residual MSE. The coding mode selected by Test Model criteria does not result in the optimal coding performance.

In attempting to achieve optimal coding performance, it is important to realize that coding modes should be determined jointly with rate control because the best coding mode depends upon the operating point for rate. In deciding which of the various coding modes is best, one should consider what the operating point is for distortion, and also consider the trade-off between spending bits for coding the prediction residuals and bits for coding motion vectors.

The amount of bits used for coding the MB is the sum of bits used for coding motion vectors and bits used for coding residuals:

$$R_{\text{MB}} = R_{\text{mv}} + R_{\text{residual}} \tag{16.19}$$

For example, in Figure 16.22, consider the decision between (a) frame-mode forward prediction and (b) field-mode bidirectional prediction. Mode (b) will almost always produce a prediction that has lower MSE than mode (a). However, mode (a) requires coding of fewer motion vectors than mode (b). Which mode is best? The answer depends on the operating point for distortion. When coding at a very coarse quant-scale, mode (a) can perform better than mode (b) because the difference in bits required for coding motion vectors between the two modes may be much greater than the difference in bits required for coding residuals between the two modes. However, when coding at a fine quant-scale, mode (b) can perform better than mode (a) because mode (b) provides a better prediction and the bits required for motion vectors would become negligible compared to bits for coding residuals.

Coding mode decisions and rate control can be determined jointly and optimally starting from the basics of constrained optimization using $R(D)$ curves. This optimal solution

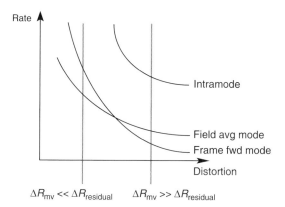

FIGURE 16.22
Rate distortion [R(D)] curves for different macroblock (MB) coding modes.

would be an a-posterior solution that assumes complete knowledge of R(D). We investigate an optimal solution for objective functions of the form:

$$D_{\text{PICT}} = \sum_{i=1}^{\text{NMB}} D_{\text{MB}i} \qquad (16.20)$$

Equation 16.20 states that the distortion for the picture, D_{PICT}, can be measured as an accumulation of individual MB distortions, D_{MB}, for all NMB number of MBs in the picture. We minimize this objective function subject to having individual MB distortions being uniform over the picture:

$$D_1 = D_2 = \cdots = D_{\text{NMB}} \qquad (16.21)$$

and having the bits generated from coding each MB, R_{MB}, sum to a target bit allocation for the entire picture, R_{PICT}

$$\sum_{i=1}^{\text{NMB}} R_{\text{MB}i} = R_{\text{PICT}} \qquad (16.22)$$

The choice for the MB distortion measure, D_{MB}, can be MSE computed over the pixels in the MB, or it can be a measure that reflects subjective distortion more accurately, such as luminance and frequency-weighted MSE. Other choices for D_{MB} may be the quantizer scale used for coding the MB, or better yet, the quantizer scale weighted by an activity masking factor. In this chapter, we select distortion for each MBi to be spatial-masking-activity-weighted quantizer scale:

$$D_{\text{MB}_i} = \text{qscale}_i / N\_\text{act}_i \qquad (16.23)$$

where $N\_\text{act}_i \in [0.5, 2.0]$ is the normalized spatial-masking-activity quantizer weighting factor, as defined in the Test Model [tm5]:

$$N\_\text{act}_i = \frac{2 * \text{act}_i + \text{avg}\_\text{act}}{\text{act}_i + 2 * \text{avg}\_\text{act}} \qquad (16.24)$$

where act$_i$ is the minimum luma block spatial variance for MB I_i and avg_act is the average value of act$_i$ over the last picture to be coded. N_act_i reflects the relative amount of quantization error that can be tolerated for MBi as compared to the rest of the MBs that compose the picture. N_act$_i$ depends strongly on whether the MB belongs to a smooth, edge, or textured region of the picture. Hence, the MB distortion metric is space variant and depends on the context of the local picture characteristics surrounding each MB. We assume that maintaining the same D_{MBi} for all MBs, or selecting the quantizer scales directly proportional to N_act$_i$ in such a manner, corresponds to maintaining uniform subjective quality throughout the picture. Masking-activity-weighted quantizer scale is somewhat a coarse measure for image quality, but it reflects subjective image quality better than MSE or peak signal to noise ratio (PSNR), and it is a practical metric to compute that lends itself to an additive form for distortion.

It is important to note that the resulting distortion measure for the picture D_{PICT} is really only meaningful as a relative comparison figure for the same identical picture (thus having the same masking activities) quantized different ways. It is not useful in comparing between two different images. PSNR is only useful in this sense too, though with poorer subjective accuracy.

In the following, a procedure for obtaining the optimal coding performance with the joint optimization of coding mode selection and rate control is discussed. As this method would be too complex to implement, a practical suboptimal heuristic algorithm is presented. Some simulation results and comparisons between the different algorithms—Test Model algorithm, Near-Optimum algorithm, and the practical Sub-Optimum algorithm—are also provided to assist the reader in understanding the differences in performance.

16.5.2 Procedure for Obtaining the Optimal Mode

16.5.2.1 Optimal Solution

The solution to the optimization problem is unique because the objective function is monotonic and the individual MB $R(D)$ functions are also monotonic. To solve for the optimal set of MB modes and quant-scales for the picture (mode and qscale), the differential encoding of motion vectors and intra dc coefficients as done in MPEG should be accounted for. According to MPEG, each slice has its own differential encoding chain. At the start of each slice, prediction motion vectors are reset to zero. As each MB is encoded in raster scan order, the MB motion vectors are encoded differentially with respect to prediction motion vectors that depend on the coding mode of the previous MB. These prediction motion vectors may be reset to zero in the case that the previous MB was coded as intra or skipped. Similarly, dc coefficients in continuous runs of intra MBs are encoded differentially with respect to the previous intra MB. The intra dc predictors are reset at the start of every slice, and at inter or skipped MBs. Slice boundaries delimit independent self-contained decodable units. Finding the optimal set of coding modes for the MBs in each slice entails a search through a trellis of dimensions S stages by M states per stage, with S being the slice size and M being the number of coding modes being considered (see Figure 16.23). This trellis structure arises because there are M^2 distinct rate distortion, $R_{mode|previous-mode}(D)$, characteristic curves corresponding to each of M coding modes, with each in turn having a different dependency for each of M coding modes of the previous MB. We now consider populating the trellis links with values by sampling the set of these M^2S. rate distortion curves at a specific distortion level. For a given fixed MB distortion level, D_{MB}, each link on the trellis is assigned a cost equal to the number of bits to code an MB in a certain mode given the mode from which the preceding MB was coded. For any group of links entering a node, the cost of these links differs only because of the difference in bits caused by the motion vector and dc coefficient coding dependency upon the earlier MB.

Digital Video Coding Standards: MPEG-1/2 Video

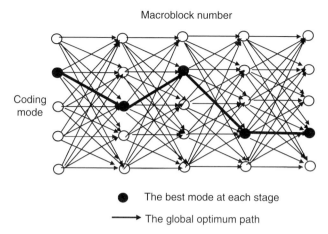

FIGURE 16.23
Full search trellis, M^S (M is number of modes at each stage and S is the length of slice) searches needed to obtain the best path.

● The best mode at each stage
→ The global optimum path

The computational requirements per slice involve

- To determine link costs in the trellis, the number of code the MB operates (i.e., DCT + quantization + RLC/VLC) is equal to $M^2 S$.
- After determining all trellis link costs, the number of path searches is equal to M^S.

A general iterative procedure for obtaining the optimal solution is as follows:

1. Initialize a guess for $D_{MB} = D_{MB0}$. However, D_{MB} is the same for every MB in the picture; this sets an initial guess for the operating distortion level of the picture.
2. Follow the given procedure for each slice in the picture:
   (a) For each MB in the slice and the mode considered, determine the quantizer scale, which yields the distortion level D_{MB}, i.e., $q_s = f(D_{MB})$, where f is the function which describes the relationship between quantizer scale q_s and distortion D_{MB}. If we use spatial-masking-activity-weighted quantizer scale as a measure of distortion (as from Equation 16.4), then q_s equals $N\_act * D_{MB}$.
   (b) Compute all the link costs in the trellis representing the slice.
   The link costs, R_{MBi} (mode k|mode j), represent the number of resulting bits (total bits for coding residual, motion vectors, and MB header) for coding MBi in mode k given that the preceding MB was coded in mode j.
   (c) Search through the trellis to find the path that has the lowest ΣR_{MBi} over the slice.
3. Compute ΣR_{MBi} for all MBs in the picture and compare to target R_{PICT}.
   (a) If $|\Sigma R_{MBi} - R_{PICT}| < \varepsilon$ then the optimal $\overrightarrow{mode}$ and $\overrightarrow{qscale}$ has been found for picture. Repeat the process for the next picture.
   (b) If $\Sigma R_{MBi} < R_{PICT}$ then decrement $D_{MB} = D_{MB} - \Delta D_{MB}$ and goto step 2.
   (c) If $\Sigma R_{MBi} > R_{PICT}$ then increment $D_{MB} = D_{MB} + \Delta D_{MB}$ and goto step 2.

16.5.2.2 Near-Optimal Greedy Solution

The solution from the full exponential-order search requires an unwieldy amount of computations. To avoid the heavy computational burden, we can use a greedy approach [lee 1994] to simplify and sidestep the dependency problems of the full search method. In the greedy algorithm, the best coding mode selection for the current MB depends only upon

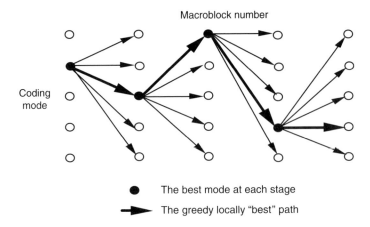

FIGURE 16.24
Greedy approach, $M \times S$ comparisons needed to obtain the locally "best" path.

the best mode of the previous coded MB. Therefore, the upper bound we obtain is a near-optimum solution instead of a global optimum. Figure 16.24 illustrates the greedy algorithm. After coding an MB in each of the M modes, the mode resulting in the least number of bits is chosen to be best. The very next MB is coded with dependencies to that chosen best mode. The computations per slice are reduced to $M \times S$ "code the MB" operations and $M \times S$ comparisons. A general iterative procedure for obtaining the greedy solution is as follows:

1. Initialize a guess for $D_{MB} = D_{MB0}$.
2. Follow the given procedure for each MB:
   (a) For each mode considered, determine the quantizer scale which yields the distortion level D_{MB}, i.e., $q_s = f(D_{MB})$, where f is the function we mentioned earlier.
   (b) For each mode, code the MB in that mode with that q_s value and record the resulting number of generated bits, $R_{MBi}(\text{mode } i|\text{mode } j)$. The MB is coded based on the earlier determined mode of the preceding MB.
   (c) The best mode for MBi is the mode for which $R_{MBi}(\text{mode } i|\text{mode } j)$, mode is smallest. This yields R_{MBi} bits for MBi.
3. Compute ΣR_{MBi} for all MBs in the picture and compare to target R_{PICT}.
   (a) If $|\Sigma R_{MBi} - R_{PICT}| < \varepsilon$ then the optimal $\overrightarrow{\text{mode}}$ and $\overrightarrow{\text{qscale}}$ has been found for picture. Repeat the process for the next picture.
   (b) If $\Sigma R_{MBi} < R_{PICT}$ then decrement $D_{MB} = D_{MB} - \Delta D_{MB}$ and goto step 2.
   (c) If $\Sigma R_{MBi} > R_{PICT}$ then increment $D_{MB} = D_{MB} + \Delta D_{MB}$ and goto step 2.

16.5.3 Practical Solution with New Criteria for the Selection of Coding Mode

It is obvious that the near-optimal solution discussed in the previous section is not a practical method because of its complexity. To determine the best mode, we have to know how many bits it takes to code each MB in every mode with the same distortion level. The total number of bits for each MB, R_{MB}, consists of three parts, bits for coding motion vectors, R_{mv}, bits for coding the predictive residue, R_{res}, and bits for coding MB header information, R_{header}, such as MB-type, quantizer scale, and coded-block-pattern.

$$R_{MB} = R_{mv} + R_{res} + R_{header} \tag{16.25}$$

Digital Video Coding Standards: MPEG-1/2 Video

FIGURE 16.25
Coding stages to find out bit count.

The number of bits for motion vectors, R_{mv}, can be easily obtained by VLC table look-up. But to obtain the number of bits for coding the predictive residue, one has to go through the three-step coding procedure: (1) DCT, (2) quantization, and (3) VLC as shown in Figure 16.25. At step 3, R_{res} is obtained with a look-up table according to the run length of zeros and the level of quantized coefficients, i.e., R_{res} depends on the pair of values of run and level:

$$R_{res} = f(\text{run, level}) \tag{16.26}$$

As stated above, to obtain the upper-bound coding performance all three steps are needed for each coding mode, and then the coding mode resulting in the least number of bits is selected as the best mode.

To obtain a much less computationally intensive method, it is preferred to use a statistical model of DCT coefficient bit usage versus variance of the prediction residual and quantizer step size. This will provide an approximation of the number of residual bits, R_{res}. For this purpose, we assume that the run and level pair in Equation 16.7 is strongly dependent on values of the quantizer scale, q_s, and the variance of the residue, V_{res}, for each MB. Intuitively, we would expect the number of bits to encode an MB is proportional to the variance of the residual and inversely proportional to the value of quantizer step size. Therefore a statistical model can be constructed by plotting R_{res} versus the independent variables V_{res} and q_s over a large set of representative MB pixels from images typical of natural video material. This results in a scatter plot showing tight correlation, and hence a surface can be fit through the data points. It was found that Equation 16.24 can be approximately expressed as

$$R_{res} \approx f(q_s, V_{res}) = [K/(Cq_s + q_s^2)]V_{res} \tag{16.27}$$

where K and C are constants found through surface fitting regression. If we assume R_{header} is a relatively fixed component that does not vary much with MB coding mode and can be ignored, then Equation 16.23 can be approximately replaced by

$$R_{MB'} = R_{mv} + [K/(Cq_s + q_s^2)]V_{res} \tag{16.28}$$

The value of $R_{MB'}$ reflects the variable portion of bit usage that is dependent on coding mode, and can be used as the measure for selecting the coding mode in our encoder. For a given quantizer step size, the mode resulting in the smallest value of $R_{MB'}$ is chosen as the best mode. It is obvious that in the use of this new measurement to select the coding mode, the computational complexity increase over the Test Model method is very slight (the same identical calculation for V_{res} is made in the Test Model).

16.6 Statistical Multiplexing Operations on Multiple Program Encoding

In this section, the strategies for StatMux operation on the multiple program encoding are introduced. This topic is an extension of rate control into the case of multiple program encoding. First, a background survey of general encoding and multiplexing modes is reviewed. Second, the specific algorithm used in some current systems has been introduced; its shortcomings are addressed and possible amendments to the basic algorithm are described. Some potential research topics such as modeling strategies and methods for solving the problem are proposed for investigation. These topics may be good research topics for the interested graduate student.

16.6.1 Background of Statistical Multiplexing Operation

In many applications, several video sources may often be combined, or multiplexed onto a single link for transmission. At the receiving end, the individual sources of data from the multiplexed data are demultiplexed and supplied to the intended receivers. For example, in an asynchronous transfer mode (ATM) network scenario many video sources originating from a local area are multiplexed onto a wide area backbone trunk. In a satellite broadcasting scenario, several video sources are multiplexed for transmission through a transponder. In a cable TV scenario, hundreds of video programs are broadcasted onto a cable bus. Because the transmission channel, such as a trunk, a transponder, or a cable, is always an expensive resource, the limited channel capacity should be exploited as much as possible. The goal of StatMux encoding is to make the best use of the limited channel capacity as possible. There are several approaches to encoding and multiplexing a plurality of video sources. In the following, we compare the methods and describe the situation where each method is applicable. The qualitative comparisons are made in terms of trade-offs among factors of computation, implementation complexity, encoded picture quality, buffering delay, and channel utilization. To understand the StatMux method, we introduce a simple case of deterministic multiplexing function of CBR encoder. The standard method for performing the encoding and multiplexing function is to independent encode source with a CBR. The CBR encoder produces an encoded bitstream, representing the video supplied to it, at a predetermined CBR. To produce CBR, the CBR encoder utilizes a rate buffer and feedback control mechanism that continually modifies the amount of quantization applied to the video signal as shown in Figure 16.26.

The CBR encoder provides a CBR with varying encoded picture quality. This means that the degree of quantization applied depends upon the current frame's coding complexity offered to the MPEG compression algorithm. Fine quantization is then applied to those

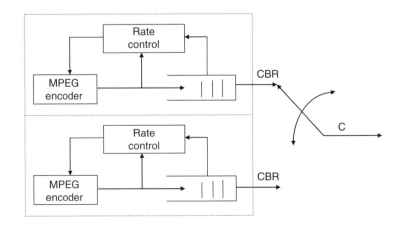

FIGURE 16.26
Independent encoding/muxing of CBR sources.

frames that have low spatial and temporal coding complexity, and conversely coarse quantization is applied to frames that possess high spatial and temporal coding complexity to meet the bit rate. However, varying the quantization level corresponds to varying the video quality. Thus, in a CBR encoder, spatial and temporal complexity tends to be encoded in such a manner that the subjective quality of the reproduced image is lower than that of less complex images. This makes any form of rate control inherently bad in the sense that control is always imposed in a direction contrary to the goal of achieving uniform image quality. Usually, bit rates for CBR encoders are chosen so that the moderately difficult scenes can be coded to an acceptable quality level. Given that moderately difficult scenes give good results, then all simpler scenes will yield even better results with the given rate, while very difficult scenes will result in noticeable degradation. As CBR encoders produce CBR, the multiplexing of a plurality of sources is very simple. The required channel capacity would simply be the sum of all the individual CBRs. Deterministic time or frequency division multiplexing of the individual CBR bitstreams onto the channel in a well-known and simple process. So with uniform CBR encoding, consistent image quality is impossible for the video sequence with varying scene complexity but the reward is the ease of multiplexing. The penalty of CBR coding with easy multiplexing may not only result in the nonuniform picture quality but also result in lower efficiency of channel bandwidth employment. Better efficiency can be gained by StatMux, whereby each source is encoded at a VBR coding approach. The VBR coding will result in uniform or consistent coded image quality by fixing the quantization scale or by modulating quantization scale to a limited extent according to activity masking attributes of the human visual system (HVS). Then the bit rates generated by VBR coding vary with the incoming video source material's coding complexity. The StatMux is referred to as StatMux in short. The coding gain of StatMux is possible through sharing of the channel resource jointly among the encoders. For example, two MPEG encoders may assign the appearance of their I-pictures at different time, this may reduce the limitation of the maximum channel bandwidth requirement because coding I-picture may generate a large number of bits. This may not be a good example for practical applications. However, this explains that the process of StatMux is not a zero-sum game whereby one encoder's gain must be exactly another encoder's loss. In the process of StatMux, one encoder's gain is obtained by using the channel bandwidth, which another encoder does not need at that time or would bring a very marginal gain for another encoder at that time. More exactly, this concept of gains through sharing arises when the limited amount of bits is dynamically appropriated toward encoders that can best utilize those bits in substantially improving its image quality during complex segments and eschewed from encoders that can improve its image quality only marginally during easy segment. It is obvious that the CBR-encoded sources do not need StatMux because the bandwidth for each encoded source is well defined. The gain of StatMux can only be possibly obtained with VBR-encoded sources. In the following section, we discuss two kinds of multiplexing with multiple VBR-encoded sources.

16.6.2 VBR Encoders in StatMux

There are two multiplexing methods for encoding multiple sources with VBR encoders, open-loop and close-loop. Each VBR encoder in open-loop multiplexing mode produces the most consistently uniform predefined image quality level regardless of the coding complexity of incoming video sources. The image quality is decided by fixing quantization scale. When the quantization scale is fixed, the SNR is fixed under assumption of white Gaussian quantization noise. Sometimes, the quantization scale is slightly modulated according to the image activity to match HVS for example in the method in MPEG-2 TM5. The resulting VBR process is generated by allowing the encoder to freely use, however, many bits needed to meet the predetermined quality level. Usually, each video source encoded by VBR encoder in open-loop mode is not geographically colocated and cannot be encoded jointly. However, the resulting VBR processes do share the channel jointly, in the sense that the total channel bandwidth is not rigidly allocated among the sources in a fixed manner such as done in CBR operation mode where each source has the fixed portion of channel bandwidth. The instantaneous combined rates of all the VBR encoders may exceed the channel capacity; especially, in the case when all the encoders generate the bursts of bits at the same time the joint buffer will overflow, thereby leading to loss of data. However, there always still exists a possibility to more efficiently utilize the channel capacity by carefully allocating the loading conditions without losing of data. But the totally open-loop VBR coding is not stationary and it is hard to achieve both good channel utilization and very limited data loss. A practical method of VBR transmission for use in ATM environment involves placing limitations to the degree of variability allowed in VBR processes. Figure 16.27 illustrates the idea of self-regulating VBR encoders.

The difference between proposed VBR encoder and totally open-loop VBR encoder is that a looser form of rate control is imposed to the VBR encoder to avoid violating transmission constraints that are agreed to by the user and the network as part of the contract negotiated during the call setup stage. The rate control will match the policing function, which is enforced by the network. Looser rate control means that the rate control is not as strict as the one in CBR case because it allows for the encoder to vary its output bit rate according to the coding complexity up to a certain degree as decided by the policing function.

In some applications such as the TV broadcasting or cable TV, the video sources may be geographically colocated at the same site. In such scenarios, additional gains can be realized by the StatMux in which the sources are jointly encoded and jointly multiplexed. By using a common rate controller, all encoders operate in VBR mode, but without contending and stepping over one another as in independent VBR encoding and multi-plexing. The joint rate controller assigns the total available channel capacity to each encoder so that a certain common quality level is maintained. The bit rates assigned to each individual encoder by joint rate control dynamically change based on the coding complexities of each video source to achieve the most uniform quality among the encoders

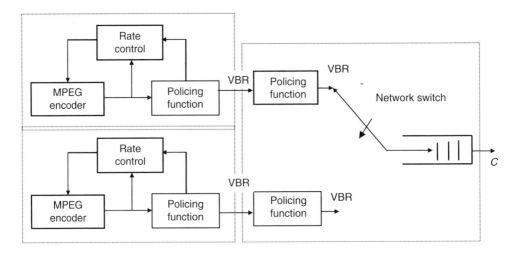

FIGURE 16.27
Independent encoding/muxing of geographically dispersed VBR sources.

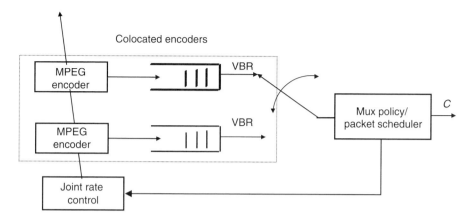

FIGURE 16.28
Method of joint rate control and multiplexing.

and along time for each encoder. In such a joint rate control method, although each encoder produces its own VBRs, the sum of bits produced by all encoders combined together is a CBR to fit the channel capacity. Such an idea is shown in Figure 16.28.

16.6.3 Research Topics of StatMux

The major problem of StatMux is how to allocate the bit rate resource among the video sources which share the common channel bit rate and are jointly encoded by a joint rate controller. This allocation should be based on the coding complexity of each source. The bit rate, $R_i(t)$, for encoder i at time t according to the normalized coding complexity of all encoders for the GOP period ending at time t such as

$$R_i(t) = \frac{X_i(t)}{\sum_{j=1}^{N} X_j(t)} \cdot C \qquad (16.29)$$

where $X_i(t)$ is the coding complexity of source for encoder i at the time t over a GOP period and C is the total channel capacity. Also the bit rate assignment has to be updated from time to time to trace the variation of source complexity. In the following, we will discuss several topics which may be the research topics for graduate students.

16.6.3.1 Forward Analysis

Without forward analysis, scene transitions are unanticipated and lead to incorrect bit allocation for a brief transient period following the scene changes. If the bit allocation of current video segment is based on the complexity of previous video segment and is adjusted by the available bit rate resource, those video segments which change from easy coding complexity to difficult coding complexity suffers the greatest degradation without pre-analysis of upcoming increased complexity. Pre-analysis could be performed with a dual set of encoders operating with a certain preprocessing delay ahead of the actual encoding process. As a simple example, we start to assign the equal portion of bit rate for each encoder, then we can obtain the average quantization scale for this GOP that can be considered as the forward analysis results of coding complexity. The real coding process can operate on the coding complexity obtained by the pre-analysis. If we choose one or two

GOPs according to the synchronous status of the input video sources to perform the pre-analysis, it will result in small buffering delay.

16.6.3.2 Potential Modeling Strategies and Methods

Several modeling strategies and methods have been investigated to find a suitable procedure for classifying sources and determining what groups of sources can appropriately be jointly encoded together for transmission over a common channel as so to meet a specified image quality level. These modeling strategies and methods include modeling of video encoding, modeling of source coding complexity, and source classification. The modeling of video encoding algorithm involves measuring the operating performance of the individual encoders or characterizing its rate distortion function for a variety of scenes. Embodied into this model are the MPEG algorithms implemented for motion estimation, mode decision, rate control, and their joint optimization issues. It has been speculated that a hyperbolic functional form of

$$\text{Rate} = X/\text{distortion} \qquad (16.30)$$

would be appropriate over the normal operating bit rate range of 3–7 Mbits/s for MPEG-2-encoded ITU-R 601-sized videos. The hyperbolic shape of rate distortion curves would be also suitable for all video scenes. Actually, we can use a set of collected rate distortion data pairs with an encoder to fit a hyperbola through the points as shown in Figure 16.29 and estimate the shape parameter X. The value of X will be used to present the coding complexity offered to an encoder. For modeling at the GOP level, the rate would be the number of bits used to code that GOP and the distortion can be chosen as the averaging quantization scale over the GOP. In some literatures, the distortion is taken as the average PSNR over the GOP or overall sequence. If it is assumed that the quantization noise is modeled by white Gaussian noise then both distortion measures are equivalent.

After obtaining the correct coding complexity parameters, we can improve the StatMux algorithm by assigning an encoding bit budget to each encoder based on the GOP level normalized complexity measure X that each encoder is encoding. The GOP level normalized complexity measure $X(n)$ is defined as

$$X(n) = \sum_{i \in \text{GOP}} T(i)Q(i) \qquad (16.31)$$

where
n is the GOP number
$T(i)$ is the total number of bits used for encoding picture i
$Q(i)$ is the average quantization scale used for encoding picture i

Some research results have shown that the $X(n)$ is insensitive to the operating bit rate; therefore, $X(n)$ is a reliable measure of a video source's loading characteristics. Therefore,

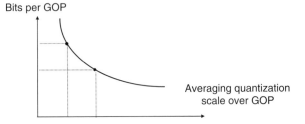

FIGURE 16.29
Rate distortion modeling of encoding algorithm and video source.

the study of accurate model of the random process of $X(n)$ is very important for improving the operations of the StatMux algorithm. The accurate model of $X(n)$ reflects the video source's loading characteristics which dictates the share of total bit budget that an encoder expects to get. Several statistical models have been proposed to describe the complexity measure, $X(n)$. For example, an auto-regressive process model is proposed for the intra-scene $X(n)$ process. This proposed model is based on the following observations; the complexity measure within a single scene has a skewed distribution by the Gamma function, and furthermore, the complexity measure within a scene displays a strong temporal correlation and the form of the correlation is essentially exponential. The definition for the Mth order auto-regressive model is

$$X(n) = \sum_{m=1}^{M} a(m) \cdot X(n-m) + e(n) \qquad (16.32)$$

where
$e(n)$ is the white noise process
$a(m)$s are the innovation filter coefficients

The statistics of the model such as the mean value, the variance, the correlation, and marginal distribution are used to match those of actual signals by adjusting $a(m)$s, $e(n)$, and M. Other cases, such as scene transition model and intercoded scene models, we leave as the project topics for the graduate students.

16.7 Summary

In this chapter, the technical detail of MPEG video was introduced. The technical detail of MPEG standards includes the decoding process of MPEG-1 and MPEG-2 video. Although the encoding process is not a standard part, it is very important for the content providers and service providers. We discussed the most important parts of encoding techniques. Some examples such as the joint optimizing of mode decision and rate control are good examples to understand how the standard is used.

Exercises

1. According to your understanding, give several reasons to explain why the MPEG standards specify only the decoding as normative part and define the encoding as informative part (Test Model).
2. Can an MPEG-2 video decoder decode a bitstream generated by an MPEG-1 video encoder? Summarize the main difference between the MPEG-1 and MPEG-2 video standards.
3. Pre-filtering may reduce the noise of original video source and increase the coding efficiency. But at the same time the pre-filtering will result in a certain information loss. Conduct a project to investigate at what bit rate range the pre-filtering may benefit the coding efficiency for some video sources.
4. Use TM5 rate control to encode several video sequences (such as Flower Garden sequence) in two ways: (a) with adaptive quantization step and (b) without adaptive

quantization step (Equation 16.16). Compare and discuss the numerical results and subjective results (observe the smooth areas carefully).
5. Why does MPEG-2 uses different quantizer matrices for intra- and intercoding? Conduct a project to use different quantization matrices to encode several video sequences and report the results.
6. Conduct a project to encode several video sequences (a) with B-picture and (b) without B-picture. Compare the numerical and subjective results. Observe what difference exists between the sequences with fast motion and the sequences with slow motion. (Typical bit rates for ITU-R 601 sequences are 4–6 Mbits/s.)

References

[cano 1983] D. Cano and M. Benard, 3-D Kalman filtering of image sequences, in *Image Sequence Processing and Dynamic Scene Analysis*, T.S. Huang (Ed.), Berlin, Springer, 1983, pp. 563–579.

[haskell 1997] B.G. Haskell, A. Puri, and A.N. Netravali, *Digital Video: Introduction to MPEG-2*, Chapman and Hall, New York, 1997.

[katsaggelos 1991] A.K. Katsaggelos, R.P. Kleihorst, S.N. Efstratiadis, and R.L. Lagendijk, Adaptive image sequence noise filtering methods, *Proceeding of SPIE Visual Communication and Image Processing*, Boston, MA, pp. 10–13, November 1991.

[katto 1994] Jiro Katto, Jun-ichi Ohki, Satoshi Nogaki, and Mutsumi Ohta, A wavelet codec with overlapped motion compensation for very low bit rate enviroment, *IEEE Transactions on Circuits and Systems for Video Technology*, 4, 3, 328–338, June 1994.

[koc 1998] U.-V. Koc and K.J.R. Liu, DCT-based motion estimation, *IEEE Transactions on Image Processing*, 7, 948–965, July 1998.

[lee 1994] J. Lee and B.W. Dickerson, Temporally adaptive motion interpolation exploiting temporal masking in visual perception, *IEEE Transactions on Image Processing*, 3, 5, 513–526, September 1994.

[mitchel 1997] J.L. Mitchell, W.B. Pennebaker, C.E. Fogg, and D.J. LeGall, *MPEG Video Compression Standard*, Chapman and Hall, New York, 1997.

[mpeg1] ISO/IEC 11172, International Standard, 1992.

[mpeg2] ISO/IEC 13818 MPEG-2 International Standard, Video Recommendation ITU-T H.262, January 10, 1995.

[ozkan 1993] M.K. Ozkan, M.I. Sezan, and A.M. Tekalp, Adaptive motion compensated filtering of noisy image sequences, *IEEE Transactions on Circuits and Systems for Video Technology*, 3, 4, 277–290, August 1993.

[ramchandran 1994] Kannan Ramchandran, Antonio Ortega, and Martin Vetterli, Bit Allocation for Dependent Quantization with Application to MPEG Video Coders, *IEEE Transactions on Image Processing*, 3, 5, 526, 533–545, September 1994.

[sezan 1991] M.I. Sezan, M.K. Ozkan, and S.V. Fogel, Temporal adaptive filtering of noisy image sequences using a robust motion estimation algorithm, *IEEE ICASSP*, 2429–2432, 1991.

[sun 1994] H. Sun, Sarnoff Internal Technical Report, May 1994.

[sun 1997] H. Sun, W. Kwok, M. Chien, and C.H. John Ju, MPEG coding performance improvement by jointly optimization coding mode decision and rate control, *IEEE Transactions on Circuits and Systems for Video Technology*, 7, 3, 449–458, June 1997.

[tm5] MPEG-2 Test model 5, ISO-IEC/JTC1/SC29/WG11, April, 1993.

[wang 1995] L. Wang, Rate control for MPEG-2 video coding, *SPIE on Visual Communications and Image Processing*, Taipei, Taiwan, pp. 53–64, May 1995.

[young 1993] R.W. Young and N.G. Kingsbury, Frequency-domain motion estimation using a complex lapped transform, *IEEE Transactions on Image Processing*, 2, 1, 2–17, January 1993.

17

Application Issues of MPEG-1/2 Video Coding

This chapter is an extension of Chapter 16. We will introduce several important application issues of MPEG-1/2 video that include the Advanced Television Standard Committee (ATSC) DTV standard which has been adopted by the Federal Communications Commission (FCC) as TV standard in the United States: transcoding, down-conversion decoder, and error concealment.

17.1 Introduction

Digital video signal processing is an area of science and engineering that has developed rapidly over the last decade. The maturity of the moving picture expert group (MPEG) video coding standard is a very important achievement for the video industry and provides a strong support for digital transmission and storage of video signals. The MPEG coding standard is now being deployed for a variety of applications, which include high definition television (HDTV), teleconferencing, direct broadcasting by satellite (DBS), interactive multimedia terminals, and digital video disc (DVD). The common feature of these applications is that the different source information such as video, audio, and data are all converted to the digital format and then mixed together to a new format that is referred to as the bitstream. This new format of information is a revolutionary change in the multimedia industry, since the digitized information format, i.e., the bitstream, can be decoded by not only the traditional consumer electronic products such as television, but also the digital computer. In this chapter, we will present several application examples of MPEG-1/2 video standards, which include the ATSC DTV standard, transcoding, down-conversion decoder, and error concealment. The DTV standard is the application extension of MPEG video standard. The transcoding and down-conversion decoders are the practical application issues which increase the features of compression-related products. The error concealment algorithms provide the tool for transmitting the compressed bitstream over noisy channels.

17.2 ATSC DTV Standards

17.2.1 A Brief History

The birth of digital television (DTV) in the United States has undergone several stages, which are the initial stage, the competition stage, the collaboration stage, and the approval stage [reitmeier 1996]. The concept of HDTV was proposed in Japan in the late 1970s and early 1980s. During that period, Japan and Europe continued to make their efforts in the development of analog television transmission systems such as MUSE and HD-MAC systems. In early 1987, U.S. broadcasters fell behind in this field and felt they should take action to catch up with the new HDTV technology and petitioned the FCC to reserve spectrum for terrestrial broadcasting of HDTV. As a result, the Advisory Committee on Advanced Television Service (ACATS) was founded in August 1987. This committee takes the role of recommending a standard to the FCC for approval. Thus, the process of selecting an appropriate HDTV system for the United States started. At the initial stage between 1987 and 1990, there were over 23 different analog systems proposed; among these systems two typical approaches were extended definition television (EDTV) that fits into a single 6 MHz channel, and HDTV approach that requires two 6 MHz channels. By 1990, ACATS had established the Advanced Television Test Center (ATTC), an official testing laboratory sponsored by broadcasters to conduct extensive laboratory tests in Virginia and field tests in Charlotte, North Carolina. Also, the industry had formed the Advanced Television Standards Committee (ATSC) to perform the task of drafting the official standard documents of the selected winning system.

As we know, the current ATSC proposed television standard is a digital system. In the early 1990s, the FCC issued a very difficult request to the industry about DTV standard. The FCC required the industry to provide full quality HDTV service in a single 6 MHz channel. Having recognized the technical difficulty of this requirement at that time, the FCC also stated that this service could be provided by a simulcast service in which programs would be simultaneously broadcasted in both NTSC and the new television system. However, the FCC decided not to assign new spectrum bands for televisions. This means that simulcasting would occur in the already crowded VHF and UHF spectrum. The new television system had to use low-power transmission to avoid excessive interference into the existing NTSC services. In addition, the new television system had to use a very aggressive compression approach to squeeze full HDTV signal into 6 MHz spectrum. One good thing was that backward compatibility with NTSC was not required. Actually, under these constraints the backward compatibility had already become impossible. Moreover, this goal could not be achieved by any of the previously proposed system and it caused most of the competing proponents to reconsider their approaches. Engineers realized that it was almost impossible to use the traditional analog approaches to reach this goal and that the solution may be in digital approaches. After a few months of considering, General Instrument announced their first digital system proposal for HDTV, DigiCigher, in June 1990. In the middle of the following year, three other digital systems were proposed: the Advanced Digital HDTV by the Advanced Television Research Consortium, which included Thomson, Philips, Sarnoff, and NBC in November 1990; Digital Spectrum Compatible HDTV by Zenith and AT&T in December 1990; and Channel Compatible Digicipher by General Instrument and the Massachusetts Institute of Technology in January 1991. Thus the competition stage started. The prototypes of four competing digital systems and the analog system, Narrow MUSE, proposed by NHK, were officially tested and extensively analyzed during 1992. After the first round of tests, they concluded that the digital systems would be continued for further improvement and would be adopted. In February 1992, the ACATS recommended digital HDTV for the U.S. standard. It also recommended that the competing systems be either further improved and retested, or be combined with a new system. In the middle of 1993, the former competitors joined a Grand Alliance. Then the DTV development entered the collaboration stage. The Grand Alliance began a collaborative effort to create the best system which combines the best features and capabilities of the formerly competing systems into a single best-of-the-best system. After one-year of joint effort by the seven Grand Alliance members, the Grand Alliance provided a new system that was prototyped and extensively tested in the laboratory and field.

The test results showed that the system is indeed the best of the best compared with formerly competing systems [ga 1994]. The ATSC then recommended this system to the FCC as the candidate HDTV standard in the United States. During the following period, the computer industry realized that DTV provides the signals that can now be used for computer applications and the TV industry was invading their terrain. They presented different opinion about the signal format and where especially opposed to the interlaced format. This reaction delayed the approval of the ATSC standard. After long-time debate, the FCC finally approved the ATSC standard in early 1997. But, the FCC did not specify the picture formats and left this issue to be decided by the market.

17.2.2 Technical Overview of ATSC Systems

The ATSC DTV system has been designed to satisfy the FCC requirements. The basic requirement is that no additional frequency spectrum will be assigned for DTV broadcasting. In other words, during a transition period, both NTSC and DTV service will be simultaneously broadcast on different channels and DTV can only use the taboo channels. This approach allows a smooth transition to DTV, such that the services of the existing NTSC receivers will remain and gradually be phased out of existence in the year 2006. The simulcasting requirement causes some technical difficulties of DTV design. First, the high-quality HDTV program must be delivered in a 6 MHz channel to make efficient use of spectrum and fit allocation plans for spectrum assigned to television broadcasting. Second, a low-power and low-interference signal must be used so that simulcasting in the same frequency allocations as current NTSC service does not cause excessive interference to the existing NTSC receiving, since the taboo channels are generally unsuitable for broadcasting an NTSC signal due to high interference. Besides satisfying the frequency spectrum requirement, the DTV standard has several important features that allow DTV to achieve interoperability with computers and data communications. The first feature is the adoption of a layered digital system architecture. Each individual layer of the system is designed to be interoperable with other systems at the corresponding layers. For example, the square pixel and progressive scan picture format should be provided to allow computers access to the compression layer or picture layer depending on the capacity of computers and the ATM-like packet format for ATM network to access the transport layer. Second, the DTV standard uses a header/descriptor approach to provide maximum flexible operating characteristics. Therefore, the layered architecture is the most important feature of DTV standards. The additional advantage of layering is that the elements of the system can be combined with other technologies to create new applications. The system of DTV standard includes four layers: picture, compression, transport, and transmission.

17.2.2.1 Picture Layer

At the picture layer the input video formats have been defined. The Executive Committee of the Advanced Television Systems Committee has approved to release the statement regarding the identification of the HDTV and SDTV transmission formats within the ATSC DTV standards. There are six video formats in the ATSC DTV standard, which are HDT (Table 17.1).

The remaining 12 video formats are not HDTV format. These formats represent some improvements over analog NTSC and are referred to as standard definition television (SDTV) (Table 17.2).

These definitions are fully supported by the technical specifications for the various formats as measured against the internationally accepted definition of HDTV established in 1989 by the International Telecommunication Union (ITU) and the definitions cited by

TABLE 17.1

HDTV Formats

Spatial Format (X × Y Active Pixels)	Aspect Ratio	Temporal Rate
1920 × 1080 (square pixel)	16:9	23.976/24 Hz progressive scan 29.97/30 Hz progressive scan 59.94/60 Hz interlaced scan
1280 × 720 (square pixel)	16:9	23.976/24 Hz progressive scan 29.97/30 Hz progressive scan 59.94/60 Hz progressive scan

the FCC during the DTV standard develop process. These formats cover a wide variety of applications, which include motion picture film, currently available HDTV production equipment, the NTSC television standard, and computers such as personal computers and workstations. However, there is no simple technique that can convert images from one pixel format and frame rate to another that achieve interoperability among film and the various worldwide television standards. For example, all low cost computers use square pixels and progressive scanning while current television uses rectangular pixels and interlaced scanning. The video industry has paid a lot of attention to developing the format converting techniques. Some techniques such as de-interlacing, down/up conversion for format conversion have already been developed. It should be noted that the broadcasters, content providers, and service providers can use any one of these DTV format. This results in a difficult problem for DTV receiver manufacturers who have to provide all kinds of DTV receivers to decode all these formats and then to convert the decoded signal to its particular display format. On the other hand, this requirement also gives receiver manufacturers the flexibility to produce a wide variety of products that have different functionality and cost, and the consumers freedom to choose among them.

17.2.2.2 Compression Layer

The raw data rate of HDTV of 1920 × 1080 × 30 × 16 (16 bits/pixel corresponds to 4:2:2 color format) is about 1 Gbits/s. The function of the compression layer is to compress the raw data from about 1 Gbits/s to the data rate of approximately 19 Mbits/s to satisfy the 6 MHz spectrum requirement. This goal is achieved by using the main profile and high level of MPEG-2 video standard. Actually, during the development of Grand Alliance HDTV system, many research results have been adopted by the MPEG-2 standard at the same time. For example, the support for interlaced video format and the syntax for data partitioning and scalability. The ATSC DTV standard is the first and important application example of the MPEG-2 standard. The use of MPEG-2 video compression fundamentally

TABLE 17.2

SDTV Formats

Spatial Format (X × Y Active Pixels)	Aspect Ratio	Temporal Rate
704 × 480 (CCIR 601)	16:9 or 4:3	23.976/24 Hz progressive scan 29.97/30 Hz progressive scan 59.94/60 Hz progressive scan
640 × 480 (VGA, square pixel)	4:3	23.976/24 Hz progressive scan 29.97/30 Hz progressive scan 59.94/60 Hz progressive scan

enables ATSC DTV devices to interoperate with MPEG-1/2 computer multimedia applications directly at the compressed bitstream level.

17.2.2.3 Transport Layer

The transport layer is another important issue for interoperability. The ATSC DTV transport layer uses the MPEG-2 system transport stream syntax. It is a fully compatible subset of the MPEG-2 transport protocol. The basic function of transport layer is to define the basic format of data packets. The purposes of packetization include:

- Packaging the data into the fixed size cells or packets for forward error correction (FEC) encoding to protect the bit-error due to the communication channel noise
- Multiplexing the video, audio, and data of a program into a bitstream
- Providing time synchronization for different media elements
- Providing flexibility and extensibility with backward compatibility.

The transport layer of ATSC DTV uses a fixed length (FL) packet. The packet size is 188 bytes consisting of 184 bytes of payload and 4 bytes of header. Within the packet header, the 13-bit packet identifier (PID) is used to provide the important capacity to combine the video, audio, and ancillary data stream into a single bitstream as shown in Figure 17.1. Each packet contains only a single type of data (video, audio, data, program guide, etc.) identified by the PID.

This type of packet structure packetizes the video, audio, and auxiliary data separately. It also provides the basic multiplexing function that produces a bitsteam including video, five-channel surround-sound audio, and an auxiliary data capacity. This kind of transport layer approach also provides complete flexibility to allocate channel capacity to achieve any mix among video, audio, and other data services. It should be noted that the selection of 188-packet length is a trade-off between reducing the overhead due to the transport header and increasing the efficiency of error correction. Also, one ATSC DTV packet can be completely encapsulated with its header within four ATM packets by using 1 AAL byte per ATM header leaving 47 usable payload bytes times 4 for 188 bytes. The details of the transport layer will be discussed in Chapter 21.

17.2.2.4 Transmission Layer

The function of transmission layer is to modulate the transport bitstream into a signal that can be transmitted over 6 MHz analog channel. The ATSC DTV system uses a trellis-coded 8-level vestigial sideband (8-VSB) modulation technique to deliver approximately 19.3 Mbits/s in the 6 MHz terrestrial simulcast channel. VSB modulation inherently requires only processing of the in-phase signal sampled at the symbol rate, thus reducing the

FIGURE 17.1
Packet structure of ATSC DTV transport layer.

complexity of receiver, and ultimately the cost of implementation. The VSB signal is organized in a data frame that provides a training signal to facilitate channel equalization for removing multipath distortion. However, from several field test results, the multipath distortion is still a serious problem of terrestrial simulcast receiving. The frame is organized into segments each with 832 symbols. Each transmitted segment consists of one-synchronization byte (four symbols), 187 data bytes, and 20 R-S parity bytes. This corresponds to a 188-byte packet, which is protected by 20-byte R-S code. Interoperability at the transmission layer is required by different transmission media applications. The different media use different modulation techniques now, such as QAM for cable and QPSK for satellite. Even for terrestrial transmission, European DVB systems use OFDM transmission. The ATV receivers will not only be designed to receive terrestrial broadcasts, but also the programs from cable, satellite, and other media.

17.3 Transcoding with Bitstream Scaling

17.3.1 Background

As indicated in the previous chapters, digital video signals exist everywhere in the format of compressed bitstreams. The compressed bitstreams of video signals are used for transmission and storage through different media such as terrestrial TV, satellite, cable, ATM network, and the Internet. The decoding of a bitstream can be implemented in either hardware or software. However, for high bit rate compressed video bitstreams of high-definition video signals, specially designed hardware is still the major decoding approach due to the speed limitation of current computer processors. The compressed bitstream as a new format of video signal is a revolutionary change of video industry since it enables many applications. For example, the coded video bitstreams can be decoded not only with digital televisions or set-top boxes, but also with computers and mobile terminals, such as cellular phones. Therefore, the problem of interactivity and integration of video data with computer, cellular, and television systems is relatively new and subject to a great deal of research worldwide. As the number of networks, types of devices, and content representation formats increase, interoperability between different systems and different networks is becoming more important. Thus, devices such as gateways, multipoint control units, and servers must be developed to provide a seamless interaction between content creation and consumption. Transcoding of video content is one key technology to make this possible. Generally speaking, transcoding can be defined as the conversion of one coded signal to another. In the earliest work on transcoding, the majority of interest focused on reducing the bit rate to meet an available channel capacity. Additionally, researchers investigated conversions between constant bit rate (CBR) streams and variable bit rate (VBR) streams to facilitate more efficient transport of video. As time moved on and mobile devices with limited display and processing power became a reality, transcoding to achieve spatial resolution reduction, as well as temporal resolution reduction, has also been studied. Furthermore, with the introduction of packet radio services over mobile access networks, error-resilience video transcoding has gained a significant amount of attention lately, where the aim is to increase the resilience of the original bit stream to transmission errors. Also in some applications, the syntax conversion is needed between different compression standards such as JPEG, MPEG-1, MPEG-2, H.261, H.263, and H.264/AVC. In this section, we will focus on the topic of bit rate conversion since it finds wide application and the readers can extend the idea for other kinds of transcoding. Also, we limit ourselves to focus on the problem of scaling an MPEG CBR encoded bitstream down to a lower CBR. A comprehensive survey of video transcoding can be found in [vetro 2003].

The basic function of bitstream scaling may be thought as a black box, which passively accepts a precoded MPEG bitstream at the input and produces a scaled or size-reduced bitstream, which meets new constraints that are not known a priori during the creation of the original precoded bitstream. The bitstream scaler is a transcoder, or filter, that provides a match between an MPEG source bitstream and the receiving load. The receiving load consists of the transmission channel, the destination decoder, and perhaps a destination storage device or display. The constraint on the new bitstream may be bound by a variety of conditions. Among them these conditions include the peak or average bit rate imposed by the communications channel, the total number of bits imposed by the storage device, and the variation of bit usage across pictures due to the amount of buffering available at the receiving decoder. While the idea of bitstream scaling has many concepts similar to those provided by the various MPEG-2 scalability profiles, the intended applications and goals differ. The scalable video coding (SVC) methods are aimed at providing encoding of source video into multiple service grades (that are predefined at the time of encoding) and multitiered transmission for increased signal robustness. The multiple bitstreams created by MPEG SVC are hierarchically dependent in such a way that by decoding an increasing number of bitstreams, higher service grades are reconstructed. Bitstream transcoding methods, in contrast, are primarily decoder/transcoder techniques for converting an existing precoded bitstream to another one that meets new rate constraints. Several applications that motivate bitstream scaling or transcoding include:

1. *Video on-demand*: Consider a video on-demand (VOD) scenario wherein a video file-server includes a storage device containing a library of precoded MPEG bitstreams. These bitstreams in the library are originally coded at a high quality (e.g., studio quality). A number of clients may request retrieval of these video programs at one particular time. The number of users and the quality of video delivered to the users are constrained by the outgoing channel capacity. This outgoing channel, which may be a cable bus or an ATM trunk, for example, must be shared among the users who are admitted to the service. Different users may require different levels of video quality, and the quality of a respective program will be based on the fraction of the total channel capacity allocated to each user. To simultaneously accommodate a plurality of users, the video file-server must scale the stored precoded bitstreams to a reduced rate before it is delivered over the channel to respective users. The quality of the resulting scaled bitstream should not be significantly degraded compared to the quality of a hypothetical bitstream so obtained by coding the original source material at the reduced rate. Complexity cost is not such a critical factor because only the file-server has to be equipped with the bitstream scaling hardware, not every user. Presumably, video service providers would be willing to pay a high cost for delivering the possible highest quality video at a prescribed bit rate.

    As an option, a sophisticated video file-server may also perform scaling of multiple original precoded bitstreams jointly and statistically multiplex the resulting scaled VBR bitstreams into the channel. By scaling the group of bitstreams jointly, statistical gains can be achieved. These statistical gains can be realized in the form of higher and more uniform picture quality for the same channel capacity. Statistical multiplexing over a DirecTv transponder [isnardi 1993] is one example for the application of video statistical multiplexing.

2. *Trick-play track on digital VTRs*: In this application, the video bitstream is reduced to create a sidetrack on video tape recorders. This sidetrack contains very coarse quality

video sufficient to facilitate trick-modes on the VTR (e.g., FF and REW at different speeds). Complexity cost for the bitstream scaling hardware is of significant concern in this application since the VTR is a mass consumer item subject to mass production.

3. *Extended-play recording on digital VTRs*: In this application, video is broadcast to users' homes at a certain broadcast quality (~6 Mbits/s for standard definition video and ~19 Mbits/s for high-definition video). With a bitstream scaling feature in their video tape recorders, users may record the video at a reduced rate, akin to extended play (EP) mode on today's VHS recorders, thereby recording a greater duration of video programs onto a tape at lower quality. Again, hardware complexity costs would be a major factor here.

4. *Universal multimedia access (UMA)*: The concept of UMA is to enable access to any multimedia content over any type of network, such as Internet and Wireless LAN, from any type of terminals with varying capabilities such as mobile phones, personal computers, and television sets [mpeg21]. The primary function of UMA services is to provide the best QoS or user experience by either selecting appropriate content formats, or adapting the content format directly to meet the playback environment, or to adapt the content playback environment to accommodate the content. Toward the above goal, video transcoder can bridge the mismatch between the large number of video data, bandwidth of communication channels, and capability of end user terminals. The concept of UMA is illustrated in Figure 17.2.

17.3.2 Basic Principles of Bitstream Scaling

As described previously, the idea of scaling an MPEG-2 compressed bitstream down to a lower bit rate is initiated by several applications. One problem is the criteria that should be used to judge the performances of an architecture that can reduce the size or rate of an MPEG compressed bitstream. Two basic principles of bitstream scaling are (1) the information in the original bitstream should be exploited as much as possible, and (2) the resulting image quality of the new bitstream with a lower bit rate should be as close as possible to a bitstream created by coding the original source video at the reduced rate. Here, we assume that for a given rate the original source is encoded in an optimal way. Of course, the implementation of hardware complexity also has to be considered. Figure 17.3 shows a simplified encoding structure of MPEG encoding in which the rate control mechanism is not shown.

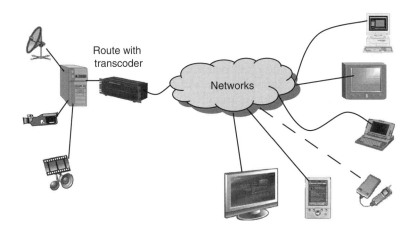

FIGURE 17.2
Concept of UMA.

Application Issues of MPEG-1/2 Video Coding

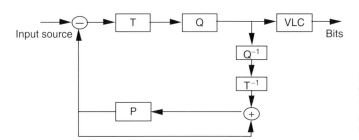

FIGURE 17.3
Simplified encoder structure. T: transform; Q: quantizer; P: motion compensated prediction; VLC: variable-length coding.

In this structure, a block of an image data is first transformed to a set of coefficients, the coefficients are then quantized with a quantizer step that is decided by the given bit rate budget, or number of bits assigned to this block. Finally, the quantized coefficients are coded in variable-length coding (VLC) to the binary format, which is called the bitstream or bits.

From this structure it is obvious that the performance of changing the quantizer step will be better than cutting higher frequencies when the same amount of rate needs to be reduced. In the original bitstream the coefficients are quantized with finer quantization steps that are optimized at the original high rate. After cutting the coefficients of higher frequencies the rest of coefficients are not quantized with an optimal quantizer. In the method of re-quantization all coefficients are re-quantized with an optimal quantizer that is determined by the reduced rate, the performance must be better than the one by cutting high frequencies to reach the reduced rate. The theoretical analysis is given in the appendix.

In the following, several different architectures that accomplish the bitstream scaling are discussed. The different methods have varying hardware implementation complexity; each having its own degree of trade-off between required hardware and resulted image quality.

17.3.3 Architectures of Bitstream Scaling

Four architectures for bitstream scaling are discussed. Each of the scaling architectures described has their own particular benefits that are suitable for a particular application.

Architecture 1: The bitstream is scaled by cutting high frequencies.

Architecture 2: The bitstream is scaled by re-quantization.

Architecture 3: The bitstream is scaled by re-encoding the reconstructed pictures with motion vectors and coding decision modes extracted from the original high-quality bitstream.

Architecture 4: The bitstream is scaled by re-encoding the reconstructed pictures with motion vectors extracted from the original high-quality bitstream, but new coding decisions are computed based on reconstructed pictures.

Architectures 1 and 2 are considered for VTR applications such as trick-play modes and EP recording. Architectures 3 and 4 are considered for VOD and other applicable StatMux scenarios.

17.3.3.1 Architecture 1: Cutting AC Coefficients

A block diagram illustrating architecture 1 is shown in Figure 17.4a. The method of reducing the bit rate in architecture 1 is based on cutting the higher frequency coefficients. The incoming precoded CBR stream enters a decoder rate buffer. Following the top branch leading from the rate buffer, a variable-length decoder (VLD) is used to parse the bits for the next frame in the buffer to identify all the variable-length code words that correspond

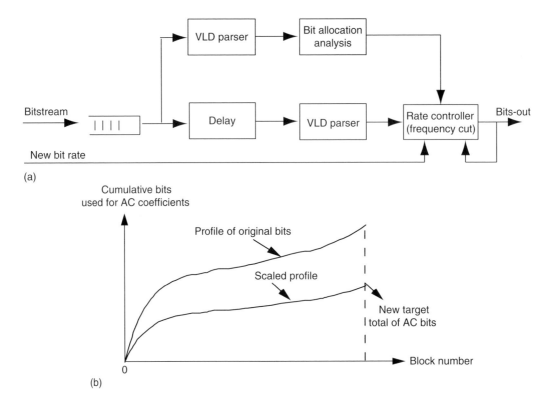

FIGURE 17.4
(a) Architecture 1: cutting high frequencies. (b) Profile map.

to AC coefficients used in that frame. No bits are removed from the rate buffer. The code words are not decoded, but just simply parsed by the VLD parser to determine code word lengths. The bit allocation analyzer accumulates these AC bit-counts for every macroblock (MB) in the frame and creates an AC bit usage profile as shown in Figure 17.4b. That is, the analyzer generates a running sum of AC DCT coefficient bits on an MB basis:

$$PV_N = \sum AC\_BITS \qquad (17.1)$$

where PV_N is the profile value of a running sum of AC code word bits until the MB N. In addition, the analyzer counts the sum of all coded bits for the frame, TB (total bits). After all MBs for the frame have been analyzed, a target value TV_{AC} of AC DCT coefficient bits per frame is calculated as

$$TV_{AC} = PV_{LS} - \alpha * TB - B_{EX} \qquad (17.2)$$

where
TV_{AC} is the target value of AC code word bits per frame
PV_{LS} is the profile value at the last MB
α is the percentage by which the pre-encoded bitstream is to be reduced
TB is the total bits
B_{EX} is the amount of bits by which the previous frame missed its desired target

The profile value of AC coefficient bits is scaled by the factor TV_{AC}/PV_{LS}. Multiplying each PV_N performs scaling by that factor to generate the linearly scaled profile shown in Figure

17.4b. Following the bottom branch from the rate buffer, a delay is inserted equal to the amount of time required for the top branch analysis processing to be completed for the current frame. A second VLD parser accesses and removes all code word bits from the buffer and delivers them to a rate controller. The rate controller receives the scaled target bit usage profile for the amount of AC bits to be used within the frame. The rate controller has memory to store all coefficients associated with the current MB it is operating on. All original code word bits at a higher level than AC coefficients (i.e., all FL header codes, motion vector codes, MB-type codes, etc.) are held in memory and will be re-multiplexed with all AC code words in that MB that have not been excised to form the outgoing scaled bitstream. The rate controller determines and flags in the MB code word memory which AC code words to keep and which to excise. AC code words are accessed from the MB code word memory in the order AC11, AC12, AC13, AC14, AC15, AC16, AC21, AC22, AC23, AC24, AC25, AC26, AC31, AC32, AC33, etc., where AC*ij* denotes the *i*th AC code words from *j*th block in the MB if it is present. As the AC code words are accessed from memory, the respective code word bits are summed and continuously compared with the scaled profile value to the current MB, less number of bits for insertion of end-of-block (EOB) code words. Respective AC code words are flagged as kept until the running sum of AC code words bits exceeds the scaled profile value less EOB bits. When this condition is met, all remaining AC code words are marked for being excised. This process continues until all MBs have their kept code words reassembled to form the scaled bitstream.

17.3.3.2 Architecture 2: Increasing Quantization Step

Architecture 2 is shown in Figure 17.5. The method of bitstream scaling in architecture 2 is based on increasing the quantization step. This method requires additional dequantizer/quantizer and VLC hardware over the first method. Like the first method, it also makes a first VLD pass on the bitstream and obtains a similar scaled profile of target cumulative code word bits versus MB count to be used for rate control.

The rate control mechanism differs from this point on. After the second pass VLD is made on the bitstream, quantized DCT (QDCT) coefficients are dequantized. A block of finely QDCT coefficients is obtained as a result of this. This block of DCT coefficients is re-quantized with a coarser quantizer scale. The value used for the coarser quantizer scale is determined adaptively by making adjustments after every MB so that the scaled target profile is tracked as we progress through the MBs in the frame:

$$Q_N = Q_{NOM} + G^* \left(\sum_{N-1} BU - PV_{N-1} \right) \quad (17.3)$$

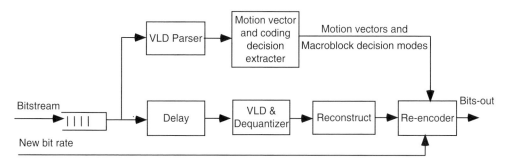

FIGURE 17.5
Architecture 2: increasing quantization step.

where

Q_N is the quantization factor for MB N
Q_{NOM} is an estimate of the new nominal quantization factor for the frame
$\Sigma_{N-1}BU$ is the cumulative amount of coded bits up to MB $N-1$
G is a gain factor that controls how tightly the profile curve is tracked through the picture

Q_{NOM} is initialized to an average guess value before the very first frame, and updated for the next frame by setting it to Q_{LS} (the quantization factor for the last MB) from the frame just completed. The coarsely re-quantized block of DCT coefficients is variable-length-coded to generate the scaled bitstream. The rate controller also has provisions for changing some MB-layer code words, such as the MB-type and coded-block-pattern to ensure a legitimate scaled bitstream that conforms to MPEG-2 syntax.

17.3.3.3 Architecture 3: Re-Encoding with Old Motion Vectors and Old Decisions

The third architecture for bitstream scaling is shown in Figure 17.6. In this architecture, the motion vectors and MB coding decision modes are first extracted from the original bitstream, and at the same time the reconstructed pictures are obtained from the normal decoding procedure. Then the scaled bitstream is obtained by re-encoding the reconstructed pictures using the old motion vectors and MB decisions from the original bitstream. The benefits obtained from this architecture compared to full decoding and re-encoding are that no motion estimation and decision computation are needed.

FIGURE 17.6
Architecture 3.

17.3.3.4 Architecture 4: Re-Encoding with Old Motion Vectors and New Decisions

Architecture 4 is a modified version of architecture 3 in which new MB decision modes are computed during re-encoding based on reconstructed pictures. The scaled bitstream created this way is expected to yield an improvement in picture quality because the decision modes obtained from the high-quality original bitstream are not optimal for re-encoding at the reduced rate. For example, at higher rates the optimal mode decision for an MB is more likely to favor of bidirectional field motion compensation (MC) over forward frame MC. But at lower rates, only the opposite decision may be true. In order for the re-encoder to have the possibility of deciding on new MB coding modes, the entire pool of motion vectors of every type must be available. This can be supplied by augmenting the original high-quality bitstream with ancillary data containing the entire pool of motion vectors during the time it was originally encoded. It could be inserted into the user data every frame. For the same original bit rate, the quality of an original bitstream obtained this way is degraded compared to an original bitstream obtained from architecture 3 because the additional overhead required for the extra motion vectors steals away bits for actual encoding. However, the resulting scaled bitstream is expected to show quality improvement over the scaled bitstream from architecture 3 if the gains from computing new and more accurate decision modes can overcome the loss in original picture quality. Table 17.3 outlines the hardware complexity savings of each of the three proposed architectures as compared to full decoding and re-encoding.

17.3.3.5 Comparison of Bistream Scaling Methods

We have described four architectures for bitstream scaling, which are useful for various applications as described in the introduction. Among the four architectures, architectures 1 and 2 neither require entire decoding and encoding loops nor frame store memory for reconstructed pictures, thereby saving significant hardware complexity. However, video quality tends to degrade through the group of pictures (GOP) until the next I-picture due to drift in the absence of decoder/encoder loops. For large scaling, say for rate reduction greater than 25%, architecture 1 produces poor quality blocky pictures, primarily because many bits were spent in the original high-quality bitstream on finely quantizing the DC and other very low-order AC coefficients. Architecture 2 is a particularly good choice for VTR applications since it is a good compromise between the hardware complexity and reconstructed image quality. Architectures 3 and 4 are suitable for video-on-demand (VOD) server applications and other StatMux applications.

TABLE 17.3

Hardware Complexity Savings Over Full Decoding/Re-Encoding

Coding Method	Hardware Complexity Savings
Architecture 1	No decoding loop, no DCT/IDCT, no frame store memory, no encoding loop, no quantizer/dequantizer, no motion compensation, no VLC, simplified rate control
Architecture 2	No decoding loop, no DCT/IDCT, no frame store memory, no encoding loop, no motion compensation, simplified rate control
Architecture 3	No motion estimation, no macroblock coding decisions
Architecture 4	No motion estimation

Appendix

In this analysis, we assume that the optimal quantizer is obtained by assigning the number of bits according to the variance or energy of the coefficients. It is slightly different from MPEG standard that will be explained later, but the principal concept is the same and the results will hold for the MPEG standard. We first analyze the errors caused by cutting high coefficients and increasing the quantizer step. The optimal bit assignment is given by [jayant 1984]:

$$R_{k0} = R_{av0} + \frac{1}{2} \log_2 \frac{\sigma_k^2}{\left(\prod_{i=0}^{N-1} \sigma_i^2\right)^{1/N}}, \quad k = 0, 1, \ldots, N-1, \tag{17.4}$$

where
  N is the number of coefficients in the block
  R_{k0} is the number of bits assigned to the kth coefficient

R_{av0} is the average number of bits assigned to each coefficient in the block, i.e., $R_{T0} = N \cdot R_{av0}$ is the total bits for this block under a certain bit rate and is the variance of kth coefficient. Under the optimal bit assignment (Equation 17.3), the minimized average quantizer error, σ_{q0}^2, is

$$\sigma_{q0}^2 = \frac{1}{N} \sum_{k=1}^{N-1} \sigma_{qk}^2 = \frac{1}{N} \sum_{k=1}^{N-1} 2^{-2R_{k0}} \cdot \sigma_k^2 \tag{17.5}$$

where σ_{qk}^2 is the quantizer error of kth coefficient. According to Equation 17.4, we have two major methods to reduce the bit rate, cutting high coefficients or decreasing the R_{av}, i.e., increasing the quantizer step. We are now analyzing the effects on the reconstructed errors caused due to the cut of bits using these methods. Assume that the number of the bits assigned to the block is reduced from R_{T0} to R_{T1}. Then the bits to be reduced, ΔR_1, is equal to $R_{T0} - R_{T1}$.

In the case of cutting high frequencies, say the number of coefficients is reduced from N to M, then

$$R_{k0} = 0 \text{ for } K < M, \text{ and } \Delta R_1 = R_{T0} - R_{T1} = \sum_{k=M}^{N-1} R_{k0} \tag{17.6}$$

the quantizer error increased due to the cutting is

$$\Delta \sigma_{q1}^2 = \sigma_{q1}^2 - \sigma_{q0}^2 = \frac{1}{N} \left(\sum_{k=0}^{M-1} 2^{-2R_{k0}} \cdot \sigma_k^2 + \sum_{k=M}^{N-1} \sigma_k^2 - \sum_{k=0}^{N-1} 2^{-2R_{k0}} \cdot \sigma_k^2 \right)$$

$$= \frac{1}{N} \left(\sum_{k=M}^{N-1} \sigma_k^2 - \sum_{k=M}^{N-1} 2^{-2R_{k0}} \cdot \sigma_k^2 \right)$$

$$= \frac{1}{N} \sum_{k=M}^{N-1} (1 - 2^{-2R_{k0}}) \cdot \sigma_k^2 \tag{17.7}$$

where σ_{q1}^2 is the quantizer error after cutting the high frequencies.

In the method of increasing quantizer step, or decreasing the average bits, from R_{av0} to R_{av2}, assigned to each coefficient, the number of bits reduced for the block is

$$\Delta R_2 = R_{T0} - R_{T2} = N \cdot (R_{av0} - R_{av2}) \tag{17.8}$$

Application Issues of MPEG-1/2 Video Coding

and the bits assigned to each coefficient become now

$$R_{k2} = R_{av2} + \frac{1}{2} \log_2 \frac{\sigma_k^2}{\left(\prod_{i=0}^{N-1} \sigma_i^2\right)^{1/N}}, \quad k = 0, 1, \ldots, N-1, \quad (17.9)$$

The corresponding quantizer error increased by the cutting bits is

$$\Delta\sigma_{q2}^2 = \sigma_{q2}^2 - \sigma_{q0}^2 = \frac{1}{N} \left(\sum_{k=0}^{N-1} 2^{-2R_{k2}} \cdot \sigma_k^2 - \sum_{k=0}^{N-1} 2^{-2R_{k0}} \cdot \sigma_k^2\right)$$

$$= \frac{1}{N} \sum_{k=0}^{N-1} (2^{-2R_{k2}} - 2^{-2R_{k0}}) \cdot \sigma_k^2 \quad (17.10)$$

where σ_{q2}^2 is the quantizer error at the reduced bit rate.

If the same number of bits is reduced, i.e., $\Delta R_1 = \Delta R_2$, it is obvious that $\Delta\sigma_{q2}^2$ is smaller than $\Delta\sigma_{q1}^2$ since σ_{q2}^2 is the minimized value at the reduced rate. This implies that the performance of changing the quantizer step will be better than cutting higher frequencies when the same amount of rate needs to be reduced. It should be noted that in the MPEG video coding, more sophisticated bit assignment algorithms are used. First, different quantizer matrices are used to improve the visual perceptual performance. Second, different VLC tables are used to code the DC values and the AC transform coefficients; and the run-length coding (RLC) is used to code the pairs of the zero-run length and the values of amplitudes. However, in general, the bits are still assigned according to the statistical model that indicates the energy distribution of the transform coefficients. Therefore, the above theoretical analysis will hold for the MPEG video coding.

17.3.4 MPEG-2 to MPEG-4 Transcoding

In this section, we are going to indicate that there is another kind of transcoding, which is to convert the bitstream between standards. An example is the transcoder between MPEG-2 to MPEG-4. The technical detail of MPEG-4 will be introduced in Chapter 18. Since we have not learnt the MPEG-4 yet, here we just introduce the concept of transcoding between MPEG-2 to MPEG-4. This concept can be extended to other standards.

The transcoding methods themselves can be applied within the same syntax format as described in the previous section or between different syntax formats. As MPEG-4 simple profile is adopted as the solution for mobile multimedia communications and a large amount of MPEG-1/2 contents is available, we focus our discussion on MPEG-2 to MPEG-4 transcoding. The transcoding from MPEG-2 to MPEG-4 is necessary and useful for allowing the mobile or PDA terminals to receive MPEG-2 compressed contents with their limited display size. Here, we describe the principles and techniques used in the MPEG-2 to MPEG-4 video transcoder. The conversions include techniques for bit rate reduction, spatial resolution down-sampling, temporal resolution down-scaling, and picture-type change. The main difficulty of MPEG-2 to MPEG-4 transcoding is to perform transcoding on both bit rate reduction and spatial resolution reduction at the same time. The transcoding on bit rate reduction and spatial resolution reduction would cause serious error drift due to the change of predictive references. To address this problem, several issues were investigated. First, an analysis of drift errors is provided to identify the sources of quality degradation when transcoding to a lower spatial resolution. Two types of drift error are considered: a reference picture error and an error due to the noncommutative property of MC and down-sampling. To overcome these sources of error, several novel transcoding architectures are then presented. One architecture attempts to compensate for the reference picture error in the reduced resolution, while another attempts to do the same in the original resolution. We present a third architecture that attempts to eliminate the second type of drift error and a final architecture that relies on an intrablock refresh method to compensate all types of errors. In all these architectures, a variety of MB level conversions are required, such as motion vector mapping and texture down-sampling. These conversions are discussed in detail. Another important issue for the transcoder is rate control. Rate contol is especially important for the intrarefresh architecture since it must find a balance between the number of intrablocks used to compensate errors and the associated rate-distortion characteristics of the low-resolution signal. The complexity and quality of the architectures are compared. Based on the results, we find that the intrarefresh architecture offers the best trade-off between quality and complexity, and is also the most flexible. After learning MPEG-4 in Chapter 18, you can refer to the technical details of MPEG-2 to MPEG-4 transcoding in [peng 2002].

17.4 Down-Conversion Decoder

17.4.1 Background

Digital video broadcasting has had a major impact in both academic and industrial communities. A great deal of effort has been made to improve the coding efficiency at the transmission side and offer cost-effective implementations in the overall end-to-end system. Along these lines, the notion of format conversion is becoming increasingly popular. On the transmission side, there are a number of different formats that are likely candidates for digital video broadcast. These formats vary in horizontal, vertical, and temporal resolution. Similarly, on the receiving side, there are a variety of display devices that the receiver should account for. In this section, we are interested in the specific problem of how to receive an HDTV bitstream and display it at a lower spatial resolution. In the conventional method of obtaining a low-resolution image sequence, the HD bitstream is fully decoded then it is simply prefiltered and subsampled [tm5]. The block diagram of this system, shown in Figure 17.7a, is referred to as a full-resolution decoder (FRD) with spatial down-conversion. Although the quality is very good, the cost is quite high due to the large memory requirements. As a result, low-resolution decoders (LRDs) have been proposed to reduce some of the costs [ng 1993; sun 1993; boyce 1995; bao 1996]. Although the quality of the picture will be compromised, significant reductions in the amount of memory can be realized; the block diagram for this system is shown in Figure 17.7b. Here, incoming blocks are subject to down-conversion filters within the decoding loop. In this way, the down-converted blocks are stored into memory rather than the full-resolution blocks. To achieve a high-quality output with the LRD, it is important to take special care in the algorithms for down-conversion and MC. These two processes are of major importance to the decoder as they have significant impact on the final quality. Although a moderate amount of complexity within the decoding loop is added, the reductions in external memory are expected to provide significant cost savings, provided that these algorithms can be incorporated into the typical decoder structure in a seamless way.

As stated above, the filters used to perform the down-conversion are an integral part of the LRD. In Figure 17.7b, down-conversion is shown to take place before the IDCT. Although filtering is not required to take place in the DCT domain, we initially assume

Application Issues of MPEG-1/2 Video Coding

FIGURE 17.7
Decoder structures. (a) Block diagram of full-resolution decoder with down-conversion in the spatial domain. The quality of this output will serve as a drift-free reference. (b) Block diagram of low-resolution decoder. Down-conversion is performed within the decoding loop and is a frequency domain process. Motion compensation is performed from a low-resolution reference using motion vectors that are derived from the full-resolution encoder. Motion compensation is a spatial domain process.

that it takes place before the adder. In any case, it is usually more intuitive to derive a down-conversion filter in the frequency domain rather than in the spatial domain [mokry 1994; pang 1996; merhav 1997]. The major drawback of these approaches is that high-frequency data is lost or not preserved very well. To overcome this, a method of down-conversion that better preserves high-frequency data within the MB has been reported in [bao 1996; vetro 1998a]; this method is referred to as frequency synthesis.

Although the above statement of the problem has mentioned only filtering based approaches to memory reduction within the decoding loop, readers should be aware that other techniques have also been proposed. For the most part, these approaches rely on methods of embedded compression. For instance, in [de with 1998], the data being written to memory is quantized adaptively using a block predictive coding scheme, then a segment of MBs is fit into a FL packet. Similarly, in [yu 1999], an adaptive minimum–maximum quantizer and edge detector is proposed. With this method, each MB is compressed to a fixed size to simplify memory access. Another more simple approach may be to truncate the 8-bit data to 7 or 6 bits. However, in this case, it is expected the drift would accumulate very fast and result in poor reconstruction quality. In [bruni 1998], a vector quantization method has been utilized, and in [lei 1999] a wavelet-based approach is described. Overall, these approaches offer exceptional techniques to reduce the memory requirements, but in most cases, the reconstructed video would still be a high-resolution signal. The reason is that compressed high-resolution data is stored in memory rather than the raw low-resolution data. For this reason, the remainder of this section emphasizes the filtering based approach, in which the data stored in memory represents the actual low-resolution picture data.

The main novelty of the system that we describe is the filtering, which is used to perform the MC from low-resolution anchor frames. It is well known that prediction drift has been difficult to avoid. It is partly due to the loss of high-frequency data from the down-conversion and partly due to the inability to recover the lost information. Although prediction drift cannot be totally avoided in an LRD, it is possible to significantly reduce the effects of drift in contrast to simple interpolation methods. The solution that we describe is optimal in the least-squares sense and is dependent on the method of down-conversion that is used [vetro 1998b]. In its direct form, the solution cannot be readily applied to a practical decoding scheme. However, it is shown that a cascaded realization is easily implemented into the FRD-type structure [vetro 1998c].

17.4.2 Frequency Synthesis Down-Conversion

The concept of frequency synthesis was first reported in [bao 1996] and later expanded in [vetro 1998b]. The basic premise is to better preserve the frequency characteristics of an MB in comparison to simpler methods that extract or cut specified frequency components of an 8×8 block. To accomplish this, the four blocks of an MB are subject to a global transformation—this transformation is referred to as frequency synthesis. Essentially, a single frequency domain block can be realized using the information in the entire MB. From this, low-resolution blocks can be achieved by cutting out the low-order frequency components of the synthesized block—this action represents the down-conversion process and is generally represented in the following way:

$$\tilde{\underline{A}} = X\underline{A} \qquad (17.11)$$

where
  $\tilde{\underline{A}}$ denotes the original DCT MB
  $\underline{A}$ denotes the down-converted DCT block
  X is a matrix that contains the frequency synthesis coefficients

The original idea for frequency synthesis down-conversion was to directly extract an 8×8 block from the 16×16 synthesized block in the DCT domain as shown in Figure 17.8a. The advantage of doing this is that the down-converted DCT block is directly applicable to an 8×8 IDCT (for which fast algorithms exist). The major drawback with regard to computation is that each frequency component in the synthesized block is dependent on all of the frequency components in each of the 8×8 blocks, i.e., each synthesized frequency component is the result of a 256-tap filter. The major drawback with regard to quality is that interlaced video with field-based predictions should not be subject to frame-based filtering [vetro 1998b]. If frame-based filtering is used, it becomes impossible to recover the appropriate field-based data that is required to make field-based predictions. In areas of large motion, severe blocking artifacts will result.

Obviously, the original approach would incur too much computation and quality degradation; so instead, the operations are performed separately and vertical down-conversion is performed on a field-basis. In Figure 17.8b, it is shown that a horizontal down-conversion can be performed. To perform this operation, a 16-tap filter is ultimately required. In this way, only the relevant row information is applied as the input to the horizontal filtering operation and the structure of the incoming video has no bearing on the down-conversion process. The reason is that the data in each row of an MB belongs to the same field, hence the format of the output block will be unchanged. It is noteworthy that the set of filter coefficients is dependent on the particular output frequency index. For

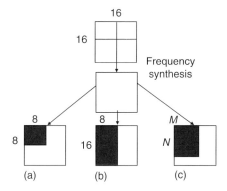

FIGURE 17.8
Concept of frequency synthesis down-conversion, (a) 256-tap filter applied to every frequency component to achieve vertical and horizontal down-conversion by a factor of 2 frame-based filtering, (b) 16-tap filter applied to frequency components in the same row to achieve horizontal down-conversion by 2, picture structure is irrelevant, (c) illustrates that the amount of synthesized frequency components which are retained is arbitrary.

1-dimensional (1-D) filtering, this means that the filters used to compute the second output index, for example, are different from those used to compute the fifth output index. Similar to the horizontal down-conversion, vertical down-conversion can also be applied as a separate process. As reasoned earlier, field-based filtering is necessary for interlaced video with field-based predictions.

However, since an MB consists of 8 lines for the even field and 8 lines for the odd field, and the vertical block unit is 8, frequency synthesis cannot be applied. Frequency synthesis is a global transformation and is only applicable when one wishes to observe the frequency characteristics over a larger range of data than the basic unit. Therefore, to perform the vertical down-conversion, we can simply cut the low-order frequency components in the vertical direction. This loss that we accept in the vertical direction is justified by the ability to perform accurate low-resolution MC that is free from severe blocking artifacts.

In the above, we have explained how the original idea to extract an 8×8 DCT block is broken down into separable operations. However, since frequency synthesis provides an expression for every frequency component in the new 16×16 block, it makes sense to generalize the down-conversion process so that decimation, which are multiples of 1/16 can be performed. In Figure 17.8c, an $M \times N$ block is extracted. Although this type of down-conversion filtering may not be appropriate before the IDCT operation and may not be appropriate for a bitstream containing field-based predictions, it may be applicable elsewhere, e.g., as a spatial domain filter somewhere else in the system or for progressive material. To obtain a set of spatial domain filters, an appropriate transformation can be applied. In this way, Equation 17.8 is expressed as

$$\tilde{\underline{a}} = x\underline{a} \qquad (17.12)$$

where the lowercase counterparts denote spatial equivalents. The expression that transforms X to x is derived in Appendix A.

17.4.3 Low-Resolution Motion Compensation

The focus of this section is to provide an expression for the optimal set of low-resolution MC filters given a set of down-conversion filters. The resulting filters are optimal in the least-squares sense as they minimize the mean squared error (MSE) between a reference block and a block obtained through low-resolution MC. The results derived in [vetro 1998a] assume that a spatial domain filter, x, is applied to incoming MBs to achieve the down-conversion. The scheme shown in Figure 17.9a illustrates the process by which reference blocks are obtained. First, full-resolution MC is performed on MBs a, b, c, and d to yield h. To execute this process, the filters $S_a^{(r)}$, $S_b^{(r)}$, $S_c^{(r)}$, and $S_d^{(r)}$ are used. Basically, these

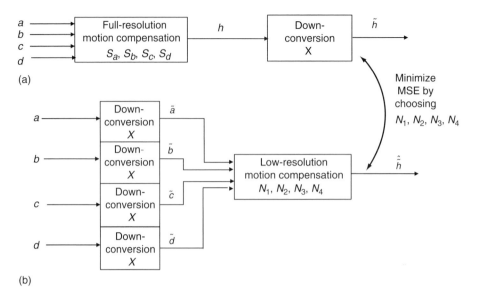

FIGURE 17.9
Comparison of decoding methods to achieve low-resolution image sequence. (a) FRD with spatial down-conversion, (b) LRD. The objective is to minimize the MSE between the two outputs by choosing N_1, N_2, N_3, and N_4 for a fixed down-conversion.

filters represent the masking/averaging operations of the MC in a matrix form. More on the composition of these filters can be found in the Appendix B. Once h is obtained, it is down-converted to $\underline{\tilde{h}}$ via the spatial filter, x:

$$\underline{\tilde{h}} = x\underline{h} \qquad (17.13)$$

The above block is considered to be the drift-free reference. On the other hand, in the scheme of Figure 17.9b, the blocks $\underline{a}$, $\underline{b}$, $\underline{c}$, and $\underline{d}$ are first subject to the down-conversion filter, x, to yield the down-converted blocks, $\underline{\tilde{a}}$, $\underline{\tilde{b}}$, $\underline{\tilde{c}}$, and $\underline{\tilde{d}}$, respectively. Using these down-converted blocks as input to the low-resolution MC process, the following expression can be assumed:

$$\underline{\hat{\tilde{h}}} = [N_1 \ N_2 \ N_3 \ N_4] \begin{bmatrix} \underline{\tilde{a}} \\ \underline{\tilde{b}} \\ \underline{\tilde{c}} \\ \underline{\tilde{d}} \end{bmatrix} \qquad (17.14)$$

where N_k, $k=1, 2, 3, 4$, are the unknown filters that are assumed to perform the low-resolution MC, $\underline{\hat{\tilde{h}}}$ is the low-resolution prediction.

As in [vetro 1998a], these filters are solved for by differentiating the following objective function:

$$J\{N_k\} = \left\| \underline{\tilde{h}} - \underline{\hat{\tilde{h}}} \right\|^2 \qquad (17.15)$$

with respect to each unknown filter and setting each result equal to zero. It can be verified that the optimal least-square solution for these filters is given by

$$N_1^{(r)} = xS_a^{(r)}x^+; \quad N_2^{(r)} = xS_b^{(r)}x^+$$
$$N_3^{(r)} = xS_c^{(r)}x^+; \quad N_4^{(r)} = xS_d^{(r)}x^+ \qquad (17.16)$$

where

$$x^+ = x^T(xx^T)^{-1} \qquad (17.17)$$

is the Moore–Penrose Inverse [lancaster 1985] for an $m \times n$ matrix with $m \leq n$. In the solution of Equation 17.16, the superscript (r) is added to the filters, N_k, due to their dependency on the full-resolution MC filters. In using these filters to perform the low-resolution MC, the MSE between $\tilde{h}$ and $\hat{\tilde{h}}$ is minimized. It should be emphasized that Equation 17.16 represents a generalized set of MC filters that are applicable to any x, which operates on a single MB. For the special case of the 4×4 cut, these filters are equivalent to the ones that were determined in [morky 1994] to minimize the drift.

In Figure 17.10, two equivalent MC schemes are shown. However, for implementation purposes, the optimal MC scheme is realized in a cascade form rather than a direct form. The reason is that the direct form filters are dependent on the matrices that perform full-resolution MC. Although these matrices were very useful in analytically expressing the full-resolution MC process, they require a huge amount of storage due to their dependency on the prediction mode, motion vector, and half-pixel accuracy. Instead, the three linear processes in Equation 17.13 are separated, so that an up-conversion, full-resolution MC, and down-conversion can be performed. Although one may be able to guess such a scheme, we have proven here that it is an optimal scheme provided the up-conversion filter is Moore–Penrose inverse of the down-conversion filter. In [vetro, 1998b], the optimal MC scheme, which employs frequency synthesis, has been compared to a nonoptimal MC scheme, which employs bilinear interpolation, and an optimal MC scheme, which employs

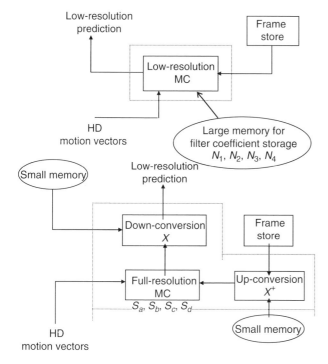

FIGURE 17.10
Optimal low-resolution MC scheme: direct form (top) versus cascade form (bottom). Both forms yield equivalent quality, but vary significantly in the amount of internal memory.

the 4 × 4 cut down-conversion. Significant reductions in the amount of drift were realized by both optimal MC schemes over the method, which used bilinear interpolation as the method of up-conversion. But more importantly, a 35% reduction in the amount of drift was realized by the optimal MC scheme using frequency synthesis over the optimal MC scheme using the 4 × 4 cut.

17.4.4 Three-Layer Scalable Decoder

In this section, we show how the key algorithms for down-conversion and MC are integrated into a three-layer scalable decoder. The central concept of this decoder is that three layers of resolution can be decoded using a decreased amount of memory for the lower resolution layers. Also, regardless of which layer is being decoded, much of the logic can be shared. Three possible decoder configurations are considered: full-memory decoder (FMD), half-memory decoder (HMD), and quarter-memory decoder (QMD). The LRD configurations are based on the key algorithms, which were described for down-conversion and MC. In the following, three possible architectures are discussed that provide equal quality, but vary in system level complexity. The first (ARCH1) is based on the LRD modeled in Figure 17.7b, the second (ARCH2) is very similar, but attempts to reduce the IDCT computation, while the third (ARCH3) is concerned with the amount of interface with an existing high-level decoder.

With regard to functionality, all of the architectures share similar characteristics. For one, an efficient implementation is achieved by arranging the logic in a hierarchical manner, i.e., employs separable processing. In this way, the FMD configuration is the simplest and serves as the logic core from which other decoder configurations are built on. In the HMD configuration, an additional horizontal down-conversion and up-conversion are performed. In the QMD configuration, all of the logic components from the HMD are utilized, such that an additional vertical down-conversion is performed after a horizontal down-conversion, and an additional vertical up-conversion is performed after a horizontal up-conversion. In summary, the logic for the HMD is built on the logic for the FMD, and the logic for the QMD is built on the logic of the HMD. The total system contains a moderate increase in logic, but HD bitstreams may be decoded to a lower resolution with a smaller amount of external memory. By simply removing external memory, lower layers can be achieved at a reduced cost.

The complete block diagram of ARCH1 is shown in Figure 17.11a. The diagram shown here assumes two things: (i) the initial system model of an LRD from Figure 17.6b is assumed and (ii) the down-conversions in the incoming branch are performed after the IDCT to avoid any confusion regarding MB format conversions in the DCT domain [vetro 1998b]. In looking at the resulting system, it is evident that full computation of the IDCT is required, and that two independent down-conversion operations must be performed. The latter is necessary so that low-resolution predictions are added to low-resolution residuals. Overall, the increase in logic for the added feature of memory savings is quite small. However, it is evident that ARCH1 is not the most cost-effective implementation, but it represents the foundation of previous assumptions, and allows us to better analyze the impact of the two modified architectures to follow.

In Figure 17.11b, the block diagram of ARCH2 is shown. In this system, realizing that the IDCT operation is simply a linear filter reduces the combined computation for the IDCT and down-conversion. In the FMD, we know that a fast IDCT is applied separately to the rows and columns of an 8 × 8 block. For the HMD, our goal is to combine the horizontal down-conversion with the horizontal IDCT. In 1-D case, an 8 × 16 matrix can represent the horizontal down-conversion, and an 8 × 8 matrix can represent the horizontal IDCT. Combining these processes such that the down-conversion operates on the incoming

Application Issues of MPEG-1/2 Video Coding

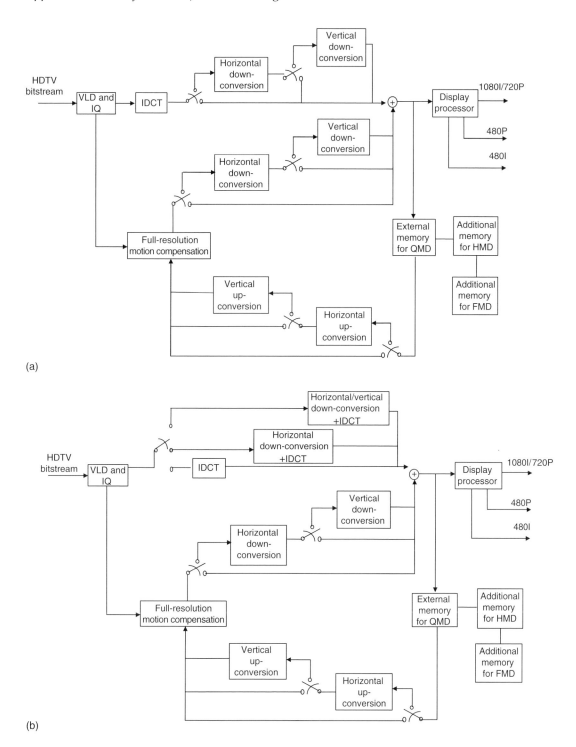

FIGURE 17.11
Block diagram of various three-layer scalable decoder architectures. All architectures provide equal quality with varying system complexity. (a) ARCH1, derived directly from block diagram of assumed low-resolution decoder. (b) ARCH2, reduced computation of IDCT by combining down-conversion and IDCT filters.

(continued)

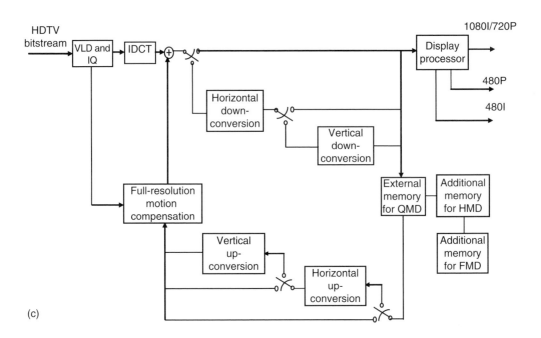

FIGURE 17.11 (continued)
(c) ARCH3, minimized interface with existing HL decoder by moving linear filtering for down-conversion outside of the adder.

DCT rows first, results in a combined 8×16 matrix. To complete the transformation, the remaining columns can then be applied to the fast IDCT. In the above description, computational savings are achieved in two places: first, the horizontal IDCT is fully absorbed into the down-conversion computation which must take place anyway, and second, the fast IDCT is utilized for a smaller amount of columns. In the case of the QMD, these same principles can be used to combine the vertical down-conversion with the vertical IDCT. In this case, one must be aware of the MB type (field-DCT or frame-DCT) so that an appropriate filter can be applied. In contrast to the previous two architectures, ARCH3 assumes that the entire front-end processing of the decoder is used; it is shown in Figure 17.11c. In this way, the adder is always a full-resolution adder, whereas in ARCH1 and ARCH2, the adder needed to handle all three-layers of resolution. The major benefit of ARCH3 is that it does not require many interfaces with the existing decoder structure. The memory is really the only place where a new interface needs to be defined. Essentially, a down-conversion filtering may be applied before storing the data, and an up-conversion filtering may be applied, as the data is needed for full-resolution MC. This final architecture is similar in principle to the embedded compression schemes that were mentioned in the beginning of this section. The main difference is that the resolution of the data is decreased rather than compressed. This allows a simpler means of low-resolution display.

17.4.5 Summary of Down-Conversion Decoder

A number of integrated solutions for a scalable decoder have been presented. Each decoder is capable of decoding directly to a lower resolution using a reduced amount of memory in comparison to the memory required by the high-level decoder. The method of frequency synthesis is successful in better preserving the high-frequency data within an MB and the filtering that is used to perform optimal low-resolution MC is capable of minimizing the drift. It has been shown that a realizable implementation can be achieved, such that the filters for optimal low-resolution MC are equivalent to an up-conversion, full-resolution MC, and down-conversion, where the up-conversion filters are determined by a Moore–Penrose inverse of the down-conversion. The amount of logic required by these processes is kept minimal since they are realized in a hierarchical structure. Since the down-conversion and up-conversion processes are linear, the architecture design is flexible in that equal quality can be achieved with varying levels of system complexity. The first architecture that we examined came from the initial assumptions that were made on the LRD, i.e., a down-conversion is performed before the adder. It was noted that a full IDCT computation was required and that a down-conversion must be performed in two places. As a result, a second architecture was presented to reduce the IDCT computation, and a third was presented to minimize the amount of interface with the existing high-level decoder. The major point here is that the advantages of ARCH2 and ARCH3 cannot be realized by a single architecture. The reason is that performing a down-conversion in the incoming branch reduces the IDCT computation, therefore a down-conversion must be performed after the full-resolution MC as well. In any case, equal quality is offered by each architecture and the quality is of commercial grade.

Appendix A: DCT-to-Spatial Transformation

Our objective in this section is to express the following DCT domain relationship:

$$\tilde{A}(k,l) = \sum_{p=0}^{M-1} \sum_{q=0}^{N-1} [X_{k,l}(p,q) A(p,q)] \quad (17.18)$$

as

$$\tilde{a}(i,j) = \sum_{s=0}^{M-1} \sum_{t=0}^{N-1} [x_{i,j}(s,t) a(s,t)] \quad (17.19)$$

where
  $\tilde{A}$ and $\tilde{a}$ are the DCT and spatial output
  A and a are the DCT and spatial input
  X and x are the DCT and spatial filters

By definition, the $M \times N$ DCT transform is defined by

$$A(k,l) = \sum_{i=0}^{M-1} \sum_{j=0}^{N-1} a(i,j) \psi_k^M(i) \psi_l^N(j) \quad (17.20)$$

and its inverse, the $M \times N$ IDCT by

$$a(i,j) = \sum_{k=0}^{M-1} \sum_{l=0}^{N-1} A(k,l) \psi_k^M(i) \psi_l^N(j) \quad (17.21)$$

where the basis function is given by

$$\psi_k^N = \sqrt{\frac{2}{N}} \alpha(k) \cos\left(\frac{2i+1}{2N} k\pi\right) \quad (17.22)$$

and

$$\alpha(k) = \begin{cases} \frac{1}{\sqrt{2}} & \text{for } k = 0 \\ 1 & \text{for } k \neq 0. \end{cases} \quad (17.23)$$

By substituting Equation 17.22 into the expression for the IDCT yields:

$$\tilde{a}(i,j) = \sum_{k=0}^{M-1}\sum_{l=0}^{N-1} \psi_k^M(i)\psi_l^N(j) \cdot \left[\sum_{p=0}^{M-1}\sum_{q=0}^{N-1} X_{k,l}(p \cdot q) A(p \cdot q)\right]$$

$$= \sum_{p=0}^{M-1}\sum_{q=0}^{N-1} A(p,q) \cdot \left[\sum_{k=0}^{M-1}\sum_{l=0}^{N-1} X_{k,l}(p,q)\psi_k^M(i)\psi_l^N(j)\right] \quad (17.24)$$

Substituting the DCT definition into Equation 17.24 gives the following:

$$\tilde{a}(i,j) = \sum_{p=0}^{M-1}\sum_{q=0}^{N-1} \left[\sum_{s=0}^{M-1}\sum_{t=0}^{N-1} a(s,t)\psi_p^M(s)\psi_q^N(t)\right] \sum_{k=0}^{M-1}\sum_{l=0}^{N-1} \left[X_{k,l}(p,q) \cdot \psi_p^M(i)\psi_q^N(j)\right] \quad (17.25)$$

Finally, Equation 17.16 can be formed with

$$x_{i,j}(s,t) = \sum_{k=0}^{M-1}\sum_{l=0}^{N-1} \psi_k^M(i) \cdot \psi_l^N(j) \left[\sum_{p=0}^{M-1}\sum_{q=0}^{N-1} X_{k,l}(p,q) \cdot \psi_p^M(s)\psi_q^N(t)\right] \quad (17.26)$$

and the transformation is fully defined.

Appendix B: Full-Resolution Motion Compensation in Matrix Form

In 2-D, a motion compensated MB may have contributions from at most 4 MBs per motion vector. As noted in Figure 17.12, MBs a, b, c, and d include four 8×8 blocks each. These subblocks are raster-scanned so that each MB can be represented as a vector. According to

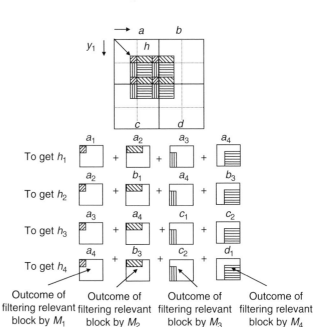

FIGURE 17.12
Relationship between the input and output blocks of the motion compensation process in the FRD.

Application Issues of MPEG-1/2 Video Coding

the motion vector, (dx, dy), a local reference, (y_1, y_2), is computed to indicate where the origin of the MC block is located; the local reference is determined by

$$y_1 = dy - 16 \cdot [\text{Integer}(dy/16) - \gamma(dy)]$$
$$y_2 = dx - 16 \cdot [\text{Integer}(dx/16) - \gamma(dx)]$$
(17.27)

where

$$\gamma(d) = \begin{cases} 1 & \text{if } d < 0 \text{ and } d \bmod 16 = 0 \\ 0 & \text{otherwise.} \end{cases}$$
(17.28)

The reference point for this value is the origin of the upper-leftmost input MB. With this, the MC prediction may be expressed as

$$\underline{h} = \begin{bmatrix} h_1 \\ h_2 \\ h_3 \\ h_4 \end{bmatrix} = \begin{bmatrix} S_a^{(r)} & S_b^{(r)} & S_c^{(r)} & S_d^{(r)} \end{bmatrix} \cdot \begin{bmatrix} \underline{a} \\ \underline{b} \\ \underline{c} \\ \underline{d} \end{bmatrix}; \quad r = 1, 2, 3, 4.$$
(17.29)

As an example, Figure 17.11 considers $(y_1, y_2) \in [0, 7]$, which implies that $r = 1$. In this case the MC filters are given by

$$S_a^{(1)} = \begin{bmatrix} M_1 & M_2 & M_3 & M_4 \\ 0 & M_1 & 0 & M_3 \\ 0 & 0 & M_1 & M_2 \\ 0 & 0 & 0 & M_1 \end{bmatrix}, \quad S_b^{(1)} = \begin{bmatrix} 0 & 0 & 0 & 0 \\ M_2 & 0 & M_4 & 0 \\ 0 & 0 & 0 & 0 \\ 0 & 0 & M_2 & 0 \end{bmatrix},$$

$$S_c^{(1)} = \begin{bmatrix} 0 & 0 & 0 & 0 \\ 0 & 0 & 0 & 0 \\ M_3 & M_4 & 0 & 0 \\ 0 & M_3 & 0 & 0 \end{bmatrix}, \quad S_d^{(1)} = \begin{bmatrix} 0 & 0 & 0 & 0 \\ 0 & 0 & 0 & 0 \\ 0 & 0 & 0 & 0 \\ M_4 & 0 & 0 & 0 \end{bmatrix}$$
(17.30)

In Equation 17.30, the M_1, M_2, M_3, and M_4 matrices operate on the relevant 8×8 blocks of a, b, c, and d. Their elements vary according to the amount of overlap as indicated by (y_1, y_2) and the type of prediction. The type of prediction may be frame-based or field-based and is predicted with half-pixel accuracy. As a result, the matrices $S_a^{(r)}, S_b^{(r)}, S_c^{(r)}$, and $S_d^{(r)}$ are extremely sparse and may only contain nonzero values of 1, 1/2, and 1/4. For different values of (y_1, y_2) the configuration of the above matrices will change: $y_1 \in [0,7]$ and $y_2 \in [8,15]$ implies $r = 2$; $y_1 \in [8,15]$ and $y_2 \in [0,7]$ implies $r = 3$; $y_1, y_2 \in [8,15]$ implies $r = 4$. The resulting matrices can easily be formed using the concepts illustrated in Figure 17.11.

17.5 Error Concealment

17.5.1 Background

Practical communication channels available for delivery of compressed digital video are characterized by occasional bit-error and packet loss, although the actual impairment mechanism varies widely with the specific medium under consideration. The class of

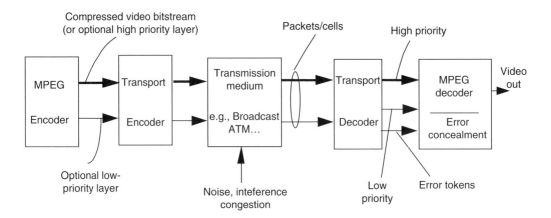

FIGURE 17.13
System block diagram of visual communication system.

decoder error concealment schemes described here is based on identification and predictive replacement of picture regions affected by bit-error or data loss. It is noted that this approach is based on conversion (via appropriate error/loss detection mechanisms) of the transmission medium into an erasure channel in which all error or loss events can be identified in the received bitstream. In a block structured compression algorithm such as MPEG, all channel impairments are manifested as erasures of video units (such as MPEG MBs or slices). Concealment at the decoder is then based on exploiting temporal and spatial picture redundancy to obtain an estimate of erased picture areas. The efficiency of error concealment depends on redundancies in pictures and on redundancies in the compressed bitstream that are not removed by source coding. Block compression algorithms do not remove a considerable amount of inter-block redundancies, such as structure, texture, and motion information about objects in the scene.

To be more specific, error resilience for compressed video can be achieved through the addition of suitable transport and error concealment methods, as outlined in the system block diagram shown in Figure 17.13.

The key elements of such a robust video delivery system are outlined below:

- The video signal is encoded using appropriate video compression syntax such as MPEG. Note that we have restricted consideration primarily to the practical case in which the video compression process itself is not modified, and robustness is achieved through additive transport and decoder concealment mechanisms (except for I-frame motion described in Section 4.3). This approach simplifies encoder design, because it separates media-independent video compression functions from media-dependent transport operations. On the receiver side, although a similar separation is substantially maintained, the video decoder must be modified to support an error token interface and error concealment functionality.

- Compressed video data is organized into a systematic data structure with appropriate headers for identification of the temporal and spatial pixel-domain location of encoded data [joseph 1992b]. When an erroneous/lost packet is detected, these video units serve as resynchronization points for resumption of normal decoding, while the headers provide a means for precisely locating regions of the picture that were not correctly received. Note that two-tier systems may require additional transport level support for high and low priority (HP/LP) resynchronization [siracusa 1993].

- The video bitstream may optionally be segregated into two layers for prioritized transport [ghanbari 1989; karlsson 1989; kishno 1989; zdepski 1989; joseph 1992a, b; siracusa 1993] when a high degree of error resilience is required. Note that separation into high and low priorities may be achieved either by using a hierarchical (layered) compression algorithm [ghanbari 1989; siracusa 1993] or by direct code word parsing [zdepski 1989, 1990]. Note that both these layering mechanisms have been accepted for standardization by MPEG-2 [mpeg2].

- Once the temporal and spatial location(s) corresponding to lost or incorrectly received packets is determined by the decoder, it will execute an error concealment procedure for replacement of lost picture areas with subjectively acceptable material estimated from available picture regions [harthanck 1986; jeng 1991; wang 1991]. Generally, this error concealment procedure will be applied to all erased blocks in one-tier (single priority) transmission systems, while for two-tier (HP/LP) channels the concealment process may optionally ignore loss of LP data.

- In the following sections, the technical detail of some commonly used error concealment algorithms is provided. Specifically, we focus on the recovery of code word errors and errors that affect the pixels within an MB.

17.5.2 Error Concealment Algorithms

In general, design of specific error concealment strategies depends on the system design. For example, if two-layered transmission is used, the receiver should be designed to conceal HP error and LP error with different strategies. Moreover, if some redundancy (steering information) could be added to the encoder the concealment could be more efficient. However, we first assume that the encoder is defined for maximum compression efficiency, and that concealment is only performed in the receiver. It should be noted that some exemptions exist for this assumption. These exemptions include the use of I-frame motion vectors, scalability concealment, and limitation of slice length (in order to perform acceptable concealment in the pixel domain the limitation of slice length exists, i.e., the length of slices cannot be longer than one row of picture). Figure 17.14 shows a block diagram of a generic one/two-tier video decoder with error concealment.

Note that Figure 17.14 shows two stages of decoder concealment in the code word domain and pixel domain, respectively. Code word domain concealment, in which locally generated decodable code words (e.g., B-picture motion vectors, EOB code, etc.) are inserted into the

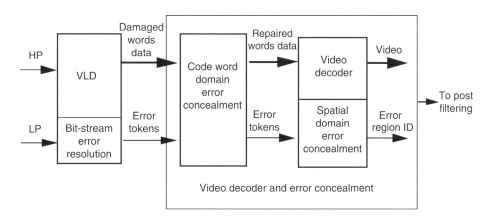

FIGURE 17.14
MPEG video decoder with error concealment.

bitstream, is convenient for implementation of simple temporal replacement functions (which in principle can also be performed in the pixel domain). The second stage of pixel-domain processing is for temporal and spatial operations not conveniently done in the code word domain. Advanced spatial processing will generally have to be performed in the pixel domain, although limited code word domain options can also be identified.

17.5.2.1 Code Word Domain Error Concealment

The code word domain concealment receives video data and error tokens from the transport processor/VLD. Under normal conditions, no action is taken and the data is passed along to the video decoder. When an error token is received, damaged data is repaired to the extent possible by insertion of locally generated code words and resynchronization codes. An error region ID is also created to indicate the image region to be concealed by subsequent pixel-domain processing. Two mechanisms have been used in code word domain error concealment: neglect the effect of lost data by declaring an EOB, or replace the lost data with a pseudo code to handle the MB-types or other VLC codes. If high-level data such as DC or MB header is lost, the code word domain concealment with pseudo codes can only provide signal resynchronization (decodability) and replaces the image scene with a fixed gray level in the error region. Obviously, further improvement is needed in the video decoder. This task is implemented with the error concealment in the video decoder. It is desirable to replace erased I- or P-picture regions with a reasonably accurate estimate to minimize the impact of frame-to-frame propagation.

17.5.2.2 Spatio-Temporal Error Concealment

In general, two basic approaches are used for spatial domain error concealment: temporal replacement and spatial interpolation. In temporal replacement, as shown in Figure 17.15, the spatially corresponding ones in the previously decoded data with MC replace the damaged blocks in the current frame if motion information is available. This method exploits temporal redundancy in the reconstructed video signals and provides satisfactory results in areas with small motion and for which motion vectors are provided. If motion information is lost, this method will fail in the moving areas. In the method of spatial interpolation as shown in Figure 17.16, the lost blocks are interpolated by the data from the adjacent non-error blocks with maximally smooth reconstruction criteria or other techniques.

In this method, the correlation between adjacent blocks in the received and reconstructed video signals is exploited. However, severe blurring will result from this method if data in adjacent blocks is also lost. In an MPEG decoder, temporal replacement outlined above is based on previously decoded anchor (I, P) pictures that are available in the frame memory. If motion vectors corresponding to pixels in the erasure region can also be estimated, this temporal replacement operation can be improved via MC. Also, in the MPEG decoder,

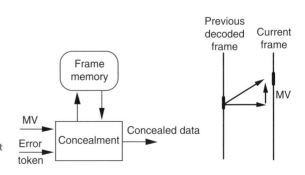

FIGURE 17.15
Error concealment uses temporal replenishment with motion compensation.

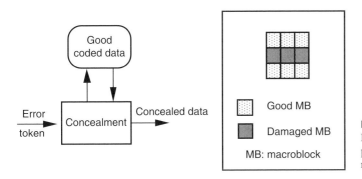

FIGURE 17.16
Error concealment uses spatial interpolation with the data from good neighbors.

groups of video pixels (blocks, macroblocks, or slices) are separately decoded, so that pixel values and motion information corresponding to adjacent picture regions are generally available for spatial concealment. However, estimation from horizontally adjacent blocks may not always be useful since cell loss tends to affect a number of adjacent blocks (due to the MPEG and ATM data structures); also differential encoding between horizontally adjacent blocks tends to limit the utility of data obtained from such neighbors. Therefore, most of the useable spatial information will be located in blocks above or below the damaged region. That is, vertical processing/concealment is found to be most useful due to the transmission order of the data.

For I-pictures, the damaged data can be reconstructed by either temporal replacement from the previously decoded anchor frame or by spatial interpolation from good neighbors. These two methods will be discussed later. For P- and B-pictures, the main strategy to conceal the lost data is to replace the region with pixels from the corresponding (and possibly motion compensated) location in the previously decoded anchor. In this replacement the motion vectors play a very important role. In other words, if good estimates of motion information can be obtained, its use may be the least noticeable correction. Since DPCM coding for motion vectors only exploited the correlations between the horizontally neighboring MBs, the redundancy between the vertical neighborhood still exists after encoding. Therefore, the lost motion information can be estimated from the vertical neighbors. In the following, three algorithms that have been developed for error concealment in the video decoder are described.

Algorithm 1: Spatial interpolation of missing I-picture data and temporal replacement for P- and B-pictures with MC [sun1 1992]

For I-pictures, DC values of damaged block are replaced by the interpolation from the closest top and bottom good neighbors; the AC coefficients of those blocks are synthesized from the DC values of the surrounding neighboring blocks.

For P-pictures, the previously decoded anchor frames with MC replace the lost blocks. The lost motion vectors are estimated by interpolation of the ones from the top and bottom MBs. If motion vectors in both top and bottom MBs are not available, zero motion vectors are used. The same strategy is used for B-pictures; the only difference is that the closest anchor frame is used. In other words, the damaged part of the B-picture could be replaced by either the forward or backward anchor frame, depending on its temporal position.

Algorithm 2: Temporal replacement of missing I-picture data and temporal replacement for P- and B-pictures with top MC

For I-pictures, the damaged blocks are replaced with the colocated ones in the previously decoded anchor frame.

For P- and B-pictures, the closest previously decoded anchor frame replaces the damaged part with MC as in the Algorithm 1. The only difference is that the motion vectors are estimated only from the closest top MB instead of interpolation of top and bottom motion vectors. This makes the implementation of this scheme much easier. If these motion vectors are not available then zero motion vectors are used.

In the above two algorithms, the damaged blocks in an I-picture (anchor frame) are concealed by two methods: temporal replacement and spatial interpolation. Temporal replacement is able to provide high-resolution image data to substitute for lost data; however, in motion areas, a big difference might exist between the current intracoded frame and the previously decoded frame. In this case, temporal replacement will produce large shearing distortion unless some motion-based processing can be applied at the decoder. However, this type of processing is not generally available since it is a computationally demanding task to locally compute motion trajectories at the decoder. In contrast, the spatial interpolation approach synthesizes lost data from the adjacent blocks in the same frame. Therefore, the intraframe redundancy between blocks is exploited, while the potential problem of severe blurring due to insufficient high order AC coefficients for active areas. To alleviate this problem, an adaptive concealment strategy can be used as a compromise; this is described in Algorithm 3.

Algorithm 3: Adaptive spatiotemporal replacement of missing I-picture data and temporal replacement with MC for P- and B-pictures

For I-pictures, the damaged blocks are concealed with temporal replacement or spatial interpolation according to the decision made by the top and bottom MBs (Figure 17.17). The decision of which concealment method to use will be based on the more cheaply obtained measures of image activity from the neighboring top and bottom MBs. One candidate for the decision processor is to make the decision based on prediction error statistics measured in the neighborhood. The decision region is shown in Figure 17.15, where

$$\begin{aligned} \text{VAR} &= E[(x - \hat{x})^2], \\ \text{VAROR} &= E[x^2] - \mu^2, \end{aligned} \tag{17.31}$$

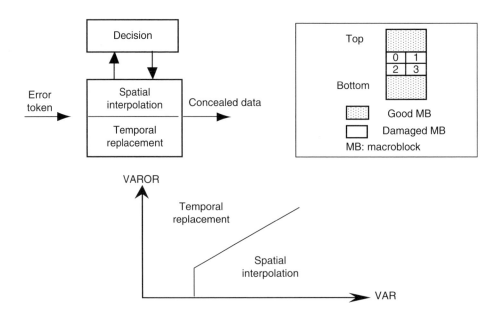

FIGURE 17.17
Adaptive error concealment strategy.

and x is the neighboring good MB data, $\hat{x}$ is the data of the corresponding MB in the previously decoded frame at the colocated position and μ is the average value of the neighboring good MB data in the current frame. One can appreciate that VAR is indicative of the local motion and VAROR of the local spatial detail. If VAR > VAROR and VAR > T, where T is a preset threshold value which is set to 5 in the experiments, the concealment method is spatial interpolation; if VAR < VAROR or VAR < T, the concealment method is temporal replacement.

It should be noted that the concealment for luminance is performed on a block basis instead of MB basis, while the chrominance is still on the MB basis. The detailed decisions for the luminance blocks are described as follows:

If both top and bottom are temporally replaced, then four blocks (0, 1, 2, and 3) are replaced by the colocated ones (colocated means no MC) in the previously decoded frame;

If top is temporally replaced and bottom is spatially interpolated, then blocks 0 and 1 are replaced by the colocated ones in the previously decoded anchor frame and blocks 2 and 3 are interpolated from the block boundaries;

If top is spatially interpolated and bottom is temporally replaced, then blocks 0 and 1 are interpolated from the boundaries, and block 2 and 3 are replaced by the colocated ones in the previously decoded anchor frame;

If both top and bottom are not temporally replaced, all four blocks are spatially interpolated.

In spatial interpolation, a maximal smoothing technique with boundary conditions under certain smoothness measures is used. The spatial interpolation process is carried out with two steps: the mean value of the damaged block is first bilinearly interpolated with ones from the neighboring blocks, then spatial interpolation for each pixel is performed with a Laplacian operator. Minimizing the Laplacian on the boundary pixels using the iterative process in [wang 1991] enforces the process of maximum smoothness.

For P- and B-pictures a similar concealment method is used as in Algorithm 2 except motion vectors from top and bottom neighboring MBs are used for top two-blocks and bottom two-blocks, respectively.

A schematic block diagram for implementation of adaptive error concealment for intra-coded frames is given in Figure 17.18. Corrupted MBs are first indicated by error tokens obtained through the transport interface. Then, a decision regarding which

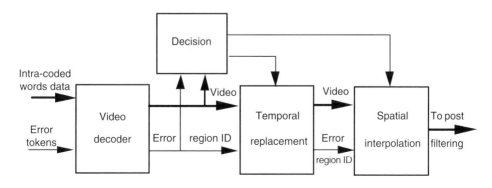

FIGURE 17.18
Two-stage error concealment strategy.

concealment method (temporal replacement or spatial interpolation) should be used is based on easily obtained measures of image activity from the neighboring top and bottom MBs. The corrupted MBs are first classified into two classes according to the local activities. If local motion is smaller than spatial detail, the corrupted MBs are defined as the first class and will be concealed by temporal replacement; when local motion is greater than local spatial detail, the corrupted MBs are defined as the second class and will be concealed by spatial interpolation. The overall concealment procedure consists of two stages. First, temporal replacement is applied to all corrupted MBs of the fist class throughout the whole frame. After the temporal replacement stage, the remaining unconcealed damaged MBs of the second class are more likely to be surrounded by valid image MBs. A stage of spatial interpolation is then performed on them. This will now result in less blurring, or the blurring will be limited to smaller areas. Therefore, a good compromise between shearing (discontinuity or shift of edge or line) and blurring can be obtained.

17.5.3 Algorithm Enhancements

As discussed above, I-picture errors, which are imperfectly concealed, will tend to propagate through all frames in the GOP. Therefore, it is desirable to develop enhancements for the basic spatiotemporal error concealment technique to further improve the accuracy with which missing I-picture pixels are replaced. Three new algorithms have been developed for this purpose. The first is an extension of the spatial restoration technique outlined earlier, and is based on processing of edge information in a large local neighborhood to obtain better restoration of the missing data. The second and third are variations, which involve encoder modifications aimed at improved error concealment performance. Specifically, information such as I-picture pseudo motion vectors, or low-resolution data in a hierarchical compression system, are added in the encoder. These redundancies can significantly benefit error concealment in the decoders that must operate under higher cell loss/error conditions, while having a relatively modest impact on nominal image quality.

17.5.3.1 Directional Interpolation

Improvements in spatial interpolation algorithms (for use with MPEG I-pictures) have been proposed in [sun 1995, kwok 1993]. In these studies, additional smoothness criteria and directional filtering are used for estimating the picture area to be replaced. The new algorithms utilize spatially correlated edge information from a large local neighborhood of surrounding pixels and perform directional or multidirectional interpolation to restore the missing block. The block diagram illustrating the general principle of the restoration process is shown in Figure 17.19.

Three parts are included in the restoration processing: edge classification, spatial interpolation, and pattern mixing. The function of the classifier is to select the top one, two or three directions that strongly characterize edge orientations in the surrounding neighborhood. Spatial interpolation is performed for each of the directions determined by the classifier. For a given direction, a series of 1-D interpolations are carried out along that direction. All of the missing pixels are interpolated from a weighted average of good neighborhood pixels. The weights depend inversely on the distance from the missing pixel to the good neighborhood pixels. The purpose of pattern mixing is to extract strong characteristic features of two or more images and merge them into one image, which is then used to replace the corrupted one. Results show that these algorithms are capable of providing subjectively better edge restoration in missing areas, and may thus be useful for I-picture processing in high error-rate scenarios. However, the computational practicality of these edge-filtering techniques needs further investigation for given application scenarios.

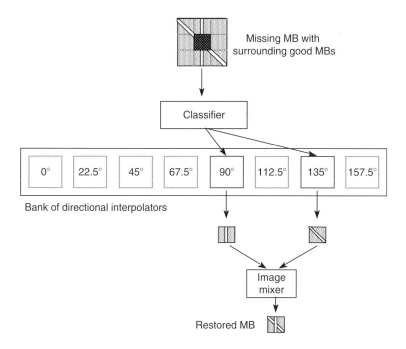

FIGURE 17.19
The multi-directional edge restoration process.

17.5.3.2 I-Picture Motion Vectors

Motion information is very useful in concealing losses in P and B frames, but is not available for I-pictures. This limits the concealment algorithm to spatial or direct temporal replacement options described above, which may not always be successful in moving areas of the picture. If motion vectors are made available for all MPEG frames (including intra-coded ones) as an aid for error concealment [sun1 1992], good error concealment performance can be obtained without the complexity of adaptive spatial processing. Therefore, a syntax extension has been adopted by the MPEG-2 where motion vectors can be transmitted in an I-picture as the redundancy for error concealment purposes [sun2 1992]. The MB syntax is not changed; however, motion vectors are interpreted in the following way: the decoded forward motion vectors belong to the MB spatially below the current MB, and describe how that MB can be replaced from the previous anchor frame in the event that the MB cannot be recovered. Simulation results have shown that subjective picture quality with I-picture motion vectors is noticeably superior to conventional temporal replacement, and that the overhead for transmitting the additional motion vectors is less than 0.7% of the total bit rate at bit rate of about 6–7 Mbits/s.

17.5.3.3 Spatial Scalable Error Concealment

This approach for error concealment of MPEG video is based on the scalability (or hierarchy) feature of MPEG-2 [mpeg2]. Hierarchical transmission provides more possibilities for error concealment, when a corresponding two-tier transmission media are available. A block diagram illustrating the general principle of coding system with spatial scalability and error concealment is shown in Figure 17.20.

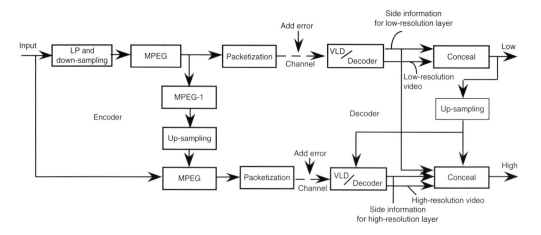

FIGURE 17.20
Block diagram of spatial scalability with error concealment.

It should be noted that the concept of scalable error concealment is different from the two-tier concept with data partitioning. Scalable concealment uses the spatial scalability feature in MPEG-2, while the two-tier case uses the data partitioning feature of MPEG-2, in which the data corresponds to the same spatial resolution layer but is partitioned to two parts with a breakpoint. In spatial scalability, the encoder produces two separate bitstreams: one for the low-resolution base layer and another for the high-resolution enhancement. The high-resolution layer is encoded with an adaptive choice of temporal prediction from previous anchor frames and compatible spatial prediction (obtained from the up-sampled low-resolution layer) corresponding to the current temporal reference. In the decoder, redundancies that exist in the scaling data greatly benefit the error concealment processing. In a simple experiment with spatial scalable MPEG-2, we consider a scenario in which losses in the high-resolution MPEG-2 video are concealed with information from the low-resolution layer. Actually, there are two kinds of information in the lower layer that can be used to conceal the data loss in the high-resolution layer: up-sampled picture data and scaled motion information. Therefore, three error concealment approaches are possible:

1. *Up-sampled substitution*: Lost data is replaced by colocated up-sampled data in the low-resolution decoded frame. The up-sampled picture is obtained from the low-resolution picture with proper up-sampling filter.
2. *Mixed substitution*: Lost MBs in I-picture are replaced by colocated up-sampled MBs in the low-resolution decoded frame, while lost MBs in P- and B-picture are temporally replaced by the previously decoded anchor frame with the motion vectors for the low-resolution layer.
3. *Motion vector substitution*: The previously decoded anchor frame with the motion-vectors replaces lost MBs for the low-resolution layer appropriately scaled.

Since motion vectors are not available for I-pictures, obviously, method 3 does not work for I-picture (unless I-picture motion vectors [concealment motion vectors] of MPEG-2 are generated in encoder). Simulation results have shown that, on average, the up-sampled substitution outperforms the other two, and mixed substitution also provides acceptable results in the case of video with smooth motion.

TABLE 17.4

Subjective Quality Comparison

Picture Material	Items	Algorithm 1	Algorithm 2	Algorithm 3	Comments
Still	Blurring	High	None	Low	Temporal replacement works
	Shearing	None	None	None	very good in no motion areas
	Artifact blocking	Medium	None	Low	
Slow motion	Blurring	High	None	Low	Temporal replacement works
	Shearing	None	Low	Low	well in slow motion areas
	Artifact blocking	Medium	None	Low	
Fast motion	Blurring	High	None	Medium	Temporal replacement causes more shearing
	Shearing	None	High	Low	Spatial interpolation results in blurring
	Artifact blocking	High	Low	Medium	Adaptive strategy limits blurring in smaller areas
Overall	The adaptive strategy of steering the temporal replacement and spatial interpolation according to the measures of local activity and local motion gives a good compromise between shearing and blurring.				

17.5.4 Summary of Error Concealment

In this section, a general class of error concealment algorithms for MPEG video has been discussed. The error concealment approaches that have been described are practical for current MPEG decoder implementations and have been demonstrated to provide significant robustness. Specifically, it has been shown that the adaptive spatiotemporal algorithm can provide reasonable picture quality at cell loss ratios (CLR) as high as 10^{-3} when used in conjunction. These results confirm that compressed video is far less fragile than originally believed when appropriate transport and concealment techniques are employed. The results can be summarized as in Table 17.4.

Several concealment algorithm extensions based on directional filtering, I-picture pseudo motion vectors and MPEG-2 scalability were also considered and shown to provide performance gains that may be useful in certain application scenarios. In view of the practical benefits of robust video delivery, it is recommended that such error-resilience functions (along with associated transport structures) be important for implementation in emerging TV, HDTV, teleconferencing, and multimedia systems if the cell loss rates on these transmission system are significant. Particularly for terrestrial broadcasting and ATM network scenarios, we believe that robust video delivery based on decoder error concealment is an essential element of a viable system design.

17.6 Summary

In this chapter, several application issues of MPEG-2 have been discussed. The most successful application of MPEG-2 is US HDTV standard. The other application issues include transcoding with bitstream scaling, down-conversion decoding, and error concealment. Transcoding is a very interesting topic that converts the bitstreams between different standards. The error concealment is very useful in the noisy communication channels such as terrestrial television broadcasting. The down-conversion decoder responds to the market requirement during the DTV transition period and long term need for displaying DTV signal on computer monitors.

Exercises

1. In DTV applications, describe the advantages and disadvantages of interlaced format and progressive format. Explain why the computer industry favors progressive format and TV manufacturers like interlaced format.
2. Do all DTV formats have square pixel format? Why is square pixel format important for digital television?
3. The bitstream scaling is one kind of transcoding, according to your knowledge, describe several other kinds of transcoding (such as MPEG-1 to JPEG) and propose a feasible solution to achieve the transcoding requirements.
4. What type of MPEG-2 frames will cause a higher degree of error propagation if errors occur? What technique of error concealment is allowed by the MPEG-2 syntax? Using this technique, perform simulations with several images to determine the penalty in the case of no errors.
5. To reduce the drift in a down-conversion decoder, what coding parameters can be chosen at the encoder? Will these actions affect the coding performance?
6. What are the advantages and disadvantages of a down-conversion decoder in the frequency-domain and spatial-domain?

References

[bao 1996] J. Bao, H. Sun, and T. Poon, HDTV down-conversion decoder, *IEEE Transactions on Consumer Electronics*, 42, 3, 402–410, August 1996.

[boyce 1995] J. Boyce, J. Henderson, and L. Pearlestien, an SDTV decoder with HDTV capability: an all-format ATV decoder, SMPTE Fall Conference, New Orleans, LA, 1995.

[bruni 1998] R. Bruni, A. Chimienti, M. Lucenteforte, D. Pau, and R. Sannino, A novel adaptive vector quantization method for memory reduction in MPEG-2 HDTV receivers, *IEEE Transactions on Consumer Electronics*, 44. 3, 537–544, August 1998.

[de with 1998] P.H.N. de With, P.H. Frencken, and M.V.d. Schaar-Mitrea, An MPEG decoder with embedded compression for memory reduction, *IEEE Transactions on Consumer Electronics*, 44. 3, 545–555, August 1998.

[ga 1994] Grand Alliance HDTV System Specification Version 2.0, December 7, 1994.

[ghanbari 1989] M. Ghanbari, Two-layer coding of video signals for VBR networks, *IEEE Journal on Selected Areas in Communications*, 7, 5, 771–781, June 1989.

[harthanck 1986] W. Harthanck, W. Keesen, and D. Westerkamp, Concealment techniques for block encoded TV-signals, Picture Coding Symposium, 1986.

[isnardi 1993] M.A. Isnardi, Consumers seek easy to-use products, IEEE Spectrum, January 1993, p. 64.

[jayant 1984] N.N. Jayant and P. Noll, *Digital Coding of Waveforms to Speech and Video*, Prentice-Hall, Englewood Cliffs, NJ, 1984.

[jeng 1991] F.-C. Jeng and S.H. Lee, Concealment of bit error and cell loss in inter-frame coded video transmission, ICC Proceeding, ICC'91, pp. 496–500.

[joseph 1992a] K. Joseph, S. Ng, D. Raychaudhuri, R.S. Girons, T. Savatier, R. Siracusa, and J. Zdepski, MPEG++: A robust compression and transport system for digital HDTV, *Signal Processing Image Communication*, 4, 307–323, 1992.

[joseph 1992b] K. Joseph, S. Ng, D. Raychaudhuri, R. Saint Girons, R. Siracusa, and J. Zdepski, Prioritization and transport in the ADTV digital simulcast system, Proceedings of the ICCE '92, June 1992.

[karlsson 1989] G. Karlsson, and M. Vetterli, Packet video and its integration into the network architecture, *IEEE Journal on Selected Areas in Communication*, June 1989, 739–751.

[kishno 1989] F. Kishino, K. Manabe, Y. Hayashi, and H. Yasuda, Variable bit-rate coding of video signals for ATM networks, *IEEE Journal on Selected Areas in Communications*, 7, 5, 801–806, June 1989.

[kwok 1993] W. Kwok and H. Sun, Multi-directional interpolation for spatial error concealment, *IEEE Transactions on Consumer Electronics*, 455–460, August 1993.

[lancaster 1985] P. Lancaster and M. Tismenetsky, *The Theory of Matrices with Application*, Academic Press, Boston, MA, 1985.

[lei 1999] S. Lei, A quadtree embedded compression algorithm for memory saving DTV decoders, Proceedings of the International Conference on Consumer Electronics, Los Angeles, CA, June 1999.

[merhav 1997] N. Merhav and V. Bhaskaran, Fast algorithms for DCT-domain image down-sampling and for inverse motion compensation, *IEEE Transaction on Circuits and Systems for Video Technology*, 7, 3, 468–476, June 1997.

[mokry 1994] R. Mokry and D. Anastassiou, Minimal error drift in frequency scalability for motion compensated DCT coding, *IEEE Transaction on Circuits and Systems for Video Technology*, 4, 4, 392–406, August 1994.

[mpeg2] MPEG-2 International Standard. Video Recommendation ITU-T H.262, ISO/IEC 13818-2, January 10, 1995.

[mpeg21] MPEG-21 Overview v.5, ISO/IEC JTC1/SC29/WG11/N5231, October 2002.

[ng 1993] S. Ng, Thompson Consumer Electronics, Low resolution HDTV receivers, U.S. Patent 5,262,854, November 16, 1993.

[pang 1996] K.K. Pang, H.G. Lim, S. Dunstan, and J.M. Badcock, Frequency domain decimation and interpolation techniques, Picture Coding Symposium, Melbourne, Australia, March 1996.

[peng 2002] P. Yin, A. Vetro, B. Lui, and H. Sun, Drift compensation for reduced spatial resolution transcoding, *IEEE Transactions on Circuits Systems for Video Technology*, 12, 1009–1020, November 2002.

[reitmeier 1996] Glenn A. Reitmeier, The U.S. Advanced Television Standard and its Impact on VLSI, Submitted to VLSI and Signal Processing, 1996.

[siracusa 1993] R. Siracusa, K. Joseph, J. Zdepski, and D. Raychaudhuri, Flexible and robust packet transport for digital HDTV, *IEEE Journal of Selected Areas in Communications*, 11, 1, 88–98, January 1993.

[sun1 1992] H. Sun, K. Challapali, and J. Zdepski, Error concealment in simulcast AD-HDTV decoder, *IEEE Transaction on Consumer Electronics*, 38, 3, 108–118, August 1992.

[sun2 1992] H. Sun, M. Uz, J. Zdepski, and R. Saint Girons, A proposal for increased error resilience, ISO-IEC/JTC1/SC29/WG11, MPEG92, September 30, 1992.

[sun 1993] H. Sun, Hierarchical decoder for MPEG compressed video data, *IEEE Transaction on Consumer Electronics*, 39, 3, 559–562, August 1993.

[sun 1995] H. Sun and W. Kwok, Restoration of damaged block transform coded image using projection onto convex sets, *IEEE Transaction on Image Processing*, 4, 4, 470–477, April 1995.

[tm5] MPEG Test Model 5, ISO/IEC JTC/SC29/WG11 Document. April, 1993.

[vetro 1998a] A. Vetro and H. Sun, On the motion compensation within a down-conversion decoder, *Journal of Electronic Imaging*, 7, 3, July 1998.

[vetro 1998b] A. Vetro and H. Sun, Frequency domain down-conversion using an optimal motion compensation scheme, *Journal of Imaging Science and Technology*, 9, 4, August 1998.

[vetro 1998c] A. Vetro, H. Sun, P. DaGraca, and T. Poon, Minimum drift architectures for three-layer scalable DTV decoding, *IEEE Trans Consumer Electronics*, 44, 3, August 1998.

[vetro 2003] A. Vetro, C. Christopolous, and H. Sun, Video transcoding architectures and techniques: An overview, *IEEE Signal Processing Magazine*, 18–29, March 2003.

[wang 1991] Y. Wang and Q.-F. Zhu, Signal loss recovery in DCT-based image and video codecs, Proceedings of SPIE on Visual Communication and Image Processing, Boston, 667–678, November 1991.

[yu 1999] H. Yu, W.-M. Lam, B. Canfield, and B. Beyers, Block-based image processor for memory efficient MPEG video decoding, Proceedings of the International Conference on Consumer Electronics, Los Angeles, CA, June 1999.

[**zdepski 1989**] J. Zdepski, et al., Packet transport of rate-free interframe DCT compressed digital video on a CSMA/CD LAN, Proceedings of the IEEE Global Conference on Communication, Dallas, TX, November 1989.

[**zdepski 1990**] J. Zdepski, et al., Prioritized packet transport of VBR CCITT H.261 format compressed video on a CSMA/CD LAN, Third International Workshop on Packet Video, Morristown, NJ, March 22–23, 1990.

18
MPEG-4 Video Standard: Content-Based Video Coding

This chapter provides an overview of the ISO MPEG-4 standard. The MPEG-4 work includes natural video, synthetic video, audio, and systems. Both natural and synthetic video has been combined into a single part of the standard, which is referred to as MPEG-4 visual [mpeg-4 visual]. It should be emphasized that neither MPEG-1 nor MPEG-2 considers synthetic video (or computer graphics) and the MPEG-4 is also the first standard to consider the problem of content-based coding. Here, we will focus on the video parts of the MPEG-4 standard.

18.1 Introduction

As discussed in the previous chapters, MPEG has completed two standards MPEG-1 that was mainly targeted for CD-ROM applications up to 1.5 Mbits/s and MPEG-2 for digital TV and HDTV applications at bit rates between 2 and 30 Mbits/s. In July 1993, MPEG started its new project, MPEG-4, which was targeted to provide technology for multimedia applications. The first working draft (WD) was completed in November 1996 and the committee draft (CD) of version 1 reached in November 1997. The draft international standard (DIS) of MPEG-4 was completed in November of 1998. The international standard (IS) of MPEG-4 version 1 was completed in February of 1999. The goal of the MPEG-4 standard is to provide the core technology that allows efficient content-based storage, transmission and manipulation of video, graphics, audio, and other data within a multimedia environment. As mentioned earlier, there exist several video coding standards such as MPEG-1/2, H.261 and H.263. Why do we need a new standard for multimedia applications? In other words, are there any new attractive features of MPEG-4 that the current standards do not have or cannot provide? The answer is yes. The MPEG-4 has many interesting features that will be described later in this chapter. Some of these features are focused on improving coding efficiency; some are used to provide robustness of transmission and interactivity with end user. However, among these features the most important one is the content-based coding. MPEG-4 is the first standard that supports content-based coding of audiovisual objects (AVO). For content providers or authors, the MPEG-4 standard can provide the greater reusability, flexibility, and manageability of the content that is produced. For network providers, MPEG-4 will offer transparent information that can be interpreted and translated into the appropriate native signaling messages of each network. This can be accomplished with the help of relevant standard bodies that have the jurisdiction. For end users, MPEG-4 can provide more functionality to make the user terminal have more capabilities of interaction with the content. To reach these goals, MPEG-4 should have the following important features.

The contents such as audio, video, or data are represented in the form of primitive AVOs. These AVOs can be natural scenes or sounds, which are recorded by video camera or synthetically generated by computers.

The AVOs can be composed together to create compound AVOs or scenes.

The data associated with AVOs can be multiplexed and synchronized so that they can be transported through network channels with certain quality requirements.

18.2 MPEG-4 Requirements and Functionalities

As the MPEG-4 standard is mainly targeted for multimedia applications, there are many requirements to ensure that several important features and functionalities are offered. These features include the allowance of interactivity, high compression, universal accessibility, and portability of audio and video content. From the MPEG-4 video requirement document, the main functionalities can be summarized by the following three aspects: content-based interactivity, content-based efficient compression, and universal access.

18.2.1 Content-Based Interactivity

In addition to provisions for efficient coding of conventional video sequences, MPEG-4 video has the following features of content-based interactivity.

18.2.1.1 Content-Based Manipulation and Bitstream Editing

The MPEG-4 supports the content-based manipulation and bitstream coding without the need for transcoding. In MPEG-1 and MPEG-2, there is no syntax and no semantics for supporting true manipulation and editing in the compressed domain. MPEG-4 provides the syntax and techniques to support content-based manipulation and bitstream editing. The level of access, editing, and manipulation can be done at the object level in connection with the features of content-based scalability.

18.2.1.2 Synthetic and Natural Hybrid Coding

The MPEG-4 supports combining synthetic scenes or objects with natural scenes or objects. This is for compositing synthetic data with ordinary video, allowing for interactivity. The related techniques in MPEG-4 for supporting this feature include sprite coding, efficient coding of 2-D and 3-D surfaces, and wavelet coding for still textures.

18.2.1.3 Improved Temporal Random Access

The MPEG-4 provides efficient method to randomly access, within a limited time and with the fine resolution, parts, e.g., video frames or arbitrarily shaped image objects from an audio-visual sequence. This includes conventional random access at very low bit rate. This feature is also important for content-based bitstream manipulation and editing.

18.2.2 Content-Based Efficient Compression

Our initial goal of MPEG-4 is to provide highly efficient coding tool with high compression at very low bit rates. But this goal has now extended to a large range of bit rates from

10 kbits/s to 5 Mbits/s, which covers from QSIF to CCIR 601 video formats. Two important items are included in this requirement.

18.2.2.1 Improved Coding Efficiency

The MPEG-4 video standard provides subjectively better visual quality at comparable bit rates comparing the existing or emerging standards, including MPEG-1/2 and H.263. MPEG-4 video contains many new tools that optimize the code in different bit rate range. Some experimental results have shown that it outperforms MPEG-2 and H.263 at the low bit rates. Also, the content-based coding reaches the similar performance of the frame-based coding.

18.2.2.2 Coding of Multiple Concurrent Data Streams

The MPEG-4 provides the capability of coding multiple views of a scene efficiently. For stereoscopic video applications, MPEG-4 allows the ability to exploit redundancy in multiple viewing points of the same scene, permitting joint coding solutions that allow compatibility with normal video as well as the ones without compatibility constraints.

18.2.3 Universal Access

Another important feature of the MPEG-4 video is the feature of universal access.

18.2.3.1 Robustness in Error-Prone Environments

The MPEG-4 video provides strong error robustness capabilities to allow access to applications over a variety of wireless and wired networks and storage media. Sufficient error robustness is provided for low bit rate applications under severe error conditions (e.g., long error bursts).

18.2.3.2 Content-Based Scalability

The MPEG-4 video provides the ability to achieve scalability with fine granularity in content, quality (e.g., spatial and temporal resolution), and complexity. These scalabilities are especially intended to result in content-based scaling of visual information.

18.2.4 Summary of MPEG-4 Features

From the above description of MPEG-4 features, it is obvious that the most important application of MPEG-4 will be in a multimedia environment. The media that can use the coding tools of MPEG-4 include computer networks, wireless communication networks, and the Internet. Although it can also be used for satellite, terrestrial broadcasting, and cable TV, these are still the territories of MPEG-2 video because MPEG-2 has already made such a large impact in the market. A large number of silicon solutions exist and its technology is more mature compared to the current MPEG-4 standard. From the viewpoint of coding theory, we can say there is no significant breakthrough in MPEG-4 video compared to MPEG-2 video. Therefore, we cannot expect to have a significant improvement of coding efficiency when using MPEG-4 video over MPEG-2. Although MPEG-4 optimized its performance in a certain range of bit rates, its major strength is that it provides more functionality than MPEG-2. Recently, MPEG-4 added the necessary tools to support interlaced material. With this addition, MPEG-4 video does the support functionalities already provided by MPEG-1 and MPEG-2, including the provision to efficiently compress standard rectangular sized video at different levels of input formats, frame rates, and bit rates.

Overall, the incorporation of an object or content-based coding structure is the feature that allows MPEG-4 to provide more functionality. It enables MPEG-4 to provide the most elementary mechanism for interactivity and manipulation with objects of images or video in the compressed domain without the need for further segmentation or transcoding at the receiver, since the receiver can receive separate bitstreams for different objects contained in the video. To achieve content-based coding, the MPEG-4 uses the concept of a video object plane (VOP). It is assumed that each frame of an input video is first segmented into a set of arbitrary shaped regions or VOPs. Each such region could cover a particular image or video object (VO) in the scene. Therefore, the input to the MPEG-4 encoder can be a VOP, and the shape and the location of the VOP can vary from frame to frame. A sequence of VOPs is refereed to as a VO. The different VOs may be encoded into separate bitstreams. MPEG-4 specifies demultiplexing and composition syntax, which provide the tools for the receiver to decode the separate VO bitstreams and composite them into a frame. In this way, the decoders have more flexibility to edit or rearrange the decoded VOs. The detailed technical issues will be addressed in the following sections.

18.3 Technical Description of MPEG-4 Video

18.3.1 Overview of MPEG-4 Video

The major feature of MPEG-4 is to provide the technology for object-based compression that is capable of separately encoding and decoding VOs. To clearly explain the idea of object-based coding, we should review the set of VO-related definitions. An image scene may contain several objects. In the example of Figure 18.1, the scene contains the background and two objects. The time instant of each VO is referred to as the VOP. The concept of a VO provides a number of functionalities of MPEG-4, which are either impossible or very difficult in MPEG-1 or MPEG-2 video coding. Each VO is described by the information of texture, shape, and motion vectors (MV). The video sequence can be encoded in a way that will allow the separate decoding and reconstruction of the objects and allow the editing and manipulation of the original scene by simple operation on the compressed bitstream domain. The feature of object-based coding is also able to support functionality such as warping of synthetic or natural text, textures, image, and video overlays on reconstructed VOs.

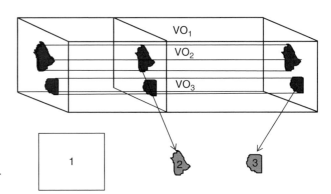

FIGURE 18.1
Video object definition and format.
(a) Video object, (b) VOPs.

As MPEG-4 aims at providing coding tools for multimedia environment, these tools not only allow one to efficiently compress natural VOs, but also compress synthetic objects, which are a subset of the larger class of computer graphics. The tools of MPEG-4 video includes:

- Motion estimation (ME) and MC
- Texture coding
- Shape coding
- Sprite coding
- Interlaced video coding
- Wavelet-based texture coding
- Generalized temporal and spatial as well as hybrid scalability
- Error resilience

The technical details of these tools will be explained in the following sections.

18.3.2 Motion Estimation and Compensation

For object-based coding, the coding task includes two parts: texture coding and shape coding. The current MPEG-4 video texture coding is still based on the combination of motion compensated (MC) prediction and transform coding (TC). MC predictive coding is a well-known approach for video coding. The MC is used to remove the interframe redundancy and the TC is used to remove the intraframe redundancy, as in the MPEG-2 video coding scheme. However, there are lots of modifications and technical details in MPEG-4 for coding a very wide range of bit rates. Moreover, MPEG-4 coding has been optimized for low bit rate applications with a number of new tools. In other words, MPEG-4 video coding uses the most common coding technologies such as MC and TC, but at the same time, it modifies some traditional methods such as advanced MC and also creates some new features such as sprite coding.

The basic technique to perform MC predictive coding for coding a video sequence is ME. The basic ME method used in the MPEG-4 video coding is still the block matching technique. The basic principle of block matching for ME is to find the best-matched block in the previous frame for every block in the current frame. The displacement of the best-matched block relative to the current block is referred to as the MV. Positive values for both MV components indicate that the best-matched block is on the bottom-right of the current block. The MC prediction difference block is formed by subtracting the pixel values of the best-matched block from the current block, pixel by pixel. The difference block is then coded by texture-coding method. In MPEG-4 video coding, the basic technique of texture coding is a discrete cosine transformation (DCT). The coded MV information and difference block information is contained in the compressed bitstream, which is transmitted to the decoder. The major issues in the ME and MC are the same as in the MPEG-1 and MPEG-2, which include the matching criterion, the size of search window (searching range), the size of matching block, the accuracy of MVs (one pixel or half pixel), and inter/intra mode decision. As we have discussed these topics already, we will focus on the new features in the MPEG-4 video coding. The feature of the advanced motion prediction is a new tool of MPEG-4 video. This feature includes two aspects: adaptive selection of 16×16 block or four 8×8 blocks to match the current 16×16 block and overlapped MC for luminance block.

18.3.2.1 Adaptive Selection of 16 × 16 Block or Four 8 × 8 Blocks

The purpose of the adaptive selection of the matching block size is to further enhance coding efficiency. The coding performance may be improved at low bit rate because the bits for coding prediction difference could be greatly reduced at the limited extra cost for increasing MVs. Of course, if the cost of coding MVs is too high, this method will not work. The decision in the encoder should be very careful. For explaining the procedure of how to make decision, we define $\{C(i,j), i,j = 0, 1, \ldots, N-1\}$ to be the pixels of the current block and $\{P(i,j), i,j = 0, 1, \ldots, N-1\}$ to be the pixels in the search window in the previous frame. The sum of absolute difference (SAD) is calculated as

$$\text{SAD}_N(x,y) = \begin{cases} \sum_{i=0}^{N-1} \sum_{j=0}^{N-1} |C(i,j) - P(i,j)| - T & \text{if } (x,y) = (0,0); \\ \sum_{i=0}^{N-1} \sum_{j=0}^{N-1} |C(i,j) - P(i+x, j+y)| & \text{otherwise} \end{cases} \quad (18.1)$$

where
(x,y) is the pixel within the range of searching window
T is a positive constant

The following steps then make the decision:

Step 1: To find $\text{SAD}_{16}(\text{MV}_x, \text{MV}_y)$
Step 2: To find $\text{SAD}_8(\text{MV1}_x, \text{MV1}_y)$, $\text{SAD}_8(\text{MV2}_x, \text{MV2}_y)$, $\text{SAD}_8(\text{MV3}_x, \text{MV3}_y)$, and $\text{SAD}_8(\text{MV4}_x, \text{MV4}_y)$;
Step 3: if

$$\sum_{i=1}^{4} \text{SAD}_8(\text{MV}_{ix}, \text{MV}_{iy}) < \text{SAD}_{16}(\text{MV}_x, \text{MV}_y) - 128$$

then choose 8 × 8 prediction; otherwise, choose 16 × 16 prediction.

If the 8 × 8 prediction is chosen, there are four MVs for the four 8 × 8 luminance blocks that will be transmitted. The MV for the two chrominance blocks is then obtained by taking an average of these four MVs and dividing the average value by a factor of two. Since each MV for the 8 × 8 luminance block has half-pixel accuracy, the MV for the chrominance block may have a 16th-pixel accuracy.

18.3.2.2 Overlapped Motion Compensation

This kind of MC is always used for the case of four 8 × 8 blocks. The case of one MV for a 16 × 16 block can be considered as having four identical 8 × 8 MVs, each for an 8 × 8 block. Each pixel in an 8 × 8 of the best-matched luminance block is a weighted sum of three prediction values specified in Equation 18.2:

$$p'(i,j) = (H_0(i,j) \cdot q(i,j) + H_1(i,j) \cdot r(i,j) + H_2(i,j) \cdot s(i,j))/8 \quad (18.2)$$

MPEG-4 Video Standard: Content-Based Video Coding

where division is with roundoff. The weighting matrices are specified as

$$H_0 = \begin{bmatrix} 4 & 5 & 5 & 5 & 5 & 5 & 5 & 4 \\ 5 & 5 & 5 & 5 & 5 & 5 & 5 & 5 \\ 5 & 5 & 6 & 6 & 6 & 6 & 5 & 5 \\ 5 & 5 & 6 & 6 & 6 & 6 & 5 & 5 \\ 5 & 5 & 6 & 6 & 6 & 6 & 5 & 5 \\ 5 & 5 & 6 & 6 & 6 & 6 & 6 & 6 \\ 5 & 5 & 5 & 5 & 5 & 5 & 5 & 5 \\ 4 & 5 & 5 & 5 & 5 & 5 & 5 & 4 \end{bmatrix}, H_1 = \begin{bmatrix} 2 & 2 & 2 & 2 & 2 & 2 & 2 & 2 \\ 1 & 1 & 2 & 2 & 2 & 2 & 1 & 1 \\ 1 & 1 & 1 & 1 & 1 & 1 & 1 & 1 \\ 1 & 1 & 1 & 1 & 1 & 1 & 1 & 1 \\ 1 & 1 & 1 & 1 & 1 & 1 & 1 & 1 \\ 1 & 1 & 1 & 1 & 1 & 1 & 1 & 1 \\ 1 & 1 & 2 & 2 & 2 & 2 & 1 & 1 \\ 2 & 2 & 2 & 2 & 2 & 2 & 2 & 2 \end{bmatrix}, \text{ and } H_2 = \begin{bmatrix} 2 & 1 & 1 & 1 & 1 & 1 & 1 & 2 \\ 2 & 2 & 1 & 1 & 1 & 1 & 2 & 2 \\ 2 & 2 & 1 & 1 & 1 & 1 & 2 & 2 \\ 2 & 2 & 1 & 1 & 1 & 1 & 2 & 2 \\ 2 & 2 & 1 & 1 & 1 & 1 & 2 & 2 \\ 2 & 2 & 1 & 1 & 1 & 1 & 2 & 2 \\ 2 & 2 & 1 & 1 & 1 & 1 & 2 & 2 \\ 2 & 1 & 1 & 1 & 1 & 1 & 1 & 2 \end{bmatrix}$$

It is noted that $H_0(i,j) + H_1(i,j) + H_2(i,j) = 8$ for all possible (i,j). The value of $q(i,j)$, $r(i,j)$, and $s(i,j)$ are the values of the pixels in the previous frame at the locations,

$$\begin{aligned} q(i,j) &= p(i + MV_x^0, j + MV_y^0); \\ r(i,j) &= p(i + MV_x^1, j + MV_y^1); \\ s(i,j) &= p(i + MV_x^2, j + MV_y^2). \end{aligned} \quad (18.3)$$

where (MV_x^0, MV_y^0) is the MV of the current 8×8 luminance block $p(i,j)$, (MV_x^1, MV_y^1) is the MV of the block either above (for $j = 0, 1, 2, 3$) or below (for $j = 4, 5, 6, 7$) the current block and (MV_x^2, MV_y^2) is the MV of the block either to the left (for $i = 0, 1, 2, 3$) or right (for $i = 4, 5, 6, 7$) of the current block. The overlapped MC can reduce the prediction noise at certain level.

18.3.3 Texture Coding

Texture coding is used to code the INTRA VOPs and the prediction residual data after MC. The algorithm for video texture coding is based on the conventional 8×8 DCT with MC. DCT is performed for each luminance and chrominance block, where the MC is performed only on the luminance blocks. This algorithm is similar to those in H.263 and MPEG-1 as well as MPEG-2. However, MPEG-4 video texture coding has to deal with the requirement of object-based coding, which is not included in the other video coding standards. Hereafter, we will focus on the new features of the MPEG-4 video coding. These new features include: the INTRA DC and AC prediction for I-VOP and P-VOP; the algorithm of ME and MC for arbitrary shape VOP; and the strategy of arbitrary shape texture coding. The definitions of I-VOP, P-VOP, and B-VOP are similar to the intra-coded (I), predictive-coded (P), and bidirectionally predictive-coded (B) pictures in Chapter 16 for MPEG-1 and MPEG-2.

18.3.3.1 INTRA DC and AC Prediction

In the intra-mode coding, the predictive coding is not only applied on the DC coefficients but also the AC coefficients to increase the coding efficiency. The adaptive DC prediction involves the selection of the quantized DC (QDC) value of the immediately left block or the immediately above block. The selection criterion is based on comparison of the horizontal and vertical DC gradients around the block to be coded. Figure 18.2 shows the three surrounding blocks A, B, and C to the current block X whose QDC is to be coded where block A, B, and C are the immediately left, immediately left and above, and immediately above block to the X, respectively. The QDC value of block X, QDC_X, is predicted by either

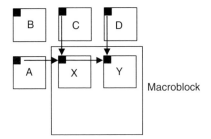

FIGURE 18.2
Previous neighboring blocks used in DC prediction.

the QDC value of block A, QDC_A, or the QDC value of block C, QDC_C, based on the comparison of horizontal and vertical gradients as follows:

$$\text{If } |QDC_A - QDC_B| < |QDC_B - QDC_C|, \quad QDC_P = QDC_C$$
$$\text{Otherwise } QDC_P = QDC_A \tag{18.4}$$

The differential DC is then obtained by subtracting the DC prediction, QDC_P, from QDC_X. If any of block A, B, or C are outside of the VOP boundary, or they do not belong to an INTRA coded block, their QDC value are assumed to take a value of 128 (if the pixel is quantized to 8 bits) for computing the prediction. The DC prediction are performed similarly for the luminance and each or the two chrominance blocks.

For AC coefficient prediction, either coefficients from the first row or the first column of a previous coded block are used to predict the co-sited (same position in the block) coefficients in the current block. On a block basis, the same rule for selecting the best predictive direction (vertical or horizontal direction) for DC coefficients is also used for the AC coefficient prediction. A difference between DC and AC prediction is the issue of quantization scale. All DC values are quantized to the 8 bits for all blocks. However, the AC coefficients may be quantized by the different quantization scales for the different blocks. To compensate for differences in the quantization of the blocks used for prediction, scaling of prediction coefficients becomes necessary. The prediction is scaled by the ratio of the current quantization step size and the quantization step size of the block used for prediction. In the cases when AC coefficient prediction results in a larger range of prediction errors as compared to the original signal, it is desirable to disable the AC prediction. The decision of AC prediction switched on or off is performed on macroblock (MB) basis instead of block basis to avoid the excessive overhead. The decision for switching on or off AC prediction is based on a comparison of the sum of the absolute values of all AC coefficients to be predicted in an MB and that of their predicted differences. It should be noted that the same DC and AC prediction algorithm is used for the INTRA blocks in the inter-coded VOP. If any blocks used for prediction are not INTRA blocks, the QDC and QAC values used for prediction are set to 128 and 0 for DC ad AC prediction, respectively.

18.3.3.2 Motion Estimation/Compensation of Arbitrary Shaped VOP

In the previous section, we discussed the general issues of ME and MC. In this section, we are going to discuss the ME and MC for coding the texture in the arbitrary shaped VOP. In an arbitrary shaped VOP, the shape information is given by either binary shape information or alpha components of a gray level shape information. If the shape information is available to both encoder and decoder, three important modifications have to be considered for the arbitrary shaped VOP. The first is for the blocks that are located in the border of VOP. For these boundary blocks, the block matching criterion should be modified.

Second, a special padding technique is required for the reference VOP. Finally, as the VOPs have arbitrary shapes rather rectangular shapes, and the shapes change from time to time, an agreement on a coordinate system is necessary to ensure the consistency of MC. At the MPEG-4 video, the absolute frame coordinate system is used for referencing all of the VOPs. At each particular time instance, a bounding rectangle that includes the shape of that VOP is defined. The position of upper-left corner in the absolute coordinate in the VOP spatial reference is transmitted to the decoder. Thus, the MV for a particular block inside a VOP is referred to as the displacement of the block in absolute coordinates.

Actually, the first and second modifications are related since the padding of boundary blocks will affect the matching of ME. The purpose of padding is aiming at more accurate block matching. In current algorithm, the repetitive padding is applied to the reference VOP for performing ME and MC. The repetitive padding process is performed as per the following steps:

- Define any pixel outside the object boundary as a zero pixel.
- Scan each horizontal line of a block (one 16×16 for luminance and two 8×8 for chrominance). Each scan line is possibly composed of two kinds of line segments: zero and nonzero segments. It is obvious that our task is to pad zero segments. There are two kinds of zero segments: (1) between an end point of the scan line and the end point of a nonzero segment and (2) between end points of two different nonzero segments. In the first case, all zero pixels are replaced by the pixel value of the end pixel of nonzero segment; whereas in the second kind, all zero pixels take the averaged value of the two end pixels of the nonzero segments.
- Scan each vertical line of the block and perform the identical procedure as described for the horizontal line.
- If a zero pixel is located at the intersection of horizontal and vertical scan lines, this zero pixel takes the average of two possible values.
- For the rest of the zero pixels, to find the closest nonzero pixel on the same horizontal scan line and the same vertical scan line (if there is a tie, the nonzero pixel on the left or the top of the current pixel is selected). Replace the zero pixel by the average of these two nonzero pixels.
- For a fast moving VOP, padding is further extended to the blocks outside the VOP but immediately next to the boundary blocks. These blocks are padded by replacing the pixel values of adjacent boundary blocks. This extended padding is performed in both horizontal and vertical directions. Since block matching is replaced by polygon matching for the boundary blocks of the current VOP, the SAD values are calculated by the modified formula:

$$\text{SAD}_N(x,y) = \begin{cases} \sum_{i=0}^{N-1} \sum_{j=0}^{N-1} |c(i,j) - p(i,j)| \cdot \alpha(i,j) - C & \text{if } (x,y) = (0,0), \\ \sum_{i=0}^{N-1} \sum_{j=0}^{N-1} |c(i,j) - p(i+x,j+y)| \cdot \alpha(i,j) - C & \text{otherwise,} \end{cases} \quad (18.5)$$

where
$C = N_B/2 + 1$ and N_B are the number of pixels inside the VOP and in this block
$\alpha(i,j)$ is the alpha component specifying the shape information and it is not equal to zero here

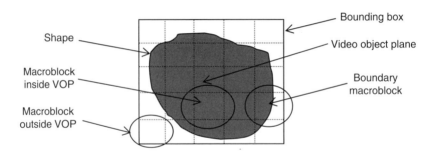

FIGURE 18.3
A VOP is represented by a bounding rectangular box.

18.3.3.3 Texture Coding of Arbitrary Shaped VOP

During encoding the VOP is represented by a bounding rectangle that is formed to completely contain the VO but with minimum number of MBs in it (Figure 18.3). The detailed procedure of VOP rectangle formation is given in MPEG-4 video VM [mpeg-4 vm12].

There are three types of MBs in the VOP with arbitrary shape: the MBs that are completely located inside of the VOP, the MBs that are located along the boundary of the VOP, and the MBs that are located outside of the boundary. For the first kind of the MB, there is no need for any particular modified technique to code them and just use normal DCT with entropy coding of quantized DCT (QDCT) coefficients such as coding algorithm in H.263. The second kind of MBs, referred to as transparent blocks are located along the boundary, contain two kinds of 8×8 blocks: the blocks lie along the boundary of VOP and the blocks do not belong to the arbitrary shape but lie inside the rectangular bounding box of the VOP. For those 8×8 blocks, which do lie along the boundary of VOP, two different methods have been proposed: low pass extrapolation (LPE) padding and shape adaptive DCT (SA-DCT). All blocks in the MB outside of the boundary are also referred to as transparent blocks. The transparent blocks are skipped and not coded at all.

18.3.3.3.1 Low Pass Extrapolation Padding Technique

This block padding technique is applied to intra-coded blocks, which are not located completely within the object boundary. To perform this padding technique we first assign the mean value of those pixels that are located in the object boundary (both inside and outside) to each pixel outside the object boundary. Then an average operation is applied to each pixel $p(i,j)$ outside the object boundary starting from the upper-left corner of the block and proceeding row by row to the lower-right corner pixel:

$$p(i,j) = [p(i,j-1) + p(i-1,j) + p(i,j+1) + p(i+1,j)]/4 \qquad (18.6)$$

If one or more of the four pixels used for filtering are outside of the block, the corresponding pixels are not considered for the average operation and the factor 1/4 is modified accordingly.

18.3.3.3.2 SA-DCT

The shape adaptive DCT is only applied to those 8×8 blocks that are located on the object boundary of an arbitrary shaped VOP. The idea of the SA-DCT is to apply one-dimensional (1-D) DCT vertically and horizontally according to the number of active pixels in the row and column of the block, respectively. The size of each vertical DCT is the same as the number of active pixels in each column. After vertical DCT is performed for all columns

FIGURE 18.4
Illustration of shape adaptive discrete cosine transformation (SA-DCT).

with at least one active pixel, the coefficients of the vertical DCTs with the same frequency index are lined up in a row. The DC coefficients of all vertical DCTs are lined up in the first row, the first-order vertical DCT coefficients are lined up in the second row, and so on. After that, horizontal DCT is applied to each row. As the same as for the vertical DCT, the size of each horizontal DCT is the same as the number of vertical DCT coefficients lined up in the particular row. The final coefficients of SA-DCT are concentrated into the upper-left corner of the block. This procedure is shown in the Figure 18.4.

The final number of the SA-DCT coefficients is identical to the number of active pixels of image. Since the shape information is transmitted to the decoder, the decoder can perform the inverse shape adapted DCT to reconstruct the pixels. The regular zigzag scan is modified so that the non-active coefficient locations are neglected when counting the runs for the run-length coding (RLC) of the SA-DCT coefficients. It is obvious that for a block with all 8×8 active pixels, the SA-DCT becomes a regular 8×8 DCT and the scanning of the coefficients is identical to the zigzag scan. All SA-DCT coefficients are quantized and coded in the same way as the regular DCT coefficients employing the same quantizers and VLC code tables. The SA-DCT is not included in MPEG-4 video version 1, but it is being considered for inclusion into version 2.

18.3.4 Shape Coding

Shape information of the arbitrary shaped objects is very useful not only in the field of image analysis, computer vision, and graphics, but also in object-based video coding. MPEG-4 video coding is the first to make effort at providing a standardized approach to compress the shape information of objects and contain the compressed results within a video bitstream. In the current MPEG-4 video coding standard, the video data can be coded on an object basis. The information in the video signal is decomposed to shape, texture, and motion. This information is then coded and transmitted within the bitstream. The shape information is provided in binary format or gray-scale format. The binary format of shape information consists of a pixel map that is generally the same size as the bounding box of the corresponding VOP. Each pixel takes on one of two possible values indicating whether it is located within the VO or not. The gray-scale format is similar to the binary format with the additional feature that each pixel can take on a range of values, i.e., times an alpha value. Alpha typically has a normalized value of 0–1. The alpha value can be used to blending two images on a pixel-by-pixel basis in such a way: new pixel = (alpha) (pixel A color) + (1-alpha)(pixel B color).

Let us now discuss how to code the shape information. As mentioned earlier, shape information is classified as binary shape or gray-scale shape. Both binary and gray-scale shapes are referred to as an alpha plane. The alpha plane defines the transparency of an object. The multilevel alpha maps are frequently used to blend different images. A binary alpha map defines whether or not a pixel belongs to an object. The binary alpha planes are encoded by modified content-based arithmetic encoding (CAE), while the gray-scale alpha planes are encoded by MC DCT coding, which is similar to texture coding. For binary shape coding, a rectangular box enclosing the arbitrary shaped VOP is formed as shown in Figure 18.3. The bounded rectangle box is then extended in both vertical and horizontal directions on the right-bottom side to the multiple of 16×16 blocks. Each 16×16 block within the rectangular box is referred to as binary alpha block (BAB). Each BAB is associated with colocated MB. The BAB can be classified to three types: transparent block, opaque block, and alpha or shape block. The transparent block does not contain any information about object. The opaque block is entirely located inside the object. The alpha or shape block is located in the area of object boundary, i.e., a part of block is inside the object and the rest of block is in the background. The value of pixels in the transparent region is zero. For shape coding, the type information will be included in the bitstream and signaled to the decoder as MB type. But only the alpha blocks need to be processed by the encoder and decoder. The methods used for each shape format contain several encoding modes. For example, the binary shape information can be encoded using either an intra or inter-mode. Each of these modes can be further divided into lossy and lossless option. Gray-scale shape information also contains intra and inter modes; however, only a lossy option is used.

18.3.4.1 Binary Shape Coding with CAE Algorithm

As mentioned previously, the CAE is used to code each binary pixel of the BAB. For a P-VOP, the BAB may be encoded in intra or inter mode. Pixels are coded in scan-line order i.e., row by row for both modes. The process for coding a given pixel includes three steps: (1) compute a context number, (2) index a probability table using the context number, and (3) use the indexed probability to drive an arithmetic encoder. In intra mode, a template of 10 pixels is used to define the causal context for predicting the shape value of the current pixel as shown in Figure 18.5. For the pixels in the top and left boundary of the current MB, the template of causal context will contain the pixels of the already transmitted MBs on the top and on the left side of the current MB. For the 2 rightmost columns of the VOP, each undefined pixel such as C_7, C_3, and C_2, of the context is set to the value of its closest neighbor inside the MB, i.e., C_7 will take the value of C_8 and C_3 and C_2 will take the value of C_4.

A 10-bit context is calculated for each pixel, X as

$$C = \sum_{k=0}^{9} C_k \cdot 2^k \qquad (18.7)$$

This causal context is used to predict the shape value of the current pixel. For encoding the state transition, a context-based arithmetic encoder is used. The probability table of the arithmetic encoder for the 1024 contexts was derived from sequences that are outside of the test set. Two bytes are allocated to describe the symbol probability for each context, the table size is 2048 bytes. To increase coding efficiency and rate control, the algorithm

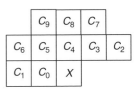

FIGURE 18.5
Template for defining the context of the pixel, X, to be coded in intra mode.

allows lossy shape coding. In lossy shape coding an MB can be down-sampled by a factor of 2 or 4 resulting in a subblock of size 8×8 pixels or 4×4 pixels, respectively. The subblock is then encoded using the same method as for full size block. The down-sampling factor is included in the encoded bitstream and then transmitted to the decoder. The decoder decodes the shape data and then up-samples the decoded subblock to full MB size according to the down-sampling factor. Obviously, it is more efficient to code shape using a high down-sampling factor, but the coding errors may occur in the decoded shape after up-sampling. However, in the case of low bit rate coding, lossy shape coding may be necessary since the bit budget may not be enough for lossless shape coding. Depending on the up-sampling filter, the decoded shape can look somewhat blocky. Several up-sampling filters were investigated. The best performing filter in terms of subjective picture quality is an adaptive nonlinear up-sampling filter. It should be noted that the coding efficiency of shape coding also depends on the orientation of the shape data. Therefore the encoder can choose to code the block as described above or transpose the MB prior to arithmetic coding. Of course, the transpose information has to be signaled to the decoder.

For shape coding in a P-VOP or B-VOP, the inter mode may be used to exploit the temporal redundancy in the shape information with MC. For MC, a 2-D integer pixel MV is estimated using full search for each MB in order to minimize the prediction error between the previous coded VOP shape and the current VOP shape. The shape MVs are predictively encoded with respect to the shape MVs of neighboring MBs. If no shape MV is available, texture MVs are used as predictors. The template for inter mode differs from the one used for intra mode. The inter mode template contains nine pixels among which five pixels are located in the previous frame and four are the current neighbors as shown in Figure 18.6.

The inter mode template defines a context of nine pixels. Accordingly, a 9-bit context or 512 contexts can be computed in a similar way to Equation 18.7.

$$C = \sum_{k=0}^{8} C_k \cdot 2^k \qquad (18.8)$$

The probability for one symbol is also described by 2 bytes giving a probability table size of 1024 bytes. The idea of lossy coding can also be applied to the inter mode shape coding by down-sampling the original BABs. For inter mode shape coding, the total bits for coding the shape consists of two parts, one part for coding MVs and other for prediction residue. The encoder may decide that the shape representation achieved by just using MVs is sufficient, thus bits for coding the prediction error can be saved. Actually, there are several modes to code the shape information of each MB: transparent, opaque, intra, inter with/without shape MVs, and prediction error coding. These different options with optional down-sampling and transposition allow for encoder implementations of different coding efficiency and implementation complexity. Again, this is a problem of encoder optimization that does not belong to the standard.

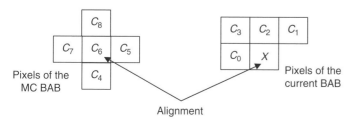

FIGURE 18.6
Template for defining the context of the pixel, X, to be coded in inter mode.

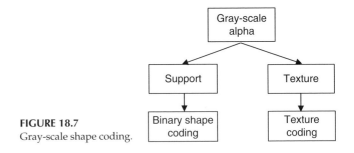

FIGURE 18.7 Gray-scale shape coding.

18.3.4.2 Gray-Scale Shape Coding

The gray-scale shape information is encoded by separately encoding the shape and transparency information as shown in Figure 18.7. For a transparent object, the shape information is referred to as the support function and is encoded using the binary shape coding method. The transparency or alpha values are treated as the texture of luminance and encoded using padding, MC and the same 8×8 block DCT approach for the texture coding. For an object with varying alpha maps, shape information is encoded in two steps. The boundary of the object is first losslessly encoded as a binary shape, and then the actual alpha map is encoded as texture coding.

The binary shape coding allows one to describe objects with constant transparency, while gray-scale shape coding can be used to describe objects with arbitrary transparency, providing more flexibility for image composition. One application example is a gray-scale alpha shape that consists of a binary alpha shape with the value around the edges tapered from 255 to 0 to provide for a smooth composition with the background. The description of each VO layer includes the information to give instruction for selecting one of six modes for feathering. These six modes include no effects, linear feathering, constant alpha, linear feathering and constant alpha, feathering filter, and feathering filter, and constant alpha. The detailed description of the function of these modes are given in the reference of VM 12.0 [mpeg4 vm12].

18.3.5 Sprite Coding

As mentioned earlier, MPEG-4 video has investigated a number of new tools, which attempt to improve the coding efficiency at low bit rates compared with MPEG-1/2 video coding. Among these tools, sprite coding is an efficient technology to reach this goal. A sprite is an especially composed VO that is visible throughout an entire piece of video sequence. For example, the sprite generated from a panning sequence contains all the visible pixels of the background throughout the video sequence. Portion of the background may not be seen in certain frames due to the occlusion of the foreground objects or the camera motion. This particular example is one of the static sprites. In other words, a static sprite is a possible still image. Since the sprite contains all visible background scenes of a segment video sequence where the changes within the background content is mainly caused by camera parameters, the sprite can be used for direct reconstruction of the background VOPs or as the prediction of the background VOPs within the video segment. The sprite coding technology first efficiently transmits this background to the receiver and then stores it in a frame at both encoder and decoder. The camera parameters are then transmitted to the decoder for each frame so that the appropriate part of the background scene can be either used as the direct reconstruction or as the prediction of the background VOP. Both cases can significantly save the coding bits and increase the coding efficiency. There are two types of sprites, static sprite and dynamic sprite, which are being considered as coding tools for MPEG-4 video. A Static sprite is used for a video sequence where the objects in a scene can be separated into foreground objects and a static background. A static sprite is a special VOP, which is generated by copying the background from a video sequence. This copying includes the appropriate warping and cropping. Therefore, a static sprite is always built off-line. In contrast, a dynamic sprite is dynamically built during the predictive coding. It can be built either online or off-line. The static sprite has shown significant coding gain over existing compression technology for certain video sequence. The dynamic sprite is more complicated in the real-time application due to the difficulty of updating the sprite during the coding. Therefore, the dynamic sprite has not been adopted by version 1 of standard. Additionally, both sprites are not easily applied to the generic scene content. Also, there is another kind of classification of sprite coding according to the method of sprite generation, namely, off-line and online sprites. Off-line is always used for static sprite generation. Off-line sprites are well suited for synthetic objects and objects that mostly undergoes rigid motion. Whereas online sprites are used only for dynamic sprites. Online sprites provide a no-latency solution in the case of natural sprite objects. Off-line dynamic sprites provide an enhanced predictive coding environment. The sprite is built with a similar way in both off-line and online methods. In particular, the same global ME algorithm is exploited. The difference is that the off-line sprite is built before starting the encoding process while in the online sprite case, both the encoder and the decoder build the same sprite from reconstructed VOPs. This is why the online dynamic sprites are more complicated in the implementation. The online sprite is not included in version 1, and will most likely not be considered for version 2 either. In sprite coding, the chrominance components are processed in the same way as the luminance components, with the properly scaled parameters according to the video format.

18.3.6 Interlaced Video Coding

Since June 1997, MPEG-4 extended its application to support interlaced video. Interlaced video consists of two fields per frame: even field and odd field. MPEG-2 has a number of tools that are used to deal with field structure of video signals. These tools include frame/field adaptive DCT coding and frame/field adaptive MC. However, the field issue in MPEG-4 has to be considered on a VOP basis instead of the conventional frame basis. When field-based MC is specified, two field MVs and the corresponding reference fields are used to generate the prediction from each reference VOP. The shape information has to be considered in the interlaced video for MPEG-4.

18.3.7 Wavelet-Based Texture Coding

In MPEG-4 there is a texture-coding mode that is used to code the texture or still image such as in JPEG. The basic technique used in this mode is the wavelet-based TC. The reason of adopting wavelet transform instead of DCT for still texture coding is not only due to high coding efficiency, but also because the wavelet can provide excellent scalability, both spatial scalability and SNR scalability. Since the principle of the wavelet-based TC for image compression has been explained in Chapter 8, we just briefly describe the coding procedure of this mode. The block diagram of the encoder is shown in Figure 18.8.

18.3.7.1 Decomposition of the Texture Information

The texture or still image is first decomposed into bands using a bank of analysis filters. This decomposition can be applied recursively on the obtained bands to yield a decomposition tree of subbands. An example of decomposition to depth 2 is shown in Figure 18.9.

FIGURE 18.8
Block diagram of encoder of wavelet-based texture coding, DWT stands for discrete wavelet transform (DWT).

18.3.7.2 Quantization of Wavelet Coefficients

After decomposition, the coefficients of the lowest band are coded independently from the other bands. These coefficients are quantized using a uniform midriser quantizer. The coefficients of high bands are quantized with a multilevel quantization. The multilevel quantization provides a very flexible approach to support the right trade-off between levels and type of scalability, complexity, and coding efficiency for any application. All quantizers for the higher bands are uniform midrise quantizer with a dead zone that is 2 times the quantizer step size. The levels and quantization steps are determined by the encoder and specified in the bitstream. To achieve scalability, a bi-level quantization scheme is used for all multiple quantizers. This quantizer is also uniform and midrise with a dead zone that is 2 times the quantization step. The coefficients outside of the dead zone are quantized to 1-bit. The number of quantizers is equal to the maximum number of bit planes in the wavelet transform representation. In this bi-level quantizer, the maximum number of bit planes instead of quantization step size is specified in the bitstream.

18.3.7.3 Coding of Wavelet Coefficients of Low–Low Band and Other Bands

The quantized coefficients at the lowest band are DPCM coded. Each of the current coefficients is predicted from three other quantized coefficients in its neighborhood as shown in Figure 18.10.

The coefficients in high bands are coded with zero-tree algorithm [Shapiro 1993], discussed in Chapter 8.

18.3.7.4 Adaptive Arithmetic Coder

The quantized coefficients and the symbols generated by the zero-tree are coded using an adaptive arithmetic coder. In the arithmetic coder three different tables that correspond to the different statistical models have been utilized. The method used here is very similar to one in Chapter 8. Further details can be found in MPEG-4 VM [mepg4 vm12].

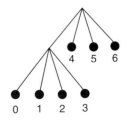

FIGURE 18.9
An example of wavelet decomposition of depth 2.

If $|W_A - W_B| < |W_B - W_C|$, $W_{Xp} = W_C$, else $W_{Xp} = W_A$

FIGURE 18.10
Adaptive DPCM coding of the coefficients in the lowest band.

18.3.8 Generalized Spatial and Temporal Scalability

The scalability framework is referred to as generalized scalability that includes the spatial and the temporal scalability similar to MPEG-2. The major difference is that MPEG-4 extends the concept of scalability to be content-based. This unique functionality allows MPEG-4 to be able to resolve objects into different VOPs. Using the multiple VOP structure, different resolution enhancement can be applied to different portions of a video scene. Therefore, the enhancement layer may be applied only to a particular object or region of the base layer instead of the entire base layer. This is a feature that MPEG-2 does not have.

In spatial scalability, the base layer and the enhancement layer can have different spatial resolution. The base layer VOPs are encoded in the same way as the non-scalable encoding technique described previously. The VOPs in the enhancement layer are encoded as P-VOPs or B-VOPs as shown in Figure 18.11. The current VOP in the enhancement layer can be predicted from either the up-sampled base layer VOP or the previously decoded VOP at the same layer as well as both of them. The down-sampling and up-sampling processing in spatial scalability is not a part of standard and can be defined by the user.

In temporal scalability, a subsequence of subsampled VOP in the time domain is coded as a base layer. The remaining VOPs can be coded as enhancement layers. In this way, the frame rate of a selected object can be enhanced so that it has a smoother motion than other objects. An example of the temporal scalability is illustrated in Figure 18.12. In Figure 18.12, the VOL_0 is the entire frame with both an object and a background, while VOL_1 is a particular object in VOL_0. VOL_0 is encoded with a low frame rate and VOL_1 is the enhancement layer. The high frame rate can be reached for the particular object by combining the decoded data from both base layer and enhancement layer. Of course, the B-VOP is also used in temporal scalability for coding the enhancement layer, which is another type of temporal scalability. As in spatial scalability, the enhancement layer can be used to improve either the entire base layer frame resolution or only a portion of the base layer resolution.

18.3.9 Error Resilience

The MPEG-4 visual coding standard provides error robustness and resilience to allow access of image and video data over a wide range of storage and transmission media. The error resilience tool development effort is divided into three major areas, which include

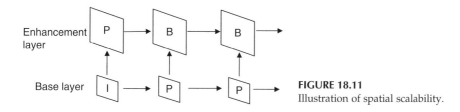

FIGURE 18.11
Illustration of spatial scalability.

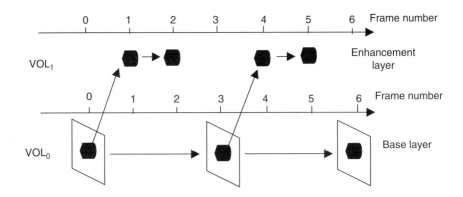

FIGURE 18.12
An example of temporal scalability.

resynchronization, data recovery, and error concealment. As with other coding standards, MPEG-4 makes heavy use of variable-length coding (VLC) to reach high coding performance. However, if even one bit is lost or damaged, the entire bitstream becomes undecodeable due to loss of synchronization. The resynchronization tools attempt to enable resynchronization between the decoder and the bitstream after a transmission error or errors have been detected. Generally, the data between the synchronization point prior to the error and the first point, where synchronization is reestablished, is discarded. The purpose of resynchronization is to effectively localize the amount of data discarded by the decoder, then the other methods such as error concealment can be used to conceal the damaged areas of a decoded picture. Currently, the resynchronization approach adopted by MPEG-4 is referred to as a packet approach. This approach is similar to the Group of Blocks (GOBs) structure used in H.261 and H.263. In the GOB structure, the GOB contains a start code that provides information on the location of the GOB. MPEG-4 adopted a similar approach where a resynchronization marker is periodically inserted into the bitstream at the particular MB locations. The resynchronization marker is used to indicate the start of new video packet. This marker is distinguished from all possible VLC code words as well as the VOP start code. The packet header information is then provided at the start of a video packet. The header contains the information necessary to restart the decoding process. These include the MB number of the first MB contained in this packet and the quantization parameter necessary to decoder the first MB. The MB number provides the necessary spatial resynchronization while the quantization parameter allows the differential decoding process to be resynchronized. It should be noted that when the error resilience is used within MPEG-4, some of the compression efficiency tools need to be modified. For example, all predictively encoded information must be contained within a video packet to avoid error propagation. In conjunction with the video packet approach to resynchronization, MPEG-4 has also adopted fixed interval synchronization method, which requires that VOP start codes and resynchronization markers appear only at legal fixed interval locations in the bitstream. This will help to avoid the problems associated with start codes emulation. In this case, when fixed interval synchronization is utilized, the decoder is only required to search for a VOP start code at the beginning of each fixed interval. The fixed interval synchronization method extends this approach to be any predetermined interval.

After resynchronization is reestablished, the major problem is recovering lost data. A new tool called reversible variable-length codes (RVLC) is developed for the purpose of data recovery. In this approach, the VLCs are designed such that the codes can be read both in the forward as well as the reverse direction. An example of such code includes code words like 111, 101, 010. All these code words can be read reversibly. However, it is obvious that this approach will reduce the coding efficiency that is achieved by the entropy coder. Therefore, this approach is used only in the case where the error resilience is important.

Error concealment is an important component of any error robust video coding. The error concealment strategy is highly dependent on the performance of the resynchronization technique. Basically, if the resynchronization method can localize the damaged data area efficiently, the error concealment strategy becomes much more tractable. Error concealment is actually a decoder issue if there is no additional information provided by the encoder. There are many approaches of error concealment (refer Chapter 17).

18.4 MPEG-4 Visual Bitstream Syntax and Semantics

The common feature of MPEG-4 with MPEG-1 and MPEG-2 is the layered structure of the bitstream. MPEG-4 defines a syntactic description language to describe the exact binary syntax of an AVO bitstream, as well as that of the scene description information. This provides a consistent and uniform way of describing the syntax in a very precise form, while at the same time it simplifies bitstream compliance testing. The visual syntax hierarchy includes the layers:

- Video session (VS)
- Video object (VO)
- Video object layer (VOL) or texture object layer (TOL)
- Group of video object plane (GOV)
- Video object plane (VOP)

A typical video syntax hierarchy is shown in Figure 18.13:

Video session is the highest syntactic structure of the coded video bitstream. A VS is a collection of one or more VOs. A VO can consist of one or more layers. Since MPEG-4 is extended from video coding to visual coding, the type of visual objects not only includes VOs, but also still texture objects, mesh objects, and face objects. These layers can be either video or texture. Still texture coding is designed for high visual quality applications in transmission and rendering of texture. The still coding algorithm supports a scalable representation of image or synthetic scene data such as luminance, color, and shape. This is very useful for progressive transmission of images or 2D/3D synthetic scenes. The

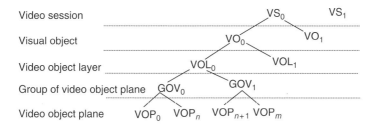

FIGURE 18.13
MPEG-4 video syntax hierarchy.

images can be gradually built up in the terminal monitor as they are received. The bitstreams for coded mesh objects are non-scalable; they define the structure and motion of a 2-D mesh. The texture of the mesh has to be coded as a separate VO. The bitstreams for face objects are also non-scalable; these bitstreams contain the face animation parameters. VOs are coded with different types of scalability. The base layer can be decoded independently, while the enhancement layers can be decoded only with the base layer. In the special case of a single rectangular VO, all of the MPEG-4 layers can be related to MPEG-2 layers. That is, VS is as same as VO since in this case a single VO is a video sequence, VOL or TOL is the same as the sequence scalable extension, GOV is like the GOP and VOP is a video frame. Visual object sequence may contain one or more visual objects coded concurrently. The visual object header information contains the start code followed by profile and level identification, and a visual object identification to indicate the type of object, which may be a VO, a still texture object, a mesh object, or a face object. The VO may contain one or more VOLs. In the VOL, the VO can be coded with spatial or temporal scalability. Also, the VO may be encoded in several layers from coarse to fine resolution. Depending on the need of the application, the decoder can choose the number of layers to decode. A VO at a specified time is called a VOP. Thus, a VO contains many VOPs. A scene may contain many VOs. Each VO can be encoded to an independent bitstream. A collection of VOPs in a VOL is called a group of VOPs (GOV). This concept corresponds to the group of pictures (GOP) in MPEG-1 and MPEG-2. A VOP is then coded by shape coding and texture coding, which is specified at lower layers of syntax, such as MB and block layer. The VOP or higher than VOP layer always commence with a start code and are followed by the data of lower layers, which are similar to the MPEG-1 and MPEG-2 syntax.

18.5 MPEG-4 Visual Profiles and Levels

In MPEG-4, many profiles have been defined. In this section, we have only introduced some visual profiles. There are totally 19 visual profiles defined for different applications:

1. Simple
2. Simple scalable
3. Core
4. Main
5. *N*-bit
6. Hybrid
7. Basic animated texture
8. Scalable texture
9. Simple face animation
10. Simple FBA
11. Advanced real-time simple
12. Core scalable
13. Advanced coding efficiency
14. Advance core profile
15. Advanced scalable texture
16. Simple studio

17. Core studio
18. Advanced simple
19. FGS

Among these visual profiles, simple profile and advanced simple profile have been extensively used by industry for mobile application and streaming on the networks.

The simple profile is used to code the rectangular video with intra (I) and predicted (P) VOPs, actually is the same as for MPEG-2 frame-based coding. The simple profile permits the use of three compression levels with bit rates from 64 kbits/s in level 1 to 384 kbits/s in level 3.

The advanced simple profile is also used to code the rectangular video with intra (I) and predicted (P) VOPs, but it is enhanced to add bidirectional (B) VOPs for better coding efficiency than the simple profile. It supports six compression levels (0–5). Levels 0–3 have bit rates from 128 to 768 kbits/s. The support for interlaced coding is added for levels 4 and 5 with bit rates from 3 to 8 Mbits/s.

18.6 MPEG-4 Video Verification Model

Since all video coding standards define only the bitstream syntax and decoding process, the use of test model (TM) to verify and optimize the algorithms is needed during the development process. For this purpose a common platform with a precise definition of encoding and decoding algorithms has to be provided. The TM of MPEG-2 took the above-mentioned role. The TM of MPEG-2 was updated continually from version 1.0 to version 5.0, until the MPEG-2 Video IS was completed. MPEG-4 video uses a similar tool during the development process; this tool in MPEG-4 is called the verification model (VM). So far, the MPEG-4 video VM has evolved gradually from version 1.0 to version 12.0 and in the process has addressed an increasing number of desired functionalities such as content-based scalability, error resilience, and coding efficiency. The material presented in this section is different from Section 18.3. Because Section 18.3 presented the technologies adopted or will be adopted by MPEG-4, while this section provides an example how to use the standard, i.e., how to encode or generate the MPEG-4 compliant bitstream. Of course, the decoder is also included in the VM.

18.6.1 VOP-Based Encoding and Decoding Process

Since the most important feature of MPEG-4 is an object-based coding method, the input video sequence is first decomposed into separate VOs, these VOs are then encoded into separate bitstream so that the user can access and manipulate (cut, past, ...) the video sequence in the bitstream domain. Instances of VO in a given time are called VOP. The bitstream contains also the composition information to indicate where and when each VOP is to be displayed. At the decoder, the user may be allowed to change the composition of the scene displayed by interactively changing the composition information.

18.6.2 Video Encoder

For an object-based coding, the encoder mainly consists of two parts: the shape coding and the texture coding of the input VOP. Texture coding is based on the DCT coding with traditional MC predictive coding. The VOP is represented by means of a bounding

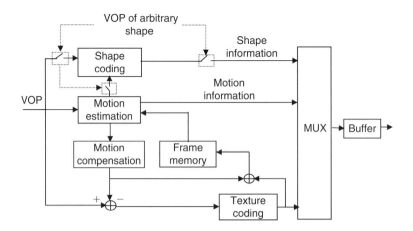

FIGURE 18.14
Block diagram of MPEG-4 video encoder structure.

rectangular as described previously. The phase between luminance and chrominance pixels of the bounding rectangular has to be correctly set to the 4:2:0 format as in MPEG-1/2. The block diagram of encoding structure is shown in Figure 18.14.

The core technologies used in VOP coding of MPEG-4 have been described previously. Here we are going to discuss several encoding issues. Although these issues are essential to the performance and application, they are not dependent on the syntax. As a result, they are not included as normative parts of the standard, but are included as informative annexes.

18.6.2.1 Video Segmentation

Object-based coding is the most important feature of MPEG-4. Therefore, the tool for object boundary detection or segmentation is a key issue in efficiently performing the object-based coding scheme. But the method of decomposing a natural scene to several separate objects is not specified by the standard since it is a preprocessing issue. There are currently two kinds of algorithms for segmentation of VOs. One kind of algorithm is an automatic segmentation algorithm. In the case of real-time applications, the segmentation must be done automatically. Real-time automatic segmentation algorithms are currently not mature. An automatic segmentation algorithm has been proposed in [mpeg96/m960]. This algorithm separates regions corresponding to moving objects from regions belonging to a static background for each frame of a video sequence. The algorithm is based on motion analysis for each frame. Motion analysis is performed along several frames to track each pixel of the current frame and to detect whether the pixel belongs to the moving objects.

Another kind of segmentation algorithms is one that is user-assisted or "semiautomatic". In non-real-time applications, the semiautomatic segmentation may be used effectively and give better results than the automatic segmentation. In the core experiments of MPEG-4, a semiautomatic segmentation algorithm was proposed in [mpeg97/m3147]. The block diagram of the semiautomatic segmentation is shown in Figure 18.15.

This technique consists of two steps. First, the intraframe segmentation is applied to the first frame, which is considered as a frame that either contains newly appeared objects or a reset frame. Then the interframe segmentation is applied to the consecutive frames. For intraframe, the segmentation is processed by a user manually or semiautomatically. The

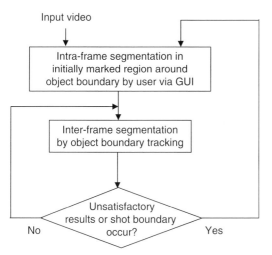

FIGURE 18.15
Block diagram of a user-assisted video object segmentation method.

user uses a graphical user interface (GUI) to draw the boundaries of interested objects. The user can mask the entire objects all the way around objects using a mouse with a predefined thickness of the line (number of pixels). A marked swath is then resulted by the mouse and this marked area is assumed to contain the object boundaries. A boundary detection algorithm is applied to the marked area to create the real object boundaries. For interframe segmentation, an object boundary-tracking algorithm is proposed to obtain the object boundaries of the consecutive frames. At first, the boundary of the previous object is extracted and the ME is performed on the object boundary. The object boundary of the current frame is initially obtained by MC and then refined by using temporal information (TI) and spatial information (SI) all the way around the object boundary. Finally, the refined object boundary can be obtained. As mentioned previously, the segmentation technique is an important tool for object-based processing in MPEG-4, but it is not defined by the standard. The method described here is just an example provided by the core experiments of MPEG-4. There are many other algorithms under investigation such as the circular Viterbi algorithm described in [lin 1998].

18.6.2.2 Intra/Inter Mode Decision

For inter-VOP coding, an MB can be coded in one of the four modes. These four modes include direct coding mode, forward coding, backward coding, and bidirectional coding. In the encoder we have to decide which mode is the best. The mode decision is an important part of encoding optimization. An example of the selection of optimized mode decision has been given in Chapter 17 for MPEG-2 encoder. The same technique can be extended to an MPEG-4 encoder. The basic idea of mode decision is to choose the coding mode that results in the best operation point on the rate distortion curve. For obtaining the best operation point on the rate distortion curve, the encoder has to compare all possible coding modes and choose the best one. This is a very complicated procedure. In the MPEG-2 case, we used a quadratic model to unify the measures of bits used to code prediction residues and the MVs. A simplified mode but near optimized mode decision method has resulted. Here, the VM.12 proposes the following steps to make coding mode decisions. First, the MC prediction error is calculated by each of the four modes. Next, the SAD of each of the MC prediction MBs is calculated and compared with the variance of the MB to be coded. Then a mode generating the smallest SAD (for direct mode, a bias is applied) is selected. For the interlaced video, more coding modes are involved. This method of mode decision is simple, but it is not optimal since the cost for coding MVs is not considered. Consequently, the mode may not lie on the best operation point on the distortion curve. But again this is an encoding issue, the encoding designers have the freedom to use their own algorithm. The VM just provides an example of encoder, which can generate the compliant bitstream.

18.6.2.3 Off-Line Sprite Generation

The sprite is a useful tool in MPEG-4 for coding certain kind of video sequences at very low bit rates. The method of generating a sprite for a video sequence is an encoder issue. The VM gives an example of off-line sprite generation. For a natural VO, sprite is referred to as a representative view collected from a video sequence. Before decoding, the sprite is transmitted to the decoder. Then the MC can be performed by using the sprite from which the video can be reconstructed. The effectiveness of video reconstruction depends on whether the motion of the object can be effectively represented by a global motion model such as translation, zooming, affine, and perspective. The key technology of the sprite generation is the ME to find perspective motion parameters. This can be implemented by many algorithms described in this book such as the three-step matching technique. The block diagram of sprite generation using the perspective ME is shown as in Figure 18.16.

The sprite is generated from the input video sequence by the following steps. First, the first frame is used as the initial value of sprite. From the second frame, the ME is applied to find the perspective motion parameters between two frames. The current frame is wrapped towards the initial sprite using the perspective MVs to get wrapped image. Then the wrapped image is blended with initial sprite to obtain a updated sprite. This procedure is continued to the entire video sequence. The final sprite is then generated.

18.6.2.4 Multiple VO Rate Control

As we know, the purpose of rate control is to obtain the best coding performance for a given bit rate in the constant bit rate (CBR) video coding. In MPEG-4 video coding, there is an additional objective for rate control, how to assign the bits among multiple VOs. In the multiple VO video coding rate control algorithm, the total target is first adjusted based on the buffer fullness, and then distributed proportional to the size of the object, the motion which the object is experiencing, and its maximum absolute differences. Based on the new individual targets and second-order model parameters [lee 1997], appropriate quantization parameters can be calculated for each VO. To compromise the trade-offs in spatial and temporal coding, two modes of operation have been introduced. With these modes, suitable decisions can be made to differentiate between low and high bit rate coding. In addition, a shape rate control algorithm has been included. The algorithm for performing

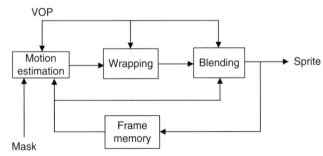

FIGURE 18.16
Block diagram of sprite generation.

MPEG-4 Video Standard: Content-Based Video Coding

the joint rate control can be decomposed into a pre-encoding stage and a post-encoding stage. The pre-encoding stage consists of (i) the target bit estimation, (ii) joint buffer control, (iii) pre-frame skip control, and (iv) the quantization level and alpha threshold calculation. Whereas the post-encoding stage consists of (i) updating the rate distortion model, (ii) post-frame skip control, and (iii) determine mode of operation. The initialization process is very similar to the single VOP initialization process. Since a single buffer is used, the buffer drain rate and initializations remain the same, but many of the parameters are extended to vector quantities. As a means of regulating the trade-offs between spatial and temporal coding, two modes of operation are introduced: low mode and high mode. When encoding at high bit rates, the availability of bits allows the algorithm to be flexible in its target assignment to each VO. Under these circumstances, it is reasonable to impose homogeneous quality among each VO. Therefore, the inclusion of $MAD^2[i]$ is essential to the target distribution and should carry the highest weighting. On the other hand, when the availability of bits is limited, it is very difficult (if not impossible) to achieve homogeneous quality among the VO. Under these conditions, it is desirable to spend less bits on the background and more bits on the foreground. Consequently, the significance of the variance has decreased and the significance of the motion has increased. Besides regulating the quality within each frame, it is also important to regulate the temporal quality as well, i.e., keep the frame skipping to a minimum. In high mode, this is very easy to do since the availability of bits is plentiful. However, in low mode, frame skipping occurs much more often. In fact, the number of frames being skipped is a good indication of which mode the algorithm should be operating. Overall, this particular algorithm is able to successfully achieve the target bit rate, effectively code arbitrarily shaped objects, and maintain a stable buffer [vetro 1999].

18.6.3 Video Decoder

The decoder mainly consists of three parts: shape, motion, and texture decoding. The decoder block diagram is shown in Figure 18.17. At the decoder the bitstream is first demultiplexed into shape and motion information as well as texture information. The reconstructed VOP is obtained by the right combination of the shape, texture, and motion information. The shape decoding is a unique feature of MPEG-4 decoder. The basic technology of shape decoding is the context-based arithmetic decoding and block-based MC.

The primary data structure denoted is the BAB. The BAB is a square block of binary pixels representing the opacity or transparency for the pixels in a specified block-shaped

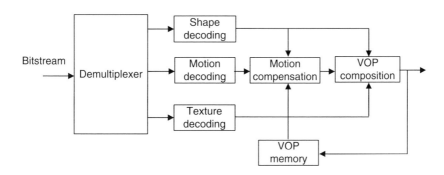

FIGURE 18.17
VOP decoder structure.

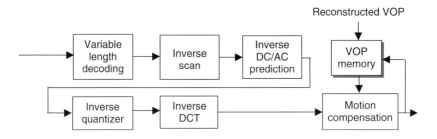

FIGURE 18.18
Block diagram of texture decoding.

spatial region of size 16 × 16 pixels which is colocated with each texture MB. Block diagram of texture decoder is shown in Figure 18.18.

The texture decoding is similar to the video decoder in MPEG-1/2 except that the inverse DC/AC prediction and more quantization methods. The DC prediction is different from the one used in MPEG-1/2. In MPEG-4 the DC coefficient is adaptively predicted from the above block or left block. The AC prediction is similar to the one used in H.263 but is not used in the MPEG-1/2. For MC, the MVs must be decoded. The horizontal and vertical MV components are decoded differentially by using a prediction from the spatial neighborhood consisting of three MVs already decoded. The final MV is obtained by adding the prediction MV values to the decoded differential motion values. Also, in MPEG-4 video coding the several advanced MC modes such as four 8 × 8 MV compensation and overlapped MC have to be handled. Another issue of MC in MPEG-4 is raised by VOP-based coding. In order to perform MC prediction on a VOP basis, a special padding technique is used to each MB that lies on the shape boundary of the VOP. The padding process defines the values of pixels, which are located outside the VOP for prediction of arbitrary shaped objects. Padding for luminance pixels and chrominance pixels is defined in the standard [mpeg4 visual]. The additional decoding issues that are special for MPEG-4 include sprite decoding, generalized scalable decoding, and still texture decoding. We do not go into further detail for these topics (interested readers can get detail from the standard documents). The outputs of decoded results are the reconstructed VOPs that are finally sent to the compositor. In the compositor, the VOPs are recursively blended in the order specified by the VOP composition order. It should be noted that the decoders could take advantage of object-based decoding. They are able to be flexible in the composition of the reconstructed VOPs such as re-allocating, rotation, or other editing actions.

18.7 Summary

In this chapter, the new video coding standard, MPEG-4, was introduced. The unique feature of MPEG-4 video is the content-based coding. This feature allows the MPEG-4 to provide much functionality, which other video coding standards do not have. The key technologies used in MPEG-4 video have been described. These technologies provide basic tools for MPEG-4 video to provide object-based coding functionality. Finally, the video VM, a platform of MPEG-4 development and an encoding and decoding example, has been described.

Exercises

1. Why is object (or content)-based coding the most important feature of MPEG-4 visual coding standard? Describe several applications for this feature.
2. What are the new coding tools in MPEG-4 visual coding that are different from MPEG-2 video coding? Is MPEG-4 backward compatible to MPEG-2?
3. MPEG-4 video coding has the feature of using either 16×16 block MV or 8×8 block MV, for what kind of video sequences will the 8×8 block motion increase coding efficiency? For what kind of video sequences will the 8×8 block MC decrease the coding efficiency?
4. What approaches for error resilience are supported by the MPEG-4 syntax? Make a comparison with the error resilience method adopted in MPEG-2 (supported by MPEG-2 syntax) and indicate their relative advantages and disadvantages.
5. Design an arithmetic coder for zero-tree coding and write a program to test it with several images.
6. Sprite is a new feature of MPEG-4 video coding. MPEG-4 specifies the syntax for sprite coding, but does not give any detail how to generate a sprite. Conduct a project to generate an off-line sprite for a video sequence and use it for coding the video sequence. Do you observe any increased coding efficiency? When do you expect to see such an increase?
7. Shape coding (binary-shape coding) is an important part of MPEG-4 due to object-based coding. Besides the shape coding method used in MPEG-4, name another shape coding method. Conduct a project to compare the method you know with the method proposed in MPEG-4. (Do not expect to get better performance, but expect to reduce the complexity.)

References

[lee 1997] H.J. Lee, T. Chiang, and Y.Q. Zhang, Scalable rate control for very low bit-rate coding, Proceedings of the International Conference on Image Processing (ICIP'97), Vol. II, Santa Barbara, CA, pp. 768–771, October 1997.

[lin 1998] Circular Viterbi: Boundary detection with dynamic programming, I-jong Lin, S.Y. Kung, Anthony Vetro and Huifang Sun, MPEG98/.

[mpeg4 visual] ISO/IEC 14496-2, Coding of audio-visual objects, Part 2, December 18, 1998.

[mpeg4 vm12] ISO/IEC 14496-2, Video Verffication Model V.12, N2552, December, 1998.

[mpeg96/m960] S. Colonnese and G. Russo, FUB results on core experiment N2: comparison of automatic segmentation techniques, MPEG96/M960, Tempere, July 1996.

[mpeg97/m3147] J.G. Choi, M. Kim, H. Lee, and C. Ahn, Partial experiments on a user-assisted segmentation technique for video object plane generation, MPEG97/M3147, San Jose meeting of ISO/IEC JTC1/SC29/WG11, February 1998.

[shapiro 1993] J. Shapiro, Embedded image coding using zero trees of wavelet coefficients, *IEEE Transactions on Signal Processing*, 3445–3462, December 1993.

[vetro 1999] A. Vetro, H. Sun, and Y. Wang, MPEG-4 rate control for multiple video objects, *IEEE Transaction Circuits and Systems for Video Technology*, 9, 1, 186–199, February 1999.

19
ITU-T Video Coding Standards H.261 and H.263

This chapter introduces ITU-T video coding standards H.261 and H.263, which are established mainly for videophony and videoconferencing. The basic technical detail of H.261 is presented. The technical improvements with which H.263 achieves high coding efficiency are discussed. Features of H.263+, H.263++, and H.26L are presented.

19.1 Introduction

Very low bit rate video coding has found many industrial applications such as wireless and network communications. The rapid convergence of standardization of digital video coding standards is the reflection of several factors: the maturity of technologies in terms of algorithmic performance, hardware implementation with VLSI technology, and the market need for rapid advances in wireless and network communications. As stated in the previous chapters, these standards include JPEG for still image coding and MPEG-1/2 for CD-ROM storage and digital television (DTV) applications. In parallel with the ISO/IEC development of the MPEG-1/2 standards, the ITU-T has developed H.261 [h261] for videotelephony and videoconferencing applications in an ISDN environment.

19.2 H.261 Video Coding Standard

The H.261 video coding standard was developed by ITU-T study group XV during 1988–1993. It was adopted in 1990 and the final revision approved in 1993. This is also referred to as the $P \times 64$ standard because it encodes the digital video signals at the bit rates of $P \times 64$ kbits/s, where P is an integer from 1 to 30, i.e., at the bit rates from 64 kbits/s to 1.92 Mbits/s.

19.2.1 Overview of H.261 Video Coding Standard

The H.261 video coding standard has many common features with the MPEG-1 video coding standard. However, as they target different applications, there exist many differences between the two standards, such as data rates, picture quality, end-to-end delay, etc. Before indicating the differences between two coding standards, we describe the major similarity between H.261 and MPEG-1/2. First, both standards are used to code the similar video format. H.261 is mainly used to code the video with common intermediate format (CIF) or quarter-CIF (QCIF) spatial resolution for teleconference application. MPEG-1 uses CIF, SIF, or higher-spatial resolution for CD-ROM application. The original motivation of developing

H.261 video coding standard is to provide a standard, which can be used, for both PAL and NTSC television signals. But later, the H.261 is mainly used for video conferencing and the MPEG-1/2 is used for DTV, VCD (video CD), and DVD (digital video disk). The two TV systems, phase alternating line (PAL) and National Television Systems Committee (NTSC), use different line and picture rate. The NTSC, which is used in North America and Japan, uses 525 lines per interlaced picture at 30 frames/s. The PAL system is used for most other countries and it uses 625 lines per interlaced picture at 25 frames/s. For this purpose, the CIF was adopted as the source video format for H.261 video coder. The CIF consists of 352 pixels per line, 288 lines per frame, and 30 frames/s. This format represents half the active lines of the PAL signal and the same picture rate of the NTSC signal. The PAL systems need only perform a picture rate conversion and NTSC systems need only perform a line numbers conversion. Color pictures consist of one luminance and two color-difference components (referred to as $Y\ Cb\ Cr$ format) as specified by the CCIR 601 standard. The Cb and Cr components are half the size on both horizontal and vertical directions and have 176 pixels per line and 144 lines per frame. Another format, QCIF is used for very low bit rate applications. The QCIF has half the number of pixels and half the number of lines of CIF. Second, the key coding algorithms of H.261 and MPEG-1 are very similar. Both H.261 and MPEG-1 use discrete cosine transform (DCT)-based coding to remove intraframe redundancy and motion compensation to remove interframe redundancy.

Now let us describe the main differences between the two coding standards with respect to coding algorithms. The main differences include:

H.261 uses only I- and P-macroblocks (MBs) but no B-MBs, whereas MPEG-1 uses three MB types, I-, P-, B-MBs (I-MB is intraframe-coded MB, P-MB is predictive-coded MB and B-MB is bidirectionally coded MB), also three picture types, I-, P-, and B-pictures as defined in Chapter 16 for MPEG-1 standard.

There is a constraint of H.261 that for every 132 interframe-coded MBs, which corresponds to four GOBs (group of blocks) or to one-third of CIF pictures; it requires at least one intraframe-coded MB. To obtain better coding performance at low bit rate applications, most encoding schemes of H.261 prefer not to use intraframe coding on all the MBs of a picture but only few MBs in every picture with a rotational scheme. MPEG-1 uses the GOP (group of pictures) structure, where the size of GOP (the distance between two I-pictures) is not specified.

The end-to-end delay is not a critical issue for MPEG-1 but is critical for H.261. The video encoder and video decoder delays of H.261 need to be known to allow audio compensation delays to be fixed when H.261 is used in interactive applications. This will allow lip synchronization to be maintained.

The accuracy of motion compensation in MPEG-1 is up to a half pixel, but only a full pixel in H.261. However, H.261 uses a loop-filter to smooth the earlier frame. This filter attempts to minimize the prediction error.

In H.261, a fixed picture aspect ratio of 4:3 is used. In MPEG-1, several picture aspect ratios can be used and the picture aspect ratio is defined in the picture header.

Finally, in H.261, the encoded picture rate is restricted to allow up to three skipped frames. This would allow the control mechanism in the encoder some flexibility to control the encoded picture quality and satisfy the buffer regulation. Although MPEG-1 has no restriction on skipped frames, the encoder usually does not perform frame skipping. Rather, the syntax for B-frames is exploited, as B-frames require much fewer bits than P-pictures.

19.2.2 Technical Detail of H.261

The key technologies used in the H.261 video coding standard are the DCT and motion compensation. The main components in the encoder include DCT, prediction,

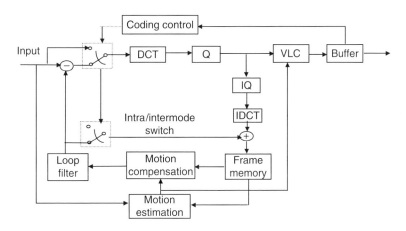

FIGURE 19.1
Block diagram of a typical H.261 video encoder.

quantization (Q), inverse DCT (IDCT), inverse quantization (IQ), loop filter, frame memory, variable-length coding (VLC), and coding control unit. A typical encoder structure is shown in Figure 19.1.

The input video source is first converted to the CIF frame and then is stored in the frame memory. The CIF frame is then partitioned into groups of blocks (GOBs). The GOB contains 33 MBs, which are one-twelfth of a CIF picture or one-third of a QCIF picture. Each MB consists of six 8×8 blocks among which four are luminance (Y) blocks and two are chrominance blocks (one of Cb and one of Cr).

For intraframe mode, each 8×8 block is first transformed with DCT and then quantized. The VLC is applied to the quantized DCT (QDCT) coefficients with a zigzag scanning order such as in MPEG-1. The resulting bits are sent to the encoder buffer to form a bitstream.

For interframe coding mode, the frame prediction is performed with motion estimation in a similar manner to that in MPEG-1 but only P-MBs and P-pictures but no B-MBs and B-pictures are used. Each 8×8 block of differences or prediction residues is coded by the same DCT coding path as for the intraframe coding. In the motion compensated (MC) predictive coding, the encoder should perform the motion estimation with the reconstructed pictures instead of the original video data, as it will be done in the decoder. Therefore, the IQ and IDCT blocks are included in the motion compensation loop to reduce the error propagation drift. However, the VLC operation is lossless; there is no need to include the VLC block into the motion compensation loop. The role of spatial filter is to minimize the prediction error by smoothing the previous frame that is used for motion compensation.

The loop filter is a separable two-dimensional (2-D) spatial filter that operates on 8×8 block. The corresponding one-dimensional (1-D) filters are nonrecursive with coefficients 1/4, 1/2, 1/4. At block boundaries, the coefficients are 0, 1, 0 to avoid that the taps fall outside the block. It should note that MPEG-1 uses sub-pixel accurate motion vectors instead of a loop filter to smooth the anchor frame. The performance comparison of two methods should be interesting.

The role of coding control includes the rate control, the buffer control, the quantization control, and the frame rate control. These parameters are intimately related. The coding control is not the part of standard; however, it is an important part for the encoding process. For a given target bit rate, the encoder has to control several parameters to reach the rate target and at the same time provide reasonable coded picture quality.

As H.261 is a predictive coder and the VLCs are used everywhere such as coding QDCT coefficients and motion vectors, a single transmission error may cause a loss of synchronization and consequently cause problems for the reconstruction. To enhance the performance of H.261 video coder in the noisy environment, the transmitted bitstream of H.261 can optionally contain a BCH (Bose, Chaudhuri, and Hocquengham) (511,493) forward error correction code.

The H.261 video decoder performs the inverse operations of the encoder. After optional error correction decoding, the compressed bitstream enters the decoder buffer and then is parsed by the variable-length decoder (VLD). The output of VLD is applied to the IQ and IDCT where the data are converted to the values in the spatial domain. For interframe coding mode, the motion compensation is performed and the data from the MBs in the anchor frame are added to the current data to form the reconstructed data.

19.2.3 Syntax Description

The syntax of H.261 video coding has a hierarchical-layered structure. From the top to the bottom, the layers are picture layer, GOB layer, MB layer, and block layer.

19.2.3.1 Picture Layer

The picture layer begins with a 20-bit picture start code (PSC). Following the PSC, there are temporal reference (5-bit), picture-type information (PTYPE, 6-bit), extra insertion information (PEI, 1-bit), and spare information (PSPARE). Then the data for GOBs are followed.

19.2.3.2 Group of Blocks Layer

A GOB corresponds to 176 pixels by 48 lines of Y and 88 pixels by 24 lines of Cb and Cr. The GOB layer contains the following data in order: 16-bit GOBs start code (GBSC), 4-bit group number (GN), 5-bit quantization information (GQUANT), 1-bit extra insertion information (GEI), and spare information (GSPARE). The number of bits for GSPARE is variable depending on the set of GEI bit. If GEI is set to 1, then 9 bits follow, consisting of 8 bits of data and another GEI bit to indicate whether a further 9 bits follow and so on. Data of GOB header is then followed by data for MBs.

19.2.3.3 Macroblock Layer

Each GOB contains 33 MBs, which is arranged as in Figure 19.2.

An MB consists of 16 pixels by 16 lines of Y that spatially correspond to 8 pixels by 8 lines of each Cb and Cr. Data in the bitstream for an MB consists of an MB header followed by data for blocks. The MB header may include MB address (MBA) (variable length), type information (MTYPE) (variable length), quantizer (MQUANT) (5 bits), motion vector data (MVD) (variable length), and coded block pattern (CBP) (variable length). The MBA information is always present and is coded by VLC. The VLC table for MB addressing is shown in Table 19.1. The presence of other items depends on MB-type information, which is shown in the VLC table (Table 19.2).

FIGURE 19.2 Arrangement of macroblocks (MBs) in a group of block (GOB).

1	2	3	4	5	6	7	8	9	10	11
12	13	14	15	16	17	18	19	20	21	22
23	24	25	26	27	28	29	30	31	32	33

TABLE 19.1

Variable-Length Coding (VLC) Table for Macroblock (MB) Addressing

MB Address (MBA)	Code	MB Address (MBA)	Code	MB Address (MBA)	Code
1	1	13	0000 1000	25	0000 0100 000
2	011	14	0000 0111	26	0000 0011 111
3	010	15	0000 0110	27	0000 0011 110
4	0011	16	0000 0101 11	28	0000 0011 101
5	0010	17	0000 0101 10	29	0000 0011 100
6	0001 1	18	0000 0101 01	30	0000 0011 011
7	0001 0	19	0000 0101 00	31	0000 0011 010
8	0000 111	20	0000 0100 11	32	0000 0011 001
9	0000 110	21	0000 0100 10	33	0000 0011 000
10	0000 1011	22	0000 0100 011	MBA stuffing	0000 0001 111
11	0000 1010	23	0000 0100 010	Start code	0000 0000 0000 0001
12	0000 1001	24	0000 0100 001		

19.2.3.4 Block Layer

Data in the block layer consists of the transformed coefficients followed by an end of block (EOB) marker (10). The data of transform coefficients (TCOEFF) is first converted to the pairs of RUN and LEVEL according to the zigzag scanning order. The RUN represents the number of successive zeros and the LEVEL represents the value of nonzero coefficients. The pairs of RUN and LEVEL are then encoded with VLCs. The DC coefficient of an intrablock is coded by a fixed-length code with 8 bits. All VLC tables can be found in the standard document [h261].

19.3 H.263 Video Coding Standard

The H.263 video coding standard [h263] is specifically designed for very low bit rate applications such as practical video telecommunication. Its technical content was completed in late 1995 and the standard was approved in early 1996.

TABLE 19.2

Variable-Length Coding (VLC) Table for Macroblock (MB) Type

Prediction	MB Quantizer (MQUANT)	Motion Vector Data (MVD)	Coded Block Pattern (CBP)	Transform Coefficients (TCOEFF)	Variable-Length Coding (VLC)
Intra				X	0001
Intra	X			X	0000 001
Inter			X	X	1
Inter	X		X	X	0000 1
Inter + MC		X			0000 0000 1
Inter + MC		X	X	X	0000 0001
Inter + MC	X	X	X	X	0000 0000 01
Inter + MC + FIL		X			001
Inter + MC + FIL		X	X	X	01
Inter + MC + FIL	X	X	X	X	0000 01

Notes: "X" means that the item is present in the MB. It is possible to apply the filter in a nonmotion compensated (MC) MB by declaring it as MC + FIL but with a zero vector.

TABLE 19.3

Number of Pixels per Line and the Number of Lines for Each Picture Format

Picture Format	Number of Pixels for Luminance (dx)	Number of Lines for Luminance (dy)	Number of Pixels for Chrominance (dx/2)	Number of Lines for Chrominance (dy/2)
Subquarter common intermediate format (QCIF)	128	96	64	48
QCIF	176	144	88	72
CIF	352	288	176	144
4CIF	704	576	352	288
16CIF	1408	1152	704	576

19.3.1 Overview of H.263 Video Coding

The basic configuration of the video source coding algorithm of H.263 is based on the H.261. Several important features that are different from H.261 include the following new options: unrestricted motion vectors, syntax-based arithmetic coding, advanced prediction, and PB-frames. All these features can be used together or separately for improving the coding efficiency. The H.263 video standard can be used for both 625-line and 525-line television standards. The source coder operates on the noninterlaced pictures at picture rate about 30 pictures/s. The pictures are coded as luminance and two color difference components (Y, Cb, and Cr). The source coder is based on a CIF. Actually, there are five standardized formats, which include sub-QCIF, QCIF, CIF, 4CIF, and 16CIF. The detail of formats is shown in Table 19.3.

It is noted that for each format, the chrominance is a quarter size of the luminance picture, i.e., the chrominance pictures are half the size of the luminance picture in both horizontal and vertical directions. This is defined by the ITU-R 601 format. For CIF, the number of pixels per line is compatible with sampling the active portion of the luminance and color difference signals from a 525- or 626-line source at 6.75 and 3.375 MHz, respectively. These frequencies have a simple relationship to those defined by the ITU-R 601 format.

19.3.2 Technical Features of H.263

The H.263 encoder structure is similar to the H.261 encoder with the exception that there is no loop filter in H.263 encoder. The main components of the encoder include block transform, MC prediction, block quantization, and VLC. Each picture is partitioned into GOBs. A GOB contains multiple number of 16 lines, $k*16$ lines, depending on the picture format ($k = 1$ for sub-QCIF, QCIF; $k = 2$ for 4CIF; $k = 4$ for 16CIF). Each GOB is divided into MBs that are the same as in H.261, each MB consists of four 8×8 luminance blocks and two 8×8 chrominance blocks. Compared to H.261, H.263 has several new technical features for the enhancement of coding efficiency for very low bit rate applications. These new features include picture-extrapolating motion vectors (or unrestricted motion vector mode), motion compensation with half-pixel accuracy, advanced prediction (which includes variable-block size motion compensation and overlapped block motion compensation), syntax-based arithmetic coding and PB-frame mode.

19.3.2.1 Half-Pixel Accuracy

In H.263 video coding, the half-pixel accuracy motion compensation is used. The half-pixel values are found using bilinear interpolation as shown in Figure 19.3.

```
A       B        ×  Integer pixel position
 ×   ○   ×
   a   b         ○  Half-pixel position

   ○   ○
   c   d         a = A
                 b = (A + B + 1)/2
                 c = (A + C + 1)/2
 ×       ×       d = (A + B + C + D + 2)/4    FIGURE 19.3
 C       D                                    Half-pixel prediction by bilinear
                 "/" indicates division by truncation    interpolation.
```

Note that H.263 uses sub-pixel accuracy for motion compensation instead of using a loop filter to smooth the anchor frames as in H.261. This is also done in other coding standards such as MPEG-1 and MPEG-2, which also use half-pixel accuracy for motion compensation. In MPEG-4 video, quarter-pixel accuracy for motion compensation has been adopted as a tool for the version 2.

19.3.2.2 Unrestricted Motion Vector Mode

Usually the motion vectors are limited within the coded picture area of anchor frames. In the unrestricted motion vector mode, the motion vectors are allowed to point outside the pictures. When the values of motion vectors exceed the boundary of anchor frame in the unrestricted motion vector mode, the picture-extrapolating method is used. The values of reference pixels outside the picture boundary will take the values of boundary pixels. The extension of motion vector range is also applied to the unrestricted motion vector mode. In the default prediction mode, the motion vectors are restricted to the range of $[-16, 15.5]$. In the unrestricted mode, the maximum range for motion vectors is extended to $[-31.5, 31.5]$ under certain conditions.

19.3.2.3 Advanced Prediction Mode

Generally, the decoder will accept no more than one motion vector per MB for baseline algorithm of H.263 video coding standard. However, in the advanced prediction mode, the syntax allows up to four motion vectors to be used per MB. The decision of using one or four vectors is indicated by the MB type and CBP for chrominance (MCBPC) code word for each MB. How to make this decision is the task of encoding process.

The following example gives the steps of motion estimation and coding mode selection for advanced prediction mode in the encoder.

Step 1: Integer pixel motion estimation

$$\mathrm{SAD}_N(x, y) = \sum_{i=0}^{N-1} \sum_{j=0}^{N-1} |\mathrm{original} - \mathrm{previous}| \tag{19.1}$$

where SAD is the sum of absolute difference, values of (x, y) is within the search range, N is equal to 16 for 16×16 block, and N is equal to 8 for 8×8 block.

$$\mathrm{SAD}_{4 \times 8} = \Sigma \mathrm{SAD}_8(x, y) \tag{19.2}$$

$$\mathrm{SAD}_{\mathrm{inter}} = \min\left(\mathrm{SAD}_{16}(x, y), \mathrm{SAD}_{4 \times 8}\right) \tag{19.3}$$

Step 2: Intra/intermode decision
  If $A < (\mathrm{SAD}_{\mathrm{inter}} - 500)$, this MB is coded as intra-MB

otherwise, it is coded as inter-MB
where SAD_{inter} is determined in step 1, and

$$A = \sum_{i=0}^{15} \sum_{j=0}^{15} |\text{original} - MB_{mean}|$$
$$MB_{mean} = \frac{1}{256} \sum_{i=0}^{15} \sum_{j=0}^{15} \text{original}$$
(19.4)

If this MB is determined to coded as inter-MB, go to step 3

Step 3: Half-pixel search

In this step, half-pixel search is performed for both 16×16 and 8×8 blocks as shown in Figure 19.3.

Step 4: Decision on 16×16 or four 8×8 (one motion vector or four motion vectors per MB)

If $SAD_{4 \times 8} < SAD_{16} - 100$, four motion vectors per MB will be used, one of the motion vectors is used for all pixels in one of the four luminance blocks in the MB, otherwise, one motion vector will be used for all pixels in the MB.

Step 5: Differential coding of motion vectors for each of 8×8 luminance blocks is performed as in Figure 19.4.

When it has been decided to use four motion vectors, the MVD_{CHR} motion vector for both chrominance blocks is derived by calculating the sum of the four luminance vectors and dividing by 8. The component values of the resulting sixteenth pixel resolution vectors are modified toward the position as indicated in the Table 19.4.

Another advanced prediction mode is overlapped motion compensation for luminance. Actually, this idea is also used by MPEG-4, which has been described in Chapter 18. In the overlapped motion compensation mode, each pixel in an 8×8 luminance block is a weighted sum of three values divided by 8 with rounding. The three values are obtained by the motion compensation with three motion vectors: the motion vector of the current

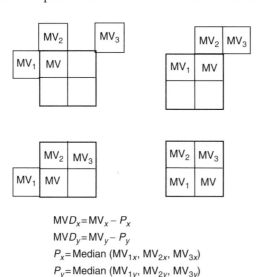

FIGURE 19.4
Differential coding of motion vectors.

TABLE 19.4

Modification of Sixteenth Pixel Resolution Chrominance Vector Components

Sixteenth pixel position	0	1	2	3	4	5	6	7	8	9	10	11	12	13	14	15	/16
Resulting position	0	0	0	1	1	1	1	1	1	1	1	1	1	1	2	2	/2

luminance block and two of four remote vectors. These remote vectors include the motion vector of the block to the left or right of the current block and the motion vector of the block above or below the current block. The remote motion vectors from other GOBs are used in the same way as remote motion vectors inside the current GOB. For each pixel to be coded in the current block, the remote motion vectors of the blocks at the two nearest block borders are used, i.e., for the upper half of the block, the motion vector corresponding to the block above the current block is used while for the lower half of the block, the motion vector corresponding to the block below the current block is used. Similarly, the left half of the block uses the motion vector of the block at the left side of the current block and the right half uses the one at the right side of the current block. To make this more clear, let (MV_x^0, MV_y^0) be the motion vector for the current block, (MV_x^1, MV_y^1) be the motion vector for the block either above or below, and (MV_x^2, MV_y^2) be the motion vector of the block either to the left or right of the current block. Then the value of each pixel, $p(x,y)$ in the current 8×8 luminance block is given by

$$p(x,y) = [q(x,y) \cdot H_0 + r(x,y) \cdot H_1 + s(x,y) \cdot H_2(x,y) + 4]/8 \quad (19.5)$$

where

$$q(x,y) = p(x + MV_x^0, y + MV_y^0); \; r(x,y) = p(x + MV_x^1, y + MV_y^1); \text{ and}$$
$$s(x,y) = p(x + MV_x^2, y + MV_y^2); \quad (19.6)$$

H_0 is the weighting matrix for prediction with the current block motion vector
H_1 is the weighting matrix for prediction with the top or bottom block motion vector
H_2 is the weighting matrix for prediction with the left or right block motion vector

This applies to the luminance block only. The values of H_0, H_1, and H_2 are shown in Figure 19.5.

H_0

4	5	5	5	5	5	5	4
5	5	5	5	5	5	5	5
5	5	6	6	6	6	5	5
5	5	6	6	6	6	5	5
5	5	6	6	6	6	5	5
5	5	6	6	6	6	5	5
5	5	5	5	5	5	5	5
4	5	5	5	5	5	5	4

H_1

2	2	2	2	2	2	2	2
1	1	2	2	2	2	1	1
1	1	1	1	1	1	1	1
1	1	1	1	1	1	1	1
1	1	1	1	1	1	1	1
1	1	1	1	1	1	1	1
1	1	2	2	2	2	1	1
2	2	2	2	2	2	2	2

H_2

2	1	1	1	1	1	1	2
2	2	1	1	1	1	2	2
2	2	1	1	1	1	2	2
2	2	1	1	1	1	2	2
2	2	1	1	1	1	2	2
2	2	1	1	1	1	2	2
2	2	1	1	1	1	2	2
2	1	1	1	1	1	1	2

FIGURE 19.5
Weighting matrices for overlapped motion compensation.

It should be noted that the above coding scheme is not optimized in the selection of mode decision since the decision depends only on the values of predictive residues. Optimized mode decision techniques that include the above possibilities for prediction have been considered in [weigand 1996].

19.3.2.4 Syntax-Based Arithmetic Coding

As in other video coding standards, H.263 uses VLC/VLD to remove the redundancy in the video data. The basic principle of VLC is to encode a symbol with a specific table based on the syntax of the coder. If the symbol is mapped to an entry of the table in a table look-up operation then the binary code word specified by the entry is sent to a bitstream buffer for transmitting to the decoder. In the decoder, an inverse operation, VLD, is performed to reconstruct the symbol by the table look-up operation based on the same syntax of the coder. The tables in the decoder must be the same as the one used in the encoder for encoding the current symbol. To obtain the better performance, the tables are generated in a statistically optimized way (such as a Huffman coder) with a large number of training sequences. This VLC/VLD process implies that each symbol must be encoded into a fixed integral number of bits. An optional feature of H.263 is to use arithmetic coding to remove the restriction of fixed integral number bits for symbols. This syntax-based arithmetic coding mode may result in bit rate reductions.

19.3.2.5 PB-Frames

The PB-frame is a new feature of H.263 video coding. A PB-frame consists of two pictures, one P-picture and one B-picture, being coded as one unit as shown in Figure 19.6. Since H.261 does not have B-pictures, the concept of a B-picture comes from the MPEG video coding standards. In a PB-frame, the P-picture is predicted from the previous decoded I- or P-picture and the B-picture is bidirectionally predicted both from the previous decoded I- or P-picture and the P-picture in the PB-frame unit, which is currently being decoded.

Several detailed issues are addressed at MB level in PB-frame mode:

If an MB in PB-frame is intracoded, the P-MB in the PB unit is intracoded and the B-MB in the PB unit is intercoded. The motion vector of intercoded PB-MB is used for the B-MB only.

An MB in PB-frame contains 12 blocks for 4:2:0 format, six (four luminance blocks and two chrominance blocks) from P-frame and six from B-frame. The data for six P-blocks is transmitted first and then for the six B-blocks.

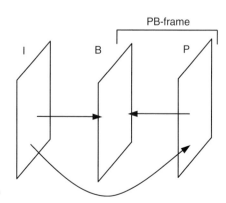

FIGURE 19.6
Prediction in PB-frames mode.

Different parts of a B-block in a PB-frame can be predicted with different modes. For pixels where the backward vector points inside of coded P-MB, bidirectional prediction is used. For all other pixels, forward prediction is used.

19.4 H.263 Video Coding Standard Version 2

19.4.1 Overview of H.263 Version 2

The H.263 version 2 [h263+] video coding standard, also known as H.263+, was approved in January of 1998 by the ITU-T. H.263 version 2 includes a number of new optional features based on the H.263 video coding standard. These new optional features are added to broaden the application range of H.263 and to improve its coding efficiency. The main features are flexible video format, scalability, and backward-compatible supplemental enhancement information. Among these new optional features, five features are intended to improve the coding efficiency and three features are proposed to address the needs of mobile video and other noisy transmission environments. The features of scalability provide the capability of generating layered bitstream which are spatial scalability, temporal scalability, and signal-to-noise ratio (SNR) scalability similar to those defined by the MPEG-2 video coding standard. There are also other modes of H.263 version 2 that provide some enhancement functions. We will describe these features in the following section.

19.4.2 New Features of H.263 Version 2

The H.263 version 2 includes a number of new features. In the following we briefly describe the key techniques used for these features.

19.4.2.1 Scalability

The scalability function allows for encoding the video sequences in a hierarchical way that partitions the pictures into one basic layer and one or more enhancement layers. The decoders have the option to decode only the base layer bitstream to obtain lower quality reconstructed pictures or further decode the enhancement layers to obtain the higher quality decoded pictures. There are three types of scalability in H.263: temporal scalability, SNR scalability, and spatial scalability.

Temporal scalability is achieved by using B-pictures as the enhancement layer. All three types of scalability are similar to the ones in MPEG-2 video coding standard. The B-pictures are predicted from either or both a previous and later decoded picture in the base layer as shown in Figure 19.7.

In the SNR scalability, the pictures are first encoded with coarse quantization in the base layer. The differences or coding error pictures between a reconstructed picture and its original in the base layer encoder are then encoded in the enhancement layer and sent to the decoder providing an enhancement of SNR. In the enhancement layer, there are two types of pictures. If a picture in the enhancement layer is only predicted from the base layer, it is referred to as an EI (enhancement I)-picture. It is a bidirectionally predicted picture if it uses both an earlier enhancement layer picture and a temporally simultaneous base layer reference picture for prediction. Note that the prediction from the reference layer uses no motion vectors. However, EP (enhancement P)-pictures use motion vectors when predicted from their temporally earlier reference picture in the same layer. Also, if more

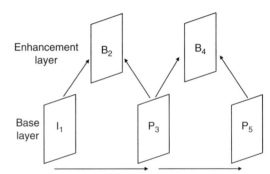

FIGURE 19.7
Temporal scalability.

than two layers are used, the reference may be the lower layer instead of the base layer (see Figure 19.8).

In the spatial scalability, lower-resolution pictures are encoded in the base layer or lower layer. The differences or error pictures between up-sampled decoded base layer pictures and its original picture are encoded in the enhancement layer and sent to decoder providing the spatial enhancement pictures. As in MPEG-2, spatial interpolation filters are used for the spatial scalability. There are also two types of pictures in the enhancement layer: EI and EP. If a decoder is able to perform spatial scalability, it may also need to be able to use a custom picture format. For example, if the base layer is sub-QCIF (128×96), the enhancement layer picture would be 256×192, which does not belong to a standard picture format (see Figure 19.9).

The scalability in H.263 can be performed with multilayers. In the case of multilayer scalability, the picture layer used for upward prediction in an EI or EP picture may be an I, P, EI, or EP picture, or may be the P part of a PB or improved PB-frame in the base layer as shown in Figure 19.10.

19.4.2.2 Improved PB-Frames

The difference between the PB-frame and the improved PB-frame is that bidirectional prediction is used for B MBs in the PB-frame, while in the improved PB-frame, B MBs can be coded in three prediction modes: bidirectional, forward, and backward predictions. This means that in forward prediction or backward prediction, only one motion vector is used for a 16×16 MB instead of using two motion vectors for a 16×16 MB in bidirectional prediction. In very low bit rate case, this mode can improve the coding efficiency by saving bits for coding motion vectors.

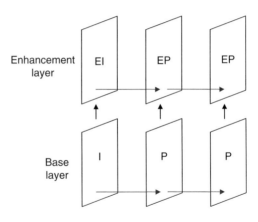

FIGURE 19.8
Signal-to-noise ratio (SNR) scalability.

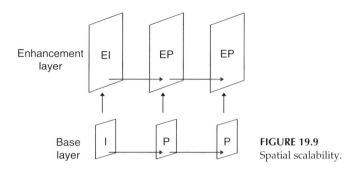

FIGURE 19.9
Spatial scalability.

19.4.2.3 Advanced Intracoding

The advantage of intracoding is to protect the error propagation because intracoding does not depend on the previous decoded picture data. However, the problem of intracoding is that more bits are needed since the temporal correlation between frames are not exploited. The idea of advanced intracoding (AIC) is used to address this problem. The coding efficiency of intracoding is improved by the use of following three methods:

1. Intrablock prediction using neighboring intrablocks for the same color component (Y, Cb, or Cr): a particular intracoded block may be predicted from the block above or left to the current block being decoded, or from both. The main purpose of these predictions tries to use the correlation between neighboring blocks. For example, the first row of AC coefficients may be predicted from those in the block above, the first column of AC coefficients may be predicted from those in the left and the DC value may be predicted as an average from the block above and left.
2. Modified IQ for intracoefficients: IQ of the intra DC coefficient is modified to allow a varying quantization step size. IQ of all intra AC coefficients is performed without a dead-zone in the quantizer reconstruction spacing.
3. A separate VLC for intracoefficients: To improve intracoding a separate VLC table is used for all intra DC and intra AC coefficients. The price paid for this modification is the use of more tables.

FIGURE 19.10
Multilayer scalability.

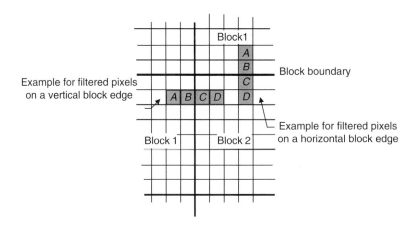

FIGURE 19.11
Positions of filtered pixels.

19.4.2.4 Deblocking Filter

The deblocking filter (DF) is used to further improve the decoded picture quality by smoothing the block artifacts. Its function on improving picture quality is similar to the overlapped block motion compensation. The filter operations are performed across 8×8 block edges using a set of four pixels on both horizontal and vertical directions at the block boundaries such as shown in Figure 19.11. In the figure, the filtering process is applied to the edges. The edge pixels, A, B, C, and D, are replaced by A_1, B_1, C_1, and D_1 by the following operations:

$$B_1 = \text{clip}(B + d_1) \tag{19.7a}$$

$$C_1 = \text{clip}(C - d_1) \tag{19.7b}$$

$$A_1 = A - d_2 \tag{19.7c}$$

$$D_1 = D + d_2 \tag{19.7d}$$

$$d = (A - 4B + 4C - D)/8 \tag{19.7e}$$

$$d_1 = f(d, S) \tag{19.7f}$$

$$d_2 = \text{clip } d_1[(A - D)/4, d_1/2] \tag{19.7g}$$

where clip is a function of clipping the value to the range of 0–255, clip $d(x, d)$ is a function that clips x to the range of from $-d$ to $+d$, and the value S is a function of quantization step QUANT that is defined in Table 19.5.

TABLE 19.5
The Value S as a Function of Quantization Step (QUANT)

QUANT	1	2	3	4	5	6	7	8	9	10	11	12	13	14	15	16
S		1	1	2	2	3	3	4	4	4	5	5	6	6	7	7	7
QUANT	17	18	19	20	21	22	23	24	25	26	27	28	29	30	31	
S	8	8	8	9	9	9	10	10	10	11	11	11	12	12	12	

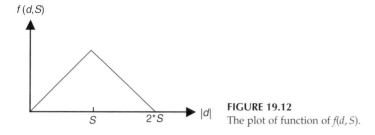

FIGURE 19.12
The plot of function of $f(d, S)$.

The function $f(d, S)$ is defined as

$$f(d, S) = \text{SIGN}(d)^* (\text{MAX}(0, \text{abs}(d)) - \text{MAX}(0, 2^* \text{abs}(d) - S)) \tag{19.8}$$

This function is described by Figure 19.12. From the figure, it is seen that this function is used to control the amount of distortion introducing by filtering. The filter has an effect only if d is smaller than $2S$. Therefore, some features such as an isolated pixel, corner, etc. would be reserved during the nonlinear filtering since for those features the value d may exceed the $2S$. The function $f(d, S)$ is also designed to ensure that small mismatch between encoder and decoder will remain small and will not allow the mismatch to be propagated over multiple pictures. For example, if the filter is simply switched on or off with a mismatch of only $+1$ or -1 for d, then this will cause the filter to be switched on at the encoder and off at the decoder, or vice versa. It should be noted that the DF proposed here is an optional selection. It is a result of a large number of simulations; it may be effective for some sequences but may not be effective for all kinds of video sequences.

19.4.2.5 Slice-Structured Mode

A slice contains a video picture segment. In the coding syntax, a slice is defined as a slice header followed by consecutive MBs in scanning order. The slice-structured (SS) mode is designed to address the needs of mobile video and other unreliable transmission environments. This mode contains two submodes: the rectangular slice (RS) submode and the arbitrary slice ordering (ASO) submode. In the rectangular submode, a slice contains a rectangular region of a picture, such that the slice header specifies the width. The MBs in this slice are in scan order within the rectangular region. In the ASO submode, the slices may appear in any order within the bitstream. The arbitrary arrangement of slices in the picture may provide an environment for obtaining better error concealment. The reason is that the damaged area caused by packet loss may be isolated from each other and can be easily concealed by the good decoded neighboring blocks. In this submode, there is usually no data dependency that can cross the slice boundaries, except for the DF mode because the slices may not be decoded in the normal scan order.

19.4.2.6 Reference Picture Selection

With optional mode of the reference picture selection (RPS), the encoder is allowed to use a modified interframe prediction method. In this method, additional picture memories are used. The encoder may select one of the picture memories to suppress the temporal error propagation due to the interframe coding. The information to indicate which picture is selected for prediction is included in the encoded bitstream that is allowed by syntax. The strategy used by encoder to select the picture to be used for prediction is open for algorithm design. This mode can use the backward channel message that is sent from a decoder to an encoder to inform the encoder which part of which pictures have been correctly decoded.

The encoder can use the message from the backward channel to decide which picture will provide better prediction. From the above description of RPS mode, it becomes evident that this mode is useful for improving the performance over unreliable channels.

19.4.2.7 Independent Segmentation Decoding

The independent segmentation decoding (ISD) mode is another option of H.263 video coding which can be used for unreliable transmission environment. In this mode, each video picture segment is decoded without the presence of any data dependencies across slice boundaries or across GOB boundaries, i.e., with complete independence from all other video picture segments and all data outside the same video picture segment location in the reference pictures. This independence includes no use of motion vectors outside of the current video picture segment for motion prediction or remote motion vectors for overlapped motion compensation in the advanced prediction mode, no DF operation and no linear interpolation across the boundaries of current video picture segment.

19.4.2.8 Reference Picture Resampling

The reference picture resampling (RPR) mode allows an earlier-coded picture to be resampled, or wrapped, before it is used as a reference picture. The idea of using this mode is similar to the idea of global motion, which is expected to obtain better performance of motion estimation and compensation. The wrapping is defined by four motion vectors for the corners of the reference picture as shown in Figure 19.13.

For the current picture with horizontal size H and vertical size V, four conceptual motion vectors, MV_{OO}, MV_{OV}, MV_{HO}, and MV_{HV} are defined for the upper left, low left, upper right, and low right corners of the picture, respectively. These motion vectors as wrapping parameters have to be coded with VLC and included in the bitstream. These vectors are used to describe how to move the corners of the current picture to map them onto the corresponding corners of the earlier decoded pictures as shown in Figure 19.13. The motion compensation is performed using bilinear interpolation in the decoder with the wrapping parameters.

19.4.2.9 Reduced-Resolution Update

When encoding a video sequence with highly active scenes, the encoder may have a problem to provide sufficient subjective picture quality at low bit rate coding. The reduced-resolution update (RRU) mode is expected to be used in this case for improving the coding performance. This mode allows the encoder to send update information for a picture that is encoded at a reduced resolution to create a final image at the higher resolution. At the encoder, the pictures in the sequence are first down-sampled to a quarter size (half in both horizontal and vertical directions) and then the resulting low-resolution pictures are encoded as shown in Figure 19.14.

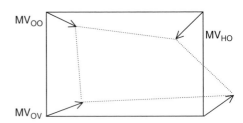

FIGURE 19.13
Reference picture resampling.

ITU-T Video Coding Standards H.261 and H.263

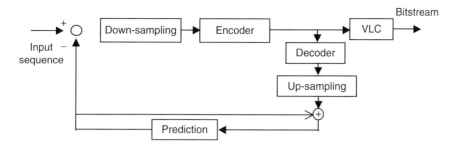

FIGURE 19.14
Block diagram of encoder with reduced-resolution update (RRU) mode.

The decoder with this mode is more complicated than one without this mode. The block diagram of decoding process with the RRU mode is shown in Figure 19.15.

The decoder with RRU mode has to deal with several new issues. First, the reconstructed pictures are up-sampled to the full size for display. However, the reference pictures have to be extended to the integer times of 32×32 MBs if it is necessary. The pixel values in the extended areas take the values of original border pixels. Second, the motion vectors for 16×16 MBs in the encoder are used for the up-sampled 32×32 MB in the decoder. Therefore, an additional procedure is needed to reconstruct the motion vectors for each up-sampled 16×16 MBs including chrominance MBs. Third, bilinear interpolation is used for up-sampling in the decoder loop. Finally, in the boundary of reconstructed picture, a block boundary filter is used along the edges of the 16×16 reconstructed blocks at the encoder as well as on the decoder. There are two kinds of block boundary filters that have been proposed. One is the DF, described in Section 19.4.2.4. The other is defined as follows. If two pixels, A and B, are neighboring pixels and A is in block 1 and B is in block 2, respectively, then the filter is designed as

$$A_1 = (3*A + B + 2)/4 \tag{19.9a}$$

$$B_1 = (A + 3*B + 2)/2 \tag{19.9b}$$

where A_1 and B_1 are the pixels after filtering and "/" is division with truncation.

FIGURE 19.15
Block diagram of decoder with reduced-resolution update (RRU) mode.

19.4.2.10 Alternative Inter VLC and Modified Quantization

The alternative inter-VLC (AIV) mode is developed for improving coding efficiency of inter-picture coding for the pictures containing significant scene changes. This efficiency improvement is obtained by allowing some VLC codes originally designed for intra-picture to be used for inter-picture coefficients. The idea is very intuitive and simple. When the rapid scene change occurs in the video sequence, the inter-picture prediction becomes difficult. This results in large prediction differences, which are similar to the intra-picture data. Therefore, the use of intra-picture VLC tables instead of using inter-picture tables may yield better results. However, there is no syntax definition for this mode. In other words, the encoder may use the intra VLC table for encoding an inter-block without informing the decoder. After receiving all coefficient codes of a block, the decoder will first decode these code words with the inter VLC tables. If the addressing of coefficients stays inside the 64 coefficients of a block, the VLD will accept the results even if some coding mismatch exists. Only if coefficients outside the block are addressed, the code words will be interpreted according to the intra VLC table. The modified quantization mode is designed for providing several features, which can improve the coding efficiency. First, with this mode more flexible control of the quantizer step can be specified in the dequantization field. The dequantization field is no longer a 2-bit fixed-length field; it is a variable-length field which can either be 2 or 6 bits depending on the first bit. Second, in this mode, the quantization parameter of the chrominance coefficients is different from the quantization parameter of the luminance coefficients. The chrominance fidelity can be improved by specifying a smaller quantization step for chrominance than that for luminance. Finally, this mode allows the extension of range of coefficient values. This provides more accuracy representation of any possible true coefficient value with the accuracy allowed by the quantization step. However, the range of quantized coefficient levels is restricted to those which can reasonably occur to improve the detectability or errors and minimize decoding complexity.

19.4.2.11 Supplemental Enhancement Information

The usage of supplemental information may be included in the bitstream in the picture layer to signal enhanced display capabilities or to provide tagging information for external usage. This supplemental enhancement information includes full-picture freeze/freeze–release request, partial-picture freeze/freeze–release request, resizing partial-picture freeze request, full-picture snapshot tag, partial-picture snapshot tag, video time segment start/end tag, progressive refinement segment start/end tag, and chroma key information. The full-picture freeze request is used to indicate that the contents of the entire earlier displayed video picture will be kept and not updated by the contents in current decoded picture. The picture freeze will be kept under this request until the full-picture freeze–release request occurs in the current or later picture-type information. The partial-picture freeze request indicates that the contents of a specified rectangular area of the earlier displayed video picture are frozen until the release request is received or timeout occurs. The resizing partial-picture freeze request is used to change the specified rectangular area for the partial picture. One use of this information is to keep the contents of picture in the corner of display unchanged for a period for commercial use or some other purpose. All information given by the tags indicates that the current picture is labeled as either a still-image snapshot or a subsequence of video data for external usage. The progressive refinement segment tag is used to indicate the display period of the pictures with better quality. The chroma keying information is used to request transparent and semitransparent pixels in the decoded video pictures [chen 1997]. One application of chroma key is to simply describe the shape information of objects in a video sequence.

19.5 H.263++ Video Coding and H.26L

H.263++ is the next version of H.263 that is considering adding more optional enhancements to H.263. It is the extension of H.263 version 2 and is currently scheduled be completed late in the year 2000. H.26L is a project to seek more efficient video coding algorithms that will be much better than the current H.261 and H.263 standards, where the L stands for long term. The algorithms for H.26L can be fundamentally different from the current DCT with motion compensation framework that is used for H.261, H.262 (MPEG-2), and H.263. The expected improvements from the current standards include several aspects: higher coding efficiency, more functionality, low complexity permitting software implementation, and enhanced error robustness. H.26L addresses very low bit rate, real-time, and low end-to-end delay applications. The potential application targets can be Internet video phones, sign language or lip reading communications, video storage and retrieval service, multipoint communication, and other visual communication systems. H.263L is currently scheduled for approval in the year 2002.

19.6 Summary

In this chapter, the video coding standards for low bit rate applications have been introduced. These standards include H.261, H.263, H.263 version 2, and the versions under development, H.263++ as well as H.26L. H.261 and H.263 are extensively used for video conferencing and other multimedia applications at low bit rates. In H.263 version 2, all new negotiable coding options are developed for special applications. Among these options, five options that include AIC mode, alternative inter VLC mode, modified quantization mode, DF mode, and improved PB-frame mode, are intended to improve coding efficiency. Three modes that include SS mode, RPS mode, and independent segment decoding mode are used to meet the need of mobile video application. The others provide the functionality of scalability, such as spatial, temporal, and SNR scalability. H.26L is a future standard to meet the requirements of very low bit rate, real-time, low end-to-end delay applications, and other advanced performance.

Exercises

1. What is the enhancement of H.263 over H.261? Describe the applications of each enhanced tool of H.263.
2. Compare with MPEG-1 and MPEG-2, which features of H.261 and H.263 are used to improve coding performance at low bit rates? Explain the reasons.
3. What is the difference between spatial scalability and RRU mode in H.263 video coding?
4. Conduct a project to compare the results by using the DFs in coding loop and out of coding loop. Which method will cause less drift if a large number of pictures are contained between two consecutive I-pictures?

References

[chen 1997] T. Chen, C.T. Swain, and B.G. Haskell, Coding of sub-regions for content-based scalable video, *IEEE Transactions on Circuits and Systems for Video Technology*, 7, 1, 256–260, February 1997.

[h261] ITU-T Recommendation H.261, Video Codec for Audiovisual Services at px64 kbit/s, March 1993.

[h263] ITU-T Recommendation H.263, Video Coding for Low Bit Rate Communication, Draft H.263, May 2, 1996.

[h263+] ITU-T Recommendation H.263, Video Coding for Low Bit Rate Communication, Draft H.263, January 27, 1998.

[weigand 1996] T. Weigand, Rate-distortion optimized mode selection for very low bit-rate video coding and the emerging H.263 standard, *IEEE Transactions on Circuits and Systems for Video Technology*, 6, 2, 182–190, April 1996.

20

A New Video Coding Standard: H.264/AVC

Many video coding standards have been developed during past two decades. In this chapter, we introduce a recently developed video coding standard, H.264 or MPEG-4 Part 10 Advanced Video Coding (AVC) [h264], which has been developed and standardized collaboratively by the joint video team (JVT) of ISO/IEC MPEG and ITU-T VCEG. The main objective of H.264 is for high coding efficiency. The test results have shown that it has made an important milestone of video coding standard at the coding efficiency improvement.

20.1 Introduction

Several video coding standards have been introduced in the previous chapters including MPEG-1/2/4 and H.261 as well as H.263. Recently, the JVT of ISO/IEC MPEG and ITU-T VCEG (Video Coding Expert Group) has developed new video coding standard, which is referred to formally as ITU-T Recommendation H.264 and ISO/IEC MPEG-4 (Part 10) Advanced Video Coding (referred in short as H.264/AVC). The work of H.264/AVC actually started in early 1998 when the VCEG issued a call for proposals for a project called H.26L. The target of H.26L is to greatly improve the coding efficiency over any existing video coding standards. The first draft of H.26L was completed in October 1999. The JVT was formed in December 2001, with the mission to finalize the new coding standard based on H.26L. The draft of new video coding standard was submitted for formal approval as the final committee draft (FCD) in March 2003 and promoted to final draft international standard (FDIS) in June 2003. The current MPEG-4 AVC standard itself is ISO/IEC 14496-10:2004. The FRExt enhancements have also received final approval as ISO/IEC 14496-10:2004/AMD 1.

The H.264/AVC mainly targets for high coding efficiency. Based on the conventional block-based motion compensated (MC) hybrid video coding concepts H.264/AVC provides approximately a 50% bit rate savings from equivalent perceptual quality relative to the performance of earlier standards. This has been shown from extensive simulation results. The superior coding performance of H.264/AVC is obtained because many new features such as enhanced prediction capability and smaller block size motion compensation are incorporated. The details of these features are described in the following sections. With high coding efficiency, the H.264/AVC can provide technical solutions for many applications, including broadcasting over different medias, video storage on optical and magnetic devices, high definition (HD) DVD, and others. Though, it will not be that easy to replace the current existing standard such as MPEG-2 with H.264/AVC in some applications such as in the area of the digital televisions. However, it may be used for new application areas, such as HD-DVD, mobile video transmission, and others. From other side, to achieve the high coding efficiency, H.264/AVC has to use a lot of new tools or modified tools from existing standards that substantially increases the complexity of the codec; it would be about four times higher for the decoder and nine times higher for the encoder compared with MPEG-2 video coding standard. However, with fast advances of semiconductor technique, the silicon solution can alleviate the problem of high complexity.

20.2 Overview of H.264/AVC Codec Structure

To address the variety of applications and networks, the H.264/AVC codec design consists of two layers: video coding layer (VCL) and network abstraction layer (NAL). The layered structure of H.264/AVC video encoder is shown in Figure 20.1.

From Figure 20.1, the input of video source is first compressed in the VCL into a bitstream. The function of the VCL is to efficiently compress the video content. The NAL is a new concept, which is designed for efficient transmission of the compressed bitstream in different network or storage environment which includes all current and future protocols and network architectures. These applications include broadcasting over terrestrial, cable, and satellite networks; streaming over IP-networks, wireless, and ISDN channels. In this layer, the head information is added to the coded bitstream for handling a variety of transport layers or storage media. The interface of NAL is designed to enable a seamless integration of the coded video data with all possible protocols and network architectures.

The bitstream can be in one of two formats: the NAL unit (NALU) stream or the byte stream. The NALU stream format consists of a sequence of syntax structures called NALUs. The format of an NALU is shown in Figure 20.2.

In the header of an NALU, the first bit is a 0 bit, and the next 2 bits are used to indicate whether the NALU contains the sequence or picture parameter set or a slice of a reference picture. The next 5 bits are used to indicate type of NALU units, which corresponds the type of data being carried in that NALU unit. There are total 32 types of NALUs allowed. The 32 types of NALUs can be classified in two categories: VCL NALUs and non-VCL NALUs. The VCL units carry the data corresponding to the VCL, whereas the non-VCL NALUs carry information like supplemental enhancement information (SEI), sequence and picture parameter set, access unit delimiter, and others. The detail can be found in the specification of H.264/AVC [h264].

In the NALU stream, the NALUs are decoded on the decoding order. The byte stream can be constructed from the NALU stream by ordering the NALUs in decoding order and adding a start code to each NALU and zero or more zero-valued bytes to form a stream of bytes. The NALU stream can be extracted from the byte stream by removing the start code which has the unique start code prefix pattern within this byte stream. The NALU is a syntax structure containing an indication of the type of data to follow

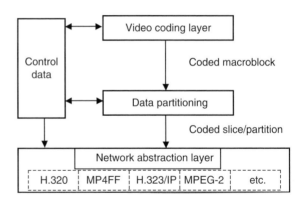

FIGURE 20.1
Layered structure of H.264/AVC video encoder.

FIGURE 20.2
Network abstraction layer NAL unit (NALU) format.

and bytes containing that data in the form of a raw byte sequence payload (RBSP) interspersed as necessary with emulation prevention bytes. The emulation prevention byte is a byte equal to 0×03 that may be present within an NALU. The presence of emulation prevention bytes ensures that no sequence of consecutive byte-aligned bytes in the NALU contains a start code prefix, which is a unique sequence of three bytes equal to 0×000001 embedded in the byte stream as a prefix to each NALU. The location of a start code prefix can be used by a decoder to identify the beginning of a new NALU and the end of a previous NALU. Emulation of start code prefixes is prevented within NALUs by the inclusion of emulation prevention bytes. An NALU specifies a generic format for use in both packet-oriented and bitstream systems. The format of NALUs for both packet-oriented transport and bitstream delivery is identical except that each NALU can be preceded by a start code prefix in a bitstream-oriented transport.

Compared to other existing video coding standards, the basic coding structure of H.264/AVC is similar, which is the structure with the MC transform coding (TC). The block diagram of H.264/AVC video encoder is shown as in Figure 20.3.

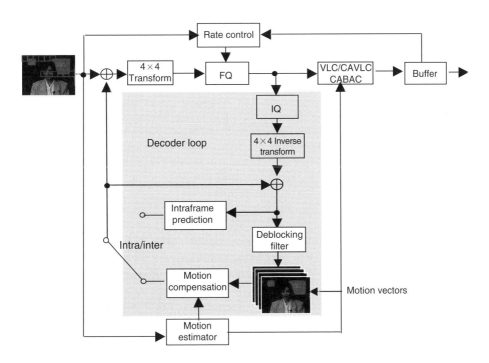

FIGURE 20.3 (See color insert following page 288.)
Block diagram of H.264 encoder.

Except many common tools, the H.264/AVC includes many highlighted features that are enabled to greatly improve the coding efficiency and increase the capability of error robustness and the flexibility for operation over a variety of network environments. Features for improving coding efficiency can be classified to two parts: the first is to improve the accuracy of prediction for the picture to be encoded and the second includes the method of transform and entropy coding. Several tools have been adopted in H.264/AVC to improve inter- and intra-prediction, which are briefly summarized as follows.

Variable block size for motion compensation with small block sizes is used, in which total seven selections of block sizes are used for motion compensation in H.264/AVC among those the smallest block size for luma motion compensation can be as small as 4×4.

The quarter-pel accurate motion compensation is adopted in H.264/AVC. The quarter-pel accurate motion compensation has been used in the advanced profile of MPEG-4 Part 2, but H.264/AVC further reduces the complexity of the interpolation process.

Multiple reference pictures for motion compensation and weighted prediction are used to predict the P- and B-pictures. The number of reference pictures can be up to 15 for level 3.0 or lower and four reference pictures for levels higher than 3.0. When the multiple reference pictures are used for motion compensation prediction, the contribution of prediction from different references should be weighted and offset by amounts specified by the encoder. This can greatly improve coding efficiency for those scenes that contain fades.

Directional spatial prediction for intracoding is adopted for further improving coding efficiency. In this technique, the intracoded regions are predicted with the references of the previously coded areas, which can be selected from different spatial directions. In such a way, the edges of the previously decoded areas of the current picture can be extrapolated into the current intracoded regions.

Skip mode in P-picture and direct mode in B-picture are used to alleviate the problem for using too many bits for coding motion vectors in the interframe coding. H.264/AVC uses the skip mode for P-picture and direct mode for B-pictures. In these modes, the reconstructed signal is obtained directly from the reference frame with the motion vectors derived from previously encoded information by exploiting either spatial (for skip mode) or temporal (for direct mode) correlation of the motion vectors between adjacent macroblocks (MBs) or pictures. In such a way, bits saving for coding motion vectors can be achieved.

The use of loop deblocking filters (DFs) is another feature that is used to reduce the block artifacts and improve both objective and subjective video quality. The difference from MPEG-1/2 is that in H.264/AVC the DF is brought within the motion compensation loop, so that it can be used for improving the interframe prediction and therefore improving the coding efficiency.

H.264/AVC uses a small transform block size of 4×4 instead of 8×8 as in most video coding standards. The merit of using the small transform block size is able to encode the picture in a more local adaptive fashion, which would reduce the coding artifacts such as ringing noise. However, the problem of using small transform block size may cause coding performance degradation due to the correlations of large area may not be exploited for certain pictures. H.264/AVC uses two ways to alleviate this problem: one is by using a hierarchical transform to extend the effective block size of non-active chroma information to an 8×8 block, and another is by allowing encoder to select a special coding type of intracoding, which enables the extension of the length of the luma transform for non-active area to a 16×16 block size. As mentioned earlier, the basic function of integer transform used in H.264/AVC does not have equal norm. To solve this problem, quantization table size has been increased.

Two very powerful entropy coding methods, content-adaptive variable-length coding (CAVLC) and content-adaptive binary arithmetic coding (CABAC), are used in H.264/AVC for further improving coding performance.

In H.264/AVC, several tools have been adopted for increasing the capability of error robustness.

Flexible slice size allows encoder to adaptively select the slice size for increasing the capability of error robustness.

Flexible macroblock ordering (FMO) allows partitioning the MBs into slices in a flexible order. Because each slice is an independently decodable unit, the FMO can significantly enhance error robustness by managing the spatial relationship between the MBs in the slice.

There are also several features, which are used to increase the flexibility for operation over a variety of network environments.

The parameter set structure is used to provide more flexible way to protect the key header information and increase the error robustness.

The NALU syntax structure allows for carrying video content in a manner appropriate for each specific network in a customized way.

Arbitrary slice ordering (ASO) is used to improve end-to-end delay in real-time application, particularly for the applications on the Internet protocol networks.

Switching P (SP)- and switching I (SI)-slices are new slice types. They are specially encoded-slices that allow efficient switching between video bitstreams and efficient random access for video decoders. This feature can be used for efficiently switching a decoder to decode different bitstreams with different bit rates, recovery from errors and trick modes.

An overview of H.264/AVC video coding standard can be found in [wiegand 2003] and the detailed specification can be found in [h264]. The technical details of above tools will be described in the following sections.

20.3 Technical Description of H.264/AVC Coding Tools

In the Section 20.2, we briefly described the features of H.264/AVC video coding standard. In this section we introduce the technical details of some of those important features.

20.3.1 Instantaneous Decoding Refresh Picture

It is well known that in the previous MPEG video coding standards, the input video sequence is organized into groups of pictures (GOPs). Each GOP consists of three types of frames or pictures, which are intracoded (I) frame or picture, predictive-coded (P) frame or picture, and bidirectionally predictive-coded (B) frame or picture. As in MPEG-2, we use the word "picture" instead of the word "frame" to provide a more general discussion, because a picture can either be a frame or a field. It should be noted that there is no I-, P-, B-picture concept in the H.264/AVC. There are only slice types which can be I-, P-, or B-slices. Therfefore, strictly speaking, there is no such thing as an I-picture in the H.264/AVC video standard. The term is not used. However, a picture can contain I-, P-, or B-slices in any combination.

To satisfy requirements for some applications, the H.264/AVC video coding standard has specified a new picture type, instantaneous decoding refresh (IDR) picture. Its exact definition is a coded picture in which all slices are I- or SI-slices that causes the decoding process to mark all reference pictures as "unused for reference" immediately after decoding the IDR picture. This means that after the decoding of an IDR picture, all following coded pictures in decoding order can be decoded without inter prediction from any picture decoded before the IDR picture. The first picture of each coded video sequence is an IDR picture.

Based on this definition, the primary difference between IDR picture of H.264 and I-picture of MPEG-2 is that for H.264 after sending an IDR picture, the encoder cannot use any pictures that preceded the IDR picture (in decoding order) as references for the inter prediction of any pictures that follow the IDR picture (in decoding order). So the presence of an IDR picture in H.264/AVC is roughly similar to the presence of an MPEG-2 GOP header in which the closed_gop flag is set to 1. The closed_gop flag is a 1-bit flag, which indicates the nature of the predictions used in the first consecutive B-pictures (if any) immediately following the first coded I-frame following the group of picture header. The closed_gop is set to "1" to indicate that these B-pictures have been encoded using only backward prediction or intracoding. The presence of a H.264/AVC "I- picture" that is not an IDR picture but contains all I-slices is similar to either an MPEG-2 I-picture without a GOP header or to the presence of an MPEG-2 GOP header in which the closed_gop flag is equal to 0. Also, the presence of an IDR picture in H.264/AVC causes a reset of the PicOrderCount and frame_num counters of the decoding process, while an I-picture that is not an IDR picture does not. Therefore, it should be noted that an IDR picture is a more severe event than an MPEG-2 I-picture, as it prohibits open GOP behavior, that means the references for inter prediction have to be within a GOP. In an open GOP, the reference pictures from the previous GOP at the current GOP boundary can be exploited. For example the GOP is open when B-pictures at the start of a GOP rely on I- or P-pictures from the immediately previous GOP.

20.3.2 Switching I-Slices and Switching P-Slices

Except the new concept of IDR picture introduced in the Section 20.3.1, H.264 has additional new types of slices: SP- and SI-slices [karczewisz 2003]. The main purpose of SP- and SI-slices is to enable efficient switching between video streams and efficient random access for video decoders. Video streaming is an important application over IP networks and 3G wireless networks. However, due to varying network conditions, the effective bandwidth to a user may vary accordingly. Therefore, the video server should scale the bit rate of the compressed video streams to accommodate the bandwidth variations. There are several ways to achieve bitstream scaling such as video transcoding, but the simplest way for real-time application is to generate several separate pre-encoded bitstreams for the same video sequence with different bit rates, of course at different quality levels at the same time. The server can then dynamically switch from higher rate bitstream to the lower rate bitstream when the network bandwidth drops. This can be described in Figure 20.4. In Figure 20.4, we assume that each frame is encoded as a single slice type and predicted from one reference. Also assume that stream A is coded with higher bit rate and stream B is coded with lower bit rate. After decoding P-slices A_1 and A_2 in stream A, the decoder wants to switch to stream B and decode B_3, B_4, and so on. In this case, it is obvious that the B_3 has to be coded as an I-slice. If B_3 is coded as a P-slice then the decoder will not have the correct decoded reference pictures required to reconstruct B_3 because B_3 is predicted from the decoded frame B_2, which does not exist in stream A. Therefore, the bitstream switching can be accomplished by inserting an I-slice at regular intervals in the coded sequence to create switching points. However, an I-slice does not exploit any temporal redundancy and it likely requires much more bits to be coded than a P-slice. This would result as a peak in the coded bitstream at each switching point. To address this problem, the SP-slices are proposed to support switching without the increased bit rate penalty of I-slices.

A New Video Coding Standard: H.264/AVC

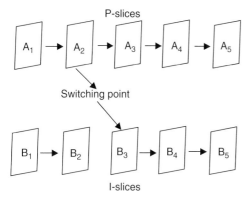

FIGURE 20.4
A decoder is decoding stream A and wants to switch to decoding stream B.

The idea behind SP-slice is that assume we encode a video sequence with different encoding parameters and generate multiple independent streams with different bit rates. For simplicity, assume we have two streams A and B as in Figure 20.4. We use the same scenario for bitstream switching. After decoding P-slices A_1 and A_2 in stream A, the decoder wants to switch to stream B and decode B_3, B_4, and so on. The SP-slices are placed at the switching points as shown in Figure 20.5.

The key point is that the SP-slice AB_3 is encoded as P-slice with B_3 as input picture and reconstructed A_2 as predictive reference. The encoding and decoding procedure of AB_3 is shown in Figure 20.6.

It is clear that the SP-slice will not result in a peak in the bitstream since it is coded using MC prediction as a P-slice, which is more efficient than intracoding. From Figure 20.5, it is shown that the SP-slice AB_3 can be decoded using reference frame A_2. It should be noted that the decoder output picture B_3 is identical whether decoding B_2 is followed by B_3 or A_2 is followed by AB_3. If we want to switch the bistream in other direction, another SP-slice, BA_3, would be required. But this is still more efficient than encoding frames A_3 and B_3 as I-slices. However, the SI-slice may be used for switching from one sequence to a completely different sequence, where the MC prediction is not efficient due to significant scene changes.

It should be indicated that the SP- and SI-slices are not only used for stream switching, they can also be used for error-resilience video coding. The feature of SP- and SI-slices can be exploited in the adaptive intra refresh mechanism.

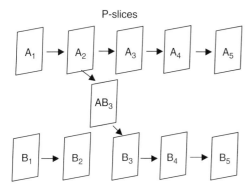

FIGURE 20.5
Switching streams using switching P (SP)-slices.

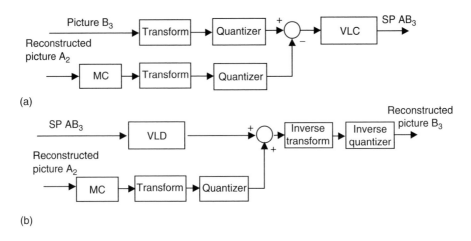

FIGURE 20.6
(a) Switching P (SP)-slice encoding and (b) switching P (SP)-slice decoding.

20.3.3 Transform and Quantization

The previous video coding standards, such as MPEG-1/2, MPEG-4 Part 2, JPEG, H.261, and H.263, all use 8×8 discrete cosine transform (DCT) as the basic transform. In H.264/AVC, the block size used in TC is 4×4. There are several questions that should be answered why the H.264/AVC chooses 4×4 integer transform. The first question is the selection of block size of 4×4 instead of 8×8 as in most previous video coding standards. In general, the larger block size could be better for exploiting the global correlations to increase the coding efficiency. From other side, the smaller block size could be better for exploiting the adaptivity accoring to the local activity in content and also it is obvious that the complexity of implementation is greatly reduced. In addition, the smaller block size could more adaptively march the motion compensation with variable block size used in H.264/AVC, the smallest block size for motion compensation is 4×4.

In the H.264 video, three transforms have been used for three different applications, which include 4×4 Hadamard transform for the 4×4 luma DC coefficients in intra MBs predicted in 16×16 mode, 2×2 transform for 2×2 of chroma DC coefficients in any MB, and 4×4 integer transform for 4×4 blocks for the lum residual data. The matrices of 4×4 (luma) and 2×2 (chroma) DC coefficients are formed as in Figure 20.7.

As shown in Figure 20.7, for the 16×16 intramode there are in total sixteen 4×4 blocks. After 4×4 transform, 16 luma DC coefficeints are extracted to form a 4×4 block, which is

FIGURE 20.7
Matrix formation for DC coefficients of luma and chroma.

A New Video Coding Standard: H.264/AVC

coded by a 4 × 4 Hadamard TC. Accordingly, the chroma DC coefficients are used to form a 2 × 2 block, which is coded by a 2 × 2 Hadamard TC. The function of these two DC TC is to remove the spatial redundancy among sixteen 4 × 4 neighboring blocks. This two-level transform is referred to as a hierarchical transformation, which aims at higher coding efficiency and lower complexity. The two transforms used to code luma and chroma DC coefficients are as follows:

$$H_L = \begin{bmatrix} 1 & 1 & 1 & 1 \\ 1 & 1 & -1 & -1 \\ 1 & -1 & -1 & 1 \\ 1 & -1 & 1 & -1 \end{bmatrix} \quad H_C = \begin{bmatrix} 1 & 1 \\ 1 & -1 \end{bmatrix}$$

The most important transform used in H.264/AVC is the 4 × 4 integer transform which is referred to as high correlation transform (HCT) [cham 1983; hallapuro 2002]. The 4 × 4 HCT is applied to 4 × 4 predicted residual blocks. The forward transform is represented in the matrix format as follows:

$$[H_f] = \cdot \begin{bmatrix} 1 & 1 & 1 & 1 \\ 2 & 1 & -1 & -2 \\ 1 & -1 & -1 & 1 \\ 1 & -2 & 2 & -1 \end{bmatrix}$$

This matrix is an integer approximation of 4 × 4 DCT. The inverse transform is represented by

$$[H_i] = \begin{bmatrix} 1 & 1 & 1 & 1/2 \\ 1 & 1/2 & -1 & -1 \\ 1 & -1/2 & -1 & 1 \\ 1 & -1 & 1 & -1/2 \end{bmatrix}$$

It can be seen that the HCT used in H.284/AVC is not orthogonal due to the approximation of DCT, which can be found from the following:

$$[H_f] \cdot [H_i] = \begin{bmatrix} 1 & 1 & 1 & 1 \\ 2 & 1 & -1 & -2 \\ 1 & -1 & -1 & 1 \\ 1 & -2 & 2 & -1 \end{bmatrix} \cdot \begin{bmatrix} 1 & 1 & 1 & 1/2 \\ 1 & 1/2 & -1 & -1 \\ 1 & -1/2 & -1 & 1 \\ 1 & -1 & 1 & -1/2 \end{bmatrix} = \begin{bmatrix} 4 & 0 & 0 & 0 \\ 0 & 5 & 0 & 0 \\ 0 & 0 & 4 & 0 \\ 0 & 0 & 0 & 5 \end{bmatrix}$$

Therefore, in the inverse transform of decoding, all scale factors resulting from this operation have to be compensated by the quantization process.

The H.264 uses a scalar quantizer. As mentioned earlier, we use integer transform to avoid division and floating-point arithmetic, we need to add rescaling function in the inverse quantization. The detailed procesure can be found in [hallapuro 2002].

20.3.4 Intraframe Coding with Directional Spatial Prediction

In the new video coding standard, H.264/AVC, a new intraframe technique based on the directional spatial prediction has been adopted. The basic idea of this technique is to predict the MBs to be coded as intra with the previously coded regions selected from proper spatial direction in the same frame. The merit of directional spatial prediction is able

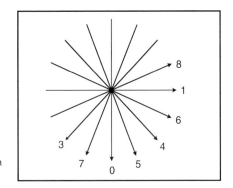

FIGURE 20.8
Eight predictive directions for intra 4×4 prediction in H.264/AVC.

to extrapolate the edges of previously decoded parts of the current picture to the MBs to be coded. This can greatly improve the accuracy of the prediction and improve the coding efficiency. For the 4×4 intramode, in addition to DC prediction, there are total eight prediction directions as shown in Figure 20.8. For the 16×16 intramode, there are four prediction modes: vertical, horizontal, DC, and plan prediction. For the technical detail, please refer to [wiegand 2003].

20.3.5 Adaptive Block Size Motion Compensation

In many video coding standards, an MB consisting of a 16×16 block of luma pixels and two corresponding blocks of chroma pixels is used as the basic processing unit of the video decoding process. An MB can be further partitioned for inter prediction. The selection of the block size using for inter prediction partitions is a compromised result between the bits saving provided by using motion compensation with smaller blocks and the increased number of bits needed for coding motion vectors. In MPEG-4 there is an advanced motion compensation mode. In this mode, the inter prediction process can be preformed with adaptive selection of 16×16 or 8×8 block. The purpose of the adaptive selection of the matching block size is to further enhance coding efficiency. The coding performance may be improved at low bit rate because the bits for coding prediction difference could be greatly reduced at the limited extra cost for increasing motion vectors. Of course, if the cost for coding motion vectors becomes too high, the mode for using small block size will not be selected. The decision made in the encoder should be very careful. If the 8×8 prediction is chosen, there are four motion vectors for the four 8×8 luminance blocks in an MB will be transmitted. The motion vectors for coding two chrominance blocks are then obtained by taking an average of these four motion vectors and dividing the average value by a factor of 2. As each motion vector for the 8×8 luminance block has half-pixel accuracy, the motion vector for the chrominance block may have a sixteenth-pixel accuracy. The issues of motion estimation process in the encoder and the selection of whether to use inter prediction for each region of the video content are not specified in the standards. The encoding issues are usually described in the informative parts of the standards. In the recent developed MPEG and ITU joint standard, H.264/AVC, the 16×16 MB, is further partitioned into even small blocks as shown in Figure 20.9.

In Figure 20.9, it can be seen that in total eight kinds of blocks can be used for adaptive selection of motion estimation/compensation. With optimal selection of motion compensation mode in the encoder, coding efficiency can be greatly improved for some sequences. Of course, this is again an encoding issue, and an optimal mode selection algorithm is needed.

A New Video Coding Standard: H.264/AVC

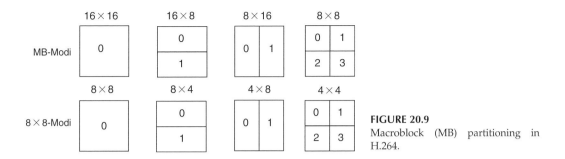

FIGURE 20.9
Macroblock (MB) partitioning in H.264.

20.3.6 Motion Compensation with Multiple References

As mentioned in Section 20.3.1, in most standards three picture types, I-, P-, and B-pictures have been defined. Also, usually no more than two reference frames have been used for motion compensation. In the recently developed new standard, H.264/AVC, a proposal for using more than two reference frames has been adopted. The comparison of H.264/AVC with MPEG-2/4 about the reference frames is shown in Figure 20.10.

The number of reference frames of H.264 can be up to 15 frames. The major reason for using multiple reference frames is to improve the coding efficiency. It is obvious that the better matching would be found by using multiple reference frames than using less or equal than two frames in the motion estimation. Such an example is shown in Figure 20.11.

MPEG-2/4 Part 2 motion estimation cannot get better reference always.

20.3.7 Entropy Coding

The H.264/AVC video standard specifies two types of entropy coding: CAVLC and CABAC. The major function of both schemes is to improve the coding performance; but among these two schemes, the former has less complexity and the latter has more complicated algorithm. As we know in the previous video coding standards, such as MPEG-2 and MPEG-4 Part 2, the fixed variable-length coding (VLC) method is used for coding each syntax element or sets of syntax elements. The VLCs are designed with the statistical characteristics of each syntax element under assumption that the statistical characteristics are closely matching the video data to be coded and also they are stationary. However, this is not true in practice; for example, the statistical behavior of the predictive residues in an MC code is nonstationary and highly depends on the video content and the accuracy of the prediction model. In the CAVLC of H.264/AVC, a total number of 32 different VLCs are used. Most of these VLCs are tables; however, some of VLCs enable simple online calculation of any code word with no need of storing the code tables.

Further, the CAVLC is simpler than CABAC, it becomes the baseline entropy coding for H.264/AVC. In the CAVLC scheme, inter-symbol redundancies are used by switching

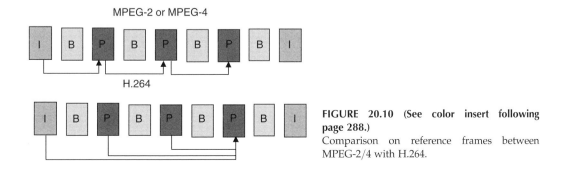

FIGURE 20.10 (See color insert following page 288.)
Comparison on reference frames between MPEG-2/4 with H.264.

FIGURE 20.11 (See color insert following page 288.)
An example to explain the benefit by using multiple reference frames, it is noted that the better reference can be obtained by using multiple reference pictures for the video sequences with periodic changes.

VLC tables for different syntax components depending on the history of transmitted coding symbols. The basic coding tool in the CAVLC is the Exp-Golomb codes (Exponential Golomb codes). The Exp-Golomb codes are VLCs, which consist of a prefix part (1, 01, 001, ...), and a suffix part that is a set of bits (x_0, x_1x_0, $x_2x_1x_0$, ...) where x_i is a binary bit. The code word structure is represented in Tables 20.1 through 20.3.

It can be seen that in Table 20.1, the structure of code word can be represented as

$$[M\ 0s][1][INFO],$$

where INFO is an M-bit suffix part carrying information. Each Exp-Golomb code word can be constructed by its index code_num as follows:

$$M = \text{Log}_2(\text{code\_num} - 1)$$
$$\text{INFO} = \text{code\_num} + 1 - 2^M$$

There are three ways of mapping in the Exp-Golomb coder. The first is the unsigned direct mapping, ue(v), which is used for coding MB type, reference frame index and others. In this

TABLE 20.1

Code Word Structure with Prefix and Suffix

Code Word	Range
1	0
$01x_1$	1–2
$001x_1x_0$	3–6
$0001x_2x_1x_0$	7–14
$00001x_3x_2x_1x_0$	15–30
$000001x_4x_3x_2x_1x_0$	31–62

TABLE 20.2

Exp-Golomb Code Words

Code Word	Code_num
1	0
010	1
011	2
00100	3
00101	4
00110	5

mapping, code_num = v. The second is the signed mapping, se(v), which is used for motion vector difference, delta quantizer parameter (QP), and others. The mapping is described in Table 20.3; the relation between syntax element value (v) and code_num is

$$\text{code\_num} = 2|v|, \quad \text{for } v < 0;$$
$$\text{code\_num} = 2|v| - 1, \quad \text{for } v > 0.$$

In the third mapping, me(v) is the mapped symbol. In this mapping, the parameter v is mapped to code_num according to the table specified in the standard [h264]. The basic principle of these mappings is to produce the code words according to the statistics, i.e., the shorter code words are used to encode the components with higher probability and longer code words are used to code the components with small probability.

The CAVLC is the method used to encode the predictive residual, zigzag ordered 4×4 (and 2×2) blocks of transform coefficients (TCOEFF). In the CAVLC, there is no end of block (EOB) code such as in MPEG-2. For a given 4×4 block after prediction, transformation, and quantization, the statistical distribution shows that only few coefficients have significant values and many coefficients have its magnitude equal to 1. An example of a typical block is shown as below:

TABLE 20.3

Mapping for Signed Exp-Golomb Code Words

Code_num	Syntax Element Value (v)
0	0
1	1
2	−1
3	2
4	−2
5	3

After zigzag reordering, the coefficients can be given as

$$0, 3, 0, 1, -1, -1, 0, 1, 0, 0, 0, 0, 0, 0, 0, 0.$$

In the CAVLC, the number of nonzero quantized coefficients (which are noted as TotalCoeff) and the actual value and position of the coefficients are encoded separately. The coefficients with value equal to 1 are called as trailing 1's (T1). For this example, the TotalCoeff = 4, T1s = 3.

At the first step of encoding, a coeff_token is used to code the number of coefficients and trailing 1's. There are four look-up tables used for encoding coeff_token. These tables are described as Num-VLC0, Num-VLC1, Num-VLC2, and Num-FLC (three VLC tables and an FLC). To use the correlation between neighboring blocks (context-based adaptive), the choice of table depends on the number of nonzero coefficients in upper and left-hand side of the previously coded blocks N_u and N_L. The parameter N is defined as follows: if blocks U and L are available (i.e., in the same coded slice), $N = (N_u + N_L)/2$; if only block U is available, $N = N_U$; if only block L is available, $N = N_L$; if neither is available, $N = 0$. After N is decided, the look-up table for coding coeff_token can be decided:

If $N = 0$ or 1, Num-VLC0 is selected; if $N = 2$ or 3, Num-VLC1 is selected; if $N = 4, 5, 6$, or 7, Num-VLC2 is selected; finally if $N \geq 8$, the Num-FLC is selected.

When you check the tables in the standard, you can find that Num-VLC0 is used for small numbers of coefficients; the short codes are assigned to low values of TotalCoeffs (0 and 1) and the long codes are used for large value of TotalCoeffs. Num-VLC1 is used for medium numbers of coefficients (TotalCoeff values around 2–4 are assigned relatively short codes), Num-VLC2 is used for higher numbers of coefficients and FLC assigns a fixed 6 bit code to every value of TotalCoeff.

The second step is to encode the sign of each T1. At this step, 1 bit is used to code the sign of each T1 with the order from highest frequency.

The third step is to encode the levels of the remaining nonzero coefficients. The choice of VLC table to encode each level adapts depending on the value of each successive coded level, and the encoding is reverse order, i.e., starting with the highest frequency toward the DC coefficient. For this step, we have seven VLC tables to choose from, starting from Level_VLC0 for coding lower level value and to Level_VLC1 for encoding slightly higher values and so on. The way how to select the look-up table depends on the threshold values such as the Level_VLC0 will be initially used unless there are more than 10 nonzero coefficients and less than three trailing ones, in which case start with Level_VLC1. Then we encode the highest-frequency nonzero coefficient. If the value of this coefficient is larger than a predefined threshold, move up to the next VLC table. In this way, the choice of level is matched to the value of the recently encoded coefficients. The thresholds are listed in Table 20.4; the first threshold is zero, which means that the table is always incremented after the first coefficient level has been encoded.

TABLE 20.4

Thresholds for Determining Whether to Increment Level Table Number

Current Variable-Length Coding (VLC) Table	Threshold to Increment Table
VLC0	0
VLC1	3
VLC2	6
VLC3	12
VLC4	24
VLC5	48
VLC6	N/A (highest table)

The fourth step is to encode the total number of zeros before the last coefficient. TotalZeros is the sum of all zeros preceding the highest nonzero coefficient in the zigzag or alternative reordered array. The reason used to separate VLC table to encode TotalZeros is that many blocks contain a number of nonzero coefficients at the start of the array and this approach means that zero-runs at the start of the array need not be encoded.

The fifth step is to encode each run of zeros, run_before. The run_before is the number of consecutive zero-valued quantized TCOEFF in the reverse scan order starting from the last nonzero-valued coefficient. For each block, run_before specifies zero-runs before the last nonzero coefficient.

Now we use the above example to explain how to perform the encoding.

- Reordered block: 0, 3, 0, 1, −1, −1, 0, 1, 0, ...
- TotalCoeffs = 5
- TotalZeros = 3
- T1s = 3 (in fact there are four trailing ones but only three can be encoded as a special case)

The encoding procedure is described in the Table 20.5.

The final result of transmitted bitstream for this block is 0000100011100101111101101. For typical test conditions and test sequences, the CAVLC can obtain 2%–7% saving in bit rate compared with conventional VLC scheme based on a single Exp-Golomb code.

The CAVLC is a simple and efficient entropy coding method. However, this cannot provide adaptation to the actually conditional symbol statistics which limits its performance. Furthermore, the symbols with probabilities higher than 0.5 cannot be efficiently coded with CAVLC as those symbols appear to be coded with a high fractional accuracy in bits, whereas VLC is limited with lower accuracy of 1 bit/symbol. As mentioned earlier, another entropy coding scheme adopted in H.264/AVC is CABAC. The CABAC achieves better performance than CAVLC with 10%–15% average bit rate saving at the cost of increasing the complexity. The main reasons for obtaining better performance include the following factors. The first reason is that CABAC is to select probability model for each

TABLE 20.5

Encoding Procedure of Content-Adaptive Variable-Length Coding (CAVLC)

Element	Value	Code
coeff_token	TotalCoeffs = 5, T1s = 3	0000100
T1 sign (4)	+	0
T1 sign (3)	−	1
T1 sign (2)	−	1
Level (1)	+1 (use Level_VLC0)	1
Level (0)	+3 (use Level_VLC1)	0010
TotalZeros	3	111
run_before(4)	ZerosLeft = 3; run_before = 1	10
run_before(3)	ZerosLeft = 2; run_before = 0	1
run_before(2)	ZerosLeft = 2; run_before = 0	1
run_before(1)	ZerosLeft = 2; run_before = 1	01
run_before(0)	ZerosLeft = 1; run_before = 1	No code required; last coefficient

syntax element according to the element's context. The second reason is to adapt probability estimation based on local statistics in CABAC. Finally, the CABAC is to use arithmetic coding, which could reach high fractional accuracy in bits.

The CABAC encoder consists of four steps: binarization, context modeling, arithmetic coding, and probability updating. In the step of binarization, only the nonbinary-valued syntax elements, such as TCOEFF and motion vectors, are uniquely mapped to a binary number, the so-called bin string. These binary-valued syntax elements will bypass this step. The reason for the need of the binarization step is to reduce the alphabet size of the syntax elements, which would result in fast and accurate estimation of conditional probabilities, subsequently minimize the computational complexity involved in performing each elementary operation of probability estimation and subsequent arithmetic coding. In this step four basic schemes of binarization and its derivatives have been used: unary, truncated unary (TU), kth-order Exp-Golomb (EGk), and fixed-length (FL) binarization schemes. From these four basic binarization schemes, three more binarization schemes are derived by concatenation. The first is a concatenation of a 4 bit FL prefix as a representation of the luminance related part of coded block pattern (CBP) and a TU suffix with $S=2$ representation of the chrominance part of CBP. The second and third concatenation schemes are derived from the TU and EGk binarization. The detail of these schemes can be found in [marpe 2003].

The context model is the conditional probability for one or more bins of the binarized symbols. At the step of context modeling selection, a context model is assigned to the given symbols from a selection of available models depending on the statistics of recently coded symbols.

The third step is the arithmetic encoding. At this step, an arithmetic coder is used to encode each bin according to the selected probability model. The encoding is performed with recursive subdivision to fractional accuracy of an existing interval; in general, the initial interval is the range from 0 to 1.

Finally, the selected context model is updated based on the actual coded value. Moreover, the statistical model determines the code and its efficiency; it is very important to choose an adequate model that explores the statistical dependencies of recently coded symbols.

20.3.8 Loop Filter

It is similar to other video coding standards, the block artifacts could be introduced in H.264/AVC video coding since its coding scheme is block based. The most significant block artifact in H.264 is caused by 4×4 integer transformation in intra- and interframe predictive residue coding followed by quantization. The course quantization of TCOEFF would result in visible discontinuities at the block boundaries. Also, for intercoded blocks, the MC reference may not be perfect and if there are not enough bits to code the predictive residues; this would cause the edge discontinuities for the blocks to be compensated. There are two ways to reduce the block artifacts. The first is the post-filtering, which operates on the display buffer and outside of the coding loop. The post-filtering is an optional for the decoder and it is not a normative part of the standard. In H.264/AVC, the DF is a normative part of the standard; it has to be in both encoder and decoder. The DF has been used in the coding loop to every decoded MB to reduce blocking artifacts. In the coding loop, the filter operation is applied after the inverse transform before reconstructing and storing the MB for future predictions in the encoder and before reconstructing and displaying the MB in the decoder. The use of DF wants to reach two main goals. The first goal is to smooth block edges and improve the appearance of decoded images, particularly

FIGURE 20.12
Boundaries in a macroblock (MB) to be filtered (dark lines represent the block boundaries to be filtered).

at higher compression ratios. The second goal is to reduce the predictive residue for MC prediction of further frames in the encoder. It should be noted that the intraprediction is carried out using unfiltered reconstructed MBs to form the prediction, though intracoded MBs are filtered. Filtering is applied to vertical or horizontal edges of 4 × 4 blocks in an MB as shown in Figure 20.12.

In Figure 20.12, the block boundaries in an MB are filtered in the following orders. First, the vertical boundaries are filtered from left to right, and then the vertical boundaries are filtered from top to bottom. This is the same for both luma and chroma. Each filtering operation is applied to a set of pixels at either side of the block boundary; total eight pixels across vertical or horizontal boundaries of the block are involved as shown in Figure 20.13. Figure 20.13 shows four pixels on either side of a vertical or horizontal boundary in adjacent blocks p and r (p_0, p_1, p_2, p_3 and q_0, q_1, q_2, q_3). Depending on the current quantizer, the coding modes of neighboring blocks, and the gradient of image samples across the boundary, several outcomes are possible, ranging from no filtering at all to filtering of all eight pixels.

The decision of whether the filtering operation would be conducted depends on the boundary strength and the gradient of image samples across the boundary. The boundary strength parameter B_S is derived based on MB type, motion vectors, reference picture ID, and MB coding parameters. B_s is defined as follows.

If pixels of p and q are intracoded and they are MB boundary, B_s is assigned to 4, which means that the strongest filtering is needed; if pixels of p and r are intracoded but they are not the MB boundary, B_s is assigned to 3. If neither pixels of p or q are intracoded but these pixels contain coded coefficients, then B_s is equal to 2; if pixels of p and q have different reference pictures or a different number of reference or different motion vector values, B_s is set to 1; finally, neither pixels p or q are intracoded; neither p or q contain coded coefficients and p and q have same reference picture as well as identical motion vectors, B_s is equal to 0.

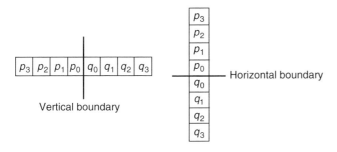

FIGURE 20.13
Boundary adjacent pixels would be involved in filtering operation.

The value of B_s indicates the strength of filtering process performed on the block boundaries including a selection between the three filtering modes. From the rule of value assignment to B_s, it can be seen that the filtering operation is stronger at places where there is likely to be significant blocking distortion, such as the boundary of an intracoded MB or a boundary between blocks that contain coded coefficients.

In cases of $B_s = 0$, the filtering process is not conducted for the current 4×4 block boundary. For $B_s > 0$, the filtering operation will be conducted and the strength of filtering depends on the difference between boundary pixels, and threshold values of α and β. As we know that the block artifacts are most visible in very smooth areas where the pixel values do not change much across block boundaries. Therefore, the filtering threshold values should be derived based on the pixel values. The values of α and β are defined in [list 2003]. In general, the values of α and β increase with the average QP of the two neighboring blocks p and q. When $B_s > 0$, and $|p_0 - q_0|$, $|p_1 - p_0|$, and $|q_1 - q_0|$ are all less than the thresholds α or β, then $(p_2, p_1, p_0, q_0, q_1, q_2)$ are filtered. The reason for switching filtering operation on when the difference or gradient is low can be described as follows. When QP is small and the difference or gradient of pixels across boundary is likely to be due to image features rather than blocking effects. In this case, the gradient should be preserved and so the thresholds α and β are low. When QP is larger, blocking distortion is likely to be more significant; α and β are higher so that the chance for switching filtering on is higher.

The detailed filtering operation is as follows. For $0 < B_s < 4$, a 4-tap linear filter is applied with inputs p_1, p_0, q_0, and q_1 producing filtered outputs P_0 and Q_0. In addition, if $|p_2 - p_0|$ is less than threshold α, a 4-tap linear filter is applied with inputs p_2, p_1, p_0, and q_0, producing filtered output P_1. If $|q_2 - q_0|$ is less than threshold β, a 4-tap linear filter is applied with inputs q_2, q_1, q_0, and p_0, producing filtered output Q_1. It should be noted that p_1 and q_1 are never filtered for chroma, but only for luma data.

For $B_s = 4$, if $|p_2 - p_0| < a$ and $|p_0 - q_0| < \mathrm{round}(b/4)$, P_0 is produced by 5-tap filtering of p_2, p_1, p_0, q_0, and q_1; P_1 is produced by 4-tap filtering of p_2, p_1, p_0, and P_2 (luma only) is produced by 5-tap filtering of p_3, p_2, p_1, p_0, and q_0; otherwise P_0 is produced by 3-tap filtering of p_1, p_0, and q_1. If $|q_2 - q_0| < a$ and $|p_0 - q_0| < \mathrm{round}(b/4)$, Q_0 is produced by 5-tap filtering of q_2, q_1, q_0, p_0, and p_1; Q_1 is produced by 4-tap filtering of q_2, q_1, q_0, and p_0, Q_2 (luma only) is produced by 5-tap filtering of q_3, q_2, q_1, q_0, and p_0; otherwise Q_0 is produced by 3-tap filtering of q_1, q_0, and p_1.

20.3.9 Error-Resilience Tools

Although coding efficiency is the most important aspect in the design of any video coding scheme, the transmission of compressed video through noisy channels has always been a key consideration. This is evident by many error-resilience tools that are available in video coding standards, such as in MPEG-2, MPEG-4 Part 2, and some only in H.264/AVC.

The first category of error-resilience tools is the localization. These tools are used to remove the spatial and temporal redundancy between segments of the video to prevent error propagation. It is well known that video compression efficiency is achieved by exploiting the redundancy in both the spatial and temporal dimensions of the video. Due to the high correlation within and among neighboring frames, predictive coding schemes are employed to exploit this redundancy. Although the predictive coding schemes are able to reach high compression ratios, they are highly susceptible to the propagation of errors. Localization techniques essentially break the predictive coding loop so that if an error does occur then it is not likely to affect other parts of the video. Obviously, a high degree of localization will lead to lower compression efficiency. There are two methods for localization of errors in a coded video: spatial localization and temporal localization. The spatial localization technique is supported in MPEG-2 and H.264/AVC using slices, and in MPEG-4 using video packets. The resynchronization marker insertion is suitable to provide a spatial localization of errors. The temporal localization is usually implemented for preventing error propagation with the insertion of intracoded MBs by decreasing the temporal dependency in the coded video sequence. Although this is not a specific tool for error resilience, the technique is widely adopted and recognized as being useful for this purpose. The higher percentage of intrablocks used for coding the video will reduce the coding efficiency, but reduce the impact of error propagation on successively coded frames. In the most extreme case, all blocks in every frame are coded as intrablocks. In this case, there will be no temporal propagation of errors, but a significant increase in bit rate could be expected. The selection of intracoded blocks may be cyclic, in which the intracoded blocks are selected according to a predetermined pattern; the intracoded blocks may also be randomly or adaptively chosen according to content characteristics.

The second category of error-resilience tools is the data partitioning. It is well known that every bit in a compressed video bitstream is not of equal importance. Some bits belong to segments defining vital information, such as picture types, quantization values, etc. When coded video bitstreams are transported over error-prone channels, errors in such segments cause a much longer lasting and severe degradation on the decoded video than that caused by errors in other segments. Therefore, data partition techniques have been developed to group together coded bits according to their importance to the decoding such that different groups may be more effectively protected using unequal protection techniques. For example, during the bitstream transmission over a single channel system, the more important partitions can be better protected with stronger channel codes than the less important partitions. Alternatively, with a multichannel system, the more important partitions could be transmitted over the more reliable channel. This kind of tool is defined in MPEG-2 and MPEG-4 Part 2 video but not in H.264/AVC.

The third category is redundant coding. This category of techniques tries to enhance error resilience by adding redundancy to the coded video. The redundancy may be added explicitly, such as the concealment motion vectors, or implicitly in the coding scheme, as in the reversible variable-length codes (RVLC) and multiple description (MD) coding.

All these strategies for error resilience indirectly lead to an increase in the bit rate and loss of coding efficiency, where the overhead with some is more than others. In the following, we describe each tool in terms of the benefit it provides for error-resilient transmission, as well as its impact on coding efficiency.

In the H.264/AVC, several new error-resilience tools, which are different from previous standards, MPEG-2 or MPEG-4 Part 2, have been adopted. These tools include FMO, ASO, and redundant slices. The idea of FMO is to specify a pattern that allocates the MBs in a picture to one or several slice groups not in normal scanning order, but in a flexible way. In such a way, the spatially consecutive MBs are assigned to different slice groups. Each slice group is transmitted separately. If a slice group is lost, the image pixels in spatially neighboring MBs that belong to other correctly received slice groups can be used for efficient error concealment. The allowed patterns of FMO range from rectangular patterns to regular scattered patterns, such as checkerboards, or completely random scatter patterns. Furthermore, the idea of FMO can be extended to the slice level. In some profiles of the H.264/AVC standard the slices can be sent in an arbitrary order to increase the capability of error resilience. The slices can also be bundled into slice groups, which may contain one or more slices. The exact number of slices is specified by a parameter in the picture parameter set.

20.4 Profiles and Levels of H.264/AVC

In this section, we would like to give brief description about H.264/AVC profiles. It is the same as in MPEG-2 and MPEG-4; a profile defines a set of coding tools or algorithms, which are used in generating a compliant bitstream with this profile. If a decoder is claimed to conform a specific profile, it must support all tools and algorithms in that profile.

20.4.1 Profiles of H.264/AVC

There are total seven profiles defined in the H.264/AVC so far, which are the baseline, main, extended, high, high 10, high 4:2:2, and high 4:4:4. The process is carried on to replace the High 4:4:4 Profile with a better one which is the Advanced High 4:4:4 Profile. In the following we briefly introduce each of these profiles.

The Baseline Profile supports all following features in H.264/AVC:

- Support I- and P-slice types, but no B-slice type.
- NALU streams do not contain the coded slices, which is a non-IDR picture.
- Sequence parameter sets contain the parameters such that every coded picture of the coded video sequence is a coded frame containing only frame MBs.
- Support 4:2:0 chroma format, 8 bit luma and chroma pixels.
- The TCOEFF decoding process and picture construction process before DF process shall not use the transform bypass operation.
- Use only flat quantization matrix; it means that all components in the matrix are 16.
- Weighted prediction shall not be applied to P- and SP-slices and the default weighted prediction specified in [h264] shall be applied to B-slices.
- Entropy coding uses Exp-Golomb codes or CAVLC and does not support CABAC.
- Support FMO.
- Use only 4×4 transformation, same quantization matrix specified in sequence level and no quantization offset.
- No interlacing support.
- No SP/SI-slices and slice data partitioning.

Also, there are several flags and their combinations used to define the conformance of a bitstreams to the Baseline Profile. The detail can be found in the specification [h264].

The Main Profile supports the following features:

- Support I-, P-, and B-slices.
- NALU streams do not contain the coded slices, which is a non-IDR picture.
- Not support ASO and FMO.
- Support 4:2:0 chroma format, 8 bit luma and chroma pixels.
- Support interlacing.
- All slices of the picture belong to the same slice group.
- No slice partitioning.
- Use only 4×4 transformation, same quantization matrix specified in sequence level and no quantization offset.

The features of the Extended Profile include

- Support I-, P-, and B-slices, and interlaced tools.
- Support 4:2:0 chroma format, 8 bit luma and chroma pixels.
- Entropy coding uses Exp-Golomb codes or CAVLC and does not support CABAC.
- Support FMO and ASO.
- Support SI-, SP-slices, and data partitioning.
- Use only 4×4 transformation, same quantization matrix specified in sequence level and no quantization offset.

In summary, the Main Profile supports all features except SP/SI-slices, slice partitioning, FMO, ASO, and redundant pictures. The Extended Profile supports all features except CABAC. Therefore, these three profiles are not the subset for each other and target for different applications. The Baseline Profile is used for videophone, mobile communication, and low delay applications. The Main Profile targets interlaced video, broadcast, and packaged media applications. The Extended Profile mainly targets streaming video and wireless transmission applications. Compared with MPEG-2 video, the Main Profile of H.264/AVC can provide 50%–100% improvement on the coding efficiency but increase about 2.5–4 times of decoder complexity.

Also several profiles have been defined in H.264/AVC for professional applications. These profiles include High Profile, High 10 Profile, High 4:2:2 Profile, and Advanced 4:4:4 Profile.

High Profile is a superset of Main Profile. The main difference with Main Profile is that the High Profile supports both the 8×8 and 4×4 transformations, which can improve the coding performance at high bit rates for high definition television (HDTV) sequences. High Profile has also the tools for enabling to change the quantization parameter offset differently for two chroma components for better subjective quality.

High 10 Profile is defined for 10 bit video sequences. The applications of this profile include medical image sequences and some other high quality image sequences encoding.

The Advanced 4:4:4 Profile is under development which is used to replace the current High 4:4:4 Profile with better coding performance and more functionality. The main features compared with the original High 4:4:4 Profile is that the three color components are encoded with the same tools which can improve the coding performance for the chromas.

20.4.2 Levels of H.264/AVC

As we discussed in the Section 20.4.1, a profile specifies a subset of the entire bitstream syntax of the standard for certain applications. However, within the specified limitations imposed by a given profile, it may still be possible to require a very large variation in the processing power and memory size of encoders and decoders. These variations depend upon the factors, such as the picture size, the frame rate, the maximum bit rate, and the maximum number of reference frames. Therefore, within a profile, we have to define the levels, which specify a set of constrains on values. The following table shows the levels specified in the standard of H.264/AVC. A decoder compliant with a specified profile and level must be able to decode the bitstreams compliant to that profile and level as well as those bitstreams with levels lower than the specified level. The detail of the level definition can be found in the H.264 specification and shown as follows.

TABLE A.20.1

Level Limits

Level Number	Typical Picture Size	Typical Frame Rate	Max Video Bit Rate (kbits/s)	Vertical MV Component Range (Luma Frame Samples)	Maximum Number of Reference Frames	Max Number of Motion Vectors per Two Consecutive MBs
1	Quarter-common intermediate format (QCIF)	15	64	[−64, +63.75]	4	—
1b	QCIF	30	128	[−64, +63.75]	4	—
1.1	Quarter video graphics array (QVGA) (320 × 240)	10	192	[−128, +127.75]	3	—
	QCIF	30			9	
1.2	Common intermediate format (CIF)	15	384	[−128, +127.75]	6	—
1.3	CIF	30	768	[−128, +127.75]	6	—
2	CIF	30	2,000	[−128, +127.75]	6	—
2.1	HHR					
	(352 × 480)	30	4,000	[−256, +255.75]	7	—
	(352 × 576)	25				
2.2	SD					
	(720 × 480)	30	4,000	[−256, +255.75]	6	—
	(720 × 576)	25				
3	SD					
	(720 × 480)	30	10,000	[−256, +255.75]	6	32
	(720 × 576)	25				
	Video graphics array (VGA) (640 × 480)	30				
3.1	1280 × 720P	30	14,000	[−512, +511.75]	5	16
	Super video graphics array (SVGA) (800 × 600)	56				
	1280 × 720P	60				
3.2	4VGA (1280 × 960)	45	20,000	[−512, +511.75]	4	16
4	HD					
	(1280 × 720P)	60	20,000	[−512, +511.75]	9	16
	(1920 × 1080I)	30			4	
	(2K × 1K)	30			4	
4.1	High definition (HD) formats					
	(1280 × 720)	60	50,000	[−512, +511.75]	9	16
	(1920 × 1080)	30			4	
4.2	1920 × 1080	60	50,000	[−512, +511.75]	4	16
5	2K × 1K	72	135,000	[−512, +511.75]	14	16
	16VGA	30			5	
5.1	2K × 1K	120	240,000	[−512, +511.75]	16	16
	4K × 2K	30			5	

HHR: half horizontal resolution.
SD: standard definition.

20.5 Summary

In this chapter, the new video coding standard, MPEG-4 Part 10 AVC standard, or H.264, which is jointly developed by JVT of MPEG and ITU-T VCEG, has been introduced. The H.264/AVC is an efficient and state-of-the-art video compression standard, where coding efficiency is about two times better than that of MPEG-2. The H.264/AVC has been planed for many applications including HD-DVD, DTV for satellite and wireless networks, IPTV, and many others.

Exercises

1. Indicate at least three new tools, which make H.264/AVC to have a better coding performance. Explain why? If it is possible, conduct computer simulations to verify it.
2. What are the entropy coding schemes used in H.264/AVC? Explain each of them.
3. Describe the principle of the DF of H.264/AVC. Conduct a simulation experiment to compare the subjective quality of decoded images between with and without the DF.
4. What are the new tools, which are different from previous MPEG video for increasing the error robustness of H.264 video coding scheme? Give explanation.
5. Describe the principle of the integer transform in H.264. Why integer transformation is adopted by H.264 video? What is the problem of integer transformation and how to solve this problem in H.264?
6. Describe the intraprediction algorithm used in H.264.

References

[cham 1983] W.K. Cham, Family of order-4 four-level orthogonal transforms, *Electronic Letters*, 19, 21, 869–871, October 1983.

[h264] ITU-T Rec. H.264/ISO/IEC 11496–10, Advanced video coding for generic audiovisual services, February 28, 2005.

[hallapuro 2002] A. Hallapuro, M. Karczewicz, and H. Malvar, Low complexity transform and quantization—Part I: Basic implementation, JVT-B38, Joint Video Team of ISO/IEC MPEG and ITUT VCEG, January 2002.

[karczewisz 2003] M. Karczewisz and R. Kurceren, The SP- and SI-frames design for H.264/AVC, *IEEE Transactions on Circuits Systems for Video Technology*, 13, 7, 637–644, July 2003.

[list 2003] P. List, A. Joch, J. Lainema, G. Bjøntegaard, and M. Karczewicz, Adaptive deblocking filter, *IEEE Transactions on Circuits Systems for Video Technology*, 13, 614–619, July 2003.

[marpe 2003] D. Marpe, H. Schwarz, and T. Wiegand, Context-adaptive binary arithmetic coding in the H.264/AVC video compression standard, *IEEE Transactions on Circuits Systems for Video Technology*, 13, 620–636, July 2003.

[wiegand 2003] T. Wiegand, G.J. Sullivan, G. Bjontegaard, and A. Luthra, Overview of the H.264/AVC video coding standard, *IEEE Transactions on Circuits and Systems for Video Technology*, 13, 7, 560–576, July 2003.

21

MPEG System: Video, Audio, and Data Multiplexing

In this chapter, we present the methods and standards of how to multiplex and synchronize the MPEG coded video, audio, and other data into a single bitstream, or multiple bitstreams for storage and transmission.

21.1 Introduction

ISO/IEC MPEG has completed work on the ISO/IEC 11172 and 13818 standards known as MPEG-1, MPEG-2, and MPEG-4 Part 2 as well as Part 10, respectively, which deal with the coding of digital audio and video signals. As mentioned in the previous chapters, the MPEG-1, 2, and 4 standards are designed as a generic standard and as such are suitable for use in a wide range of audiovisual applications. The coding part of the standards convert the digital visual, audio, and data signals to the compressed formats that are represented as binary bits. The task of MPEG system is focused on multiplexing and synchronizing the coded audio, video, and data into a single bitstream or multiple bitstreams. In other words, the digital compressed video, audio, and data all are first represented as binary formats which are referred to as bitstreams, and then the function of system is to mix the bitstreams from video, audio, and data together. For this purpose, several issues have to be addressed by the system part of the standard:

- Distinguishing different data, such as audio, video, or other data
- Allocating bandwidth during muxing
- Reallocating or decoding the different data during demuxing
- Protecting the bitstreams in error-prone media and detecting the errors
- Dynamically multiplexing several bitstreams

Additional requirements for the system should include extensibility issues such as

- New service extensions should be possible
- Existing decoders should recognize and ignore data they cannot understand
- The syntax should have extension capacity

It should also be noted that all system-timing signals are included in the bitstream. This is the big difference between digital systems and traditional analog systems in which the timing signals are transmitted separately. In this chapter, we introduce the concept of systems and give detailed explanations for existing standards such as MPEG-2. However, we will not go through the standard page by page to explain the syntax; we pay more attention to those core parts of the standard and the parts, which always cause confusion during the implementation. One of the key issues is system timing. For MPEG-4, we give a presentation of the current status of the system part of the standards.

21.2 MPEG-2 System

The MPEG-2 system standard is also referred to as ITU-T Rec. H.222.0/ISO/IEC 13818-1 [mpeg2 system]. The ISO document gives very detailed description of this standard. A simplified overview of this system is shown in Figure 21.1.

The MPEG-2 system coding is specified in two forms: the transport stream and the program stream. Each form is optimized for a different set of applications. The audio and video data are first encoded by audio and video encoder, respectively. The coded data is the compressed bitstreams, which follow the syntax rules specified by the video coding standard 13818-2 and audio coding standard 13818-3. The compressed audio and video bitstreams are then packetized to the packetized elementary streams (PES). The video PES and audio PES are coded by system coding to the transport stream or program stream according to the requirements of the application.

The system coding provides a coding syntax which is necessary and sufficient to synchronize the decoding and presentation of the video and audio information, at the same time it also has to ensure that data buffers in the decoders do not overflow and underflow. Of course, the buffer regulation is also considered by the buffer control or rate control mechanism in the encoder. The video, audio, and data information are multiplexed according to the system syntax by inserting time stamps for decoding, presenting, and delivering the coded audio, video, and other data. It should be noted that both program stream and transport stream are packet-oriented multiplexing. Before we explain these streams, we first give a set of parameter definitions used in the system documents. Then, we describe the overall picture regarding the basic multiplexing approach for single video and audio elementary streams.

FIGURE 21.1
Simplified overview of system layer scope.

21.2.1 Major Technical Definitions in MPEG-2 System Document

In this section, the technical definitions that are often used in the system document are provided. The major packet- and stream-related definitions are given.

Access unit: a coded representation of a presentation unit. In the case of audio, an access unit is the coded representation of audio frame. In the case of video, an access unit indicates all the coded data for a picture, and any stuffing that follows it, up to but excluding the start of the next access unit. In other words, the access unit begins with the first byte of the first start code. Except for the end of sequence, all bytes between the last byte of the coded picture and the sequence end code belong to the access unit.

DSM-CC: digital storage media command and control.

Elementary stream (ES): a generic term for one of the coded video, coded audio, or other coded bitstreams in PES packets. One ES is carried in a sequence of PES packets with one and only one stream identification. This implies that one ES can only carry the same type of data such as audio or video.

Packet: a packet consists of a header followed by a number of contiguous bytes from an elementary data stream.

Packet identification (PID): a unique integer value used to associate ESs of a program in a single or multiprogram transport stream. It is a 13 bit field, which indicates the type of data stored in the packet payload.

PES packet: the data structure used to carry ES data. It contains a PES packet header followed by PES packet payload.

PES: a PES consists of PES packets, all whose payloads consist of data from a single ES, and all which have the same stream identification. Specific semantic constraints apply.

PES packet header: the leading fields in the PES packet up to and excluding the PES packet data byte fields. Its function will be explained in the section of syntax description.

System target decoder (STD): a hypothetical reference model of a decoding process used to describe the semantics of the MPEG-2 system-multiplexed bitstream.

Program-specific information (PSI): PSI includes normal data that will be used for demultiplexing of programs in the transport stream by decoders. One case of PSI, the nonmandatory network information table, is privately defined.

System header: the leading fields of program stream packets.

Transport stream packet header: the leading fields of program stream packets.

The following definitions are related to the timing information:

Time stamp: a term that indicates the time of a specific action such as the arrival of a byte or the presentation of presentation unit.

System clock reference (SCR): a time stamp in the program stream from which decoder timing is derived.

Elementary stream clock reference (ESCR): a time stamp in the PES from which decoders of PES may derive timing information.

Decoding time stamp (DTS): a time stamp that may be presented in a PES packet header used to indicate the time when an access unit is decoded in the system target decoder.

Program clock reference (PCR): a time stamp in the transport stream from which decoder timing is derived.

Presentation time stamp (PTS): a time stamp that may be presented in PES packet header used to indicate the time that a presentation unit is presented in the system target decoder.

21.2.2 Transport Streams

The transport stream is a stream definition that is designed for communicating or storing one or more programs of coded video, audio, and other kind of data in lossy or noisy environments where significant errors may occur. A transport stream combines one or more programs with one or more time bases into a single stream. However, there are some difficulties with constructing and delivering a transport stream containing multiple programs with independent time bases such that the overall bit rate is variable. As in other standards, the transport stream may be constructed by any method that results in a valid stream. In other words, the standards just specify the system coding syntax. In this way, all compliant decoders can decode bitstreams generated according to the standard syntax. However, the standard does not specify how the encoder generates the bitstreams. It is possible to generate transport streams containing one or more programs from elementary-coded data streams, from program streams, or from other transport streams, which may themselves contain one or more program. An important feature of transport stream is that the transport stream is designed in such a way that makes the following operations become possible with minimum effort. These operations include several transcoding requirements, which include:

- Retrieve the coded data from one program within the transport stream, decode it, and present the decoded results. In this operation, the transport stream is directly demultiplexed and decoded. The data in the transport stream is constructed in two layers: a system layer and a compression layer. The system decoder decodes the transport streams and demultiplexes them to the compressed video and audio streams that are further decoded to the video and audio data by the video decoder and the audio decoder, respectively. It should be noted that non-audio/video data is also allowed. The function of transport decoder included demultiplexing, depacketization, and other such as error detection that will be later explained in detail. This procedure is shown in Figure 21.2.
- Extract the transport stream packets from one program within the transport stream and produce as the output a new transport stream that contains only that one program. This operation can be seen as system layer transcoding that converts a

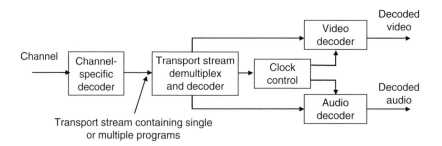

FIGURE 21.2
Example of transport demultiplexing and decoding.

transport stream containing multiple programs to a transport stream containing only a single program. In this case, the re-multiplexing operation may need the correction of PCR values to account for changes in the PCR locations in the bitstream.

- Extract the transport stream packets of one or more programs from one or more transport streams and produce as output of a new transport stream. This is another kind of transcoding that converts selected programs of one transport stream to a different one.
- Extract the contents of one program from the transport stream and produce as output another program stream. This is a transcoding that converts the transport program to a program stream for certain applications.
- Convert a program stream to a transport stream that can be used in a lossy communication environment.

To answer the question of how to define the transport stream and then make the above transcoding become simpler and more efficient, we describe the technical detail of the systems specification in the following section.

21.2.2.1 Structure of Transport Streams

As described earlier, the task of the transport stream coding layer is to allow one or more programs to be combined into a single stream. Data from each ES is multiplexed together with timing information, which is used for synchronization and presentation of the ES during decoding. Therefore, the transport stream consists of one or more programs such as audio, video, and data ES access units. The transport stream structure is a layered structure. All the bits in the transport stream are packetized to the transport packets. The size of transport packet is chosen to be 188 bytes; among those, 4 bytes are used as the transport stream packet header. In the first layer, the header of transport packets indicates whether the transport packet has an adaptation field. If there is no adaptation field, the transport payload may either consist of only PES packets or consist of both PES packets and PSI packets. Figure 21.3 illustrates the case of containing only PES packets.

If the transport stream carries both PES and PSI packets then the structure of transport stream as shown Figure 21.4 would result.

If the transport stream packet header indicates that the transport stream packet includes the adaptation field then the construction is shown in Figure 21.5.

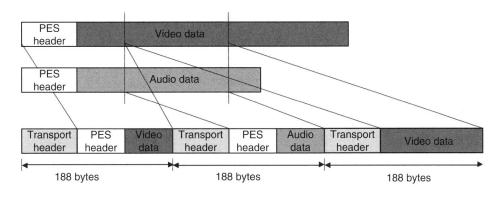

FIGURE 21.3
Structure of transport stream containing only packetized elementary streams (PES) packets.

512 *Image and Video Compression for Multimedia Engineering*

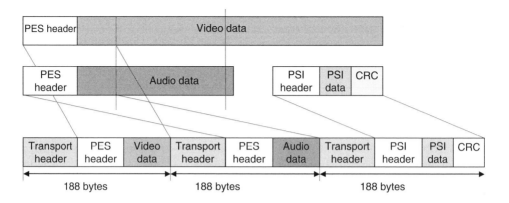

FIGURE 21.4
Structure of transport stream containing both packetized elementary streams (PES) packets and program specific information (PSI) packets.

In Figure 21.5, the appearance of the optional field depends on the flag settings. The function of adaptation field will be explained in the syntax section. Before we go ahead though, we give a little explanation regarding the size of the transport stream packet. More specifically, why is a packet size of 188 bytes chosen? Actually, there are several reasons. First, the transport packet size needs to be large enough so that the overhead due to the transport headers is not too significant. Second, the size should not be so large that the packet-based error correction code becomes inefficient. Finally, the size 188 bytes is also compatible with ATM packet size which is 47 bytes, then one transport stream packet is equal to the four ATM packets. So the size of 188 bytes is not a theoretical solution but a practical and compromised solution.

21.2.2.2 Transport Stream Syntax

As we indicated, the transport stream is a layered structure. To explain the transport stream syntax, we start from the transport stream packet header. Because the header part is very important, it is the highest layer of the stream. We describe it in more detail. For the rest, we do not repeat the standard document and just indicate the important parts that we think there may cause some confusion for the readers. The detail of other parts that are not covered here can be found from the MPEG standard document [mpeg2 system].

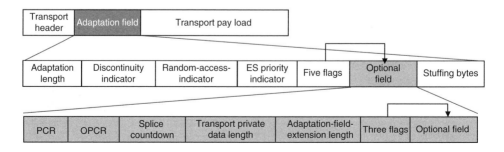

FIGURE 21.5
Structure of transport stream header, which contains adaptation field.

21.2.2.2.1 Transport Stream Packet Header

This header contains four bytes that are assigned as eight parts:

Syntax	No of Bits	Mnemonic
sync_byte	8	bslbf
transport_error_indicator	1	bslbf
payload_unit_start_indicator	1	bslbf
transport_priority	1	bslbf
PID	13	uimsbf
transport_scrambling_control	2	bslbf
adaptation_field_control	2	bslbf
continuity_counter	4	uimsbf

The mnemonic in the above table means
bslbf—bitstream left bit first
unimsbf—unsigned integer, most significant bit first

21.2.2.2.2 Syntax No of Bits Mnemonic

- The sync_byte is a fixed 8 bit field whose value is 0100 0111 (hexadecimal 47 = 71).
- The transport_error_indicator is a 1 bit flag, when it is set to 1, it indicates that at least 1 uncorrectable bit error exists in the associated transport stream packet. It will not be reset to 0 unless the bit values in error have been corrected. This flag is useful for error concealment purpose, because it indicates the error location. When an error exists, either resynchronization or other concealment method can be used.
- The payload_unit_start_indicator is a 1 bit flag that is used to indicate whether the transport stream packets carry PES packets or PSI data. If it carries PES packets then the PES header starts in this transport packet. If it contains PSI data then a PSI table starts in this transport packet.
- The transport_priority is a 1 bit flag, which is used to indicate that the associated packet is of greater priority than other packets having the same PID which do not have the flag bit set to 1. The original idea of adding a flag to indicate the priority of packets comes from video coding. The video elementary bitstream contains mostly bits that are converted from discrete cosine transform (DCT) coefficients. The priority indicator can set a partitioning point that can divide the data into a more important part and a less important part. The important part includes the header information and low frequency coefficients and the less important part includes only the high frequency coefficients that have less affect on the decoding and quality of reconstructed pictures.
- PID is a 13 bit field that provides information for multiplexing and demultiplexing by uniquely identifying which packet belongs to a particular bitstream.
- The transport_scrambling_control is a 2 bit flag. Here 00 indicates that the packet is not scrambled, the other three (01, 10, and 11) indicate that the packet is scrabbled by a user-defined scrambling method. It should be noted that the transport packet header and adaptation field (when it is present) should not be scrambled. In other words, only the payload of transport packets can be scrambled.
- The adaptation_field_control is a 2 bit indicator, which is used to inform whether there is an adaptation field present in the transport packet. The value 00 is reserved for future use. 01 indicates no adaptation field, 10 indicates that there is only an

adaptation field and no payload. Finally, 11 indicates that there is an adaptation field followed by a payload in the transport stream packet.
- The continuity_counter is a 4 bit counter which increases with each transport stream packet having the same PID.

From the header of the transport stream packet we obtain the information about future bits. There are two possibilities; if the adaptation field control value is 10 or 11 then the bits following the header are adaptation field, otherwise the bits are payload. The information contained in the adaptation field is described as follows.

21.2.2.2.3 Adaptation Field

The structure of the adaptation field data is shown in Figure 21.5. The functionality of these headers is basically related to the timing and decoding of the elementary bitsteam. Some important fields are explained below:

- Adaptation-field-length is an 8 bit field specifying the number of bytes immediately following it in the adaptation field including stuffing bytes.
- Discontinuity indicator is 1 bit flag which when it is set to 1 indicates that the discontinuity state is true for the current transport packet. When this flag is set to 0, the discontinuity is false. This discontinuity indicator is used to indicate two types of discontinuities, system time base discontinuities and continuity counter discontinuities. In the first type, this transport stream packet is the packet of a PID designed as a PCR-PID. The next PCR represents a sample of a new system time clock (STC) for the associated program. In the second type, the transport stream packet could be any PID type. If the transport stream packet is not designated as a PCR-PID, the continuity counter may be discontinuous with respect to the previous packet with the same PID or when a system time base discontinuity occurs. For those PIDs that are not designated as PCR-PIDs, the discontinuity indicator may be set to 1 in the next transport stream packet with the same PID, but will not be set to 1 in three consecutive transport stream packet with the same PID.
- Random-access-indicator is 1 bit flag that indicates the current and later transport stream packets with the same PID, containing some information to aid with random access at this point. Specifically, when this flag is set to 1, the next PES packet in the payload of transport stream packet with the current PID will contain the first byte of a video sequence header or the first byte of an audio frame.
- ES priority indicator is used for data partitioning application in the ES. If this flag is set to 1, the payload contains high-priority data such as the header information or low-order DCT coefficients of the video data. This packet will be highly protected.
- PCR-flag and OPCR-flag: if these flags are set to 1, it means that the adaptation field contains the PCR data and original PCR data. These data are coded in two parts.
- Splicing-point-flag: when this flag is set to 1, it indicates that a splice-countdown field will be present to specify the occurrence of a splicing point. The splice point is used to smoothly splice two bitstreams into one stream. SMPTE has developed a standard for seamless splicing of two streams [smpte pt20]. We will describe the function of splicing later.

- Transport-private-flag: this flag is used to indicate whether the adaptation field contains private data.
- Adaptation-field-extension-flag: this flag is used to indicate whether the adaptation field contains the extension field that gives more detailed splicing information.

21.2.2.2.4 Packetized Elementary Stream

It is noted that the ES data is carried in PES packets. A PES packet consists of a PES packet header followed by packet data, or payload. The PES packet header begins with a 32 bit start code that also identifies the stream or stream type to which the packet data belongs. The first byte of each PES packet header is located at the first available payload location of a transport stream packet. The PES packet header may also contain decoding time stamps (DTS), PTS, ES clock reference (ESCR), and other optional fields such as DSM trick mode information. The PES packet data field contains a variable number of contiguous bytes from one ES. Readers can learn this part of syntax in the same way as described for the transport packet header and adaptation field.

21.2.2.2.5 Program-Specific Information

PSI includes both MPEG-2 system compliant data and private data. In the transport streams, the PSI is classified into four table structures: program association table, program map table, conditional access (CA) table, and network information table. The network information table is private data and other three are MPEG-2 system compliant data. The program associate table provides the information of program number and the PID value of the transport stream packets. The program map table specifies PID values for components of one or more programs. The CA table provides the association between one or more CA systems, their entitlement management messages (EMM), and any special parameters associated with them. The EMM are private CA information that specifies the authorization levels or the services of specific decoders. They may be addressed to a single decoder or groups of decoders. The network information table is optional and its contents are private. Its contents provide physical network parameters, such as FDM frequencies, transponder numbers, etc.

21.2.3 Transport Streams Splicing

The operation of bitstream splicing is switching form one source to another according to the requirements of the applications. Splicing is the most common operation performed in TV stations today [hurst 1997]. The examples include inserting commercials into programming, editing, inserting, or replacing a segment into a existing stream, and inserting local commercials or news into a network feed. The most important problem for bitstream splicing is managing the buffer fullness at the decoder. Usually, the encoded bitstream satisfies the buffer regulation with a buffer control algorithm at the encoder. During decoding, this bitstream will not cause the decoder buffer to suffer from buffer overflow and underflow. A typical example of buffer fullness trajectory at the decoder is shown in Figure 21.6. However, after bitstream splicing, the buffer regulation is not guaranteed depending on the selection of splicing point and the bit rate of new bitstream. It is necessary to have a rule for selecting the splicing point.

The committee on packetized television technology, PT20 of SMPTE (Society of Motion Picture and Television Engineers), has proposed a standard that deals with the splice point for MPEG-2 transport streams [smpte pt20]. In this standard, two techniques have been proposed for selecting splicing points. One is the seamless splicing and other is non-seamless splicing. The seamless splicing approach can provide clean and instant switching of bitstreams, but it requires careful selection of splicing points on video bitstreams. The

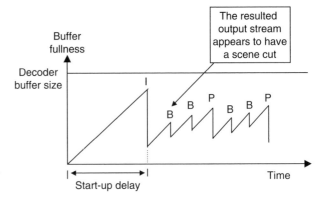

FIGURE 21.6
Typical buffer fullness trajectory at the decoder.

non-seamless splicing approach inserts a drain-time that is a period of time between the end of an old stream and the start of a new stream to avoid overflow in the decoder buffer. The drain-time assures that the new stream begins with an empty buffer. However, the decoder has to freeze the final presented picture of old stream and wait for a period of start-up delay while the new stream is initially filling the buffer. The difference between the seamless splicing and the non-seamless splicing is shown in Figure 21.7.

In the SMPTE-proposed standard [smpte pt20], the optional indicator data in the PID streams (all the packets with the same PID within a transport stream) is used to provide

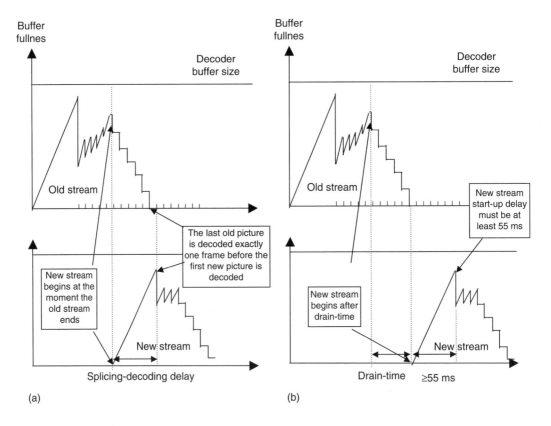

FIGURE 21.7
Difference between seamless splicing and non-seamless splicing (a) the video buffer verifier (VBV) buffer behavior of seamless splicing, (b) the VBV buffer behavior of non-seamless buffer behavior.

important information about the splice for the applications such as inserting commercial programs. The proposed standard defines a syntax that may be carried in the adaptation field in the packets of the transport stream. The syntax provides a way to convey two kinds of information. One type of information is splice point information that consists of four splicing parameters: drain-time, in-point-flag, ground-id, and picture-param-type. The other types of information are splice point indicators that provide a method to indicate application-specific information. One such application example is the insertion indicator for commercial advertisement. This indicator includes flags to indicate that the original stream is obtained from the network and that the splice point is the time point where the network is going out or going in. Other fields give information about whether it is scheduled, how long it is expected to last as well as an ID code. The detail about splicing can be found in the proposed standard [smpte pt20].

Although the standard provides a tool for bitstream splicing, there are still some difficulties for performing bitstream splicing in practice. One problem is that the selection of a splicing point has to consider that the bitstream contains video that has been encoded by a predictive coding scheme. Therefore, the new stream should begin from the anchor picture. Other problems include uneven timing frames and splicing of bitstreams with different bit rates. In such cases, one needs to be aware of any consequences related to buffer overflow and underflow.

21.2.4 Program Streams

The program stream is defined for the multiplexing of audio, video, and other data into a single stream for communication or storage application. The essential difference between the program stream and transport stream is that the transport stream is designed for applications with noisy media, such as in terrestrial broadcasting. Because the program stream is designed for applications in the relatively error-free environment, such as in the digital video disk (DVD) and digital storage applications, the overhead in the program stream is less than in the transport stream.

A program stream contains one or more elementary streams. The data from ESs is organized in the form of PES packets. The PES packets from different ESs are multiplexed together. The structure of a program stream is shown in the Figure 21.8.

A program stream consists of packs. A pack begins from a pack header followed by PES packets. The pack header is used to carry timing and bit rate information. It begins with a 32 bit start code followed by SCR information, program muxing rate, and stuffing bits. The SCR indicates the intended arrival time of the byte that contains the last bit of SCR base at the input of the decoder. The program muxing rate is a 22 bit integer that specifies the rate at the decoder. The value of this rate may vary from pack to pack. The stuffing bits are inserted by the encoder to meet channel requirements. The pack header may contain a system header that may be repeated optionally. The system header contains the summary of the system parameters such as header length, rate bound, audio bound, video bound, stream id, and other system parameters. The rate bound is used to indicate the maximum rate in any pack of the program stream, and it may be used to assess whether the decoder is capable of decoding the entire stream. The audio bound and video bound are used to indicate the maximum values of audio and video in the program stream. There are some other flags that are used to give some system information. A PES packet consists of a PES packet header followed by packet data. The PES packets have the same structure as in the transport stream.

A special type of PES packet is the program stream map; it is present when the stream id value is $0 \times BC$. The program stream map provides a description of the ESs in the program

FIGURE 21.8
Structure of program stream.

stream and their relationship to one another. The data structure of program stream map is shown in Figure 21.9.

Other special types of PES packets include program stream directory and program element descriptors. The major information contained in the program stream directory includes the number of access units, packet stream id, and PTS. The program and program descriptors provide the coding information about the ESs. There are total of 17 descriptors, including video descriptor, audio descriptor, and hierarchy descriptor. For the detail on these descriptors, the reader is referred to the standard document [mpeg2 system].

21.2.5 Timing Model and Synchronization

The principal function of the MPEG system is to define the syntax and semantics of the bitstreams that allows the system decoder to perform two operations among multiple ESs: demultiplexing and resynchronization. Therefore, the system encoder has to add the timing information to the program streams or transport streams during the process of

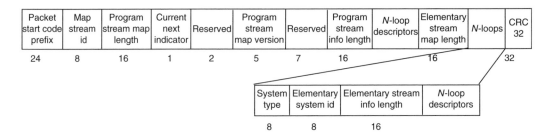

FIGURE 21.9
Data structure of program stream map.

multiplexing the coded video, audio, and data ESs to a single stream or multiple streams. System, video, and audio all have a timing model in which the end-to-end delay from the signal input to an encoder to the signal output from a decoder is a constant. The delay is the sum of encoding, encoder buffering, multiplexing, transmission or storage, demultiplexing, decoding buffering, decoding, and presentation delays. The buffering delays could be variable, while the sum of total delays should be constant.

In the program stream, the timing information for a decoding system is the SCR, whereas in the transport stream, the timing information is given by the PCR. The SCR and PCR are time stamps that are used to encode the timing information of the bitstream itself. The 27 MHz SCR is the kernel time base for the entire system. The PCR is 90 kHz, which is 1/300 of the SCR. In the transport stream, the PCR is encoded with 33 bits and is contained in the adaptation field of the transport stream. The PCR can be extended to the SCR with an additional 9 bits in the adaptation field. For the program stream, the SCR is directly encoded with 42 bits and it is located in the pack header of the program stream. The synchronization among multiple ESs is accomplished with a PTS in the program and transport streams. The PTS is 90 kHz and represented with a 33 bit number coded in three separate parts contained in the PES packet header. In the case of audio, if a PTS is present, it will refer to the first access unit commencing in the PES packet. An audio access unit starts in a PES packet if the first byte of the audio access unit is present in the PES packet. In the case of video, if a PTS occurs in the PES packet header, it refers to the access unit containing the first picture start code (PSC) that commences in this PES packet. A PSC commences in the PES packet if the first byte of the PSC is present in the PES packet. In an MPEG-2 system, the SCR is specified to satisfy the following conditions:

$$27 \text{ MHz} - 810 \text{ Hz} \leq \text{SCR} \leq 27 \text{ MHz} + 810 \text{ Hz}$$

$$\text{Change rate of SCR} \leq 75 \times 10^{-3} \text{ Hz/s}$$

In the encoder, the SCR or PCR are encoded in the bitstream at intervals up to 100 ms in the transport stream and up to 700 ms in the program stream. As such, they can be used to reconstruct the STC in the decoder with sufficient accuracy for all identified applications. The decoder has its own STC with the same frequency, 90 kHz for the transport stream and 27 MHz for the program stream. In a correctly constructed MPEG-2 system bitstream, each SCR arrives at the decoder, precisely at the time indicated by the value of that SCR. If the decoder's clock frequency matches the one in the encoder, the decoding, and presentation of video and audio will automatically have the same rate as those in the encoder, then the end-to-end delay will be constant. However, the STC in the decoder may not exactly match the one in the encoder due to the independent oscillators. Therefore, a decoder's system clock frequency may not match the encoder's system clock frequency that is sampled and indicated in the SCR. One method is to use a free run 27 MHz in the decoder. The mismatch between the encoder's STC and the decoder's STC is handled by skipping or repeating frames. Another method to handle the mismatch is to use the received SCRs (which occur at least once in the intervals of 100 ms for transport stream and 700 ms for the program stream). In this way, the decoder's STC is a slave to the encoder's STC. This can be implemented with a phase-locked loop (PLL) as shown in Figure 21.10.

The synchronization among multiple ESs can be achieved by adjusting the decoding of streams to a common master time base rather than by adjusting the decoding of one stream to match that of another. The master time base may be one of the many decoder clocks, the clock of the data source, or some external clock. Each program in a transport stream, which may contain multiple programs, may have its own time base. The time bases of different programs within a transport stream may be different.

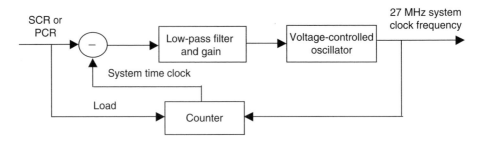

FIGURE 21.10
System time clock (STC) recovery using phase-locked loop (PLL).

In the digital video systems, the 13.5 MHz sampling rate of the luminance signal and 6.25 MHz chrominance signals of the CCIR601 digital video are all synchronized to 27 MHz system time clock. The National Television Systems Committee (NTSC) or phase alternating line (PAL) TV signals are also phase-locked to the same 27 MHz clock such that the horizontal and vertical synchronous signals and the color burst clock are all locked to the 27 MHz system time clock.

In the TV studio applications, the entire TV studio equipment is synchronized to the same time base of a composite horizontal and vertical synchronization signals to perform the seamless video source switching and editing. It should be noted that this time base is definitely not synchronized to the PCRs from various remote encoder sites. The 27 MHz local decoder's STC is locked to the same studio composite horizontal and vertical synchronization signal. The 33 bits of video STC counter is initialized by the latest video PTS then calibrated using the 90 kHz clock derived from the 27 MHz system clock in the decoder. If the 27 MHz system clock in the decoder is synchronized with the system clock on the transmitting end, the STC counter will be always the same as the incoming PTS numbers. However, there may be some mismatch between the system clocks. As each new PTS arrives, the PTS will be compared with the STC counter. If the PTS is larger than the STC plus half of the duration of the PTS, it means that the 27 MHz decoder clock is too slow and the bit buffer may overflow. In this case, the decoder should skip some of current data to search for the next anchor frame so that decoding can be continued. If the PTS is less than the STC minus half of the duration of the PTS, the bit buffer may underflow. The decoding will halt and repeatedly display the current frame. The audio decoder will also be locked on the same 27 MHz system clock, where similar skipping and repeating of audio data is used to handle the mismatch.

In the low cost consumer set-top box (STB) applications, a simple free run 27 MHz decoder system clock with the skipping and repeating frame scheme can provide pretty good results. In fact, the skipping or repeating frame may happen once in a 2 or 4 h period with a free run 27 MHz crystal clock. The STC counter will be set by the latest PTS, then count on the 90 kHz STC derived from the free run 27 MHz crystal clock. The same skipping or repeating display control as the TV studio will be used.

For a complex STC solution, a PLL with VCXO (voltage-controlled crystal oscillator) in the decoder is used to synchronize the incoming PCR data. The 33 bit decoder's PCR counter is initialized by the latest PCR data, then the 27 MHz system clock is calibrated. If the decoder's system clock is synchronized with the e encoder's remote 27 MHz system clock, every incoming PCR data will be the same as the decoder's PCR counter or have small errors from PCR jitter. The difference between the decoder's PCR counter and the incoming PCR data indicates this frequency jitter or drift. As long as the decoder's

27 MHz system clock is locked on the PCR data, the STC counter will be initialized by the latest PTS then calibrated using the 90 kHz clock. The similar skipping and repeating frame scheme will be used again, but the 27 MHz system clock in the decoder is synchronized with the incoming PCR. As long as the decoder's 27 MHz is locked on the encoder's 27 MHz, there will be no skipping or repeating of frames. However, if the PCR PLL is not working properly, the skipping or repeating of frames will occur more often than when the free run 27 MHz system clock is used.

Finally, it should be noted that the PTS-DTS-flag is used to indicate whether the PTS and DTS or both of them will be present in the PES packet header. The DTS is a 33 bit number coded in three separate fields in the PES packet header. It is used to indicate the decoding time.

21.3 MPEG-4 System

This section describes the specification of the MPEG-4 system or ISO/IEC 14496-1.

21.3.1 Overview and Architecture

The specification of the MPEG-4 system [mpeg4 system] is used to define the requirements for the communication of interactive audiovisual scenes. An example of such a scene is shown in Figure 21.11 [mpeg4system]. The overall operation of this system can be summarized as follows. At the encoder, the audio, visual, and other data information is first compressed and supplemented with synchronization timing information. The compressed data with timing information is then passed to a delivery layer that multiplexes these data into one or more coded binary streams for storing or transmission. At the decoder, these streams are first demultiplexed and decompressed. The reconstructed audio and visual objects are then composed according to the scene description and synchronization information. The composed audiovisual objects (AVO) are then presented to the end user. The important feature of the MPEG-4 standard is that the end user may have the option to interact with this presentation because the compression is performed on the object or content basis. The interaction information can be processed locally or transmitted back to the encoder. The scene information is contained in the bitstreams and used in the decoding processes.

The system part of the MPEG-4 standard specifies the overall architecture of a general receiving terminal. Figure 21.12 shows the basic architecture of the receiving terminal. The major elements of this architecture are delivery layer, sync layer (SL), and compression layer. The delivery layer consists of the FlexMux and TransMux. At the encoder, the coded ESs, which include the coded video, audio, and other data with the synchronization and scene description information, are multiplexed to the FlexMux streams. The FlexMux streams are transmitted to the TansMux of the delivery layer from the network. The function of TransMux is not within the scope of the system standard and it can be any of the existing transport protocols such as MPEG-2 transport stream, RTP/UDP/IP, AAL5/ATM, and H223/Mux.

Only the interface to the TransMux layer is part of the standard. Usually, the interface is the DMIF application interface (DAI), which is not specified in the system part, but in part 6 of the MPEG-4 standard. The DAI specifies the data that needs to be exchanged between the SL and the delivery layer. The DAI also defines the interface for signaling information required for session and channel set up as well as tear down. For some simple applications,

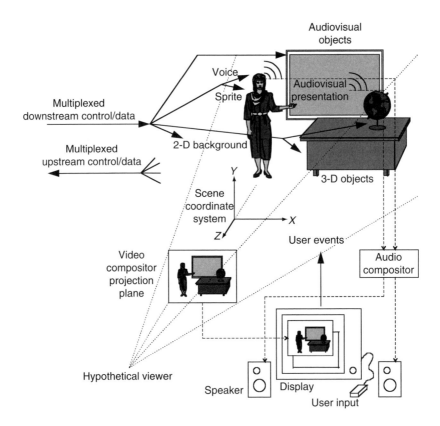

FIGURE 21.11
An example of MPEG-4 audiovisual scene [mpeg4 system].

it does not require full functionality of the system specification. A simple multiplexing tool, FlexMux, with low delay and low overhead is defined in the system part of MPEG-4. The FlexMux tool is a flexible multiplexer that accommodates for the interleaving of SL-packetized streams with varying instantaneous bit rate. The FlexMux packet has a variable length which may contain one or more SL packets. Also, the FlexMux tool provides the identification for the SL packets to indicate which ES it comes from. FlexMux packets with data from different SL-packetized streams can therefore be arbitrary interleaved.

The SL specifies the syntax for packetizing the ESs to the SL packets. The SL packets contain a SL packet header and a SL packet payload. The SL packet header provides the information for continuity checking in case of data loss and also carries the timing and synchronization information as well as fragmentation and random access information. The SL packet does not contain its length information. Therefore, SL packets must be framed by the FlexMux tool. At the decoder, the SL packets are demultiplexed to the ESs in the SL. At the same time, the timing and the synchronization information as well as fragmentation and random access information are also extracted for synchronizing the decoding process and subsequently for composition of the ESs.

At the compression layer, the encoded ESs are decoded. The decoded information is then used for the reconstruction of audiovisual information. The operation of the reconstruction includes composition, rendering, and presentation with the timing synchronization information.

FIGURE 21.12
The MPEG-4 system terminal architecture.

21.3.2 Systems Decoder Model

The systems decoder model (SDM) is a conceptual model that is used to describe the behavior of decoders complying with MPEG-4 systems. It may be used for the encoder to predict how the decoder or receiving terminal will behave in terms of buffer management and synchronization during the process of decoding, reconstructing, and composing of AVO. The SDM includes a system timing model and a system buffer model. These models specify the interfaces for accessing demultiplexed data streams, decoding buffers for each ES, the behavior of ES decoder, composition memory for decoded data from each decoder, and the output behavior of composition memory toward the compositor. The SDM is shown in Figure 21.13.

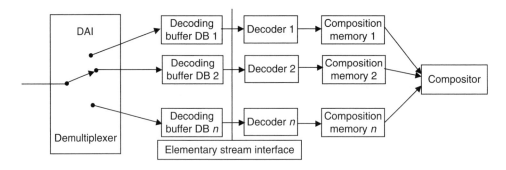

FIGURE 21.13
Block diagram of systems decoder model.

The timing model defines the mechanisms that allow a decoder or receiving terminal to process time-dependent objects. This model also allows the decoder or receiving terminal to establish mechanisms to maintain synchronization both across and within particular media types as well as with user interaction events. To facilitate these functions at the decoder or receiving terminal, the timing model requires that the transmitted data streams contain implicit or explicit timing information. There are two sets of timing information that are defined in the MPEG-4 system. One indicates the periodic values of the encoder clock that is used to convey the encoder's time base to the decoder or the receiving terminal, whereas the other is the desired presentation timing for each audiovisual object. For real-time applications, the end-to-end delay from the encoder input to the decoder output is constant. The delay is equal to the sum of the delay due to the encoding process, buffering, multiplexing at the encoder, the delay due to the delivery layer and demultiplxing, decoder buffering, and decoding process at the decoder.

The buffer model is used for the encoder to monitor and control the buffer resources that are needed for decoding each ES at the decoder. The information of the buffer requirements is transmitted to the decoder by descriptors at the beginning of the decoding process. The decoder can then decide whether it is capable of handling this particular bitstream or not. In summary, the buffer model allows the encoder to schedule data transmission and to specify when the bits may be removed from these buffers. Then the decoder can choose proper buffers so that the buffers will not overflow or underflow during the decoding process.

21.3.3 Scene Description

In multimedia applications, a scene may consist of AVO that include the objects of natural video, audio, texture, two-dimensional (2-D) or three-dimensional (3-D) graphics and synthetic video. As MPEG-4 is the first object-based coding standard, reconstructing or composing a multiple audiovisual scene is quite new. The decoder not only needs the ESs for the individual AVO but also synchronization timing information and the scene structure. This information is called the scene description and it specifies the temporal and spatial relationships between the objects or scene structures. The information of the scene description can be defined in the encoder or interactively determined by the end user and transmitted with the coded objects to the decoder. The scene description only describes the scene structure. The action of assembling these AVO to a scene is called composition.

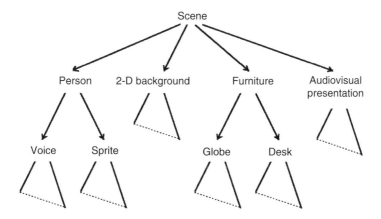

FIGURE 21.14
Hierarchical graph representation of an audiovisual scene [mpeg4system].

The action of transmitting these objects from a common representation space to a specific presentation device is called rendering.

The MPEG-4 system defines the syntax and semantics of a bitstream that can be used to describe the relationships of the objects in space and time. However, for visual data, the system standard does not specify the composition algorithms. Only for audio data, the composition process is specified in a normative manner. To allow the operations of authoring, editing, and interaction of visual objects at the decoder, the scene descriptions are coded independent from the audiovisual media. This allows the decoder to modify the scene according to the requirements of the end user. Two kinds of user interaction are provided in the system specification. One is client-side interaction that involves object manipulations requested in the end user's terminal. The manipulation includes the modification of attributes of scene objects according to the specified user actions. The other type of manipulation is the serve-side interaction that the standard does not deal with.

The scene description is a hierarchical structure that can be represented as a graph. The example of the audiovisual scene in Figure 21.11 can be represented as in Figure 21.14. The scene description is represented by a parametric approach, the binary format for scenes (BIFS). The description consists of an encoded hierarchical tree of nodes with attributes and other information. In this tree, the leaf nodes correspond to the elementary AVO and information for grouping, transformation, and other operations.

21.3.4 Object Description Framework

The ESs carry data for audio or visual objects as well as for the scene description itself. The purpose of the object description framework is to provide the link between the ESs and the audiovisual scene description. The object description framework consists of a set of descriptors that allow identifying, describing, and appropriately associating ESs to each other and to AVO used in the scene description. Each object descriptor is a collection of one or more ES descriptors that are associated to a single audiovisual object or a scene description. Object descriptors are themselves conveyed in ESs. Each object descriptor is assigned an identifier (object descriptor id), which is unique within a defined name scope. This identifier is used to associate AVO in the scene description with a particular object descriptor, and thus the ESs related to that particular object. ES descriptors include information about the source of the stream data, in the form of a unique numeric identifier (the ES id) or a URL pointing to a remote source for the stream. ES descriptors also include information about the encoding format, configuration information for the decoding process, and the SL packetization, as well as quality of service requirements for the transmission of the stream and intellectual property identification. Dependencies between streams can also be signaled within the ES descriptors. This functionality may be used, for example, in scalable audio or visual object representations to indicate the logical dependency of an enhancement layer stream to a base layer stream. It can also be used to describe alternative representations for the same content (e.g., the same speech content in various languages).

The object description framework provides the hooks to implement intellectual property management and protection (IPMP) systems. IPMP descriptors are carried as part of an object descriptor stream. IPMP ESs carry time variant IPMP information that can be associated to multiple object descriptors. The IPMP system itself is a nonnormative component that provides IPMP functions for the terminal. The IPMP system uses the information carried by the IPMP ESs and descriptors to make protected IS 14496 content available to the terminal. An application may choose not to use an IPMP system, thereby offering no management and protection features.

21.4 Summary

In this chapter, the MPEG system issues have been discussed. Two typical systems, MPEG-2 system and MPEG-4 system, have been introduced. The major task of system layer is to multiplex and demultiplex video, audio, and other data to a single bitstream with synchronization timing information. For MPEG-4 systems, there are some more issues being addressed such as interface with network applications.

Exercises

1. What are two major system streams provided by MPEG-2 system? Describe some application examples for these two streams and explain the reasons.
2. The MPEG-2 system bitstream is a self-contained bitstream to facilitate synchronous playback of video, audio, and related data. Describe what kinds of timing signals are contained in the bitstream to achieve the goal of synchronization?
3. How does MPEG-2 system deal with different system clocks between the encoder and decoder? Describe what a system may do when the decoder clock is running too slow or too fast?
4. Why is the 27 MHz system clock in MPEG-2 is represented in two parts: 33 bit + 9 bit extension?
5. What is bitstream splicing of a transport stream? Give several application examples of bitstream splicing and indicate the problems that may arise.
6. Describe the difference between the MPEG-2 system and MPEG-4 system.

References

[**hurst 1997**] Norm Hurst, Splicing—high definition broadcasting technology, year 1 demonstration, Meeting talk, 1997.

[**mpeg2 system**] ISO/IEC 13818-1: 1996 Information Technology—Generic Coding of Moving Pictures and Associated Audio Information.

[**mpeg4 system**] ISO/IEC 14496-1: 1998 Information Technology—Coding of Audio-Visual Objects.

[**smpte pt20**] Proposed SMPTE Standard for Television—Splice Points for MPEG-2 Streams, PT20.02, April 4, 1997.

Index

A

Access unit, definition of, 509
Adaptation field, 511–515, 517, 519
Adaptive arithmetic coder, 450–451
Adaptive block size motion compensation, 492
Adaptive error concealment strategy, 426
Adaptive quantization
    backward, 52–53
    forward, 52
    one-word memory, 53
    switched quantization, 53–54
    types of, 51
Additive white Gaussian noise (AWGN), 12–13, 235
Advanced High 4:4:4 Profile, 502–503
Advanced intracoding (AIC), 475
Advanced Television Standard Committee (ATSC), 395–399
Advanced Television Test Center (ATTC), 396
Advisory Committee on Advanced Television Service (ACATS), 396
Affine transformations, 212
Algorithm enhancements
    directional interpolation, 428–429
    I-picture motion vectors, 429
    spatial scalable error concealment, 429–430
Alpha plane, 445–446
Alternative inter VLC (AIV), 480
Anandan's algorithm, in optical flow determination, 298
Aperture problem, 327
    in optical flow, 285, 287–288
Arbitrary slice ordering (ASO), 487
Arithmetic coding, 111
    advantages over Huffman coding, 134, 137
    development history, 136
    double recursion, 138
    implementation problems
        carry-over problem, 136
        finite precision, 136
        increments, 135
        recursive division of subintervals, 136
    need for, 129; see also Huffman codes, limitations
    principles of
        decoding, 132–134
        encoding, 132
        first in first out (FIFO), 134
        last in first out (LIFO), 134
        subintervals division, 131
AR model, see Autoregressive (AR) model
ATSC, 352–353
ATSC DTV Standards
    compression layer, 398–399
    history of, 395–397
    picture layer, 397–398
    transmission layer, 399–400
    transport layer, 399
Automatic segmentation algorithm, see Video segmentation
Autoregressive (AR) model, 143–144
AWGN, see Additive white Gaussian noise

B

Bergmann's (1982) pel recursive algorithm, 281–282
Bidirectionally predictive-coded (B) pictures, 361
Bilinear interpolation, 233
Binary alpha block (BAB), 446
Binary format for scenes (BIFS), 525
Binary image coding, 149
Binary linear block code, 211
Binary shape coding, see Shape coding
Binary tree-search quantization, 207
Bit allocation, 86, 101
Bitstream scaling
    architectures of, 403–407
    principles of, 402–403
    transcoding, 399–402
Bitstream syntax and semantics, 453–454
Block matching, 241–243
    block diagram of multigrid, 261
    criteria, 243
    error, 244
    limitations of, 256
    multiresolution, 248
    optimal multigrid, 262
    technique, 439
    three-level hierarchical, 258–260

thresholding multigrid, 260
thresholding multiresolution, 250–251
Block pixels, transformed coefficients, 205
Block quantization, 83
Block truncation coding (BTC), 207, 209
Brightness function, 222
Brightness invariant equation, 290–292, 321

C

CABAC, see Content-adaptive binary arithmetic coding
Cafforio and Rocca's pel recursive algorithm, 282
CAVLC, see Content-adaptive variable-length coding
CBR encoder
    encoded bitstream, 388
    quantization, 388–389
    spatial and temporal complexity, 389
CDF(2,2) wavelet transform, 186
Channel coding, 346
Chebyshev polynomials, 95
Coarse-fine three-step search, 246
Codebook generation, 204
Codes
    characteristics
        block code, 115
        compact codes, 120
        instantaneous codes, 118–119
        nth extension of block code, 118
        uniquely decodable codes, 116, 118
    discrete memoryless source
        definition, 120
        entropy, 121
        noiseless source coding theorem, 121–122
    information source, 115–116
Code word, 205
Common intermediate format (CIF), 352
Compressed bitstream structure, 365
Computational complexity, 208
Concealment motion vector (CMV), 370–371
Conjugate direction search, 246–247
Conservation information, 298, 300, 315, 321–322, 328–329
    Neighborhood information, 298–301, 316, 321–322
Constant bit rate (CBR) coding, 377–378
Constant Q-factor, 181, 184
Content-adaptive binary arithmetic coding, 487, 493, 497–498, 503
Content-adaptive variable-length coding, 487, 493–497

Content-based
    efficient compression, 436–437
    interactivity, 436
    scalability, 437
Content-based arithmetic encoding (CAE), 446
Continuous-time wavelet transform, 180–182
Contraction, 180
Convergence rate, 282–283
Convergence speed
    linear, 276
    order of, 275–276
Correlation, 243
    interframe
        redundancy, 226
        temporal redundancy, 225
    intraframe
        redundancy, 225
        spatial redundancy, 225
    window, 243
        sizes of blocks, 258
        subsampling in, 248
Correlation-based approach, 297–316, 321–322
Correlation-feedback algorithm, 302, 305–306, 311–313, 315
Correlation-feedback technique, block diagram of, 304

D

Data partitioning, bitstreams, 372
DCT based method, for motion estimation, 338–340
Decoder
    block diagram of, 367
    decoding process, 365
    full-memory decoder (FMD), 416
    half-memory decoder (HMD), 416
    quarter-memory decoder (QMD), 416
Decoding time stamp (DTS), 509, 515
Deformation, 330
Delta modulation (DM)
    adaptive, 72
    block diagram of, 70
    features of, 70
    input–output characteristic of two-level quantization in, 71
Descent methods, of Pel recursive technique
    convergence speed, 275–276
    first-order necessary conditions, 273
    Newton–Raphson's method, 277–279
    second-order sufficient conditions, 273–274
    steepest descent method, 277
    underlying strategy, 274–275
Detelecine process, 373
DFD, see Displaced frame difference

Index

Dictionary coding
   adaptive, 151–152
   applications of, 150–151
   formulation and classification of, 151
   LZ77 algorithms, 152
   LZ78 algorithms
      decoding in, 157–158
      encoding in, 156–157
   LZW algorithm
      applications of, 160
      decoding in, 159
      encoding in, 158
   parsing strategy, 152
   sliding window (LZ77) algorithms
      decoding in, 154–155
      encoding in, 152–153
   static, 151
Differential pulse code modulation (DPCM), 224, 226; *see also* Quantization
   adaptive, 68
   block diagram of, 63
   channel encoding, 70
   definition of, 59
   delta modulation (DM)
      adaptive, 72
      block diagram of, 70
      features of, 70
      input-output characteristic of two-level quantization in, 71
   image and video coding, 64
   information-preserving differential coding, 76
   interframe differential coding, 73–76
   motion compensated predictive coding, 75–76
   one, two and three-dimensional, 67–68
   optimum linear prediction
      formulation, 65
      mean square prediction error, 65–66
      orthogonality condition, 66
      Yule–Walker equations, 66
   pixel-to-pixel
      block diagram of, 62
      histogram of, 61
      quantization error accumulation, 60
   prediction error, 64
   transmission error effects, 68–69
Digital image processing, 221
Digital versatile disk, 353, 356
Digital video coding standards
   MPEG-1, 359
      compressed bitstream structure, 365
      decoding process, 365–366
      layered structure based on group of pictures, 360–361
      video encoder structure, 361–365
      video sequence, group of pictures (GOPs) of, 360–361
      video signal coding, 360
   MPEG-2, 359
      basic coding structure of, 366
      concealment motion vector (CMV), 370–371
      field/frame DCT coding syntax, 369
      field/frame prediction mode, 367–369
      non-compression enhancements, 367
      pan–scan parameters, 370
      quantizer matrices and scan order, 369–370
      scalable modes, 371–372
Digital Video Processing, 224
Digram coding, 151
Dilation, 180, 183–184
Directional spatial prediction, 486, 491
Discrete Cosine Transform (DCT), 224–225, 360–363, 365–367, 369–370, 372, 377, 382, 385, 387, 439
   advantages, 98
   basis images, 96
   coding, 360
   N-point sequence, 98
   transformation kernel, 95
   two-dimensional (2-D), 362
   vs. DFT, 96–97
Discrete Cosine Transform (DCT) coefficients
   Huffman coding of, 171–172
   in progressive DCT-based coding, 174
   quantized, 170
   transformed, 169
   zigzag scanning order of, 172
Discrete Fourier transform (DFT), 92
Discrete Hadamard transform (DHT)
   ordered, 94–95
   transformation kernel, 93
   vs. DWT, 94
Discrete Karhunen–Loeve transform (KLT), 100; *see also* Hotelling transform
Discrete Markov source model
   extensions of
      definition, 143
      entropy, 143
   source symbols interdependence, 142
   state diagram, 142
   zero memory source, 141
Discrete Walsh transform (DWT), 93
Discrete wavelet transform (DWT), 183–185, 450
Displaced frame difference (DFD), 271–272, 280–281, 302, 313, 328, 334, 337
Displacement vector, 242
Dissimilarity measure, 244
Dithering, 55

Diverging Tree image sequence, 307, 311
Down conversion decoder, 410–421
Dynamic sprite, see Sprite coding

E

Early wavelet image coding algorithms, 191
Eigenvectors
    covariance matrix, 85
    orthonormal, 82
    Toeplitz matrices, 95
    transform, 83
Electromagnetic spectrum, 223
Elementary stream (ES), definition of, 509
Embedded block coding with optimized truncation of the embedded bit streams (EBCOT), 192
Embedded zerotree wavelet coding, 192–194
Encoding function, 212
End of block (EOB), 107
End-of-line (EOL), 144
Energy-based approach, 321
Energy compaction, 98–99
Entitlement management messages (EMM), 515
Entropy, 254, 347–348, 357
    average information content per symbol, 27–28
    information measurement, 27
Entropy coding, 111; see also Huffman coding
    in H.264/AVC, 493–498
EOL, see End-of-line
Error concealment, 421–431
Error resilience, 451–453, 455
    tools, in H.264/AVC, 500–501
Euclidean plane, 213
Exponential Golomb codes, 494–495

F

Facsimile coding, 144–145, 149–150
Fixed step size, 284
Flexible macroblock ordering (FMO), 487, 501–503
Forward transformation kernel; see also Inverse transformation kernel
Fractal dimension, 215
Fractal image coding
    IFS-based, 214
    mathematical foundation, 212
    segmentation-based coding, 215
Frame memories, 361–362
Frame replenishment
    buffer saturation, 237
    conditional, 226–227
    dirty window effect, 227–228

    frame-difference predictive coding, 227
    picture breakup, 227
Frequency domain method, for motion estimation, 330–331
Frequency domain motion estimation, 377
Frequency synthesis, 411–413, 415–416, 418
Frequency synthesis down-conversion, 412–413
Full-resolution decoder (FRD), 236, 410–411
Full search algorithm, 375
Full search quantization, 207

G

Gaussian mask, 301
Gaussian pyramid, 249
Generalized block matching, 377
General switched telephone network (GSTN), 144
Gradient-based approach, 290–297, 305–307, 309, 321
Gradient method, see Steepest descent method
Grand Alliance, 396, 398
Graphical user interface (GUI), 457
Gray-scale shape coding, see Shape coding
Group of Blocks (GOBs) structure, 452
Group of pictures (GOP), 360–361, 365–366, 378–379, 382, 391–392
Group of video object plane (GOV), 454
GSTN, see General switched telephone network

H

H.263+, 355
H.264, 353, 355
Haar wavelet, 181
Half-pixel accuracy, 468–469
Hamburg taxi, role in Pan, Shi, and Shu's method, 312–313
Hamburg taxi sequence, needle diagram of, 313
H.264/AVC
    block artifacts in, 498
    directional spatial prediction in, 491–492
    error-resilience tools in, 500–501
    levels of, 503–504
    overview of, 484–487
    profiles of, 502–503
HDTV formats, 397–398
Hessian matrix, 274–275, 278–279, 281–282
Hierarchical block matching, 257–260, 376
Hierarchical DCT-based encoding algorithm, 177
High correlation transform (HCT), 491
High definition television (HDTV), 236, 395–397

Horn and Schunck's algorithm
  brightness invariance equation, 290–292
  iterative algorithm, 293–295
  minimization of, 292–293
  smoothness constraint, 292
Hotelling transform
  covariance matrix, 82
  definition, 83
  inverse, 83–85
Huffman coding, 103, 106–107, 110–111
  for AC coefficients, 173
  algorithm
    image and video coding application, 125
    principle, 123–124
    process, 124
    properties, 125
  DCT coefficients, 171–172
  two-dimensional (2-D) array, 173
Human visual perception; *see also* Objective visual quality measurement, Psychovisual redundancy
  differential sensitivity, 11, 20
  luminance masking, 11
  two-unit cascade model of, 10
  Weber's law, 12
H.263++ video coding, 481
H.261 video coding standard, 353, 355
  developmental history, 463
  overview of, 463–464
  syntax of, 466–467
  technologies used in, 464–466
H.263 video coding standard [h263], 353, 355, 467
  advanced prediction mode, 469–472
  half-pixel accuracy, 468–469
  overview of, 468
  PB-frames, 472–473
  syntax-based arithmetic coding, 472
  unrestricted motion vector mode, 469
H.263 video coding standard version 2
  advanced intracoding (AIC), 475
  alternative inter-VLC (AIV) mode, 480
  deblocking filter (DF), 476–477
  independent segmentation decoding (ISD) mode, 478
  modified quantization mode, 480
  overview of, 473
  PB-frame, improved, 474
  reduced-resolution update (RRU), 478–479
  reference picture resampling (RPR) mode, 478
  reference picture selection, 477–478
  scalability in, 473–474
  slice-structured (SS) mode, 477
  supplemental enhancement information, 480–481
Hybrid coding, 109, 112

I

IDR picture, *see* Instantaneous decoding refresh picture
Ill-posed inverse problem, 327–328
Ill-posed problem, 289, 296
Image and video compression
  applications of, 4
  feasibility of
    interpixel redundancy, 5–10
    psychovisual redundancy, 10–20
    statistical redundancy, 5–10
  information theory
    entropy, 27–28
    information transmission theorem, 29–30
    Shannon's source coding theorem, 28–29
  objective visual quality measurement, 22–26
  source decoder and source encoder, 35
  subjective visual quality measurement, 21–22
  visual communication system
    block diagram of, 34
    digital format, 33
  visual storage system
    applications of, 33
    block diagram of, 34
Image attribute, 315–317, 322
Image coding, 209–210, 214
Image flow, 286
Image partitioning, 207
Image processing, 215
Image quality, 215
Image resolution, 210
Image sequences, 221
Imaging space, 223
  hierarchical structure, 224
  image, 223
  image sequence, 223
  spatial image sequence, 221
  temporal image sequence, 223
  video, 223
Independent segmentation decoding (ISD), 478
Information-preserving differential coding, 76
Instantaneous decoding refresh picture, 487–488
Integer wavelet transform (IWT), 188–189
Intellectual property management and protection (IPMP), 526
Interframe differential coding
  conditional replenishment coding technique, 73–74
  motion compensated predictive coding, 75–76
  three-dimensional DPCM, 74–75
Interframe redundancy, *see* Temporal redundancy
Interlaced, 350–352
Interlaced video coding, 449
Interpixel redundancy, *see* Statistical redundancy

Interpolation
    bilinear, 233
    polynomial, 233
    weighted linear, 233–234
    zero-order, 233
Intrablock encoding, 363–364
Intracoded (I) pictures/frames
    coding, 361
    P-frame and B-frame, 367–368
INTRA DC and AC prediction, 441–442
Intraframe coding, 360, 363
Intraframe redundancy, see Spatial redundancy
Intra quantizer weighting matrix, 363–364
Inverse problem, 289, 321
Inverse transform, Hotelling transform and, 83–85
Inverse transformation kernel
    DFT, 92
    2-D image, 87–91
    DWT, 93
Inverse wavelet transform, 186
I-picture motion vector, 429
Iterated function system (IFS)
    fractal image coding, 212
    parameters of, 213
Iterative algorithm, 293–295, 301
ITU-R, 349, 351

J

Joint Bilevel Image Experts Group (JBIG) coding, 161
Joint Photographic Experts Group (JPEG), 81, 353–354
    hierarchical DCT-based encoding algorithm, 177
    lossless spatial-based encoding algorithm, 176
    lossless still image compression
        bilevel, 161
        multilevel, 161–162
    luminance-chrominance quantization, 106
    progressive DCT-based encoding algorithm
        spectral selection method, 174–175
        successive approximation method, 174–175
    sequential DCT-based encoding algorithm, 169–174
Joint photographic (image) experts group coding, 144
Joint rate controller
    channel capacity, 390
    coding complexity, 391
Joint video team (JVT), 483

JPEG2000, 195
    parts of, 197
    requirements of, 196–197
    verification model (VM) of, 197–199
JPEG compression, 224
Just noticeable distortion (JND), 349

K

Kalman filter, 329
Karhunen–Loeve transform (KLT)
    discrete version of, 83, 100
    reconstruction error, 99

L

Labeling, method for, 212
Laplacian operator, 293
Laplacian pyramid, 299
Lattice labeling algorithm, 210
Lattice points, 210
Layer
    block layer, 467
    compression layer, 398–399
    group of blocks layer, 466
    macroblock layer, 466
    picture layer, 397–398
    transport layer, 399
LBG algorithm, 206
8-level vestigial sideband (8-VSB) modulation, 399
Levinson recursive algorithm, 66
Lifting scheme, 185–189
Limit superior, concept of, 276
Lloyd–Max quantizer, 45
Logarithmic quantization, see Nonuniform quantization
Logarithmic search algorithm, 375–376
2-D Logarithm search, 245
Loop filter, 464–465, 469, 498–500
Lossless spatial-based encoding algorithm, 176
Low pass extrapolation (LPE) padding technique, 444
Low-resolution decoder (LRD), 236, 410–411
Low-resolution motion compensation, 413–416
Lucas and Kanade's algorithm, 296

M

Macroblocks (MBs), 361, 363, 365, 367, 369, 371, 378–379, 381, 383–386
    of frame, 361
    matching criterion, 374–375
    for motion compensation, 362

searching algorithms, 375–376
Markov source model
    autoregressive (AR) model, 143–144
    discrete Markov source
        entropy of, 142–143
        extensions of, 143
        state diagrams of, 142
Maximum theoretical compression factor, 145
Mean absolute difference (MAD), 244, 334
Mean absolute error (MAE), 244
Mean square approximation error, 99–101
Mean squared error (MSE), 206, 244, 298
Mean square reconstuction error, 84, 99–101
MH, see Modified Huffman code
Mildly ill-posed problem, 289, 296, 321
Minimization, 272, 274, 278, 281
Minimum scan line time (MSLT), 149
MMR, see Modified modified READ coding
Mode
    advanced prediction mode, 469–472
    unrestricted motion vector mode, 469
Model-based coding
    image modeling, 216
Modern wavelet image coding, 191–192
Modified Horn and Schunck algorithm, 322
Modified Horn and Schunck method, 295–296
Modified Huffman code, 145
Modified modified READ (MMR) coding, 150
Modified READ code, 146
Morlet wavelet, 180
Motion
    analysis
        biological vision perspective, 230
        block matching, 232
        computer vision perspective, 230–231
        differential method, 232
        2-D translational model, 232
        feature correspondence, 231
        moving objects, image planes, 228
        optical flow, 231
        pel recursion, 237
        region matching, 232
        signal processing perspective, 232–233
    disocclusion, 328–329
    nonrigid motion, 329–330
    occlusion, 328–329
    rigid motion, 329–330
3-D Motion and optical flow, 286–287
Motion compensation (MC), 221, 285, 359–363, 366–367, 373–374, 377, 382, 407, 411, 413, 417–418, 420, 439–441
    coding, 227–230
    down-conversion, 236
    enhancement, 235
    for image sequence processing, 233
    interframe redundancy and, 360
    interpolation, 233–234
    MPEG-1 video compression technique, 362
    restoration, 236
    temporal filtering, 235
    temporal redundancy, 373
Motion estimation
    advanced
        generalized block matching, 377
        overlapped and frequency domain, 377
        using reduced set of image data, 376–377
    advanced motion estimation, 376–377
    backward motion estimation, 335–336
    DCT based method, 338–340
    deterministic methods, 330
    forward motion estimation, 335–336
    frequency domain method, 330–331
    frequency domain motion estimation, 377
    Gabor energy filter, 331–332
    gradient-based method, 333–335
    hierarchical motion estimation, 376–377
    overlapped motion estimation, 377
    region-based method, 333–335
    spatial domain method, 330
    stochastic methods, 330
Motion estimation (ME), 439, 442; see also Motion
2-D motion field, 286–287, 321
Motion interpretation, see Motion
Motion vectors
    experimental results, 253
    motion compensation, 252
    thresholding process, 253
Moving picture expert group (MPEG), 395
MPEG-1, 353–354, 359–368, 370–371, 393
    bit rates for NTSC/PAL, 359
    compressed bitstream structure, 365
    decoding process, 365–366
    layered structure based on group of pictures, 360–361
    video compression technique, 362
    video encoder structure, 361–365
    video sequence, group of pictures (GOPs) of, 360–361
    video signal coding, 360
MPEG-2, 353–356, 359–360, 365–366, 368–374, 376, 378, 382, 389, 392–394
    areas of research, designing optimized, 373
    basic coding structure of, 366
    coding in, 508
    concealment motion vector (CMV), 370–371
    field/frame DCT coding syntax, 369
    field/frame prediction mode
        coding modes, 368
        frame coding, 367
    matching criterion, 374–375

motion estimation and motion
  compensation, 374
non-compression enhancements, 367
optimum mode decision
  bit allocation, 382
  coding mode selection criteria, 386–387
  greedy approach for, 385–386
  iterative procedure for, 385
  MB distortion measure, 383
  problem formation, 381
  trellis structure, 384–385
pan–scan parameters, 370
preprocessing
  and frame rates, 374
  telecine and detelecine process, 373–374
program streams, 517
quantizer matrices, downloadable, 369
scalable modes, 371–372
searching algorithm, 375–376
splice point in, 515–517
technical definitions in, 509–510
timing model and synchronization, 518–521
transport streams, 510
zigzag scan and alternative scan, 370
MPEG-4, 353–356
  audiovisual objects (AVO), 435–436, 453
  coding tools, 439
  and content providers, 435
  and end users, 435–436
  features summary, 437–438
  FlexMux tool in, 522
  goal of, 435
  and network providers, 435
  object-based coding, 438–439
  object description framework, 525–526
  project history, 435
  requirements and functionalities, 436–437
  scene description, 524–525
  sprite coding technology, 448–449
  systems decoder model, 523–524
  terminal architecture, 521–523
  verification model (VM), 455
  video object plane (VOP), 438, 453
  visual profiles, 454–455
  visual syntax hierarchy, 453–454
MPEG-2 to MPEG-4 transcoding, 409–410
MPEG-4 visual coding, 217
MR code, see Modified READ code
MSE, see Mean square error
MSLT, see Minimum scan line time
Multigrid block matching, 260
Multiple attribute approach, 285, 315–321
Multiple references, 486, 493–494
Multiresolution analysis (MRA), 184
Multiresolution block matching, 248–250

N

Nagel's algorithm, 296
NAL unit (NALU) stream format, 484–485
National television system commission
  (NTSC), 233
National Television Systems Committee (NTSC),
  346, 349–352
National Television Systems Committee/phase
  alternating line (NTSC/PAL)
  bit rates for, 359
Natural binary code (NBC), 156
Neighborhood information, 328
Netravali–Robbins' pel recursive algorithm,
  271–272
  inclusion of neighborhood area, 280
  interpolation of, 280
  performance of, 281
  simplification of, 281
Network abstraction layer (NAL), 484–485
Newly exposed area, 329
Newly visible area, 329
Newton–Raphson's method, 275, 277, 281–283
  convergence speed, 279
  formulae, 278–279
Noise reduction, procession of, 373
Noise remove, 235
Non-intra quantizer weighting matrix, 363
Nonuniform quantization
  companding quantization
    consists of, 48
      A-law compression characteristic, 50
      μ-law compression characteristic, 49
    optimum quantization, 45, 48
      Gaussian distribution, 45
      $vs.$ optimum uniform quantization, 45, 48
Normalized activity factor, 381
Normal optical flow, 287

O

Object-based coding, 438–439
Objective visual quality measurement
  methodology, 24–25
  motivation, 22–23
  objective estimator, 26
  signal-to-noise ratio (SNR), 22–23
  spatial and temporal information, 25–26
Off-line sprite generation, 458
Off-line sprites, 458; see also Sprite coding
Online sprites, see Sprite coding
Optical flow
  computation, 289–290, 297–298, 315, 321–322
  correlation-based approach, 297
    Anandan's method, 298

Pan, Shi, and Shu's method, 302–315
  Singh's method, 298–302
 determination, 285, 287, 289–290, 292, 296–298
 field, 285–286, 290, 303–305, 307–308, 311, 320–321
 fundamentals of, 285–286
  aperature problem, 287–288
  classification of, 289–290
  2-D motion and optical flow, 286–287
  Ill-posed problem, 289
 gradient-based approach
  Horn and Schunck's method, 290–295
  Lucas and Kanade's method, 296
  modified Horn and Schunck method, 295–296
  Nagel's method, 296
  Uras, Girosi, Verri, and Torre's method, 297
 multiple attributes, 315–316
  Weng, Ahuja, and Huang's method, 316
  Xia and Shi's method, 317–321
 technique, 285, 289, 292, 306, 322
Optical flow vectors, 231
Optimal mode decision, 359
Optimal multigrid block matching, 262–263
Optimal quantization, 206
Optimization, 272–273, 275, 279, 283
Optimum code (minimum redundancy code), 120, 122–123
Optimum uniform quantizer
 conditions of, 41, 44
 mean square quantization error, 40
 quantization noise, 41
 signal-to-noise ratio (SNR), 41
 for uniform, Gaussian, Laplacian, and Gamma distributions, 42–45
Orthonormal eigenvectors, 82
Overlapped block matching, 265–267
Overlapped motion compensation, 440–441, 470, 478
Overlapped motion estimation, 377

P

Packet identification (PID), 509, 513–516
Packetized elementary streams (PES), 508, 511–512, 515
Pan, Shi, and Shu's algorithm, 313–315
 experiment I, 305
 experiment II, 305–306
 experiment III, 306–312
 experiment IV, 312
 framework of, 302–304
 implementation, 304–305

Partitioned IFS (PIFS), 214
PB-frames, 472–474
Peak signal-to-noise ratio (PSNR), 252
Pel recursive technique
 Bergmann's algorithm, 281–282
 Cafforio and Rocca's algorithm, 282
 classification of, 283
 comparison of, 283
 descent methods
  convergence speed, 275–276
  first-order necessary conditions, 273
  Newton–Raphson's method, 277–279
  second-order sufficient conditions, 273–274
  steepest descent method, 277
  underlying strategy, 274–275
 Netravali–Robbins, 280–281
 problem formulation, 271–272
 Walker and Rao algorithm, 282
PES packet, 515, 517
Phase-based approach, 321
Picture start code (PSC), 519
Predictive-coded (P) pictures, 361
Predictive coding, 111–112
Predictive motion field segmentation, 263–265
Preprocessing, 359, 373–374, 381, 391
 and frame rates, 374
 telecine and detelecine process, 373–374
Presentation time stamp (PTS), 510, 515, 519–521
Program-specific information (PSI), 509, 511–513, 515
Program stream, 508–511, 517–519
Progressive, 350–352
Progressive DCT-based encoding algorithm
 spectral selection method, 174–175
 successive approximation method, 174–175
Pseudophase
 DCT, 338–339
 DST, 338–339
Psychophysical redundancy, 346, 352
Psychovisual redundancy
 color masking
  applications in video compression, 19–20
  gamma-correction, 17
  HSI model, 17
  RGB model, 16–17
  YCbCr model, 19
  YDbDr model, 18
  YIQ model, 18
  YUV model, 17–18
 frequency masking
  description of, 15
  discrete cosine transform (DCT) domain, 16
 human visual system (HVS), 10–11

luminance masking
    contrast sensitivity function, 11
    quantization, 12–13
    Weber's law, 11–12
temporal masking, 16
texture masking
    definition, 13
    quantization levels, 13–15
Pulse amplitude modulation (PAM), 54
Pulse code modulation (PCM)
    description of, 54
    in digital image processing, 55
Pulse position modulation (PPM), 54
Pulse width modulation (PWM), 54
Pyramid, 249
Pyramid boundary, 211

Q

Q-coder and QM-coder, 136
Quadrature mirror filters (QMFs), 191
Quantization
    adaptive, 50–54, 110
    block, 83
    description of, 33–34
    optimum nonuniform, 45–50
    optimum uniform, 40–45
    scalar and vector, 35
    step size, 36–37, 39–40, 105, 110
    switched, 53–54
    table, 106
    uniform, 35–37
Quantization matrix, 362, 366, 369
    function of, 362
    for intra- and non-intracoding, 363–364

R

Raster algorithms, 149
Rate control, 359, 363, 373, 377–382, 384, 387–393
    encoding
        classification of, 377
        multiple VO rate control, 458–459
        purpose of, 373
    TM5 (Test Model 5) encoder
        adaptive quantization, 380–381
        target bit allocation, 378–379
        virtual buffer, fullness of, 379–380
Rate distortion function, 346–348
Rate distortion theory, 346–347
Raw byte sequence payload (RBSP), 485
READ, see Relative element address
    designate code
RealVideo, 355–356

Recursions, types of, 272
Reduced-resolution update (RRU), 478–479
Reference line, 146
Reference picture resampling (RPR), 478
Reference picture selection (RPS), 477–478
Relative element address designate (READ)
    code, 146
Reversible variable-length codes (RVLC),
    452–453
RLC, see Run-length coding
Robust techniques, 232
Root mean square (RMS), 305
Root mean square (RMS) error, 215
Rotation, 329–330
Run-length coding (RLC)
    1-D RLC, 144
    one-dimensional (1-D)
        EOL code word, 144
        MH code table, 145–146
        run-length calculation, 145
        transmission error effects of, 148–149
    two-dimensional (2-D)
        changing pixels, 146–147
        coding modes, 147–148
        DCT coefficients, 144
        MR code, 146
        transmission error effects of, 149
    uncompressed mode of, 149
    uses of, 144

S

Scalability, 371–372
    signal-to-noise ratio (SNR) scalability,
        473–474
    SNR-scalability, 371–372
    spatial scalability, 371, 449, 451, 474
    temporal scalability, 371–372, 451–452, 454,
        473–474
Scalable video coding (SVC), 401
Scalar quantizer, 210
Searching algorithm
    full search, 375–376
    logarithm search, 375–376
    three, 375–376
Search window, 243
Segmentation, 262
Semiautomatic segmentation algorithm, see
    Video segmentation
Sensor
    optical axis of, 222
    temporal image sequence, 243
Sequential DCT-based encoding algorithm
    block diagram of, 169
    quantized coefficients of, 170

Index

Set partitioning in hierarchical trees (SPIHT), 192, 194
Set partitioning sorting algorithm, 194
Shannon's first coding theorem, 121–122
Shape adaptive DCT (SA-DCT), 444–445
Shape coding, 445
    binary shape coding, 446–448
    gray-scale shape coding, 448
Short-time Fourier transform (STFT)
    continuous wavelet transform, 180–182
    digital wavelet transform, 189–190
    discrete wavelet transform, 183–185
    embedded zerotree wavelet image coding (EZW), 191
    wavelet transform, 179–183
Signal-to-noise ratio (SNR) scalability, 372
Similarity measure, 243
Singh's algorithm, 298–302
    conservation information, 300
    minimization and iterative algorithm, 301–302
    neighborhood information, 300–301
Sinusoidal orthogonal principle, 339
SI-slices, *see* Switching I-slices
Slice-structure (SS), 477
Smoothness constraint, 292, 296, 299, 303, 316
Source alphabet
    nth auxiliary, 124–125
    nth extension of, 120–122, 135, 139
Source coding, 346
Source input format (SIF), 351–352
Spatial domain method, for motion estimation, 330
Spatial image sequence, 243
Spatial redundancy
    consequences of, 6
    definition of, 5
    Fourier transform of autocorrelation, 5–6
Spatial scalability, 451
Spatial scalability decoder
    block diagram of, 371
Spatiotemporal domain, 331–333
Splicing, 514–517
Sprite coding, 448–449
SP-slices, *see* Switching P-slices
SSD, *see* Sum of squared difference
Stability range, 283
Static sprite, *see* Sprite coding
Statistical mean value, 206
Statistical models, 393
Statistical multiplexing operation (StaMux), 359, 387–393
    bit rate allocation problem, 391
    coding gain of, 389
    video sources coding complexity

    forward analysis, 391–392
    modeling strategies and methods, 392–393
    video sources encoding, 388
Statistical redundancy, 346
    classification of, 5
    coding redundancy, 9–10
    interpixel redundancy
        spatial redundancy, 5–7
        temporal redundancy, 7–9
Steepest descent method, 275, 277, 279–280, 282–283
Step size, 275, 283–284
    in Cafforio and Rocca algorithm, 282
    selection of, 277–278
Still image coding, *see* JPEG
Stochastic approach, 290, 321
Subjective visual quality measurement, 21–22
Sum of squared difference, 298, 300, 303
Sum of squared error (SSE), 244
Supplemental enhancement information, 480–481
Supremum, concept of, 276
Switching I-slices, 487–489, 503
Switching P-slices, 487–489, 503
Syntax-based arithmetic coding, 472
Synthetic and natural hybrid coding (SNHC), 436
System clock reference (SCR), 509, 517, 519
Systems decoder model (SDM), 523–524
System time clock (STC), 514, 519–520

T

Telecine material, 373
Television signals, 225
Temporal filtering, 235
Temporal image sequence, 243
Temporal redundancy
    definition of, 7
    motion compensated (MC) predictive coding, 9
    relative power, 8
Temporal scalability, 372, 451–452
Test model 5 (TM5), 363, 378–383, 389, 393
    rate control
        adaptive quantization, 380–381
        target bit allocation, 378–379
        virtual buffer, fullness of, 379–380
Texture coding
    INTRA DC and AC prediction, 441–442
    Motion estimation/Compensation of arbitrary shaped VOP, 442–444
    texture coding of of arbitrary shaped VOP, 444–445
Three-layer scalable decoder, 416–418

Threshold coding
  block diagram of, 102
  Huffman coding, 106
  input–output characteristic of thresholding
      and shifting, 103–104
  normalization factor, 104–105
  rate buffer feedback, 107–108
  roundoff process, 105
Thresholding multiresolution block
      matching, 250
Threshold value, 252
Toeplitz matrices, 95
Transcoding, 400–401, 409–410
Transformation kernel
  DCT
    coefficient, 103–104, 106–107
    vs. DFT, 96–98
  forward transform
    DFT, 92
    2-D image, 87–91
    DWT, 93
  inverse transform
    DCT, 103
    DFT, 92
    2-D image, 87–91
    DWT, 93
    Hotelling transform, 83–85
  orthogonal, 89
  separable, 87
  symmetric, 88
Transform coding (TC); see also Block
      quantization
  adaptive, 91
  basis vector expansion, 85, 89–91
  block diagram of, 86
  DCT
    coefficient, 103–104, 106–107
    transformation kernel, 95–96
    vs. DFT, 96–98
  DFT, 92
  DHT, 93–94
  DWT, 93
  error types, 101, 108
  gain, 99
  geometrical interpretation, 84–85, 99
  Hotelling transform
    covariance matrix of, 82
    definition, 83
    inverse, 83–85
  hybrid coding, 109
  procedures of, 86
  statistical interpretation, 83–84
  unitary transform
    definition, 89
    functions, 86

  vs. differential pulse code modulation
      (DPCM), 109
Transform coefficients
  in blocks, 103
  definition, 89
  energy compaction, 98–99
  truncation
    threshold coding, 102–107
    zonal coding, 101–102
  variance, 100
Transform coefficients (TCOEFF), 467
Transformed random vector, 82
Translating Tree 2-D velocity, 311, 320
Translating Tree image sequence,
      307, 311
Translation, 329–330, 338–339
Transport stream, 508, 510, 519
  splicing, 515–517
  structure of, 511–512
  syntax, 512–515
Tree 2-D sequence, 310
Trellis coder, 209
TV frames, 226
Two-stage error concealment
      strategy, 427

U

Uniform quantization
  input–output characteristics
    midrise uniform quantizer, 37
    midtread uniform quantizer, 36
  optimum uniform quantizer
    conditions of, 41, 44
    mean square quantization error, 40
    signal-to-noise ratio (SNR), 41
    for uniform, Gaussian, Laplacian, and
        Gamma distributions, 42–45
  quantization distortion, 38
  quantizer design, 39
Unitary transform
  definition, 89
  functions, 86
Universal access, 436–437
Universal multimedia access (UMA), 402
Uras, Girosi, Verri and Torre's algorithm, 297

V

Variable length coding (VLC), 452, 465–467,
      493–494
  arithmetic codes
    need for, 129
    principles of, 129–130

Huffman codes
    coding algorithm, 123–124
    limitations, 129, 137
    modified Huffman codes
        algorithm, 126–127
        codebook creation, 125–126
        codebook memory, 127–128
        code word length, 128
    tables, 365
Variable size block matching, 260
Variable step size, 282
VBR coding, 377
    image quality, 389
    VBR encoder, 390
VBR encoder
    comparison with totally open-loop VBR encoder, 390
    in open-loop multiplexing mode
        image quality, 389
        video source encoded by, 390
Vector dimension, 210
Vector quantization (VQ)
    basic principle of, 204
        quantization, 207
        training set generation, 205
        vector formation, 205
    block diagram of lattice, 210
    classified, 208
    finite state, 208
    image coding schemes with, 207
    lattice, 207
        construction-A, 211–212
        construction-B, 211–212
        labeling efficiency, 210
        two-dimensional labeling, 211
    predictive, 208
    residual, 207
    transform domain, 208
Video coding expert group (VCEG), 483
Video coding layer (VCL), 484–485
Video compression, 224
Video decoder, 459–460
Video encoder, 455–459
Video formats
    ATSC digital television format, 352
    common intermediate format, 352
    ITU-R, 351
    source input format, 351–352
Video object layer (VOL), 454
Video object plane (VOP), 438, 453
Video object (VO), 438, 453

Videophone service, 225
Video segmentation, 456–457
Video sequences
    coding complexity
        forward analysis, 391–392
        modeling strategies and methods, 392–393
    compression, 230
    consecutive frames of, 228
    geographical colocation of, 390
    group of pictures (GOPs) of, 360–361
    in television broadcast, 223
Video session (VS), 453
Visible frequency band, 223
Visual profiles, in MPEG-4, 454–455
Von Koch curve, construction of, 212
Voronoi cells, 211
VQ, *see* Vector quantization

W

Walker and Rao's pel recursive algorithm, 282
Wavelet-based texture coding, 449–450
Weber's law, 11
Weighted linear interpolation, 234
Weng, Ahuja and Huang's algorithm, 315–316, 321
Wire frame model, 330

X

Xia and Shi's algorithm
    algorithm steps used in, 319–320
    conservation stage, 318–319
    multiple image attributes, 317–318
    propagation stage, 319
    results of, 320–321

Y

Yosemite 2-D velocity, 312, 321
Yosemite image sequence, 307, 309–310
Yule–Walker equation, 66

Z

Zerotree, 193
Zonal coding, 101–102, 110